K-theory, originally the study of vector bundles in topology, has become a powerful tool in the subject of operator algebras and has led to profound and unexpected applications throughout mathematics. This book develops K-theory and its deep bivariant version, Kasparov's KK-theory, from scratch to its most advanced aspects and describes important applications in topology and geometry. Numerous exercises ranging from elementary to research level supplement the text.

The second edition includes a new section on E-theory and coverage of other key recent developments in the subject, as well as more than sixty new and updated references.

From reviews of the first edition:
"This book gives a comprehensive survey of 'operator' K-theory or 'noncommutative' algebraic topology. Since its inception in the early 1970s, this field has grown rapidly, until a deep and elaborate machinery has evolved. This book is the first to consolidate this material and does an excellent job of presenting the path of least resistance to the key results while keeping the reader informed about the many important sidetracks."

– Mathematical Reviews

Mathematical Sciences Research Institute
Publications

5

K-Theory for Operator Algebras

Mathematical Sciences Research Institute
Publications

Volumes 1–4 and 6–27 are available from Springer-Verlag

K-Theory for Operator Algebras

Second Edition

Bruce Blackadar

University of Nevada, Reno

CAMBRIDGE
UNIVERSITY PRESS

Bruce Blackadar
Department of Mathematics
University of Nevada, Reno
Reno, NV 89557
USA

Mathematical Sciences Research
   Institute
1000 Centennial Drive
Berkeley, CA 94720

MSRI Editiorial Committee
Hugo Rossi (chair)
Alexandre Chorin
Silvio Levy (series editor)
Jill Mesirov
Robert Osserman
Peter Sarnak

The Mathematical Sciences Research Institute wishes to acknowledge
support by the National Science Foundation.

Published by the Press Syndicate of the University of Cambridge
The Pitt Building, Trumpington Street, Cambridge CB2 1RP
40 West 20th Street, New York, NY 10011–4211, USA
10 Stamford Road, Oakleigh, Melbourne 3166, Australia

First edition published 1986 by Springer-Verlag

Second edition published 1998

Printed in the United States of America

Library of Congress cataloging-in-publication data is available.

A catalogue record for this book is available from the British Library.

ISBN 0-521-63532-2 Paperback

For Martha and Gabriel

PREFACE TO SECOND EDITION

I was pleased to learn that MSRI and Cambridge University Press have decided to issue a second edition of this book. The first edition sold out its press run rather quickly, and for the last several years I have had regular inquiries from colleagues and students about where to obtain a copy.

The first edition was successful because it filled a need: there was nothing available at the time which even remotely covered the subject satisfactorily. Even now, while there are many more references available for this and related subjects, there is no other comprehensive treatment of the topics covered in this book, and so the second edition will (I hope) still fill an important need. Some of the newer references I can recommend to the reader are: [Fillmore 1996] and [Murphy 1990] as general references for C^*-algebras, including some K-theory; [Davidson 1996] for a deep study of examples of important C^*-algebras, including many treated superficially in my book; [Wegge-Olsen 1993] for a leisurely treatment of basic K-theory in more detail than we have included here; [Higson 1990] for a detailed survey of KK-theory and its applications, written primarily for nonspecialists in operator algebras; [Rosenberg 1994] for an excellent treatment of algebraic K-theory; [Loday 1992] for cyclic homology/cohomology; and, of course, [Connes 1994] (which was already partially available at the time of the first edition)—its introduction gives a marvelous overview of the subject, and the book contains a vast supply of important applications. Several more books on operator K-theory and related subjects are forthcoming. There is also a compilation of all Mathematical Reviews on K-theory from 1940 to 1985 [Magurn 1985], and even a K-theory preprint archive on the web (www.math.uiuc.edu/K-theory/).

Time and mathematics have marched on since the book was completed, and the treatment of the book is somewhat out of date in many areas and badly out of date in a few. Despite my overly optimistic statement in the original preface that "it appears the basic theory has more or less reached a final form," the subject has continued to evolve. Even before the book was published, important new approaches and results of Cuntz and Skandalis made some sections somewhat obsolete. Since then, such advances as the E-theory of Connes and Higson have become core material for the subject. Profound new applications of K-theory and its generalizations, both within operator algebras (for example, the classification programs of Elliott *et al.* and of Kirchberg) and in topology and geometry (see [Connes 1994]) have attracted much attention.

In preparing the second edition, I had to make the most important decision right at the outset: how extensively to revise the book. If I were writing a book now from scratch, it would be very different from the first edition, taking into account both the evolution of the subject itself and of my understanding of it. I basically had three options:

(1) Extensively rewrite almost every section.
(2) Do a minimal revision, correcting errors, updating references, and adding a few comments here and there about the most important developments.
(3) Do something in between, rewriting some parts but leaving the essence of the manuscript intact.

I immediately rejected option (3): I was unlikely to be completely satisfied with the outcome, and if I was going to do that much work I should really do the job right and choose option (1). Option (1) was seriously considered, but ultimately rejected, partly because I did not feel I had the time or energy to do it properly, and partly because I think there are others far more qualified than I (some of whom are seriously considering a book of their own which I very much hope to see.) So option (2) was chosen. I felt that even though the treatment in the book might no longer be optimal, it could be relatively easily updated into a reference still of value. I have corrected those errors, gaps, and obscurities I was aware of, both typographical and mathematical (although I certainly would not be willing to assert that the book is now error-free!) I have also updated references, and added new comments and references where needed. I have tried to mention the most important new developments, usually in brief comments or exercises/problems; the reader should not infer that subjects treated only briefly here are not worthy of expanded coverage.

The one major addition to the book is a new section ($\S\,25$) on E-theory. I felt that no contemporary book on K-theory would be complete without this topic, and it warranted more than a brief treatment since there is at present no other adequate reference. A few other additions of essential topics are scattered throughout the book.

I have taken care not to change the numbering of any section, paragraph, or result, in order not to make references ambiguous. The only exceptions to this are Theorem 11.4.2 and problem 16.4.8, which were erroneously numbered 11.4.1 and 16.4.7 respectively in the first edition. All added material has numbers not used in the first edition. This policy has occasionally led to new material being inserted in a less than optimal location, but I believe the advantages of the policy outweigh the disadvantages.

I am grateful to all those who have given or sent me comments or corrections to the first edition. Colleagues who have made substantive comments include P. Baum, J. Cuntz, E. Kirchberg, A. Kumjian, L. Lehmann, N. C. Phillips,

M. Rørdam, J. Rosenberg, and C. Schochet. Unfortunately, recollection of the source of some of the comments has been lost in time, and I apologize to anyone I missed in this list.

The first edition was produced with what was then a state-of-the-art word processing system, UNIX troff with eqn, and I received compliments on the appearance. Today the printing looks crude. I am grateful to Silvio Levy of MSRI who spent a great deal of time converting the troff files to TEX and making the book look good by current standards.

PREFACE TO FIRST EDITION

K-Theory has revolutionized the study of operator algebras in the last few years. As the primary component of the subject of "noncommutative topology," K-theory has opened vast new vistas within the structure theory of C^*-algebras, as well as leading to profound and unexpected applications of operator algebras to problems in geometry and topology. As a result, many topologists and operator algebraists have feverishly begun trying to learn each others' subjects, and it appears certain that these two branches of mathematics have become deeply and permanently intertwined.

Despite the fact that the whole subject is only about a decade old, operator K-theory has now reached a state of relative stability. While there will undoubtedly be many more revolutionary developments and applications in the future, it appears the basic theory has more or less reached a "final form." But because of the newness of the theory, there has so far been no comprehensive treatment of the subject.

It is the ambitious goal of these notes to fill this gap. We will develop the K-theory of Banach algebras, the theory of extensions of C^*-algebras, and the operator K-theory of Kasparov from scratch to its most advanced aspects. We will not treat applications in detail; however, we will outline the most striking of the applications to date in a section at the end, as well as mentioning others at suitable points in the text.

There is little in these notes which is new. They represent mainly a consolidation and integration of previous work. I have borrowed freely from the ideas and writings of others, and I hope I have been sufficiently conscientious in acknowledging the sources of my presentation within the text and in the notes at the end of sections. There are some places where I have presented new arguments or points of view to (hopefully) make the exposition cleaner or more complete.

These notes are an expanded and refined version of the lecture notes from a course I gave at the Mathematisches Institut, Universität Tübingen, West Germany, while on sabbatical leave during the 1982-83 academic year. I taught the course in an effort to learn the material of the later sections. I am grateful to the participants in the course, who provided an enthusiastic and critical audience: A. Kumjian, B. Kümmerer, M. Mathieu, R. Nagel, W. Schröder, J. Vazquez, M. Wolff, and L. Zsido; and to all the others in Tübingen who made my stay pleasant and worthwhile. I am also grateful to the Alexander von Humboldt-Stiftung for their financial support through a Forschungsstipendium.

I have benefited greatly from numerous lectures and discussions at the Mathematical Sciences Research Institute, Berkeley, during the 1984–85 academic year. I am particularly indebted to P. Baum, L. Brown, A. Connes, J. Cuntz, R. Douglas, N. Higson, J. Kaminker, C. Phillips, M. Rieffel, J. Rosenberg, and C. Schochet for sharing their knowledge and insights.

In addition, I want to thank J. Cuntz, P. Julg, G. Kasparov, C. Phillips, M. Rieffel, J. Roe, D. Voiculescu, and especially J. Rosenberg, for taking the time to review a preliminary draft of the manuscript, pointing out a number of minor (and a few major) errors, and suggesting improvements.

I am also grateful to Ed Wishart, Jana Dunn, Bill Rainey, Mark Schank, and Ron Sheen for patiently helping me master the UNIX[1] system and managing to keep the UNR system operating long enough to produce the manuscript.

Since these notes are primarily written for specialists in operator algebras, we will assume familiarity with the rudiments of the theory of Banach algebras and C^*-algebras, such as can be found in the first part of [Dixmier 1969], [Pedersen 1979], or [Takesaki 1979]. Some of the sections, particularly later in the book, require more detailed knowledge of certain aspects of C^*-algebra theory.

Most of the notation we use will be standard, and will be explained as needed. Some basic notation used throughout: $\mathbb{N}$, $\mathbb{Z}$, $\mathbb{Q}$, $\mathbb{R}$, $\mathbb{C}$ will denote the natural numbers, integers, and rational, real, and complex numbers respectively; $\mathbb{M}_n$ will denote the $n \times n$ matrices over C; $\mathbb{H}$ will denote a Hilbert space, separable and infinite-dimensional unless otherwise specified; and $\mathbb{B}(\mathbb{H})$ and $\mathbb{K}(\mathbb{H})$, or often just $\mathbb{B}$ and $\mathbb{K}$, will respectively denote the bounded operators and compact operators on $\mathbb{H}$. $\mathrm{diag}(x_1,\ldots,x_n)$ will denote the diagonal matrix with diagonal elements $x_1,\ldots,x_n$. If A and B are C^*-algebras, $A \otimes B$ will always denote the minimal (spatial) C^*-tensor product of A and B.

This work was supported in part by NSF grant no. 8120790.

---

[1] UNIX is a Trademark of Bell Laboratories.

CONTENTS

K-Theory for Operator Algebras
MSRI Publications
Volume **5**, second edition, 1998

CHAPTER I

INTRODUCTION TO *K*-THEORY

1. Survey of Topological *K*-Theory

This expository section is intended only as motivation and historical perspective for the theory to be developed in these notes. See [Atiyah 1967; Karoubi 1978] for a complete development of the topological theory.

K-theory is the branch of algebraic topology concerned with the study of vector bundles by algebraic means. Vector bundles have long been important in geometry and topology. The first notions of *K*-theory were developed by Grothendieck in his work on the Riemann–Roch theorem in algebraic geometry. *K*-theory as a part of algebraic topology was begun by Atiyah and Hirzebruch [1961].

1.1. Vector Bundles

Informally, a vector bundle over a base space X (which we will usually take to be a compact Hausdorff space) is formed by attaching a finite-dimensional vector space to each point of X and tying them together in an appropriate manner so that the bundle itself is a topological space. More specifically:

DEFINITION 1.1.1. A *vector bundle* over X is a topological space E, a continuous map $p : E \to X$, and a finite-dimensional vector space structure on each $E_x = p^{-1}(x)$ compatible with the induced topology, such that E is *locally trivial*: for each $x \in X$ there is a neighborhood U of x such that $E|_U = p^{-1}(U)$ is isomorphic to a trivial bundle over U.

An *isomorphism* of vector bundles E and F over X is a homeomorphism from E to F which takes E_x to F_x for each $x \in X$ and which is linear on each fiber.

A *trivial bundle* over X is a bundle of the form $X \times V$, where V is a fixed finite-dimensional vector space and p is projection onto the first coordinate (the topology is the product topology).

One can consider real vector bundles or complex vector bundles (or even quaternionic vector bundles), according to whether the vector spaces are real or complex. We will later restrict attention to complex bundles since it is the complex theory which generalizes to (complex) Banach algebras, but for much of the basic theory either kind can be considered.

The local triviality implies that the dimension of the fibers is locally constant, and hence is globally constant if X is connected (really the only interesting case).

If the dimension of each fiber is n, we say the bundle is n-dimensional. A one-dimensional bundle is sometimes called a *line bundle*.

Every X has at least one bundle of each dimension, namely the trivial bundle. Many spaces have only trivial bundles.

EXAMPLES 1.1.2. (a) The simplest example of a nontrivial (real) vector bundle is the Möbius strip, formed from $[0, 1] \times \mathbb{R}$ by identifying $(0, x)$ with $(1, -x)$. It is a vector bundle over the circle S^1.

(b) Another interesting nontrivial vector bundle is the tangent bundle TS^2 of the 2-sphere S^2. More generally, many vector bundles are naturally associated to differentiable manifolds: tangent and cotangent bundles, normal bundles associated with immersions, bundles associated with foliations, etc.

(c) There is a general "clutching" construction which yields many bundles. If $X = X_1 \cup X_2$, and if E_i is a bundle on X_i with $E_1|_Y \cong E_2|_Y$, where $Y = X_1 \cap X_2$, then E_1 and E_2 can be glued together over Y to give a vector bundle over X. (Clutching can also be done more generally.) The resulting bundle depends on which isomorphism is taken between $E_1|_Y$ and $E_2|_Y$. For example, let X be S^2, X_1 the upper hemisphere, X_2 the lower hemisphere, and E_1 and E_2 trivial complex line bundles. Then $Y \cong S^1$, and $E_i|_Y$ is a trivial complex line bundle. Let σ_n be the map which sends $(z, w) \in E_1|_Y$, for $z \in S^1$ and $w \in \mathbb{C}$, to $(z, z^n w) \in E_2|_Y$. Then the bundles on S^2 corresponding to the σ_n are all mutually nonisomorphic, and all complex line bundles arise in this manner. σ_0, of course, gives the trivial bundle. In fact, any complex vector bundle on S^n is formed in a similar way by clutching over the "equator", which is an $(n-1)$-sphere.

There is an alternate way of viewing the bundle arising from σ_1, which helps motivate the algebraic reformulation of K-theory described in 1.7: identify S^2 with $\mathbb{CP}^1$, the projective space of one-dimensional (complex) subspaces of $\mathbb{C}^2$, and set $V = \{(x, v) \in \mathbb{CP}^1 \times \mathbb{C}^2 : v \in x\}$. This is a subbundle of the trivial two-dimensional bundle. It is probably the most important bundle, and is called the *Bott bundle*.

It is not always easy to tell whether a given bundle is trivial. In cases (a) and (b) above, the nontriviality can most easily be established by looking at the sections of the bundle. A *section* of a bundle E over X is a continuous function $s : X \to E$ with $s(x) \in E_x$ for each x, i.e. a continuous choice of a vector in each fiber. A trivial bundle has many globally nonvanishing sections, for example the constant sections. Neither the Möbius strip nor the tangent bundle to S^2 has a globally nonvanishing section. (For a tangent bundle, a section corresponds to a vector field on the manifold. It is well known that S^2 has no globally nonvanishing vector fields.) We will denote the set of all sections of E by $\Gamma(E)$.

It must be emphasized that we are considering vector bundles as topological objects only. In case (b) above, the bundles have a natural differentiable structure, and it is this differentiable structure which is crucial in many of the applications of vector bundle theory to differential geometry and differential topology. We will ignore the differentiable structure completely, as it is irrelevant for K-theory. (This is actually not a serious loss, since a theorem of differential topology [Hirsch 1976, 4.3.5] says that every topological vector bundle over a differentiable manifold has an essentially unique differentiable structure compatible with the manifold.) So for us, differentiable structures on X serve only to give a way of constructing interesting topological vector bundles over X. (This is not to say that differentiable structures are irrelevant for operator algebras. On the contrary, some of the deepest and most fascinating work now being done is the noncommutative differential geometry of Connes [1994]. There are some close connections between this work and K-theory, which we will unfortunately only be able to very briefly touch on in these notes.)

1.1.3. If E is a vector bundle over X, and $\phi : Y \to X$ is a continuous function, then E can be pulled back via ϕ to a vector bundle $\phi^*(E)$ over Y. $\phi^*(E)$ can be defined formally as the fibered product $Y \times_X E = \{(y, e) \in Y \times E \mid \phi(y) = p(e)\}$. The fiber $\phi^*(E)_y$ is just $E_{\phi(y)}$. The pullback of a bundle is a bundle of the same dimension, and the pullback of a trivial bundle is trivial.

1.2. Whitney Sum

There are a number of operations which can be used to combine vector bundles over X into new ones. The most important one for our purposes is the "direct sum" or Whitney sum. Given bundles E and F over X, the Whitney sum $E \oplus F$ is formed by taking the fiberwise direct sum and tying the fibers together in a way compatible with the topologies on E and F. More precisely, if p and q are the projection maps of E and F onto X, we have

$$E \oplus F = \{(e, f) \in E \times F \mid p(e) = q(f)\}.$$

The sum of an n-dimensional bundle and an m-dimensional bundle is an $(m + n)$-dimensional bundle, and the sum of two trivial bundles is a trivial bundle. Whitney sums commute with pullbacks.

The Whitney sum makes the set of isomorphism classes of complex vector bundles over X into a commutative monoid (semigroup with identity) denoted $V_{\mathbb{C}}(X)$. The trivial bundles form a submonoid isomorphic to the additive monoid of nonnegative integers. The identity is the (class of the) 0-dimensional trivial bundle. A continuous map $\phi : Y \to X$ induces a homomorphism $\phi^* : V_{\mathbb{C}}(X) \to V_{\mathbb{C}}(Y)$ (1.1.3), so $V_{\mathbb{C}}$ gives a contravariant functor from topological spaces to abelian monoids. Similarly, the real vector bundles give a monoid $V_{\mathbb{R}}(X)$.

One can also form tensor products and exterior powers of bundles. These give additional algebraic structure to $V(X)$. Unfortunately, there is no known way of extending this additional structure to the noncommutative case, so these operations remain unique to topological K-theory.

EXAMPLES 1.2.1. (a) The Whitney sum of two Möbius strips is a two-dimen-
sional trivial bundle. $V_{\mathbb{R}}(S^1)$ is isomorphic to $\{0\} \cup (\mathbb{N} \times \mathbb{Z}_2)$.

(b) Although TS^2 is a nontrivial 2-dimensional bundle over S^2, its sum with a
trivial line bundle is a trivial 3-dimensional bundle. So $V_{\mathbb{R}}(S^2)$ is a complicated
monoid which does not have cancellation, i.e. $x + z = y + z$ does not imply $x = y$.
There are compact manifolds X (e.g. the 5-torus $\mathbb{T}^5$) for which $V_{\mathbb{C}}(X)$ does not
have cancellation.

(c) Every complex vector bundle over S^2 is a sum of line bundles, and $V_{\mathbb{C}}(S^2) \cong
\{0\} \cup (\mathbb{N} \times \mathbb{Z})$.

1.2.2. An important theorem of Swan says that if E is a vector bundle over a
compact Hausdorff space X, then there is a bundle F such that $E \oplus F$ is a trivial
bundle.

1.3. The Grothendieck Group

If H is an abelian semigroup, then there is a universal enveloping abelian
group $G(H)$ called the Grothendieck group of H. $G(H)$ can be constructed in a
number of ways. For example, $G(H)$ may be defined to be the quotient of $H \times H$
under the equivalence relation $(x_1, y_1) \sim (x_2, y_2)$ if and only if there is a z with
$x_1 + y_2 + z = x_2 + y_1 + z$. $G(H)$ may be thought of as the group of (equivalence
classes of) formal differences of elements of H, thinking of (x, y) as $x - y$. The
prototype example of this construction is the construction of $\mathbb{Z}$ from $\mathbb{N}$. $G(H)$
may also be defined by generators and relations, with generators $\{\langle x \rangle : x \in H\}$
and relations $\{\langle x \rangle + \langle y \rangle = \langle x + y \rangle : x, y \in H\}$.

There is a canonical homomorphism from H into $G(H)$ which sends x to
$[(x+x, x)]$. This homomorphism is injective if and only if H has cancellation.
$G(H)$ has the universal property that any homomorphism from H into an abelian
group factors through $G(H)$. G gives a covariant functor from abelian semigroups
to abelian groups.

1.4. The K-Groups

DEFINITION 1.4.1. If X is a compact Hausdorff space, $K(X) = K_{\mathbb{C}}(X)$ is the
Grothendieck group of $V_{\mathbb{C}}(X)$. $K_{\mathbb{R}}(X)$ is the Grothendieck group of $V_{\mathbb{R}}(X)$. So
$K(X)$ may be thought of as the group of equivalence classes of formal differences
of vector bundles over X.

$K_{\mathbb{C}}$ and $K_{\mathbb{R}}$ are sometimes written KU and KO respectively (U for "unitary",
O for "orthogonal"). $K_{\mathbb{C}}$ and $K_{\mathbb{R}}$ are contravariant functors from compact Hausdorff spaces to abelian groups.

EXAMPLES 1.4.2. (a) If X is a one-point space or $[0, 1]$, then every bundle is
trivial; $V_{\mathbb{R}}(X) \cong V_{\mathbb{C}}(X)$ is the nonnegative integers, so $K_{\mathbb{R}}(X) \cong K_{\mathbb{C}}(X) \cong \mathbb{Z}$.
The same is true for any contractible space.

(b) $K_{\mathbb{R}}(S^1) \cong \mathbb{Z} \times \mathbb{Z}_2$.

(c) $K_{\mathbb{C}}(S^2) \cong \mathbb{Z}^2$.

1.5. Locally Compact Spaces

Before developing the fundamental exact sequence of K-theory, we must extend the definition of $K(X)$ to locally compact Hausdorff spaces. The reason is that the exact sequence relates the K-theory of X to that of a closed subspace Y and the complement $X \setminus Y$; even if X is compact $X \setminus Y$ will not be in general.

The same definition of $K(X)$ makes sense if X is not compact, but turns out not to be appropriate. The correct approach is to define a relative K-group for a compact pair (X, Y), where X is compact and Y is a closed subspace of X. $K(X, Y)$ may be defined to be the Grothendieck group of the semigroup consisting of triples (E, F, α), where E and F are vector bundles over X whose restrictions to Y are isomorphic and α is a fixed isomorphism from $E|_Y$ to $F|_Y$; (E, F, α) and (E', F', α') are identified if $E \cong E'$ and $F \cong F'$ under isomorphisms whose restrictions to Y intertwine α and α'. (E, F, α) is also identified with $(E \oplus G, F \oplus G, \alpha \oplus id)$. (The semigroup operation is coordinatewise Whitney sum.) We then define $K(X) = K(X^+, +)$ for X locally compact, where X^+ is the one-point compactification of X and $+$ is the point at infinity. It is easy to see that this definition agrees with the previous one if X is compact. We extend the definition of the relative group $K(X, Y)$ to the locally compact case by $K(X, Y) = K(X^+, Y^+)$. We may also define $K_{\mathbb{R}}(X, Y)$ and $K_{\mathbb{R}}(X)$ in the same manner.

$K_{\mathbb{C}}$ and $K_{\mathbb{R}}$ then give functors from the category of locally compact Hausdorff spaces and *proper* maps to abelian groups. It is usually better to think of the functors as being defined on the equivalent category of pointed compact Hausdorff spaces.

PROPOSITION 1.5.1. *If X is locally compact, Y is a closed subspace, and $U = X \setminus Y$, then the map q from X^+ to U^+ which is the identity on U and which sends $X^+ \setminus U$ to the point at infinity induces an isomorphism between $K(X, Y)$ and $K(U)$. So the sequence*

$$K(U) \xrightarrow{q^*} K(X) \xrightarrow{i^*} K(Y)$$

is exact in the middle (i.e. $\ker i^* = \operatorname{im} q^*$.) An analogous statement is true for $K_{\mathbb{R}}$.

1.6. Exact Sequences

It is not true that q^* is injective and i^* surjective in general. However, this exact sequence can be put into a longer exact sequence. The extension uses higher K-groups defined by suspension:

DEFINITION 1.6.1. *If X is locally compact, the (reduced) suspension SX of X is defined to be $X \times \mathbb{R}$.*

The unreduced suspension of a compact space X is the quotient of $X \times [0, 1]$ obtained by collapsing $X \times \{0\}$ and $X \times \{1\}$ to single points. The reduced suspension is the analogous construction in the category of pointed spaces: form the unreduced suspension of X^+, collapse $\{+\} \times [0,1]$ to a single point, and use this as base point.

One can also suspend a map, so suspension gives a functor from the category of locally compact spaces (or pointed compact spaces) to itself.

DEFINITION 1.6.2. Set $K^0(X) = K(X)$, $K^{-n}(X) = K(S^n X) = K(X \times \mathbb{R}^n)$ for $n > 0$. Similarly, set $K_{\mathbb{R}}^{-n}(X) = K_{\mathbb{R}}(S^n X)$. [The use of negative indices is a convention intended to exhibit K-theory as a cohomology theory. Because of Bott periodicity it is irrelevant in complex K-theory, but is necessary for real K-theory.]

The situation of 1.5.1 yields a short exact sequence $K^{-n}(U) \to K^{-n}(X) \to K^{-n}(Y)$ for each n, and similarly for $K_{\mathbb{R}}^{-n}$.

We now come to the two fundamental results of K-theory:

THEOREM 1.6.3 (LONG EXACT SEQUENCE OF K-THEORY). *Let X be locally compact, Y a closed subspace, $U = X \setminus Y$. Then there is a natural connecting homomorphism $\partial : K^{-n}(Y) \to K^{-n+1}(U)$ which makes the following long sequence exact:*

$$\cdots \xrightarrow{\partial} K^{-n}(U) \xrightarrow{q^*} K^{-n}(X) \xrightarrow{\iota^*} K^{-n}(Y) \xrightarrow{\partial} K^{-n+1}(U) \xrightarrow{q^*} \cdots \xrightarrow{\iota^*} K^0(Y)$$

and similarly for $K_{\mathbb{R}}$.

THEOREM 1.6.4 (BOTT PERIODICITY). *There is a natural isomorphism between $K(X)$ and $K^{-2}(X)$, hence between $K^{-n}(X)$ and $K^{-n-2}(X)$. ("Natural" is in the sense of category theory, i.e. a natural transformation between the functors K^{-n} and K^{-n-2}.) So the long exact sequence of complex K-theory becomes a cyclic 6-term exact sequence*

$$
\begin{array}{ccccc}
K^0(U) & \xrightarrow{q^*} & K^0(X) & \xrightarrow{\iota^*} & K^0(Y) \\
\partial \uparrow & & & & \downarrow \partial \\
K^{-1}(Y) & \xleftarrow{\iota^*} & K^{-1}(X) & \xleftarrow{q^*} & K^{-1}(U)
\end{array}
$$

Bott Periodicity is the place where complex K-theory begins to differ from real K-theory. There is also periodicity in real K-theory, but the period is 8 (i.e. $K_{\mathbb{R}}(X) \cong K_{\mathbb{R}}^{-8}(X)$), so the long exact sequence of real K-theory is a cyclic 24-term exact sequence. Bott Periodicity can be understood and proved using Clifford algebras, and the difference between the real and complex cases is reflected by the greater complexity of real Clifford algebras. (See [Karoubi 1978, III.3] for an exposition of Clifford algebras and their relationship to Bott Periodicity.)

The 6-term exact sequence is one of the primary tools which allow the K-groups of standard spaces to be computed.

THEOREM 1.6.5.   *Real or complex K-theory is an extraordinary cohomology theory, i.e. it is a sequence of homotopy-invariant contravariant functors from compact spaces and compact pairs to abelian groups, with a long exact sequence (1.6.3), and satisfying the excision and continuity axioms (but not the dimension axiom).*

See [Spanier 1966] or [Taylor 1975] for an explanation of these terms.

THEOREM 1.6.6 (CHERN CHARACTER).   *Let X be compact.   Then there are isomorphisms*

$$\chi^0 : K^0(X) \otimes \mathbb{Q} \to \bigoplus_{n \text{ even}} H^n(X;\mathbb{Q}),$$

$$\chi^1 : K^{-1}(X) \otimes \mathbb{Q} \to \bigoplus_{n \text{ odd}} H^n(X;\mathbb{Q}),$$

where $H^n(X;\mathbb{Q})$ denotes the n-th ordinary (Alexander or Čech) cohomology group of X with coefficients in $\mathbb{Q}$.

So, at least rationally, $K^0(X)$ is just the direct sum of the even cohomology groups of X, and $K^{-1}(X)$ the sum of the odd ones.

1.7. Algebraic Formulation of K-Theory

We now describe a way of translating K-theory into an algebraic form which admits a generalization to Banach algebras and, to a lesser extent, to general rings.

1.7.1. Let E be a complex vector bundle over a compact space X. Then the set $\Gamma(E)$ of sections of E has a natural structure as a module over the ring (algebra) $C(X)$ of all complex-valued continuous functions on X. If E is a real vector bundle, then $\Gamma(E)$ is a module over $C_{\mathbb{R}}(X)$, the ring of real-valued continuous functions. If E is a trivial bundle of dimension n, then $\Gamma(E)$ is a free module of rank n. We have $\Gamma(E \oplus F) \cong \Gamma(E) \oplus \Gamma(F)$. So since every bundle is a direct summand of a trivial bundle by Swan's Theorem (1.2.2), $\Gamma(E)$ is always a *projective* module, a direct summand of a free module. Using the compactness of X, the local triviality of E, and the finite-dimensionality of the fibers, it is easy to see that the module $\Gamma(E)$ is *finitely generated*.

Conversely, any finitely generated projective module over $C(X)$ occurs as the module of sections of a bundle. This can be seen most easily by identifying projective modules with idempotents as follows.

The finitely generated projective modules over a unital ring R are exactly the direct summands of R^n for some n. The endomorphism ring of the free module R^n is $M_n(R)$. If V and W are R-modules with $V \oplus W \cong R^n$, then the projection of $V \oplus W$ onto $V \oplus 0$ gives an idempotent in $M_n(R)$. The idempotent so constructed depends on the choice of W and n, and on the identification of $\text{End}(V \oplus W)$ with $M_n(R)$; but it is not difficult to see that it is uniquely determined by V up to similarity. (We identify $x \in M_n(R)$ with $\text{diag}(x,0) \in$

$M_{n+k}(R)$.) Conversely, if e is an idempotent in $M_n(R) \cong \text{End}(R^n)$, then the range of e is a finitely generated projective module.

If $R = C(X)$, we may identify $M_n(R)$ with the algebra $C(X, \mathbb{M}_n)$ of continuous functions from X to $\mathbb{M}_n$. If e is an idempotent in $C(X, \mathbb{M}_n)$, then e is a continuous function from X into the set of idempotents in $\mathbb{M}_n$. We form a bundle E by attaching the range of $e_x \subseteq \mathbb{C}^n$ to x, i.e. $E = \{(x, v) \in X \times \mathbb{C}^n \mid v \in \text{range } e_x\}$. (It is a somewhat nontrivial fact that a bundle defined this way is locally trivial.) Thus every idempotent, and hence every finitely generated projective module, over $C(X)$ comes from a bundle. The Bott bundle of 1.1.2(c) arises this way; the corresponding projection in $M_2(C(S^2))$, called the *Bott projection*, is easily described by identifying $\mathbb{CP}^1$ with the set of rank-one projections in $\mathbb{M}_2$.

Since E and F are isomorphic as bundles if and only if $\Gamma(E) \cong \Gamma(F)$ as modules, we get an isomorphism of the monoid $V(X)$ with the monoid of isomorphism classes of finitely generated projective modules over $C(X)$, with ordinary direct sum. Alternatively, $V(X)$ is isomorphic to the monoid of equivalence classes of idempotents in $M_\infty(C(X)) = \varinjlim M_n(C(X))$.

1.7.2. If R is any unital ring, we may define the monoid $V(R)$ of isomorphism classes of finitely generated projective R-modules, or of equivalence classes of idempotents in matrix algebras over R. (If R is noncommutative, we must specify left modules or right modules; but since the categories are equivalent, the resulting monoid is the same.) Thus we can define $K_0(R)$ to be the Grothendieck group of $V(R)$. K_0 is a *covariant* functor from unital rings to abelian groups (this is why we write K_0 rather than K^0). We have $K^0(X) = K_0(C(X))$ and $K_\mathbb{R}^0(X) = K_0(C_\mathbb{R}(X))$.

If R is nonunital, we define $K_0(R)$ to be the kernel of the homomorphism from $K_0(R^+)$ to $K_0(\mathbb{Z}) \cong \mathbb{Z}$, where R^+ is R with identity adjoined. (If R is a complex algebra, adjoining an identity is usually done by adding a copy of $\mathbb{C}$. For K-theory the results are the same.)

The basic properties of K^0 carry over to this algebraic situation. If $0 \to J \to R \to R/J \to 0$ is an exact sequence of rings, we have a short exact sequence $K_0(J) \to K_0(R) \to K_0(R/J)$ which is an exact generalization of 1.5.1.

1.7.3. Although K_0 works fine in a purely algebraic setting, difficulties arise in trying to define higher algebraic K-groups due to the lack of any reasonable notion of suspension. There is a way of defining higher algebraic K-groups (due to Bass for K_1, Milnor for K_2, and Quillen for general K_n), which is satisfactory in the sense that a long exact sequence is obtained, and the groups have identifiable algebraic significance (at least for n small). However, $K_1^{alg}(C(X))$ does not agree with $K^{-1}(X)$, and the definition of $K_n^{alg}(R)$ becomes successively more complicated and technical at each step. Algebraic K-theory has nonetheless become an important branch of ring theory. See [Rosenberg 1994] for a complete development of algebraic K-theory. The algebraic K-theory of C^*-algebras is of interest in certain contexts (cf. [Connes 1994]), but is not a well-developed theory at present. (The algebraic and topological K-theories coincide for stable C^*-algebras: see [Rosenberg 1997] for an account of this and related results about the algebraic K-theory of C*-algebras.)

1.7.4. If A is a Banach algebra, there is a natural notion of suspension: SA is the Banach algebra of all continuous functions from $\mathbb{R}$ to A which vanish at infinity. We have $S(C_0(X)) \cong C_0(SX)$ if X is a locally compact Hausdorff space. We then define $K_n(A) = K_0(S^n A)$. Remarkably, all of the results of topological K-theory described in this section carry over to the Banach algebra case, notably the long exact sequence and Bott Periodicity (hence the cyclic 6-term exact sequence). Not only do the results hold true, but even the proofs, when expressed in Banach algebra language, carry over verbatim in almost all cases. In fact, the Banach algebra approach is probably the most elegant and natural way to develop topological K-theory. (This approach to topological K-theory actually appeared before Banach algebra K-theory was developed, due primarily to work of Atiyah, Karoubi, Swan, and Wood.)

1.7.5. Not everything in topological K-theory carries over to the noncommutative case, however. The tensor product of vector bundles defines a multiplication on $K^0(X)$ making it into a ring. (Actually one gets a graded ring structure on $K^*(X)$.) This ring structure can be extended to $K_0(R)$ whenever R is a commutative ring, using tensor products of modules; but there is no obvious way of putting a ring structure on $K_0(R)$ for noncommutative R. Similarly, the exterior power operations on modules have no noncommutative analog.

2. Overview of Operator K-Theory

In this section, we will give an overview of the topics to be covered in these notes. The point of view taken here is considerably different than that of Section 1, and is much more in keeping with the traditional ideas of the theory of operator algebras. No knowledge of topological K-theory is assumed; however, it is beneficial to have some understanding of the material of Section 1, particularly 1.7, to appreciate how the ideas developed.

2.1. Noncommutative Topology

The theory we will develop is the heart of the subject of *noncommutative topology*, which is the process of taking a concept from topology, rephrasing it using the (contravariant) equivalence between the category of locally compact Hausdorff spaces and the category of commutative C^*-algebras, and extending the concept in a meaningful way to the category of all C^*-algebras (or some suitable subcategory). The goal of noncommutative topology is to bring ideas, techniques, and results from topology into the study of operator algebras, and vice versa; both areas have already richly benefited from this process.

One of the motivations for developing the theory of noncommutative topology (although not the only one) is that in many instances in ordinary topology the natural object of study is a "singular space" which cannot be defined and studied in purely topological terms. Good examples are the orbit space of a group action or the leaf space of a foliation. Although the singular space X may not really exist topologically, there is often a (noncommutative) C^*-algebra which plays the role of $C_0(X)$ in an appropriate sense. Most of the applications of noncommutative topology to ordinary topology and geometry exploit this point of view.

Most of the noncommutative topology done so far has been noncommutative algebraic topology, the process of extending the functors of ordinary algebraic topology, regarded as functors from commutative C^*-algebras to abelian groups, to more general C^*-algebras. There has been little success so far in extending the standard homotopy, cohomotopy, homology, or cohomology functors (perhaps with good reason—see 22.4.2 and 22.4.3); but the functors of complex K-theory extend very nicely. We will describe these functors below purely in operator algebra terms, referring to Section 1 for the connections with topology.

2.2. The K_0-Functor

The first functor we will consider is the K_0-functor. The goal here is to define a group-valued "universal dimension function," a function D from the projections of a C^*-algebra A to an abelian group G, with the properties that $D(p) = D(q)$ whenever $p \sim q$ and $D(p + q) = D(p) + D(q)$ whenever $p \perp q$, such that any function from the projections of A to an abelian group with these properties factors through D. The prototype is the Murray–von Neumann comparison theory for finite factors, which may be regarded as the K_0-theory of factors. In this theory, the dimension function takes real-number values and completely determines equivalence of projections. In the II_1 case, the dimension group is $\mathbb{R}$, since the values of the dimension function fill up an entire interval in $\mathbb{R}$, and all of $\mathbb{R}_+$ if matrix algebras are considered.

In other cases, however, the group of real numbers is either too large or inappropriate to serve as the range group of the universal dimension function. For example, for $\mathbb{C}$, $\mathbb{M}_n$, or $\mathbb{K}$, the appropriate range group is $\mathbb{Z}$. For the CAR algebra, the proper range group is the dyadic rationals. For a direct sum of two II_1 factors, the group should be $\mathbb{R}^2$. And for an infinite factor, the requirement that the function be group-valued forces the range group to be $\{0\}$.

For any C^*-algebra A, we will define an abelian group $K_0(A)$ which is the appropriate range group for the universal dimension function on A (and on all matrix algebras over A). For A unital, the elements of $K_0(A)$ are formal differences of equivalence classes of projections in matrix algebras over A. (If A is nonunital, the proper definition is less obvious.) Matrix algebras over A must be considered, and the equivalence relation on projections "stabilized", in order to get a group structure on $K_0(A)$. The definition of K_0 is functorial, i.e. any homomorphism $\phi : A \to B$ induces a homomorphism $\phi_* : K_0(A) \to K_0(B)$. In many cases, the group $K_0(A)$ has a natural partial ordering which determines comparability of projections in A; in the particularly nice case of AF algebras, this partially ordered group (with an additional piece of structure called the "scale," the actual range of the dimension function on A) is a complete isomorphism invariant for A.

2.3. The K_1-Functor

The second functor we will examine is K_1. The goal here is to define a "universal index group" or "universal determinant function," a continuous homomorphism from the group of invertible elements of A, or of matrix algebras over A (if A is nonunital, we add an identity), to a (discrete) abelian group. The continuity requires such a homomorphism to be locally constant. The prototype is the Fredholm index map from the invertible elements of the Calkin algebra $\mathbb{B}/\mathbb{K}$ to $\mathbb{Z}$. As with K_0, the group $K_1(A)$ will be the range group of this universal index function, and K_1 will be a covariant functor from C^*-algebras to abelian groups.

There are two connections between K_0 and K_1. First, we have that $K_1(A) \cong K_0(SA)$ for any A, where $SA = C_0(\mathbb{R}, A) \cong A \otimes C_0(\mathbb{R})$, and the more difficult result that $K_0(A) \cong K_1(SA)$ (Bott periodicity). And secondly, if J is a (closed two-sided) ideal in A, there are connecting homomorphisms from $K_1(A/J)$ to $K_0(J)$ and from $K_0(A/J)$ to $K_1(J)$, called the *index* and *exponential* maps respectively (both denoted ∂), making the following six-term sequence exact everywhere:

$$
\begin{array}{ccccc}
K_0(J) & \xrightarrow{\ \iota_*\ } & K_0(A) & \xrightarrow{\ \pi_*\ } & K_0(A/J) \\[4pt]
\partial \uparrow & & & & \downarrow \partial \\[4pt]
K_1(A/J) & \xleftarrow{\ \pi_*\ } & K_1(A) & \xleftarrow{\ \iota_*\ } & K_1(J)
\end{array}
$$

This is called the fundamental exact sequence of K-theory, and is the principal tool used in the calculation of the K-groups of standard C^*-algebras.

2.4. Extensions of C^*-Algebras

The next topic we consider is the theory of extensions of C^*-algebras. Given C^*-algebras A and J, we want to find all C^*-algebras E containing J as an ideal, with $E/J \cong A$, i.e., we want to classify all exact sequences of the form

$$ 0 \longrightarrow J \longrightarrow E \longrightarrow A \longrightarrow 0 $$

(up to a suitable notion of equivalence). Such an algebra or exact sequence is called an *extension of A by J*.

Busby [1968], building on work of Hochschild, discovered that such extensions are classified by homomorphisms from A into $Q(J) = M(J)/J$, the outer multiplier algebra of J. The homomorphism $\tau : A \to Q(J)$ is called the Busby invariant of the extension; an extension is often identified with its Busby invariant.

The first systematic study of extensions was done by Brown, Douglas, and Fillmore [Brown et al. 1977a], who examined the case $J = \mathbb{K}$. They were originally interested in classifying essentially normal operators, but soon realized the problem of classifying the essentially normal operators with essential spectrum X is equivalent to the classification of extensions of $C(X)$ by $\mathbb{K}$. The most natural notion of equivalence was unitary equivalence of the corresponding Busby invariants. The theory turned out to work almost equally well in the case of noncommutative A.

In the case $J = \mathbb{K}$, there is a binary operation on the set $Ext(A)$ of equivalence classes, which makes $Ext(A)$ into an abelian monoid if A is separable (a theorem of Voiculescu); in nice cases, e.g. when A is nuclear, $Ext(A)$ is an abelian group. Ext is a contravariant functor from C^*-algebras to abelian semigroups. There are analogs of Bott periodicity and the six-term cyclic exact sequence of K-theory.

BDF showed that if X is compact, then $Ext(C(X)) \cong K_1(X)$, the first K-homology group of X (K-homology is the homology theory which is dual to complex K-theory). This result attracted great attention from topologists, and may fairly be regarded as the beginning of noncommutative topology as a discipline (although work such as Murray-von Neumann dimension theory, Fredholm index theory, and the Atiyah–Karoubi approach to topological K-theory can in retrospect be considered to be pioneering work in noncommutative topology).

Building on the BDF work, and intermediate work of Pimsner, Popa, and Voiculescu [Pimsner et al. 1979; 1980], Kasparov [1975] studied the case where A is separable and $J = B \otimes \mathbb{K}$ for B a C^*-algebra with a strictly positive element. One obtains a binary operation on the set of unitary equivalence classes as in the BDF case; and when the subsemigroup of trivial extensions (ones whose Busby invariant lifts to a homomorphism from A to $M(J)$) is divided out, an abelian monoid $Ext(A, B)$ is obtained, which is a group in nice cases (e.g. when A is nuclear). Kasparov's Ext is a bifunctor from pairs of (suitably nice) C^*-algebras to abelian groups, contravariant in the first variable and covariant in the second. $Ext(A, \mathbb{C})$ is the BDF semigroup $Ext(A)$; and $Ext(\mathbb{C}, B)$ is isomorphic to $K_1(B)$. There is Bott periodicity and a six-term cyclic exact sequence in each variable separately.

2.5. KK-Theory

The final step (for us) in the development of operator K-theory is the KK-theory of Kasparov [1980b]. Given a pair of C^*-algebras (A, B), with A separable and B containing a strictly positive element, we define an abelian group $KK(A, B)$. There are two standard ways of viewing the elements of $KK(A, B)$, as Fredholm modules or as quasihomomorphisms; each is useful in certain applications. The quasihomomorphism approach (due to Cuntz) is perhaps more intuitive. A quasihomomorphism from A to B is a pair $(\phi, \overline{\phi})$ of homomorphisms from A to $M(B \otimes \mathbb{K})$ which agree modulo $B \otimes \mathbb{K}$. Then $KK(A, B)$ is the set of equivalence classes of quasihomomorphisms from A to B, under a suitable notion of equivalence; the group operation, similar to that on Ext, is essentially orthogonal direct sum. KK is a homotopy-invariant bifunctor from pairs of C^*-algebras to abelian groups, contravariant in the first variable and covariant in the second. $KK(\mathbb{C}, B) \cong K_0(B)$.

The central tool in the theory is the intersection product, a method of combining an element of $KK(A, B)$ with one in $KK(B, C)$ to yield an element of $KK(A, C)$. In the quasihomomorphism picture, the intersection product gives a way of "composing" quasihomomorphisms; if one of the quasihomomorphisms is an actual homomorphism, then the intersection product is really just composition, defined in a straightforward way. The technical details of establishing the properties of the intersection product are formidable.

Armed with the intersection product, it is fairly easy to prove all the important properties of the KK-groups. First, $KK(SA, B) \cong KK(A, SB)$; call this $KK^1(A, B)$. Then we have Bott periodicity $KK^1(SA, B) \cong KK^1(A, SB) \cong KK(A, B) \cong KK(SA, SB)$. Moreover $KK^1(\mathbb{C}, B) \cong K_1(B)$; more generally, if A is nuclear, then $KK^1(A, B) \cong Ext(A, B)$. There are cyclic six-term exact sequences in each variable separately under some mild restrictions.

There are some situations where the six-term exact sequences do not hold in KK-theory, however. This shortcoming and some important potential applications led Connes and Higson to develop E-theory, which may be regarded as a "variant" of KK-theory in which the six-term exact sequences hold in complete generality (for separable C^*-algebras). The basic objects of study in E-theory are *asymptotic morphisms*, paths of maps which become asymptotically *-linear and multiplicative. E-theory is developed in Section 25.

2.6. Further Developments

Work of Cuntz, Higson, Rosenberg, and Schochet shows that K-theory can be characterized (at least for suitably nice C^*-algebras) by a simple set of axioms analogous to the Steenrod axioms of cohomology, and has established generalizations for KK-theory of the Universal Coefficient Theorem and Künneth Theorem of topology, which provide powerful tools for the calculation of the KK-groups of many C^*-algebras.

Operator K-theory has led both to spectacular advances within the subject of operator algebras and to deep applications to problems in geometry and topology. Besides the classification of AF algebras, probably the greatest achievement of K-theory within operator algebras has been to give some insight into the previously mysterious (and still rather mysterious) internal structure of crossed products and free products of C^*-algebras. The two most notable applications to geometry and topology so far have been the various generalizations of the Atiyah–Singer Index Theorem due to Connes, Skandalis, Kasparov, Moscovici, Miščenko, Fomenko, and Teleman, and the work on the homotopy invariance of higher signatures of manifolds by Kasparov and Miščenko and vanishing of "higher $\hat{A}$-genera" of manifolds of positive scalar curvature by Rosenberg (par-

allel to work of Gromov and Lawson). It would appear certain that the work
so far has only scratched the surface of the possibilities. Until recently it would
have been difficult to conceive of theorems in differential topology whose proofs
require the use of operator algebras in an essential way.

2.6.1. The most elegant way to approach operator K-theory would be to plunge
right into KK-theory, beginning in Chapter VIII. Indeed, this would be possible,
since chapters VIII–IX are largely independent of the earlier chapters, and contain all the earlier results (except those on the ordering of K_0 and the structure
of AF algebras) as a special case. I prefer to take a more pedestrian approach
to K-theory, since I feel the "traditional" point of view of chapters III–V has
independent merit, both historical and conceptual, which is nearly impossible to
dig out of the elegant and high-powered approach of the later sections.

We do most of the basic K-theory for general Banach algebras (and even for
"local Banach algebras"), since the theory is identical and requires no more work
except for a few preliminaries, and because the K-theory of some Banach algebras
such as function algebras [Taylor 1975], non-self-adjoint operator algebras [Lance
1985], and group convolution algebras [Rosenberg 1984, 2.1], is of interest. Since
the principal objects of study are C^*-algebras, we restrict to this case whenever
necessary. Beginning with Chapter V, all algebras are C^*-algebras (or sometimes
local C^*-algebras). Even in the earlier sections all examples are C^*-algebras. It
would be a challenging project (one which I will leave to others) to generalize
Extension Theory and Kasparov Theory to more general Banach algebras.

K-Theory for Operator Algebras
MSRI Publications
Volume **5**, second edition, 1998

CHAPTER II

PRELIMINARIES

This chapter is devoted to a development of the basic facts about equivalence of idempotents and projections, inductive limits, and quotients. All readers are urged to at least browse through the chapter, since much of our basic notation is established here.

3. Local Banach Algebras and Inductive Limits

3.1. Local Banach Algebras

While we will be primarily concerned with the study of Banach algebras, in fact almost entirely with C^*-algebras, it is sometimes useful to deal with a dense $*$-subalgebra which is "complete enough" to make the theory work.

DEFINITION 3.1.1. A local Banach algebra is a normed algebra A which is closed under holomorphic functional calculus (i.e. if $x \in A$ and f is an analytic function on a neighborhood of the spectrum of x *in the completion of A,* with $f(0) = 0$ if A is nonunital, then $f(x) \in A$.) For technical reasons we will also require that all matrix algebras over A have the same property. If A is a $*$-algebra, it will be called a *local Banach $*$-algebra*; if the norm is a pre-C^*-norm, A will be called a *local C^*-algebra*.

Note that our definition of a local C^*-algebra does not agree with the definitions in [Behncke and Cuntz 1976] and [Blackadar and Handelman 1982].

EXAMPLES 3.1.2. (a) If A is any C^*-algebra, then the Pedersen ideal [Pedersen 1979, 5.6] $P(A)$ is a local C^*-algebra. In particular, if X is a locally compact Hausdorff space, then $C_c(X)$ is a local C^*-algebra.
(b) Generalizing (a) [Pedersen 1979, 5.6.1], an algebraic direct limit of Banach algebras (see Theorem 3.3.2) is a local Banach algebra.
(c) If X is a C^∞-manifold, then $C_c^\infty(X)$ is a local C^*-algebra.
(d) The domain of a closed $*$-derivation on a C^*-algebra is a local C^*-algebra.
(e) Local C^*-algebras arise naturally in many applications of K-theory to noncommutative differential geometry [Connes 1994].

PROPOSITION 3.1.3. *Let A be a unital local Banach algebra. If $z \in A$ is invertible in the completion $\bar{A}$ of A, then z is invertible in A. So in any local Banach algebra, the spectrum of an element is the same as its spectrum in the completion.*

PROOF. If z is invertible in $\bar{A}$, then the $\bar{A}$-spectrum of z is contained within the region of analyticity of $f(\lambda) = \lambda^{-1}$.   □

COROLLARY 3.1.4. *If A is a unital local Banach algebra, then the invertible elements of A are dense in the invertible elements of $\bar{A}$.*

COROLLARY 3.1.5 [Pedersen 1979, 1.5.7]. *If A is a local C^*-algebra, and ϕ is a *-homomorphism from A into a C^*-algebra B, then ϕ is norm-decreasing.*

3.1.6. Unlike the case of C^*-algebras, an injective homomorphism from a local C^*-algebra A into a C^*-algebra B need not be an isometry, and the image need not be a local C^*-algebra. For example, take for A the subalgebra of $C([0,2])$ consisting of restrictions of functions analytic in a neighborhood of $[0,2]$, take $B = C([0,1])$, and consider the homomorphism given by restriction. The difficulty is the lack of partitions of unity. If A is closed under C^∞ functional calculus of normal elements, then any injective homomorphism from A to a C^*-algebra will be an isometry, so A has a unique pre-C^*-norm. It is true in general that a local C^*-algebra has a unique local C^*-norm.

3.1.7. A local C^*-algebra admits polar decomposition of invertible elements (if z is invertible in A, then $(z^*z)^{-1/2}$ is defined in A and $z = u(z^*z)^{1/2}$ with $u = z(z^*z)^{-1/2}$), so for local C^*-algebras 3.1.4 remains true with "unitary" in place of "invertible element".

3.1.8. If A is a normed algebra with the property that whenever $x \in A$ and f is analytic on a disk centered at 0 of radius greater than $\|x\|$ (with $f(0) = 0$ if A is nonunital), then $f(x) \in A$, one can show that A is a local Banach algebra [Schmitt 1991, 2.1]. It follows that if A is a local Banach algebra and J is a closed two-sided ideal, then A/J is also a local Banach algebra.

3.2. Unitization

DEFINITION 3.2.1. $A^+ = \{(a, \lambda) : a \in A, \lambda \in \mathbb{C}\}$ is A with identity adjoined (even if A already has one.) The multiplication is given by $(a, \lambda) \cdot (b, \mu) = (ab + \lambda b + \mu a, \lambda\mu)$.

A^+ is a local Banach algebra under the norm $\|(a, \lambda)\| = \|a\| + |\lambda|$ (or other equivalent norm.) If A is complete, then A^+ is also complete. A is a two-sided ideal in A^+. Let $\tilde{A}$ be the smallest unital subalgebra of A^+ containing A, i.e. $\tilde{A} = A$ if A has an identity and $\tilde{A} = A^+$ otherwise.

If A is a C^*-algebra, then A^+ can be made into a C^*-algebra in a unique way [Pedersen 1979, 1.1.3].

We denote the $n \times n$ matrix algebra over A by $M_n(A)$. $M_n(A)$ can be made into a local Banach algebra in many equivalent ways (e.g. with the operator norm on $\tilde{A}^n$); if A is a local C^*-algebra, then $M_n(A)$ can be made into a local C^*-algebra in a unique way.

3.3. Inductive Limits

We now describe the construction of the inductive limit of a directed set of Banach algebras.

Given a set $\{X_i : i \in I\}$ of semigroups, groups, rings, algebras, etc., where I is a directed set, and a coherent family of morphisms $\phi_{ij} : X_i \to X_j$ for $i < j$ (coherent means that $\phi_{ik} = \phi_{jk} \circ \phi_{ij}$ for $i < j < k$), one can form the direct limit $X = \varinjlim(X_i, \phi_{ij})$, usually denoted $\varinjlim X_i$ when the ϕ_{ij} are understood. When all the ϕ_{ij} are injective, we can (modulo logical technicalities) regard X as the union of the X_i. X has the usual universal property: any coherent family of morphisms from the X_i to Y induces a unique morphism from X to Y. A complete discussion can be found in many standard texts such as [Grätzer 1968].

An inductive system (A_i, ϕ_{ij}) of local Banach algebras is called a *normed inductive system* if each ϕ_{ij} is bounded and $\limsup_j \|\phi_{ij}\| < \infty$ for each i. If each A_i is complete, it follows easily from the Banach–Steinhaus theorem that (A_i, ϕ_{ij}) is a normed inductive system if and only if each ϕ_{ij} is bounded and for each i and each $x \in A_i$ we have $\limsup_j \|\phi_{ij}(x)\| < \infty$. Any inductive system of local C^*-algebras and $*$-homomorphisms is a normed inductive system by 3.1.5.

If (A_i, ϕ_{ij}) is a normed inductive system, then the algebraic direct limit has a natural seminorm $\|\|x\|\| = \limsup_j \|\phi_{ij}(x)\|$. The *normed inductive limit* is the quotient of the algebraic direct limit by the elements of seminorm 0, and the *Banach algebra inductive limit* is the completion of the normed inductive limit. Unless otherwise qualified, "inductive limit" will mean "Banach algebra inductive limit". We will reserve the notation $\varinjlim A_i$ for the inductive limit, denoting the algebraic direct limit by $\underrightarrow{\mathrm{alglim}}\, A_i$. There is a canonical homomorphism ϕ_i from A_i to the normed inductive limit, which is continuous (bounded).

A particularly important case occurs when each ϕ_{ij} is an isometric embedding (an injective $*$-homomorphism between C^*-algebras is automatically an isometry.) In this case, the normed inductive limit coincides with the algebraic direct limit and may be thought of as $\bigcup A_i$.

Note that if $A = \varinjlim A_i$, then $A^+ = \varinjlim A_i^+$ under unital maps, and $M_n(A) = \varinjlim M_n(A_i)$.

LEMMA 3.3.1. *If A is the normed inductive limit of (A_i, ϕ_{ij}) under unital maps, and $z \in A_i$ with $\phi_i(z)$ invertible in $\bar{A}$, then for sufficiently large $j > i$, $\phi_{ij}(z)$ is invertible in A_j.*

PROOF. For sufficiently large $k > i$ there is a $w \in A_k$ with $\phi_i(z)\phi_k(w)$ and $\phi_k(w)\phi_i(z)$ both close to 1. Then for sufficiently large $j > k$ both $\phi_{ij}(z)\phi_{kj}(w)$ and $\phi_{kj}(w)\phi_{ij}(z)$ are close to 1, so $\phi_{ij}(z)$ is left and right invertible in A_j. $\square$

THEOREM 3.3.2. *If A is the normed inductive limit of (A_i, ϕ_{ij}), then A is a local Banach algebra.*

PROOF. We may assume that the inductive limit is unital. Let $x \in A$ and f analytic on a bounded neighborhood U of $\sigma_{\bar{A}}(x)$. Let $z \in A_i$ with $\phi_i(z) = x$,

and let D be a closed disk containing U of radius greater than $\limsup_j \|\phi_{ij}(z)\|$.
For each $\lambda \in D \setminus U$ there is a $j > i$ with $\lambda - \phi_{ij}(z)$ invertible in A_j by 3.3.1,
so there is a neighborhood V of λ such that $\mu - \phi_{ij}(z)$ is invertible in A_j for all
$\mu \in V$. Using the compactness of $D \setminus U$, for sufficiently large $j > i$, $\lambda - \phi_{ij}(z)$ is
invertible in A_j for all $\lambda \in D \setminus U$. This implies $\sigma_{A_j}(\phi_{ij}(z))$ is contained in U, so
$f(\phi_{ij}(z))$ is defined in A_j and its image is $f(x)$.   □

The argument in 3.3.2 can be greatly simplified in the case of an inductive limit
of C^*-algebras and injective connecting maps.

Combining 3.1.4, 3.3.1, and 3.3.2, we get:

PROPOSITION 3.3.3. *Let $A = \varinjlim A_i$ with unital maps, and let z be invertible
in A. Then, for any $\varepsilon > 0$, for all sufficiently large i there is an invertible w in
A_i with $\|\phi_i(w) - z\| < \varepsilon$.*

It follows from 3.1.7 that if the A_i are local C^*-algebras, then 3.3.3 remains true
with "invertible" replaced by "unitary".

3.4. Invertible Elements

We conclude this section with two results about the connected component of
the identity in the group of invertible elements in a unital local Banach algebra,
and one additional anomalous result which will be needed later.

PROPOSITION 3.4.1. *Let A be a unital local Banach algebra. If x and y are invertible in A, then there is a path of invertible elements in $M_2(A)$ from $\operatorname{diag}(xy, 1)$
to $\operatorname{diag}(x, y)$. If A is a local C^*-algebra and x and y are unitary, the path may
be chosen to consist of unitaries. So if z is invertible in A, then there is a path
of invertibles in $M_2(A)$ from 1 to $\operatorname{diag}(z, z^{-1})$.*

PROOF. Set $w_t = \operatorname{diag}(x, 1) \cdot u_t \cdot \operatorname{diag}(y, 1) \cdot u_t^{-1}$, where

$$u_t = \begin{bmatrix} \cos \frac{\pi}{2} t & -\sin \frac{\pi}{2} t \\ \sin \frac{\pi}{2} t & \cos \frac{\pi}{2} t \end{bmatrix}. \qquad \qquad □$$

Since $\operatorname{diag}(x, y)$ and $\operatorname{diag}(y, x)$ are connected by a similar path, 3.4.1 implies that
$\operatorname{diag}(xy, 1)$ and $\operatorname{diag}(yx, 1)$ are connected by a path of invertibles.

3.4.1 is a variant of the "Whitehead lemma" [Whitehead 1941; Bass 1968;
Rosenberg 1994], which says that $\operatorname{diag}(xy, 1)$, $\operatorname{diag}(yx, 1)$, and $\operatorname{diag}(x, y)$ are
all obtainable from one another by left and right multiplication by products of
elementary matrices. The form of the proof of 3.4.1 (cf. [Atiyah 1967, 2.4.6]) is
more suitable for our purposes.

3.4.2. If A is a local Banach algebra, we write $\operatorname{GL}_n(A)$ to denote the group of
invertible elements in $M_n(A^+)$ which are congruent to 1_n mod $M_n(A)$. If A is
unital, then $\operatorname{GL}_n(A)$ is isomorphic to the group of invertible elements of $M_n(A)$.
If A is a local C^*-algebra, we let $U_n(A)$ denote the group of unitaries in $M_n(A^+)$
which are congruent to 1_n mod $M_n(A)$. These are topological groups with the
norm topology. Let $\operatorname{GL}_n(A)_0$ and $U_n(A)_0$ denote the connected components of

the identity. Since the groups are locally path-connected, these are also the path components of the identity; and they are open subgroups.

PROPOSITION 3.4.3. *Let A be a unital local Banach algebra. Then $\mathrm{GL}_1(A)_0$ is the subgroup of $\mathrm{GL}_1(A)$ generated algebraically by $\{e^x : x \in A\}$.*

PROOF. e^x is invertible (with inverse e^{-x}), and $\{e^{tx} : 0 \leq t \leq 1\}$ is a path to 1, so $e^x \in \mathrm{GL}_1(A)_0$. On the other hand, if $\|z - 1\| < 1$, then $x = \log z$ is defined and satisfies $e^x = z$. So the group generated algebraically by $\{e^x : x \in A\}$ is open. $\square$

COROLLARY 3.4.4. *If A and B are unital local Banach algebras and $\phi : A \to B$ is a continuous surjective homomorphism, then $\phi(\mathrm{GL}_1(A)_0) = \mathrm{GL}_1(B)_0$.*

PROOF. If $x \in A$ with $\phi(x) = y$, then $\phi(e^x) = e^y$, so all exponentials must lie in $\phi(\mathrm{GL}_1(A)_0)$, which is a connected subgroup of $\mathrm{GL}_1(B)$. $\square$

If A and B are Banach algebras, a (simpler) proof of 3.4.4 can be given using the Open Mapping Theorem instead of 3.4.3.

Thus, although invertible elements in quotients cannot in general be lifted to invertible elements, ones in the connected component of 1 can always be lifted (with the lifted invertible also in the connected component of the identity.)

As usual, for local C^*-algebras there is a unitary analog of 3.4.3 and 3.4.4:

PROPOSITION 3.4.5. *If A is a unital local C^*-algebra, then $U_1(A)_0$ is generated algebraically by $\{e^{ix} : x = x^* \in A\}$. If B is also a unital local C^*-algebra and $\phi : A \to B$ is surjective and unital, then $\phi(U_1(A)_0) = U_1(B)_0$.*

The final result concerns lifting homotopies from quotient algebras. This result is probably true in greater generality and may be known, but I could find no reference.

PROPOSITION 3.4.6. *Let A be a C^*-algebra, J a closed two-sided ideal in A, and $\pi : A \to A/J$ the quotient map. Let $(\bar{x}_t)$, with $0 \leq t \leq 1$, be a norm-continuous path of elements in A/J, and let x_0 and x_1 be elements of A with $\pi(x_i) = \bar{x}_i$ for $i = 0, 1$. Then there is a norm-continuous path (x_t) of elements of A from x_0 to x_1 with $\pi(x_t) = \bar{x}_t$ for all t.*

PROOF. Set $\bar{y}_t = \bar{x}_t - (1-t)\bar{x}_0 - t\bar{x}_1$. Then $\bar{y}_0 = \bar{y}_1 = 0$. Using the surjectivity of the map $C_0((0,1), A)/C_0((0,1), J) \to C_0((0,1), A/J)$ (a fact about C^*-tensor products), we can find a continuous path $(y_t) \subseteq A$ with $y_0 = y_1 = 0$ and $\pi(y_t) = \bar{y}_t$ for all t. Set $x_t = y_t + (1-t)x_0 + tx_1$. $\square$

4. Idempotents and Equivalence

In this section, we will study the basic properties of idempotents in local Banach algebras, and the various notions of equivalence. Throughout, A will denote a local Banach algebra.

4.1. Idempotents

DEFINITION 4.1.1. An *idempotent* in A is an element e with $e^2 = e$. Two idempotents e and f are *orthogonal* ($e \perp f$) if $ef = fe = 0$. If $e \perp f$, then $e + f$ is an idempotent. We have $e \perp (1 - e)$. We say that $e \leq f$ if $ef = fe = e$. This implies that $f - e$ is an idempotent.

If A is a local C^*-algebra, this is not the usual notion of orthogonality, which also requires that $e^*f = fe^* = 0$. The two notions coincide for projections. The one given here is more useful for general idempotents.

4.2. Equivalence of Idempotents

There are several notions of equivalence of idempotents:

DEFINITION 4.2.1. Let e and f be idempotents in A.
$e \sim f$ if there are $x, y \in A$ with $xy = e$ and $yx = f$. (algebraic equivalence)
$e \sim_s f$ if there is an invertible z in $\tilde{A}$ with $zez^{-1} = f$. (similarity)
$e \sim_h f$ if there is a norm-continuous path of idempotents in A from e to f. (homotopy)

PROPOSITION 4.2.2. *If $e \sim f$, then there exist $x, y \in A$ with $xy = e$, $yx = f$, $x = ex = xf = exf$, $y = fy = ye = fye$.*

PROOF. Replace x by exf, y by fye. Then $(exf)(fye) = exfye = exyxye = e^4 = e$, and similarly for $(fye)(exf)$. □

COROLLARY 4.2.3. $\sim$ *is an equivalence relation.*

PROOF. Let $e = xy$, $f = yx = zw$, $g = wz$, with $x = exf$, $y = fye$, $z = fzg$, $w = gwf$. Then $(xz)(wy) = xfy = e$, and $(wy)(xz) = wfz = g$. □

PROPOSITION 4.2.4. *If $e_1 \sim f_1$, $e_2 \sim f_2$, $e_1 \perp e_2$, $f_1 \perp f_2$, then $e_1 + e_2 \sim f_1 + f_2$.*

PROOF. For $i = 1, 2$, let $x_i y_i = e_i$, $y_i x_i = f_i$, $x_i = e_i x_i f_i$, $y_i = f_i y_i e_i$. Then $(x_1 + x_2)(y_1 + y_2) = e_1 + e_2$ and $(y_1 + y_2)(x_1 + x_2) = f_1 + f_2$. □

PROPOSITION 4.2.5. *$e \sim_s f$ if and only if $e \sim f$ and $1 - e \sim 1 - f$.*

PROOF. If $zez^{-1} = f$, set $x = ez^{-1}, y = ze$. Similarly, $z(1 - e)z^{-1} = 1 - f$, so $1 - e \sim 1 - f$. Conversely, if $xy = e$, $yx = f$, $x = exf$, $y = fye$, and $ab = 1 - e$, $ba = 1 - f$, $a = (1 - e)a(1 - f)$, $b = (1 - f)b(1 - e)$, then $z = x + a$ is invertible with $z^{-1} = y + b$, and $zez^{-1} = f$. □

4.3. Algebraic Equivalence, Similarity, and Homotopy

It is not true that $e \sim f$ implies $e \sim_s f$. For example, if p is an infinite-rank projection in $\mathbb{B}$, then $p \sim 1$, but p is not similar to 1. However:

PROPOSITION 4.3.1. *If $e \sim f$, then* $\begin{bmatrix} e & 0 \\ 0 & 0 \end{bmatrix} \sim_s \begin{bmatrix} f & 0 \\ 0 & 0 \end{bmatrix}$ *in $M_2(A)$.*

PROOF. Let x, y be as in 4.2.2, and

$$z = \begin{bmatrix} 1 - f & y \\ x & 1 - e \end{bmatrix} \begin{bmatrix} 1 - e & e \\ e & 1 - e \end{bmatrix}.$$

Then

$$z^{-1} = \begin{bmatrix} 1 - e & e \\ e & 1 - e \end{bmatrix} \begin{bmatrix} 1 - f & y \\ x & 1 - e \end{bmatrix}. \qquad \square$$

PROPOSITION 4.3.2. *If $\|e - f\| < 1/\|2e - 1\|$, then $e \sim_s f$. In fact, there is a $z \in \tilde{A}$ with $\|z - 1\| \le \|2e - 1\| \|e - f\|$ and $z^{-1}ez = f$. Also, $e \sim_h f$.*

PROOF. Set $v = (2e - 1)(2f - 1) + 1$. We have $2 - v = 2(2e - 1)(e - f)$, so if $2\|2e - 1\|\|e - f\| < 2$, then v is invertible; and $ev = vf = 2ef$. Set $z = v/2$ and $w_t = tz + (1 - t)$. Then w_t is invertible. Set $e_t = w_t^{-1}ew_t$. $\qquad \square$

PROPOSITION 4.3.3. *If $e \sim_h f$ with path e_t, there is a path z_t of invertibles with $z_0 = 1$ and $z_t^{-1}ez_t = e_t$ for all t. So $e \sim_s f$.*

PROOF. Choose K with $\|2e_t - 1\| \le K$ for all t, and find $0 = t_0 < t_1 < \cdots < t_n = 1$ with $\|e_t - e_s\| < 1/K$ if s and t are in the same interval. Set $v_t = (2e_{t_i} - 1)(2e_t - 1) + 1$ for $t_i \le t \le t_{i+1}$, and $u_t = v_t/2$. Let $z_t = u_t$ for $0 \le t \le t_1$, $z_t = u_t u_{t_1}$ for $t_1 \le t \le t_2$, ..., $z_t = u_t u_{t_{n-1}} \ldots u_{t_1}$ for $t_{n-1} \le t \le 1$. $\qquad \square$

4.3.2 can be rephrased into a result which can be regarded as the fundamental technical theorem of K-theory. We let $\mathrm{Ip}(A)$ be the set of idempotents in A, and if A is unital A^{-1} denotes the set of invertible elements in A. Give these sets the norm topology; A^{-1} is a topological group.

THEOREM 4.3.4. *Let A be a local Banach algebra. Consider the action γ of $\tilde{A}^{-1}$ on $\mathrm{Ip}(A)$ given by $\gamma(z)(e) = zez^{-1}$. Then:*

(a) *The orbits of γ are open and closed; the orbit containing e contains the open ball B_e of radius $\|2e - 1\|^{-1}$ around e in $\mathrm{Ip}(A)$*

(b) *For any $e \in \mathrm{Ip}(A)$, there is a continuous cross section $w_e : B_e \to \tilde{A}^{-1}$ for γ (i.e. $w_e(f)ew_e(f)^{-1} = f$) with $w_e(e) = 1$.*

(c) *The association $e \to w_e$ is functorial, i.e. if $\phi : A \to B$ is a (bounded) homomorphism, then $w_{\phi(e)}(\phi(f)) = \phi(w_e(f))$ for f in a sufficiently small neighborhood of e (precisely, $\|e - f\| < (\|\phi\|\|2e - 1\|)^{-1}$).*

PROOF. This is mostly contained in the proof of 4.3.2. The formula for $w_e(f)$ is $\frac{1}{2}((2e - 1)(2f - 1) + 1)$. The assertions not spelled out in the proof of 4.3.2 are routine, and are left to the reader. $\qquad \square$

4.4. Similarity vs. Homotopy

It is not true that $e \sim_s f$ implies $e \sim_h f$. There is a counterexample in $M_2(C(S^3))$. However, as with $\sim$ and $\sim_s$, there is a 2×2 matrix result:

PROPOSITION 4.4.1. *If $e \sim_s f$, then* $\begin{bmatrix} e & 0 \\ 0 & 0 \end{bmatrix} \sim_h \begin{bmatrix} f & 0 \\ 0 & 0 \end{bmatrix}$.

PROOF. Let $zez^{-1} = f$, and let w_t be a path from 1 to $\mathrm{diag}(z, z^{-1})$ as in 3.4.1. Set $e_t = w_t \, \mathrm{diag}(e, 0) w_t^{-1}$.   □

4.5. Equivalence and Completion

4.3.2 also implies that idempotents and equivalence in the completion of A can be approximated in A:

PROPOSITION 4.5.1. *Let A be a local Banach algebra, e an idempotent in $\bar{A}$, and $\varepsilon > 0$. Then there is an idempotent e_0 in A with $\|e - e_0\| < \varepsilon$. If f is another idempotent in $\bar{A}$ with $e \sim_s f$, then there is an f_0 in A with $\|f - f_0\| < \varepsilon$ and $e_0 \sim_s f_0$ within A.*

PROOF. If $x \in A$ with $\|x - e\|$ small enough, then $\|x - x^2\|$ will be small, so the spectrum of x will be contained in the union of a small disk around 0 and a small disk around 1; so e_0 can be made from x by holomorphic functional calculus. For the second part, let $z \in \bar{A}$ with $zez^{-1} = f$. Approximate z closely by an invertible $w \in A$ (3.1.4), and set $f_0 = we_0w^{-1}$.   □

The similarity of 4.5.1 and 3.1.4 is no accident. In fact, one of the philosophical tenets of K-theory is the close relationship between idempotents and invertible elements (or between projections and unitaries in a C^*-algebra.)

PROPOSITION 4.5.2. *Let A be the normed inductive limit of (A_i, ϕ_{ij}), and e an idempotent in A. Then for sufficiently large i there is an idempotent $e_0 \in A_i$ with $\phi_i(e_0) = e$. If $e, f \in A$ with $e \sim_s f$, then there are idempotent preimages e_0 and f_0 in A_i for sufficiently large i with $e_0 \sim_s f_0$ in A_i.*

PROOF. Let x be a preimage of e in A_i for some i. We have $\bigcap_{j>i} \sigma(\phi_{ij}(x)) = \sigma(e) = \{0, 1\}$ by 3.3.1, so for sufficiently large j 0 and 1 lie in different components of $\sigma(\phi_{ij}(x))$, and we may find disjoint open sets $U, V \subseteq \mathbb{C}$ with $0 \in U$, $1 \in V$, and $\sigma(\phi_{ij}(x)) \subseteq U \cup V$. Set $g = 0$ on U, $g = 1$ on V, and $e_0 = g(\phi_{ij}(x))$.

For the second part, let $z \in \bar{A}$ with $zez^{-1} = f$. Choose i large enough that e and f have idempotent preimages e_0 and f_0 in A_i, and z has an invertible preimage w in $\tilde{A}_i$. Set $f_1 = we_0w^{-1}$. Since $\phi_i(f_1) = \phi_i(f_0) = f$, by increasing i if necessary we may make $\|f_1 - f_0\|$ small, so $e_0 \sim_s f_1 \sim_s f_0$.   □

4.6. Projections

We now let A be a local C^*-algebra, and show that all relevant aspects of idempotent theory can be done with projections. Some of the results are true for more general local Banach *-algebras.

DEFINITION 4.6.1. A *projection* is a self-adjoint idempotent. A *partial isometry* is an element u with u^*u a projection. If A has a unit, then u is an *isometry* if $u^*u = 1$, and u is *unitary* if $u^*u = uu^* = 1$.

If p and q are projections, the following conditions are equivalent: (i) $p \leq q$ as idempotents; (ii) $p \leq q$ in the sense of positive elements of A; (iii) $p \leq \lambda q$ (in the sense of positive elements of A) for some $\lambda > 0$; (iv) $q - p$ is a projection.

We will usually use p and q to denote projections and e and f to denote general idempotents.

PROPOSITION 4.6.2. *Every idempotent in A is similar to a projection. In fact, any idempotent is homotopic to a projection.*

PROOF. Let e be an idempotent. Set $z = 1 + (e - e^*)(e^* - e)$. Then z is an invertible positive element of $\tilde{A}$; set $r = z^{-1}$. We have $ze = ez = ee^*e$, so $er = re, re^* = e^*r$. Set $p = ee^*r$. Then $p = p^*$, and $p^2 = ee^*ree^*r = ree^*ee^*r = rzee^*r = ee^*r = p$. Moreover $ep = p, pe = e$. For any $t \in \mathbb{R}$, $(1 - te + tp)$ is invertible, with inverse $(1 - tp + te)$; $(1 - p + e)e(1 - e + p) = p$. [Actually p is the range projection of e, or the support projection of ee^*, in A^{**}; this argument shows that $p \in A$.] □

A similar proof shows:

PROPOSITION 4.6.3. *If $p \sim_h q$, then there is a path of projections from p to q.*

PROOF. Let e_t be a path of idempotents from p to q. As in the proof of 4.6.2, let $z_t = 1 + (e_t - e_t^*)(e_t^* - e_t)$, $p_t = e_t e_t^* z_t^{-1}$. □

The proofs of 4.6.2 and 4.6.3 work in any regular local Banach *-algebra, i.e. one in which $1 + x^*x$ is always invertible.

PROPOSITION 4.6.4. *If $p \sim q$, then there is a partial isometry u with $u^*u = p, uu^* = q$.*

PROOF. Let $p = xy$, $q = yx$ with $x = pxq$, $y = qyp$. Then $p = p^*p = y^*x^*xy \leq \|x\|^2 y^*y$, so y^*y is invertible in pAp. Let $(y^*y)^{1/2}r = p$ with $r = prp$, and set $u = yr$. [This u is the partial isometry in the polar decomposition of y in A^{**}. The argument shows that $u \in A$.] Then $y = u(y^*y)^{1/2}$. $u^*u = p$. Also, $q(uu^*) = uu^*$, so $uu^* \leq q$. But $q = yxx^*y^* \leq \|x\|^2 yy^* = \|x\|^2 u(y^*y)u^* \leq \|x\|^2\|y\|^2 uu^*$. □

The u in 4.6.4 is called a *partial isometry from p to q*, and p and q are said to be *Murray-von Neumann equivalent.*

PROPOSITION 4.6.5. *If $p \sim_s q$, then there is a unitary u with $upu^* = q$.*

PROOF. Let $zpz^{-1} = q$. Then $zp = qz$ and $pz^* = z^*q$, so $pz^*z = z^*qz = z^*zp$, and so $p(z^*z)^{1/2} = (z^*z)^{1/2}p$. Set $u = z(z^*z)^{-1/2}$; then u is unitary and $upu^* = q$. □

If the conclusion of 4.6.5 holds, we say p and q are *unitarily equivalent*, written $p \sim_u q$.

There are exact analogs of 4.3.2 through 4.5.2 with "idempotent" replaced by "projection" and "invertible element" by "unitary". Details are left to the reader. Note that if p is a projection, then $2p - 1$ is a (self-adjoint) unitary [and conversely], so $\|2p - 1\| = 1$. Thus for projections 4.3.2 becomes:

PROPOSITION 4.6.6. *If p and q are projections with $\|p - q\| < 1$, then $p \sim_h q$, so $p \sim_u q$.*

In this case, there is an explicit path as follows. Let $u = 2p - 1$, $v = 2q - 1$, $r = uv + vu = 2 - 4(p - q)^2$ in $\tilde{A}$. Then $u^2 = v^2 = 1$, and r commutes with u and v. We have $r > -2$, so $1 + \lambda r$ is positive and invertible for $0 \le \lambda \le 1/2$. Set $u_t = (1 + r \cos t \sin t)^{-1/2}(u \cos t + v \sin t)$ for $0 \le t \le \pi/2$. u_t is a self-adjoint unitary; set $p_t = (u_t + 1)/2$. Then (p_t) is a path of projections in A from p to q. (We have $p_t \in A$ since $u = v = -1$ and $r = 2$ mod A, so $u_t = -1$ mod A.)

4.3.4 is also worth stating explicitly for projections:

THEOREM 4.6.7. *Let A be a local C^*-algebra. Consider the action γ of $U(\tilde{A})$ on the set $\mathrm{Pr}(A)$ of projections in A given by $[\gamma(u)](p) = upu^*$. Then:*

(a) *The orbits of γ are open and closed; the orbit containing p contains the open ball B_p of radius 1 around p in $\mathrm{Pr}(A)$.*

(b) *For any $p \in \mathrm{Pr}(A)$, there is a continuous cross section $v_p : B_p \to U(\tilde{A})$ for γ (i.e. $v_p(q)pv_p(q)^* = q$) with $v_p(p) = 1$.*

(c) *The association $p \to w_p$ is functorial, i.e. if $\pi : A \to B$ is a $*$-homomorphism, then $v_{\pi(p)}(\pi(q)) = \pi(v_p(q))$ for $q \in B_p$.*

The unitary $v_p(q)$ is the unitary in the polar decomposition of $w_p(q)$ of 4.3.4.

4.7. EXERCISES AND PROBLEMS

4.7.1. [Blackadar 1985b] (a) A separable C^*-algebra A is *semiprojective* if, for any C^*-algebra B, any increasing sequence $\langle J_n \rangle$ of (closed two-sided) ideals of B, and any $*$-homomorphism $\phi : A \to B/J$, where $J = \left(\bigcup J_n\right)^-$, there is an n and a $*$-homomorphism $\psi : A \to B/J_n$ such that $\phi = \pi \circ \psi$, where $\pi : B/J_n \to B/J$ is the natural quotient map. A ϕ for which such a ψ exists is said to be *partially liftable*. If there is a $\psi : A \to B$ with $\phi = \pi \circ \psi$, then ϕ is *liftable*; if every homomorphism from A is liftable, A is said to be *projective*.

(b) $C_0((0, 1])$ is the universal C^*-algebra generated by a positive element of norm 1. Since such elements can be lifted from quotients, $C_0((0, 1])$ is projective.

(c) Let B, J_n, and J be as in (a), and let $q_1, \ldots, q_k$ be mutually orthogonal projections in B/J. Then for sufficiently large n, there are mutually orthogonal projections $p_1, \ldots, p_k$ in B/J_n with $\pi(p_k) = q_k$. If B (and hence B/J) is unital and $q_1 + \cdots + q_k = 1$, then we may choose the p_j so that $p_1 + \cdots + p_k = 1$. [Prove this by induction on k; use functional calculus to do the case $k = 1$ (see the proof of 4.5.1).]

(d) Let B, J_n, J be as in (a). Let v be a partial isometry in B/J, and set $q_1 = v^*v$, $q_2 = vv^*$. Suppose there are projections $p_1, p_2 \in B/J_n$ for some n with $\pi(p_j) = q_j$. Then, after increasing n if necessary, there is a partial isometry

$u \in B/J_n$ with $\pi(u) = v$ and $p_1 = u^*u$, $p_2 = uu^*$. [Use functional calculus and 4.3.4.]

(e) Use (c) and (d) to show that the following C^*-algebras are semiprojective: $\mathbb{C}$, $\mathbb{M}_n$, $C(\mathbb{T})$, $S = C_0(\mathbb{R})$, the Toeplitz algebra T (9.4.2), the Cuntz algebras O_n (6.3.2), and the Cuntz–Krieger algebras O_A (10.11.9). (O_∞ is also semiprojective [Blackadar 1998], but this does not follow from (c) and (d).) [Write each of these as a universal C^*-algebra on a finite set of partial isometries satisfying simple algebraic relations.] None of these C^*-algebras are projective.

(f) [Blackadar 1985b, 2.19; Loring 1997] A finite direct sum of semiprojective C^*-algebras is semiprojective.

(g) [Blackadar 1985b, 2.28–2.29; Loring 1997] If A is semiprojective, then $M_n(A)$ is semiprojective for all n. If A is unital and semiprojective, then any unital C^*-algebra strongly Morita equivalent to A is also semiprojective.

(h) [Blackadar 1985b, 3.1] Let A be a semiprojective C^*-algebra, and $(B_n, \beta_{m,n})$ be an inductive system of C^*-algebras with $B = \varinjlim(B_n, \beta_{m,n})$. If $\phi : A \to B$ is a homomorphism, then for all sufficiently large n there are homomorphisms $\phi_n : A \to B_n$ such that $\beta_n \circ \phi_n$ is homotopic to ϕ and converges pointwise to ϕ as $n \to \infty$, where β_n is the standard map from B_n to B.

The notions of projectivity and semiprojectivity were introduced in the development of shape theory for C^*-algebras [Effros and Kaminker 1986; Blackadar 1985b] as noncommutative analogs of the topological notions of absolute retract (AR) and absolute neighborhood retract (ANR) respectively. Semiprojective C^*-algebras have rigidity properties which make them conceptually and technically important in several aspects of C^*-algebra theory; this is reflected especially in the work of Loring [1997] and his coauthors. It is not too easy for a C^*-algebra to be semiprojective, but there does seem to be a reasonable supply of such algebras.

Notes for Chapter II

Many of the results in this chapter are ancient folklore, and it is difficult to trace the origin of some. Most of the results about idempotents, such as 4.6.2, are due to Kaplansky [1968]; see also [Berberian 1972]. 4.2.5 is due to Sz.-Nagy; the proof given here is taken from [Kasparov 1980b, § 6, Lemma 4].

K-Theory for Operator Algebras
MSRI Publications
Volume **5**, second edition, 1998

CHAPTER III

K_0-THEORY AND ORDER

5. Basic K_0-Theory

Throughout this section, A will denote a local Banach algebra. (In fact, for much of the section A could just be any ring.)

5.1. Basic Definitions

The definition of $K_0(A)$ requires simultaneous consideration of all of the matrix algebras over A. The most elegant way to do this is the following:

DEFINITION 5.1.1. $M_\infty(A)$ is the algebraic direct limit of $M_n(A)$ under the embeddings $a \to \mathrm{diag}(a,0)$.

$M_\infty(A)$ may be thought of as the algebra of all infinite matrices over A with only finitely many nonzero entries. Whenever it is convenient, we will identify $M_n(A)$ with its image in the upper left-hand corner of $M_{n+k}(A)$ or $M_\infty(A)$. When it is necessary to topologize $M_\infty(A)$, any topology inducing the natural topologies on $M_n(A)$ will do. For example, $M_\infty(A)$ may be given the inductive limit topology if desired. Better yet, one may choose the norms on $M_n(A)$ so that the embeddings are isometries; $M_\infty(A)$ is a local Banach algebra with the induced norm.

If A is a (local) C^*-algebra, the embeddings are isometries, so $M_\infty(A)$ has a natural norm. The completion is called the *stable algebra* of A, denoted $A \otimes \mathbb{K}$ (it is the C^*-tensor product of A and $\mathbb{K}$).

DEFINITION 5.1.2. $\mathrm{Proj}(A)$ is the set of algebraic equivalence classes of idempotents in A. We set $V(A) = \mathrm{Proj}(M_\infty(A))$.

There is a binary operation (orthogonal addition) on $V(A)$: if $[e], [f] \in V(A)$, choose $e' \in [e]$ and $f' \in [f]$ with $e' \perp f'$ (this is always possible by "moving down the diagonal"), and define $[e] + [f] = [e' + f']$. This operation is well defined by 4.2.4, and makes $V(A)$ into an abelian semigroup with identity $[0]$.

$V(A)$ can also be described as the set of isomorphism classes of finitely generated projective (left or right) A-modules as in 1.7. The binary operation on $V(A)$ corresponds to direct sum of modules.

Because of the results of Section 4, one obtains exactly the same semigroup starting with $\sim_s$ or $\sim_h$ instead of $\sim$ as the equivalence, since the three notions coincide on $M_\infty(A)$. If A is a local C^*-algebra, one can restrict to projections and use $\sim$, $\sim_u$, or $\sim_h$; and one can use $A \otimes \mathbb{K}$ in place of $M_\infty(A)$. More generally, we have $V(A) \cong V(\bar{A})$ for any local Banach algebra A.

NOTE. If A is a normed algebra, it is not true in general that the inclusion of A into $\bar{A}$ induces an isomorphism of $V(A)$ with $V(\bar{A})$. For example, consider the polynomials as a dense *-subalgebra of $C([0,1] \cup [2,3])$. One needs some weak type of completeness of A. It is a difficult and interesting problem, for example, to relate the K-theory of a Banach *-algebra with that of its enveloping C^*-algebra; cf. 9.4.3, 11.1.2, [Rosenberg 1984, 2.1].

$V(A)$ depends on A only up to stable isomorphism: if $M_\infty(A) \cong M_\infty(B)$, or more generally in the local C^* case if $A \otimes \mathbb{K} \cong B \otimes \mathbb{K}$, then $V(A) \cong V(B)$. In particular, $V(M_n(A)) \cong V(A)$.

If A is separable, then $V(A)$ is countable (4.3.2).

EXAMPLES 5.1.3. (a) $V(\mathbb{C}) \cong V(\mathbb{M}_n) \cong V(\mathbb{K}) \cong \mathbb{N} \cup \{0\}$.
(b) $V(\mathbb{B}) \cong \{0\} \cup \mathbb{N} \cup \{\infty\}$. If A is a II_1 factor, then $V(A) \cong \mathbb{R}_+ \cup \{0\}$; if A is a countably decomposable II_∞ factor, then $V(A) \cong \{0\} \cup \mathbb{R}_+ \cup \{\infty\}$; if A is a countably decomposable type III factor, then $V(A) = \{0, \infty\}$. In each case the operation is the ordinary one with $\infty + x = \infty$ for all x. (If A is not countably decomposable, then $V(A)$ will have other infinite cardinals.)
(c) If X is a compact Hausdorff space, then $V(C(X)) \cong V_{\mathbb{C}}(X)$ (1.7).
(d) If X is a connected locally compact noncompact Hausdorff space, then $V(C_0(X)) = 0$.
(e) Let $A = \{f : [0,1] \to \mathbb{M}_2 \mid f(0) = \mathrm{diag}(x,0), f(1) = \mathrm{diag}(y,y)$ for some $x, y \in \mathbb{C}\}$. Then

$$A^+ = \{f : [0,1] \to \mathbb{M}_2 \mid f(0) = \mathrm{diag}(x,z), f(1) = \mathrm{diag}(y,y) \text{ for some } x, y, z \in \mathbb{C}\}.$$

A contains no nonzero projections, but $M_2(A)$ contains nontrivial projections. We have $V(A) \cong \mathbb{N} \cup \{0\}$ and $V(A^+) \cong \{(m,n) \in \mathbb{Z}^2 \mid m, n \geq 0, m+n \text{ even}\}$.

Example 5.1.3(e) shows the necessity of considering idempotents in matrix algebras over A, since unexpected idempotents sometimes appear which have nothing to do with idempotents in A. $C(S^2)$ gives another example of this phenomenon (1.1.2(c), 1.4.2(c)). This can even happen in simple unital C^*-algebras [Blackadar 1981, 4.11] (but not in a factor). It turns out to be necessary for K-theory to take such idempotents into account. Consideration of matrix algebras is more natural if one associates idempotents with projective modules (1.7).

Example 5.1.3(b) shows that the semigroup $V(A)$ can fail to have cancellation. Even $V(C(X))$ can fail to have cancellation for certain compact differentiable manifolds X (1.2.1(b)). We will treat cancellation in more detail in Section 6.

5.2. Properties of $V(A)$

5.2.1. FUNCTORIALITY. If $\phi : A \to B$ is a homomorphism (not necessarily continuous!) of local Banach algebras, then ϕ induces a map $\phi_* : \mathrm{Proj}(A) \to$

Proj(B). ϕ extends to a homomorphism from $M_\infty(A)$ to $M_\infty(B)$, which induces a semigroup homomorphism, also denoted ϕ_*, from $V(A)$ to $V(B)$. So V is a covariant functor from the category of local Banach algebras to the category of abelian semigroups.

Actually, the construction of $V(A)$ works equally well for any ring, so we get a functor from rings to abelian semigroups.

5.2.2. HOMOTOPY INVARIANCE. We say two (continuous) homomorphisms $\phi, \psi : A \to B$ are homotopic if there is a path of (continuous) homomorphisms $\omega_t : A \to B$ for $0 \le t \le 1$, continuous in t in the topology of pointwise norm-convergence, with $\omega_0 = \phi$, $\omega_1 = \psi$. This is equivalent to the existence of a (continuous) homomorphism $\omega : A \to C([0,1], B)$ with $\pi_0 \circ \omega = \phi$, $\pi_1 \circ \omega = \psi$, where $\pi_t : C([0,1], B) \to B$ is evaluation at t; and is the exact Banach algebra analog of the ordinary notion of homotopy of continuous functions between locally compact Hausdorff spaces.

If $\phi, \psi : A \to B$ are homotopic, then $\phi(e) \sim_h \psi(e)$ for any idempotent $e \in M_\infty(A)$, and hence $\phi_* = \psi_*$, i.e. V is a *homotopy-invariant functor*.

5.2.3. DIRECT SUMS. If $A = A_1 \oplus A_2$, then $M_\infty(A) = M_\infty(A_1) \oplus M_\infty(A_2)$, and equivalence is coordinatewise; hence $V(A) \cong V(A_1) \oplus V(A_2)$.

5.2.4. INDUCTIVE LIMITS. If $A = \varinjlim(A_i, \phi_{ij})$, then $V(A)$ is the algebraic direct limit of $(V(A_i), \phi_{ij*})$. This follows easily from 4.5.1 and 4.5.2.

5.3. Preliminary Definition of K_0

We are tempted to define $K_0(A)$ to be the Grothendieck group (1.3) of $V(A)$; but it turns out that this is not the proper definition for A nonunital. It is customary to give separate definitions of $K_0(A)$ in the unital and nonunital cases, but we will unify the treatment.

DEFINITION 5.3.1. $K_{00}(A)$ is the Grothendieck group of $V(A)$.

K_{00} is a covariant functor from local Banach algebras (or even rings) to abelian groups satisfying the properties of 5.2. Elements of $K_{00}(A)$ may be pictured as formal differences $[e] - [f]$, where $[e_1] - [f_1] = [e_2] - [f_2]$ if there are orthogonal idempotents e_i', f_i', g in $M_\infty(A)$ with $e_i' \sim e_i$, $f_i' \sim f_i$, and $e_1' + f_2' + g \sim e_2' + f_1' + g$.

EXAMPLES 5.3.2. (a) $K_{00}(\mathbb{C}) \cong K_{00}(\mathbb{M}_n) \cong K_{00}(\mathbb{K}) \cong \mathbb{Z}$.
(b) If A is a II$_1$ factor, then $K_{00}(A) \cong \mathbb{R}$. If A is an infinite factor, then $K_{00}(A) = 0$.
(c) $K_{00}(C(S^1)) \cong \mathbb{Z}$; $K_{00}(C(S^2)) \cong \mathbb{Z}^2$.
(d) If X is connected and noncompact, then $K_{00}(C_0(X)) = 0$.
(e) If A is as in 5.1.3(e), then $K_{00}(A) \cong \mathbb{Z}$, $K_{00}(A^+) \cong \mathbb{Z}^2$.

5.4. Relative K-Groups

Let A be unital, J a closed two-sided ideal in A, and $\pi : A \to A/J$ the quotient map. We will define a relative K_0-group $K_0(A, J)$ in analogy with the relative group $K^0(X, Y)$ of topology (1.9). $K_0(A, J)$ has generators (e, f, z), where e, f are idempotents in $M_n(A) \subseteq M_\infty(A)$, and z is an invertible element of $M_n(A/J)$ with $z\pi(e)z^{-1} = \pi(f)$; and relations

$$(e_1, f_1, z_1) + (e_2, f_2, z_2) = (\mathrm{diag}(e_1, e_2), \mathrm{diag}(f_1, f_2), \mathrm{diag}(z_1, z_2))$$

and $(e_1, f_1, z_1) = (e_2, f_2, z_2)$ if there are idempotents $g_1, g_2 \in M_k(A)$ and invertible elements $u, v \in M_{n+k}(A)$ with

$$u \, \mathrm{diag}(e_1, g_1) \, u^{-1} = \mathrm{diag}(e_2, g_2),$$
$$v \, \mathrm{diag}(f_1, g_1) \, v^{-1} = \mathrm{diag}(f_2, g_2),$$
$$\pi(v) \, \mathrm{diag}(z_1, 1) \, \pi(u)^{-1} = \mathrm{diag}(z_2, 1).$$

Strictly speaking, we should write $K_0(A, A/J)$ instead of $K_0(A, J)$ to be in complete analogy with the notation of topology; but our notation seems less clumsy to work with.

It is not necessary to understand the relative K-group construction in full generality for the development of K-theory; it is enough to deal with the case where $A = J^+$, where the details are much simpler. However, there is some theoretical merit in the general construction, particularly in dealing with K-theory axiomatically as a cohomology theory (22.4.3, 22.4.4).

PROPOSITION 5.4.1. $K_0(A^+, A) \cong \ker \pi_* \subseteq K_{00}(A^+)$.

PROOF. It is immediate from the definition of the relative group that if (e_1, f_1, z_1) is equivalent to (e_2, f_2, z_2), then $[e_1] - [f_1] = [e_2] - [f_2]$ in $K_{00}(A)$, and the corresponding element of $K_{00}(A)$ lies in $\ker \pi_*$. On the other hand, if $[e] - [f] \in \ker \pi_*$, then $(e+g, f+g, z)$ is a generator of the relative group for some g and z, and the element of the relative group so obtained depends only on $[e]$ and $[f]$. If $[e_1] - [f_1] = [e_2] - [f_2]$, we may assume (e_1, f_1, z_1) and (e_2, f_2, z_2) are generators, and that $e_1 + f_2 \sim e_2 + f_1$. Then, choosing $g_1 = e_2$, $g_2 = e_1$, we can find a u and v implementing the equivalence between (e_1, f_1, z_1) and (e_2, f_2, z_2) except for intertwining z_1 and z_2. But invertible elements in the quotient lift to A^+, so one can correct the v, f_2, e_1 by multiplying by the inverse of the error. □

THEOREM 5.4.2. *For any A and J, the natural homomorphism from $K_0(J^+, J)$ to $K_0(A, J)$ is an isomorphism.*

A direct proof of this theorem is possible, but it will follow easily from the exact sequence of K-theory. Since we will not use this theorem, we omit the details.

Theorem 5.4.2 is called the Strong Excision Theorem of K-theory. It is really the result which allows K-theory to be developed without reference to relative K-groups.

5.5. Definition of $K_0(A)$

DEFINITION 5.5.1. $K_0(A) = K_0(A^+, A)$.

The conclusion of 5.4.1 may be taken as the definition of $K_0(A)$ if desired. Thus $K_0(A)$ may be viewed as formal differences $[e] - [f]$, where $e, f \in M_\infty(A^+)$ with $e \equiv f \bmod M_\infty(A)$, with the usual notion of equivalence of formal differences in $K_{00}(A^+)$. In fact, any element of $K_0(A)$ may be written $[e] - [p_n]$, where $p_n = \text{diag}(1, \ldots, 1, 0, \ldots)$ (with n ones on the diagonal) and $e \equiv p_n \bmod M_\infty(A)$: if n is large enough, $f \leq p_n$, and $[e] - [f] = [e' + (p_n - f)] - [p_n]$, where $e' \sim e$ and $e' \perp p_n$. This is the *Standard Picture* of $K_0(A)$ for general A.

K_0 is a covariant functor from local Banach algebras (general rings) to abelian groups which has the properties of 5.2.

5.5.2. There is a homomorphism from $K_{00}(A^+)$ to $K_0(A)$, given by $[e] \to [e] - [p_n]$, where $\pi(e)$ is a rank n projection in $M_\infty(\mathbb{C})$. Composing this with the canonical map from $K_{00}(A)$ to $K_{00}(A^+)$ yields a homomorphism $\omega_A : K_{00}(A) \to K_0(A)$. The map ω_A is injective if $V(A^+)$ has cancellation (and perhaps in general), but it is usually not surjective if A is nonunital.

EXAMPLE 5.5.3. $K_0(C_0(\mathbb{R}^2)) \cong \mathbb{Z}$ because of 5.3.2(c); but $K_{00}(C_0(\mathbb{R}^2)) = 0$.

DEFINITION 5.5.4. A local Banach algebra A is *stably unital* if $M_\infty(A)$ has an approximate identity of idempotents.

If A is unital, or more generally if A has an approximate identity of idempotents, then A is stably unital. If A is a local C^*-algebra, then A is stably unital if and only if $A \otimes \mathbb{K}$ has an approximate identity of projections.

PROPOSITION 5.5.5. *If A is stably unital, then $\omega_A : K_{00}(A) \to K_0(A)$ is an isomorphism.*

PROOF. If A is stably unital, then there is a dense local Banach subalgebra of $M_\infty(A)$ which is an algebraic direct limit of unital local Banach algebras. Since both K_{00} and K_0 respect inductive limits, it suffices to show the result for A unital. But then $A^+ \cong A \oplus \mathbb{C}$, so $K_{00}(A^+) \cong K_{00}(A) \oplus \mathbb{Z}$. The map π_* is projection onto the second coordinate. The image of ω_A is the subgroup $K_{00}(A) \oplus 0$.   □

So when A is stably unital, we may identify $K_0(A)$ with $K_{00}(A)$. This will be the *Standard Picture* of $K_0(A)$ for A unital.

5.6. Exactness of K_0

THEOREM 5.6.1. *If J is a (closed two-sided) ideal in A, then the sequence $K_0(J) \xrightarrow{\iota_*} K_0(A) \xrightarrow{\pi_*} K_0(A/J)$ is exact in the middle, i.e. $\ker(\pi_*) = \text{im}(\iota_*)$.*

PROOF. If $x \in K_0(J)$, then $x = [e] - [p_n]$ for $e \in M_\infty(J^+)$ with $e \equiv p_n \bmod M_\infty(J)$. The image of x in $K_0(A)$ is again $[e] - [p_n]$, so $\pi_*(x) = [\pi(e)] - [p_n] = 0$. Conversely, if $[e] - [p_n] \in \ker \pi_* \subseteq K_0(A) \subseteq K_{00}(A^+)$, then $\text{diag}(\pi(e), p_k) \sim \text{diag}(p_n, p_k)$ in $M_\infty((A/J)^+)$ for some k, so for large r there is an invertible $z \in M_r((A/J)^+)$ with

$$z \, \text{diag}(\pi(e), p_k, 0) z^{-1} = \text{diag}(p_n, p_k, 0).$$

We can lift $\mathrm{diag}(z, z^{-1})$ to an invertible element $w \in M_{2r}(A^+)$. Set $f = w \, \mathrm{diag}(e, p_k, 0) w^{-1}$; then $\pi(f) = p_{n+k}$, so $f \in M_\infty(J^+)$, and $[e] - [p_n] = [f] - [p_{n+k}]$ is in $\mathrm{im}\, \iota_*$. $\qquad\square$

Theorem 5.6.1 is one of the most important reasons for working with K_0 instead of K_{00} in general. The corresponding statement for K_{00} is false. For example, the result fails for the exact sequence $0 \to C_0(\mathbb{R}^2) \to C(S^2) \to \mathbb{C} \to 0$. Thus the more complicated definition of K_0 is necessary to make the desired exact sequences work.

5.6.2. It is important to realize that an exact sequence $0 \longrightarrow J \xrightarrow{\iota} A \xrightarrow{\pi} A/J \longrightarrow 0$ does *not* yield an exact sequence $0 \longrightarrow K_0(J) \xrightarrow{\iota_*} K_0(A) \xrightarrow{\pi_*} K_0(A/J) \longrightarrow 0$ in general, i.e. ι_* is not always injective (for example, $A = \mathbb{B}$, $J = \mathbb{K}$) and π_* not always surjective (for example, $A = C([0,1])$, $J = C_0((0,1))$). The problem with π_* is that idempotents in a quotient do not in general lift to idempotents. The exact sequence of 5.6.1 can be expanded to a larger exact sequence, but K_1 and Bott periodicity are needed. This will be done in Chapter IV.

5.7. EXERCISES AND PROBLEMS

5.7.1. Show that if B is a C^*-algebra, then $K_0(B)$ can be described as the set of equivalence classes of pairs $(p^{(0)}, p^{(1)})$, where the $p^{(n)}$ are projections in $\tilde{B} \otimes \mathbb{K}$ which agree mod $B \otimes \mathbb{K}$. The equivalence relation is generated by homotopy (continuous paths $(p_t^{(0)}, p_t^{(1)})$ where $p_t^{(0)} = p_t^{(1)}$ mod $B \otimes \mathbb{K}$ for each t) and orthogonal direct sum with "degenerate" elements of the form (p, p). Compare with 12.5.1 and 17.5.4.

6. Order Structure on K_0

6.1. Introduction

If A is a local Banach algebra, the semigroup $V(A)$ really contains the information we want about the idempotents of $M_\infty(A)$; however, semigroups (particularly ones without cancellation) can be nasty algebraic objects, and for technical reasons (e.g. 5.6.1) it is necessary to pass to the group $K_0(A)$ in order to apply techniques from topology and homological algebra to the study of idempotents over A. But it is desirable to keep the original semigroup in the picture as much as possible. One way to do this is to try to put an ordering on $K_0(A)$ by taking the image $K_0(A)_+$ of $V(A)$ in $K_0(A)$ to be the positive cone. (Even at this point we may lose information, since the map from $V(A)$ into $K_0(A)$ will be injective only if $V(A)$ has cancellation.) Just as the elements of $K_0(A)$ determine the (stable) equivalence of idempotents in $M_\infty(A)$, the ordering will determine the (stable) comparability of idempotents.

We also define the *scale* $\Sigma(A)$ to be the image of $\mathrm{Proj}(A)$ in $K_0(A)$. If A is unital, the scale is simply the elements of $K_0(A)_+$ which are $\leq [1_A]$, so the scale can be described by simply specifying $[1_A]$, and by slight abuse of terminology we will frequently do so. We will mostly be concerned with the unital case. The triple $(K_0(A), K_0(A)_+, \Sigma(A))$ is called the *scaled preordered K_0-group* of A.

The preordered group $(K_0(A), K_0(A)_+)$ depends on A only up to stable isomorphism (5.1); but the scale gives a finer invariant, which can be used to distinguish between algebras in the same stable isomorphism class.

If $\phi : A \to B$ is a homomorphism, then $\phi_* : K_0(A) \to K_0(B)$ is a homomorphism of scaled preordered groups, i.e. $\phi_*(K_0(A)_+) \subseteq K_0(B)_+$ and $\phi_*(\Sigma(A)) \subseteq \Sigma(B)$.

6.2. Ordered Groups

DEFINITION 6.2.1. An *ordered group* (G, G_+) is an abelian group G with a distinguished subsemigroup G_+ containing the identity 0, called the *positive cone* of G, having these properties:

(1) $G_+ - G_+ = G$.
(2) $G_+ \cap (-G_+) = \{0\}$.

G_+ induces a translation-invariant partial ordering on G by $y \leq x$ if $x - y \in G_+$. By $y < x$ we will mean that $y \leq x$ and $y \neq x$.

An element $u \in G_+$ is called an *order unit* if for any $x \in G$ there is an $n > 0$ with $x \leq nu$ (in other words, the order ideal [hereditary subgroup] generated by u is all of G.) A triple (G, G_+, u) consisting of an ordered group (G, G_+) with a fixed order unit u is called a *scaled ordered group*. We say G is a *simple* ordered group if G has no proper order ideals, i.e. if every nonzero positive element is an order unit.

EXAMPLES 6.2.2. (a) On $\mathbb{Z}^n$ or $\mathbb{R}^n$, there are two standard orderings: the *ordinary ordering*, with positive cone $\{(x_1, \ldots, x_n) \mid x_1, \ldots, x_n \geq 0\}$, and the *strict ordering*, with positive cone $\{0\} \cup \{(x_1, \ldots, x_n) \mid x_1, \ldots, x_n > 0\}$. These orderings coincide for $n = 1$.

(b) More generally, if X is a set and G is a positively generated additive group of real-valued functions on X, then G can be given the ordinary ordering with $G_+ = \{f \mid f \geq 0 \text{ everywhere}\}$ or the strict ordering with $G_+ = \{0\} \cup \{f \mid f > 0 \text{ everywhere}\}$. The positive cone in the strict ordering is sometimes denoted G_{++}. Even more generally, if ρ is a homomorphism from G into the additive group of real-valued functions on X whose image is positively generated, then G may be given the *strict ordering from ρ*, with $G_+ = \{0\} \cup \{a \in G \mid \rho(a) > 0 \text{ everywhere}\}$. There is no analog of the ordinary ordering if ρ is not injective.

(c) If ρ is a homomorphism from G into an ordered group H whose range is positively generated, then G can be given the *strict ordering from ρ* by taking $G_+ = \{0\} \cup \{x \mid \rho(x) > 0\}$.

A group G with the strict ordering in the sense of (b), or in the sense of (c) with H simple, is a simple ordered group.

6.3. $K_0(A)$ as an Ordered Group

6.3.1. The set $K_0(A)_+$ does not satisfy (1) of 6.2.1 in general. For example, if $A = C_0(\mathbb{R}^2)$, then $K_0(A) \cong \mathbb{Z}$ and $K_0(A)_+ = 0$. However, if $\omega_A : K_{00}(A) \to K_0(A)$ is surjective (in particular, if A is stably unital), then $K_0(A)_+$ does satisfy condition (1).

From now on in this section, we will assume A is *unital*, and we will identify $K_0(A)$ with $K_{00}(A)$. Almost everything carries through (with appropriate technical modifications) to the stably unital case.

EXAMPLE 6.3.2. The set $K_0(A)_+$ does not satisfy (2) of 6.2.1 in general. Let O_n be the C^*-algebra generated by n isometries $s_1, \ldots, s_n$ with $s_i^* s_i = 1$, $s_i s_i^* = p_i$, and $p_1 + \cdots + p_n = 1$. The O_n were first studied by Cuntz [1977a], who showed that up to isomorphism O_n is independent of the choice of the s_i (and is therefore simple). In $K_0(O_n)$, we have $[1] = [p_1] = \cdots = [p_n]$, so $n[1] = [p_1] + \cdots + [p_n] = [1]$, i.e. $(n-1)[1] = 0$. Thus $(n-2)[1] = -[1]$. It is shown in [Cuntz 1981b] (see 10.11.8) that $K_0(O_n) \cong \mathbb{Z}/(n-1)$, with [1] as generator, so if $n > 2$ then $K_0(O_n)_+$ does not satisfy (2). In fact, in this case $K_0(O_n)_+ = \Sigma(O_n) = K_0(O_n)$. Actually, we have $K_0(A)_+ = \Sigma(A) = K_0(A)$ whenever A is a simple unital C^*-algebra containing a nonunitary isometry (6.11.8).

Recall that a local Banach algebra A is *finite* if $e \leq f$, $e \sim f$ implies $e = f$. If A is unital, this is equivalent to the property that no proper idempotent is algebraically equivalent to 1; if A is a unital local C^*-algebra, A is finite if and only if every isometry in A is unitary. A is *stably finite* if $M_n(A)$ is finite for all n. Not every finite C^*-algebra is stably finite (6.10.1); it is an open question whether every finite simple C^*-algebra is stably finite.

PROPOSITION 6.3.3. *If A is stably finite, then $(K_0(A), K_0(A)_+)$ is an ordered group.*

PROOF. We must show $K_0(A)_+$ satisfies (2). If $[e] - [f]$ and $[f] - [e]$ are both in $K_0(A)_+$, then $[e] - [f] = [g]$, $[f] - [e] = [h]$, so $[e] = [f] + [g] = [e] + [h] + [g]$, so there are orthogonal representatives e', h', g', k' with $e' + k' \sim e' + h' + g' + k'$, so $g' = h' = 0$, $[e] - [f] = 0$. □

EXAMPLES 6.3.4.   (a) The ordering on $K_0(\mathbb{C})$ and $K_0(\mathbb{M}_n)$ is the ordinary ordering on $\mathbb{Z}$. $\Sigma(\mathbb{M}_n) = \{0, \ldots, n\}$.

(b) If A is a II_1 factor, then the ordering on $K_0(A)$ is the ordinary ordering on $\mathbb{R}$, and $\Sigma(A) = [0, 1]$.

(c) If $A = \mathbb{C}^2$, then $K_0(A) \cong \mathbb{Z}^2$ with the ordinary ordering. $\Sigma(A) = \{(0,0), (0,1), (1,0), (1,1)\}$.

(d) $K_0(C(S^2))$ is $\mathbb{Z}^2$ with the strict ordering from the first coordinate, i.e. $K_0(C(S^2))_+ = \{(0,0)\} \cup \{(m,n) \mid m > 0\}$. $\Sigma(C(S^2)) = \{(0,0), (1,0)\}$.

(e) If A^+ is as in 5.1.3(e), then $K_0(A^+) = \{(m,n) \mid m, n \in \mathbb{Z}, m+n \text{ even}\} \cong \mathbb{Z}^2$ and $K_0(A^+)_+ = \{(m,n) \mid m, n \geq 0, m+n \text{ even}\}$. $\Sigma(A^+) = \{(0,0), (1,1)\}$.

Examples (c), (d), and (e) show that $K_0(A)$ and $K_0(B)$ can be isomorphic as groups without being isomorphic as ordered groups. Thus the order structure can be used to distinguish between algebras.

$K_0(A)$ is frequently a simple ordered group:

PROPOSITION 6.3.5. *If A is stably finite, and every nonzero idempotent in $M_\infty(A)$ is full (not contained in any proper two-sided ideal), then $K_0(A)$ is a simple ordered group.*

PROOF. [Cuntz 1977b] Let e and f be idempotents in $M_n(A)$, with $f \neq 0$. Then e is in the ideal of $M_n(A)$ generated algebraically by f, i.e. there are elements $x_1, \ldots, x_k, y_1, \ldots, y_k \in M_n(A)$ with $e = \sum_{i=1}^k x_i f y_i$. We may assume $x_i = ex_i f$ and $y_i = fy_i e$ for all i. Set

$$X = \begin{bmatrix} x_1 & \cdots & x_k \\ 0 & & 0 \\ \cdots & \cdots & \cdots \\ 0 & & 0 \end{bmatrix}, \qquad Y = \begin{bmatrix} y_1 & 0 & 0 \\ \cdots & \cdots & \cdots & \cdots \\ y_k & 0 & 0 \end{bmatrix},$$

$E = \mathrm{diag}(e, 0, \ldots, 0)$, $F = \mathrm{diag}(f, f, \ldots, f)$ in $M_k(M_n(A))$. Then $E = XFY = EXFYE$. Set $G = FYEXF$. Then G is an idempotent since

$$G^2 = FYEXFFFYEXF = FY(EXFYE)XF = FYEXF = G,$$

and G is equivalent to E via EXF and FYE. F is a unit for G, so $G \leq F$. Thus $[e] \leq k[f]$ in $K_0(A)$. $\square$

COROLLARY 6.3.6. *If A is a stably finite C^*-algebra and $\mathrm{Prim}(A)$ contains no nontrivial compact open subsets, then $K_0(A)$ is a simple ordered group. So if $\mathrm{Prim}(A)$ is Hausdorff and connected, $K_0(A)$ is simple. In particular, if A is simple or if $A = C(X)$, X connected, then $K_0(A)$ is simple.*

PROOF. If p is a nonzero projection in $M_n(A)$, write $\pi_J : M_n(A) \to M_n(A)/J$ for $J \in \mathrm{Prim}(M_n(A)) \cong \mathrm{Prim}(A)$. Then $\{J \in \mathrm{Prim}(M_n(A)) : \pi_J(p) \neq 0\} = \{J \in \mathrm{Prim}(M_n(A)) : \|\pi_J(p)\| \geq 1\}$ is a compact open set in $\mathrm{Prim}(M_n(A))$ by [Dixmier 1969, 3.3.2 and 3.3.7]. $\square$

6.3.7. The order structure on K_0 seems to have played only a minimal role in topological K-theory; the order on $K_0(C(X))$ is usually either rather trivial or else badly behaved (6.10.2). But the ordering is crucial in many of the applications of K-theory to C^*-algebras, particularly to the structure of AF algebras. The significance of the ordering is partial compensation for the absence (so far) in noncommutative K-theory of additional algebraic structure, such as the product and exterior power operations, which are very important in topological K-theory (1.7.5).

6.4. Cancellation

We say that A has *cancellation of idempotents* if, whenever e, f, g, h are idempotents in A with $e \perp g, f \perp h, e \sim f, e + g \sim f + h$, then $g \sim h$. A has *cancellation* if $M_n(A)$ has cancellation of idempotents for all n. A has cancellation if and only if the semigroup $V(A)$ has cancellation.

PROPOSITION 6.4.1. *Let A be a unital local Banach algebra. Then the following conditions are equivalent*:

(1) *A has cancellation of idempotents.*
(2) *If $e \sim f$, then $1 - e \sim 1 - f$.*
(3) *If $e \sim f$, then $e \sim_s f$.*

PROOF. (2) $\Longleftrightarrow$ (3) is 4.2.5. (1) $\Longrightarrow$ (2): Take $g = 1 - e, h = 1 - f$. (2) $\Longrightarrow$ (1): If e, f, g, h are as in the definition of cancellation, we have $1 - e - g \sim 1 - f - h$; so by 4.2.4 we have $1 - g = (1 - e - g) + e \sim (1 - f - h) + f = 1 - h$. So again by (2) $g \sim h$. $\square$

So a local Banach algebra with cancellation must be stably finite. There are many stably finite C^*-algebras which do not have cancellation (e.g. some commutative C^*-algebras). It is still an open question whether every stably finite simple (unital) C^*-algebra has cancellation, but the examples of Villadsen [1995; 1997] suggest that there are ones that do not.

6.5. Stable Rank

Cancellation is related to *stable rank*. A complete discussion of Bass' stable rank and Rieffel's topological stable rank (shown to coincide for C^*-algebras in [Herman and Vaserstein 1984]) is beyond the scope of these notes; see [Rieffel 1983b] for a complete treatment, or [Blackadar 1983a] for a survey. We will mention only the simplest case, of stable rank 1.

We say that A has *stable rank* 1, written $\mathrm{sr}(A) = 1$, if the invertible elements of A are dense in A. It can be shown [Rieffel 1983b, 3.3] that $\mathrm{sr}(A) = 1$ if and only if $\mathrm{sr}(M_n(A)) = 1$ for some (hence all) n. If $\mathrm{sr}(A) = 1$, then A is stably finite.

PROPOSITION 6.5.1. *Let A be a unital local C^*-algebra with $\mathrm{sr}(A) = 1$. Then A has cancellation.*

PROOF. Suppose $p \sim q$. Let $u^*u = p, uu^* = q$. Approximate u closely by an invertible element x, and write $x = v(x^*x)^{1/2}$, with v unitary. Then $x^*x \approx p$, so $vpv^* \approx v(x^*x)v^* = xx^* \approx q$, so $p \sim_u vpv^* \sim_u q$. $\square$

There is a partial converse to 6.5.1. Recall that a local C^*-algebra A has the *(HP) property* if every hereditary *-subalgebra of A has an approximate identity of projections.

PROPOSITION 6.5.2. *If A is a local C^*-algebra with (HP) and cancellation of idempotents, then $\mathrm{sr}(A) = 1$.*

DEFINITION 6.5.3. An element x of a C^*-algebra A is *well-supported* if there is a projection $p \in A$ with $x = xp$ and x^*x invertible in pAp.

An element x is well-supported if and only if either x^*x is invertible or 0 is an isolated point of the spectrum $\sigma(x^*x)$, i.e. if and only if $\sigma(x^*x) \subseteq \{0\} \cup [\varepsilon, \infty)$ for some $\varepsilon > 0$. So x is well-supported if and only if x^* is well-supported. Invertible elements and partial isometries are well-supported.

If x is well-supported, then x can be written as ua, where u is a partial isometry with $u^*u = p$ and a is an invertible positive element of pAp. $q = uu^*$ is a left support projection for x, and xx^* is invertible in qAq.

PROPOSITION 6.5.4. *If A has (HP) and p and q are projections of A, then the well-supported elements of qAp are dense in qAp.*

PROOF. Let $x \in qAp$. Let r be a projection in x^*Ax which is almost a unit for $(x^*x)^{1/2}$, and set $y = xr$. Then $y \in qAr \subseteq qAp$, and since $r \leq n(x^*x)$ for some sufficiently large n, $y^*y = rx^*xr \leq (1/n)r$, so y^*y is invertible in rAr. Thus y is well-supported, and closely approximates x. $\square$

PROPOSITION 6.5.5. *If A is a C^*-algebra, then A has cancellation if and only if the invertible elements of $M_n(A)$ are dense in the well-supported elements of $M_n(A)$ for all n.*

PROOF. Because of polar decomposition and the fact that invertible positive elements are dense in A_+, the invertible elements are dense in the well-supported elements if and only if every partial isometry can be arbitrarily approximated by invertibles. If A has cancellation and u is a partial isometry in $M_n(A)$, set $p = u^*u$, $q = uu^*$. By cancellation there is a partial isometry v with $v^*v = 1_{M_n(A)} - p$, $vv^* = 1_{M_n(A)} - q$; then $u + \varepsilon v$ is an invertible closely approximating u. Conversely, suppose $p \sim q$, and let u be a partial isometry in $M_n(A)$ with $u^*u = p$, $uu^* = q$. Approximate u closely by an invertible element x, and write $x = v(x^*x)^{1/2}$. Then v is unitary, and $vpv^* \approx v(x^*x)v^* = xx^* \approx q$, so p is unitarily equivalent to a projection close to q and hence also unitarily equivalent to q. $\square$

PROOF OF 6.5.2. By 6.5.4 with $p = q = 1$ the well-supported elements are dense in A. Apply 6.5.5. To pass to matrix algebras use the fact that $sr(A) = 1$ implies $sr(M_n(A)) = 1$. $\square$

There are some other characterizations of (HP), which underscore the importance and naturality of the axiom:

THEOREM 6.5.6. *Let A be a C^*-algebra. Then the following are equivalent:*

(1) *A has (HP).*
(2) *The well-supported self-adjoint elements of A are dense in A_{sa}.*
(3) *A has real rank zero: the invertible self-adjoint elements of A are dense in A_{sa}.*

(4) *A has (FS): the self-adjoint elements of A of finite spectrum are dense in* A_{sa}.

OUTLINE OF PROOF. (1) $\implies$ (2) is almost identical to the proof of 6.5.4: if x is self-adjoint and r is a projection in xAx which is an approximate unit for x^*x, then rxr is a well-supported self-adjoint element closely approximating x.

(2) $\implies$ (3): if x is well-supported, then $x + \lambda 1$ is invertible for all sufficiently small nonzero λ.

(3) $\implies$ (4): suppose $x = x^*$ is given. We may assume $0 \leq x \leq 1$. Let $\{\lambda_1, \lambda_2, \ldots\}$ be the rationals in $[0, 1]$. Set $x_1 = x$. For each n, let y_n be a well-supported self-adjoint element closely approximating $x_n - \lambda_n 1$, and set $x_{n+1} = y_n + \lambda_n 1$. Then x_n approximates x and its spectrum has gaps around $\lambda_1, \ldots, \lambda_n$; the approximate to x with finite spectrum can then be made from x_n by functional calculus.

(4) $\implies$ (1) is the trickiest part; see [Pedersen 1980]. $\qquad\square$

COROLLARY 6.5.7. *Let A be a C^*-algebra with (HP). Then A has cancellation if and only if* $\mathrm{sr}(A) = 1$.

There is a theory of real rank for C^*-algebras, developed by Brown and Pedersen [1991], which formally resembles the theory of stable rank. The use of the term "real rank zero" for the condition in (3) comes from this theory.

It is obvious that (FS) is preserved in inductive limits, and it is not hard to show that real rank zero passes to matrix algebras. (HP) obviously passes to hereditary subalgebras.

Until recently, no stably finite simple C^*-algebra was known to have stable rank greater than 1. Examples have recently been constructed by Villadsen [1997].

The exact general relationship between stable rank and cancellation is not known. There are some significant generalizations of 6.5.1, however; see [Rieffel 1983a] and [Blackadar 1983b].

Cancellation questions are part of what is known as *nonstable K-theory*, which is concerned with relating the K-theory data of A (which is "stable" data) to the actual structure of A. See [Husemoller 1966, Chapter 8] and [Rieffel 1983b] for a further discussion of questions from nonstable K-theory.

6.6. Classification of Stably Isomorphic C^*-Algebras

Suppose that A is a unital C^*-algebra with cancellation. Then the scale $\Sigma(A)$ is a hereditary subset of $K_0(A)_+$; in fact, $\Sigma(A)$ is the closed interval $[0, [1_A]] = \{x \in K_0(A)_+ \mid x \leq [1_A]\}$. Although $\Sigma(A)$ does not always generate $K_0(A)$ as a group (6.3.4(d), or, for a simple example, [Blackadar 1981, 4.11]), $[1_A]$ is always an order unit, so the order ideal generated by $\Sigma(A)$ is $K_0(A)$. If B is a unital C^*-algebra stably isomorphic to A, then $K_0(B)$ is order-isomorphic to $K_0(A)$, and the image of $\Sigma(B)$ in $K_0(A)$ will be an interval $[0, u]$ for some order unit u. Conversely, if u is an order unit in $K_0(A)$, then there is a unital C^*-algebra

B stably isomorphic to A with $\Sigma(B) = [0, u]$: let $u = [p]$ for some projection $p \in M_n(A)$, and take $B = pM_n(A)p$. So one can nearly classify all unital C^*-algebras stably isomorphic to A by the order units in $K_0(A)$. The correspondence is, however, not one-to-one in general: the algebras corresponding to u and v may be isomorphic if there is an order-automorphism of $K_0(A)$ taking u to v. (But the existence of such an order-automorphism does not guarantee that the algebras are isomorphic; not every order-automorphism of $K_0(A)$ is induced from an isomorphism on the algebra level in general.)

One can extend the above classification to certain nonunital C^*-algebras as follows. If $u_1, u_2, \ldots$ is an increasing sequence of elements of $K_0(A)_+$, then one can find an increasing sequence of projections $p_1, p_2, \ldots$ in $A \otimes \mathbb{K}$ with $[p_n] = u_n$. If $\{u_1, u_2, \ldots\}$ generates $K_0(A)_+$ as an order ideal, then the C^*-algebra $B = (\bigcup p_n(A \otimes \mathbb{K})p_n)^-$ is stably isomorphic to A and corresponds naturally to the interval $\bigcup [0, u_n] \subseteq K_0(A)$. A hereditary subset Σ of $K_0(A)_+$ is of this form if and only if it generates G as an order ideal and is countably generated and upward directed, i.e. if $x, y \in \Sigma$ there is a $z \in \Sigma$ with $x \le z$ and $y \le z$. The C^*-algebra corresponding to Σ by the above construction depends up to isomorphism only on Σ and not on the choice of the u_n or p_n. Conversely, every C^*-algebra which is stably isomorphic to A and which has an approximate identity consisting of a sequence of projections is obtained in this way. So if A is separable with (HP) [i.e. with real rank zero], one obtains a complete classification (modulo the possible identifications through order-automorphisms).

6.7. Perforation

One difficulty which can occur in ordered groups is perforation.

DEFINITION 6.7.1. An ordered group (G, G_+) is *unperforated* if $nx \ge 0$ for some $n > 0$ implies $x \ge 0$; G is *weakly unperforated* if $nx > 0$ for some $n > 0$ implies $x > 0$.

An unperforated group must be torsion-free. A weakly unperforated group can have torsion: for example, $\mathbb{Z} \oplus \mathbb{Z}_2$ with strict ordering from the first coordinate. If (G, G_+) is weakly unperforated, H is the torsion subgroup of G, and $\pi : G \to G/H$ the quotient map, then $(G/H, \pi(G_+))$ is an unperforated ordered group. Hence a weakly unperforated group is "unperforated up to torsion." Conversely, if (K, K_+) is a (weakly) unperforated ordered group and $\rho : G \to K$ is a homomorphism with positively generated image, then G is weakly unperforated if given the strict ordering from ρ. A weakly unperforated group is unperforated if and only if it is torsion-free.

EXAMPLES 6.7.2. (a) Let $G = \mathbb{Z}$, $G_+ = \{0\} \cup \{n \mid n \ge 2\}$. Then (G, G_+) is not weakly unperforated.

(b) $K_0(C(\mathbb{RP}^2)) = \mathbb{Z} \oplus \mathbb{Z}_2$ with strict ordering from the first coordinate [Karoubi 1978, IV.6.47]. $K_0(C(\mathbb{T}^4)) \cong \mathbb{Z}^8$ is perforated, where $\mathbb{T}^4$ is the 4-torus (6.10.2). There are stably finite simple unital C^*-algebras A with torsion in $K_0(A)$ (10.11.2). There are even stably finite simple C^*-algebras whose K_0 is not weakly unperforated [Villadsen 1995] (cf. 6.10.2(d)).

(c) Perforation in K_0 can be eliminated by "rationalizing": if R is the (unique) UHF algebra with $K_0(R) = \mathbb{Q}$, then for any A we have $K_0(A \otimes R) \cong K_0(A) \otimes \mathbb{Q}$ with $K_0(A \otimes R)_+ = K_0(A)_+ \otimes \mathbb{Q}_+$, and $K_0(A \otimes R)$ is unperforated.

6.8. States on Ordered Groups

The order structure on an ordered group is at least partially (and in good cases completely) determined by the states, which in the K_0 case are closely related to the tracial states on the algebra.

DEFINITION 6.8.1. A *state* on a scaled ordered group (G, G_+, u) is an order-preserving homomorphism f from G to $\mathbb{R}$ with $f(u) = 1$.

The set $S(G, G_+, u)$ (or just denoted $S(G)$ when there is no confusion) of all states on (G, G_+, u) is a compact convex set in the topology of pointwise convergence. $S(G)$ is called the *state space* of G.

We now develop some properties of $S(G)$ due to Goodearl and Handelman [1976], including a Hahn–Banach type existence theorem.

LEMMA 6.8.2. *Let (G, G_+, u) be a scaled ordered group. Let H be a subgroup of G containing u, and f a state on $(H, H \cap G_+, u)$ (note we do not assume H is positively generated). Let $t \in G_+$, and $p = \sup\{f(x)/m \mid x \in H, m > 0, x \le mt\}$, $q = \inf\{f(y)/n \mid y \in H, n > 0, nt \le y\}$. Then:*

(a) $0 \le p \le q < \infty$.

(b) *If g is a state on $(H + \mathbb{Z}t, u)$ which extends f, then $p \le g(t) \le q$.*

(c) *If $p \le r \le q$, then there is a unique state g on $(H + \mathbb{Z}t, u)$ which extends f with $g(t) = r$.*

PROOF. (a) Clearly $p \ge 0$. Next, $t \le ku$ for some $k > 0$, so $q \le f(ku)/1 = k < \infty$. If $x, y \in H$ and $m, n > 0$ with $x \le mt$ and $nt \le y$, then $nx \le mnt \le my$, so $nf(x) \le mf(y)$, so $f(x)/m \le f(y)/n$, and thus $p \le q$.

(b) If $x \in H$, $m > 0$, and $x \le mt$, then $f(x) = g(x) \le mg(t)$ and so $f(x)/m \le g(t)$. Thus $p \le g(t)$. Similarly, $g(t) \le q$.

(c) The formula for g, if it exists, must be $g(z + kt) = f(z) + kr$ for $z \in H, k \in \mathbb{Z}$, so we must show this function is well defined and order-preserving. It suffices to show that $z + kt \ge 0$ (for $z \in H$ and $k \in \mathbb{Z}$) implies $f(z) + kr \ge 0$. This is obvious if $k = 0$. If $k > 0$, then $-z \le kt$, so $f(-z)/k \le p \le r$. If $k < 0$, then $-kt \le z$, so $r \le q \le f(z)/(-k)$. □

THEOREM 6.8.3. *Let (G, G_+, u) be a scaled ordered group, and let H be a subgroup of G containing u. If f is any state on $(H, H \cap G_+, u)$, then f extends to a state on (G, G_+, u).*

PROOF. This follows from 6.8.2 by a straightforward Zorn's Lemma argument.

□

COROLLARY 6.8.4. *Let (G, G_+, u) be a scaled ordered group, and let $t \in G_+$. Let $f_*(t) = p, f^*(t) = q$ defined as in 6.8.2 with $H = \mathbb{Z}u$. Then:*

(a) $0 \leq f_*(t) \leq f^*(t) < \infty$.
(b) *If f is any state on G, then $f_*(t) \leq f(t) \leq f^*(t)$.*
(c) *If $f_*(t) \leq r \leq f^*(t)$, then there is a state g on G with $g(t) = r$.*

$f_*(t)$ and $f^*(t)$ can be more elegantly described as $\sup\{n/m \mid nu \leq mt\}$ and $\inf\{n/m \mid mt \leq nu\}$ respectively.

THEOREM 6.8.5. *Let (G, G_+, u) be a simple weakly unperforated scaled ordered group. Then G has the strict ordering from its states, i.e. $G_+ = \{0\} \cup \{x \mid f(x) > 0 \text{ for all } f \in S(G)\}$.*

PROOF. If $x > 0$, then x is an order unit, so $u \leq mx$ for some $m > 0$. Then $0 < 1/m \leq f_*(x)$, so $f(x) > 0$ for all $f \in S(G)$. Conversely, suppose $f(x) > 0$ for all $f \in S(G)$. By compactness, we have $f_*(x) = \inf_{f \in S(G)} f(x) > 0$, so there are positive integers n and m with $0 < nu \leq mx$, and therefore by weak unperforation $x > 0$. (Note that this implication does not require G to be simple.) $\qquad\square$

There is an alternate way to view 6.8.5. If $x \in G$, then x induces a continuous affine function $\hat{x}$ on $S(G)$ by $\hat{x}(f) = f(x)$. Thus there is a homomorphism ρ from G to $\mathrm{Aff}(S(G))$, the group of all continuous real-valued affine functions on $S(G)$. 6.8.5 then says that G has the strict ordering from ρ, in the sense of 6.2.2(b).

6.9. Dimension Functions and States on $K_0(A)$

It is easy to identify at least some (in fact, all) of the states on $(K_0(A), K_0(A)_+, [1_A])$. If τ is a tracial state on A, or more generally a *quasitrace* on A (a function from $M_\infty(A)$ to $\mathbb{C}$ which is linear on commutative subalgebras and satisfies $0 \leq \tau(x^*x) = \tau(xx^*)$ for all x), then τ induces a state on $K_0(A)$ in an obvious way. Let $T(A)$ and $QT(A)$ denote respectively the tracial states and normalized quasitraces on A. We have $T(A) \subseteq QT(A)$; it is a very important and difficult question whether $T(A) = QT(A)$ for all A. (In a remarkable advance, Haagerup [1992] has shown that a quasitrace on an exact [e.g. nuclear] C^*-algebra must be a trace.) It is at least true that $QT(A)$, like $T(A)$, is always a Choquet simplex, which is metrizable if A is separable [Blackadar and Handelman 1982, II.4.4]. For the ordering on $K_0(A)$ the set $QT(A)$ is the more natural and important set to consider, due to the fact that the quasitraces on A are in one-one correspondence with the lower semicontinuous dimension functions on A.

A *dimension function* on A is a function $D : M_\infty(A) \to [0, \infty)$ such that $D(1) = 1$, $D(a) \leq D(b)$ whenever there are sequences x_n, y_n with $x_n b y_n \to a$, and $D(a + b) = D(a) + D(b)$ whenever $a \perp b$. One can construct an ordered group $K_0^*(A)$ (called $D(A)$ in [Cuntz 1982a]) whose states are exactly the dimension functions on A [Cuntz 1978]. $K_0^*(A)$ is constructed in a manner very similar to $K_0(A)$ except that one starts with all elements of $M_\infty(A)$, not just the projections; roughly, one pretends that each element has a well-defined (left or right) support projection, and then $K_0^*(A)$ is the Grothendieck group of the set of stable equivalence classes of these "support projections". Since a dimension function on A measures the "size" of the "support projections" of the elements of A, the correspondence between dimension functions and states of $K_0^*(A)$ is natural.

The correspondence described above gives a continuous affine map

$$\chi : QT(A) \to S = S(K_0(A)).$$

This map is not injective in general. For example, if $A = C(S^1)$, then $QT(A) = T(A)$ is the state space of A, while S is a singleton. Also, if $A = C^*(\mathbb{Z}_2 * \mathbb{Z}_2)$ (6.10.4), then S is a square and not even a simplex. There is a simple (unital, stably finite) C^*-algebra A for which χ is not injective [Blackadar and Kumjian 1985].

The map χ is always surjective [Blackadar and Rørdam 1992]. In other words, if A is any C^*-algebra, then every state on $K_0(A)$ comes from a quasitrace on A; if A is nuclear (or exact), then every state on $K_0(A)$ comes from a tracial state. Thus there are enough tracial states on A to completely determine the order structure on $K_0(A)$ (up to perforation).

The following results were proved in [Blackadar and Handelman 1982, III]:

THEOREM 6.9.1. *If A is a stably finite unital C^*-algebra with real rank zero, then χ is a bijection, hence a homeomorphism.*

COROLLARY 6.9.2. *If A is a stably finite simple unital C^*-algebra with real rank zero, such that $K_0(A)$ is weakly unperforated, then the state space of $K_0(A)$ is the simplex $QT(A)$, and $K_0(A)$ has the strict ordering induced from $\rho : K_0(A) \to \mathrm{Aff}(QT(A))$, i.e. $[p] < [q]$ if and only if $\tau(p) < \tau(q)$ for every quasitrace τ. If A has cancellation, and p and q are projections in $M_\infty(A)$ with $\tau(p) < \tau(q)$ for all τ, then $p \preceq q$ (p is equivalent to a subprojection of q).*

It is not true that $\tau(p) \leq \tau(q)$ for all τ implies $p \preceq q$, even in a simple unital AF algebra (7.6).

THEOREM 6.9.3. *Let A be simple, unital, stably finite, with real rank zero and cancellation, and with $K_0(A)$ weakly unperforated. Then the range of $\rho : K_0(A) \to \mathrm{Aff}(QT(A))$ is uniformly dense.*

6.10. EXERCISES AND PROBLEMS

6.10.1. There is a finite unital C^*-algebra which is not stably finite. Let $A = \tau(S^3)$ be the Toeplitz algebra on the unit ball of $\mathbb{C}^2$ [Coburn 1973/74]. Then A is an extension of $C(S^3)$ by $\mathbb{K}$. Since $\pi^1(S^3) \cong \pi_3(S^1) = 0$ [Spanier 1966, 7.2.12], the unitary group of $C(S^3)$ is connected. Every isometry in A has unitary image in $C(S^3)$, hence has index 0, i.e. A contains no nonunitary isometries. But $M_2(A)$

does contain a nonunitary isometry; in fact, the image of the multiplication operator with symbol

$$f(z, w) = \begin{bmatrix} z & -\bar{w} \\ w & \bar{z} \end{bmatrix}$$

is a Fredholm operator of nonzero index.

This example works because, although $K_1(C(S^3))$ is nontrivial, the nontrivial elements first appear in the 2×2 matrix algebra.

By considering the Toeplitz algebra on the unit ball of $\mathbb{C}^n$, one can make the nonunitary isometries first appear in the $n \times n$ matrix algebra.

Although $M_2(A)$ is infinite, it is in some sense "not very infinite" (for example, it is type I, an extension of a finite homogeneous C^*-algebra by $\mathbb{K}$). A C^*-algebra like this, or the usual Toeplitz algebra (9.4.2), might be called "poorly infinite." An infinite simple unital C^*-algebra must be "much more infinite" (or properly infinite: see 6.11).

This example was independently obtained by N. Clarke [1986] and the author [Blackadar 1983a].

M. Rørdam [1998] has recently constructed an example of a finite unital C^*-algebra A for which $M_2(A)$ is *properly* infinite (6.11.1).

6.10.2. (a) If $\mathbb{T}^4$ is the 4-torus, then the ring $K_*(C(\mathbb{T}^4))$ can be identified with the exterior algebra with four generators $e_1, \ldots, e_4$ over $\mathbb{Z}$, as in [Elliott 1984]. K_0 corresponds to the terms with even degree. Line bundles correspond to the positive elements with constant term 1. The exterior product corresponds to tensor product of bundles, and the sum in the exterior algebra corresponds to Whitney sum.

(b) The element $(1 + e_1 \wedge e_2)(1 + e_3 \wedge e_4)$ is positive, being the product of two line bundles [note that $(1 + e_1 \wedge e_2)$ is the pullback of the standard nontrivial line bundle on $\mathbb{T}^2$ to $\mathbb{T}^4 = \mathbb{T}^2 \times \mathbb{T}^2$ via projection onto the first coordinate]. Similarly $(1 - e_1 \wedge e_2)$, etc., are positive. So

$$2 + 2e_1 \wedge e_2 \wedge e_3 \wedge e_4 = (1 + e_1 \wedge e_2)(1 + e_3 \wedge e_4) + (1 - e_1 \wedge e_2)(1 - e_3 \wedge e_4)$$

is positive. But $1 + e_1 \wedge e_2 \wedge e_3 \wedge e_4$ is not positive, since if it were it would correspond to a line bundle. But a line bundle is completely determined by its second-order terms (the Chern character implies that every line bundle is of the form $1 + v + \frac{1}{2}v \wedge v$ for some v in the second exterior power); the only line bundle with zero second-order terms is the trivial bundle. Thus $K_0(C(\mathbb{T}^4))$ is not weakly unperforated.

(c) Show by direct calculation that any element of K_0 of constant term at least 3 can be written as a sum of line bundles and is therefore positive. Conclude from [Husemoller 1966, 8.1.2] that every element of constant term 2 is also positive, although such an element cannot in general be written as a sum of line bundles. Characterize which elements of constant term 1 come from line bundles. Show that this implies that K_0 has a unique state (cf. 6.10.3).

(d) Generalize these results to $\mathbb{T}^n$ for $n > 4$. Specifically, show that $k + e_1 \wedge \cdots \wedge e_{2m} \geq 0$ if and only if $k \geq m$ [Villadsen 1995]. This was the key technical tool in Villadsen's construction of a stably finite simple unital C^*-algebra A in which $K_0(A)$ is not weakly unperforated [Villadsen 1995].

This example is due to G. Elliott and the author (with (c) due to Rieffel [1988]), although the result may be known to topologists.

6.10.3. Let X be a connected compact Hausdorff space.

(a) All states (traces) on $C(X)$ induce the same state on $K_0(C(X))$, the one which assigns to each vector bundle its dimension. This state is called the *geometric state* on $K_0(C(X))$.

(b) If X has finite dimension d, set $s = \lceil d/2 \rceil$. Then, for any r, by [Husemoller 1966, 8.1.2] every vector bundle of dimension at least $r + s$ contains a trivial bundle of dimension r as a summand.

(c) With notation as in (b), let V be an n-dimensional bundle over X, and let W be a bundle of dimension at least s such that $V \oplus W$ is trivial. Write $W = W_1 \oplus W_2$, with W_1 s-dimensional and W_2 trivial. Thus $V \oplus W_1 \oplus W_2$ is trivial, so $V \oplus W_1$ is trivial by [Husemoller 1966, 8.1.5] since it has dimension at least s. Thus V can be embedded in a trivial bundle of dimension at most $n + s$.

(d) Combining (b) and (c), if X is d-dimensional and V and W are vector bundles over X of dimension n and m respectively, with $m - n \geq 2s$, then V is isomorphic to a subbundle of W. So if f is the geometric state on $K_0(C(X))$, and $x \in K_0(C(X))$ with $f(x) \geq 2s$, then $x \geq 0$. Conclude from 6.8.4 that f is the only state on $K_0(C(X))$.

(e) A general X can be written $X = \varprojlim X_i$, where X_i is a finite complex [Eilenberg and Steenrod 1952, X.10.1], so $K_0(C(X)) = \varinjlim K_0(C(X_i))$ as an ordered group; hence the geometric state is the only state on $K_0(C(X))$ in general (for X connected).

(f) Even though $K_0(C(X))$ is simple for X connected (6.3.6), it may be perforated (6.10.2), and so may not have the strict ordering from its (unique) state (cf. 6.8.5).

(g) Use a similar argument to show that if X is not connected, it is still true that every state on $K_0(C(X))$ comes from a state (trace) on $C(X)$, and hence the map $\chi : QT(A) \to S(K_0(A))$ of 6.9 is always surjective for A commutative. If X is not connected, then $K_0(C(X))$ no longer has a unique state (there is one extremal state for each connected component).

This result is due to A. Sheu and the author (possibly known to topologists). The result also follows from [Blackadar and Rørdam 1992].

6.10.4. Let $A = \{f : [0,1] \to \mathbb{M}_2 \mid f(0), f(1) \text{ are diagonal}\}$. Then $A \cong C^*(\mathbb{Z}_2 * \mathbb{Z}_2)$, the group C^*-algebra of the free product of $\mathbb{Z}_2$ with itself. A is also the universal unital C^*-algebra generated by two projections. There are four equivalence classes of minimal projections in A, corresponding to p_{ij} $(i, j = 0, 1)$, where $p_{ij}(0) = \text{diag}(1-i, i)$, $p_{ij}(1) = \text{diag}(1-j, j)$. We have $[p_{00}] + [p_{11}] = [p_{10}] + [p_{01}]$. $K_0(A) \cong \mathbb{Z}^4/\langle(1, 1, -1, -1)\rangle \cong \mathbb{Z}^3$, with $K_0(A)_+$ the monoid generated by $(1, 0, 0)$, $(0, 1, 0)$, $(0, 0, 1)$, and $(1, 1, -1)$. The state space of $K_0(A)$ is a square.

By doing an inductive construction using these algebras, a simple unital C^*-algebra can be constructed whose state space is a square [Elliott 1996].

6.10.5. Let A be the Choi algebra $C_r^*(\mathbb{Z}_2 * \mathbb{Z}_3)$ [Choi 1979]. A is a stably finite simple unital C^*-algebra. It is shown in [Cuntz 1983b] (cf. 10.11.11) that $K_0(A) \cong \mathbb{Z}^5/\langle(1, 1, -1, -1, -1)\rangle$, with the spectral projections of the finite-order unitaries as generators and the one obvious relation. Since A has a unique trace, there is one obvious corresponding state f_0 on $K_0(A)$. Each of the generators is in $K_0(A)_+$, and f_0 takes the value $1/2$ on the first two, $1/3$ on the last three. One could more generally consider $C_r^*(\mathbb{Z}_n^* \mathbb{Z}_m)$ for $n, m \geq 2$, $n + m \geq 5$, or even more generally reduced free products of finite-dimensional commutative C^*-algebras.

These algebras were studied in [Anderson et al. 1991] with partial results about the order structure on K_0; complete results were then obtained in [Dykema and Rørdam 1998]:

(a) f_0 is the only state on $K_0(A)$.
(b) $K_0(A)$ is unperforated. Thus $K_0(A)$ has the strict ordering from f_0, i.e. for $x \in K_0(A)$, $x > 0$ if and only if $f_0(x) > 0$.
(c) A has stable rank 1 and hence cancellation. Thus, for example, if p is a projection of trace $1/2$ and q a projection of trace $1/3$, then $q \preceq p$ and $p \preceq q \oplus q$.

6.11. Properly Infinite C^*-Algebras

In stark contrast to the stably finite C^*-algebras are the properly infinite ones. In this section, we develop ordered K-theory type results for these algebras; these results are due to Cuntz [1981b].

DEFINITION 6.11.1. A unital C^*-algebra is *properly infinite* if it contains two orthogonal projections equivalent to the identity (i.e. it contains two isometries with mutually orthogonal range projections).

PROPOSITION 6.11.2. *Let A be a unital C^*-algebra. Then A is properly infinite if and only if A contains a sequence of isometries with mutually orthogonal range projections (i.e. A contains a unital copy of O_∞ (10.11.8)).*

PROOF. Let s_1, s_2 be isometries in A with orthogonal ranges. Then the isometries $\{s_2^n s_1 : n \geq 0\}$ have mutually orthogonal ranges. □

A properly infinite C^*-algebra is infinite, but an infinite C^*-algebra need not be properly infinite (6.10.1, 9.4.2). Any quotient of a properly infinite C^*-algebra is properly infinite.

PROPOSITION 6.11.3. *Let A be a simple unital C^*-algebra. Then the following conditions are equivalent*:

(i) *A is infinite.*
(ii) *A is properly infinite.*
(iii) *A contains a sequence of mutually orthogonal equivalent nonzero projections.*

PROOF. (ii) $\implies$ (i) is trivial.

(i) $\implies$ (iii): If s is a proper isometry in A, and $p = 1 - ss^*$, then $\langle s^n p s^{*n} \rangle$ is a sequence of mutually orthogonal equivalent nonzero projections.

(iii) $\implies$ (ii): Let $\langle p_n \rangle$ be a sequence of mutually orthogonal nonzero equivalent projections. Let $p = p_1$, and u_n a partial isometry with $u_n^* u_n = p$ and $u_n u_n^* = p_n$ for each n; so $p = u_n^* p_n u_n$. Then there are elements x_i with $1 = \sum_{i=1}^n x_i^* p x_i = \sum_{i=1}^n x_i^* u_i^* p_i u_i x_i$ for some n. If $q = \sum_{i=1}^n p_i$ and $v = \sum_{1=1}^n u_i x_i$, then $1 = v^* q v$, so qv is an isometry with range projection $qvv^* q \leq q$. Similarly, there is a projection equivalent to 1 under $\sum_{i=kn+1}^{(k+1)n} p_i$ for every k. $\qquad\square$

There is an even stronger notion of infiniteness:

DEFINITION 6.11.4. *A unital C^*-algebra A is* purely infinite *if $A \neq \mathbb{C}$ and, for every nonzero $a \in A$, there are $x, y \in A$ with $xay = 1$.*

A purely infinite C^*-algebra is obviously simple. A corner in a purely infinite C^*-algebra is clearly purely infinite. If A is purely infinite, then any matrix algebra over A is isomorphic to a corner in A, and hence is also purely infinite.

Examples of purely infinite C^*-algebras are (countably decomposable) type III factors, the Calkin algebra, the Cuntz algebras (6.3.2, 10.11.8), and the simple Cuntz–Krieger algebras (10.11.9).

PROPOSITION 6.11.5. *If A is a simple unital C^*-algebra, then A is purely infinite if and only if every nonzero hereditary C^*-subalgebra of A contains an infinite projection. In particular, every nonzero projection in a purely infinite C^*-algebra is infinite.*

PROOF. Let B be a nonzero hereditary C^*-subalgebra of A, and $0 \leq b \in B$ a nonzero positive element. Find $x, y \in A$ with $xby = 1$. Then $1 = y^* b x^* x b y$. If $u = (bx^* xb)^{1/2} y$, then u is an isometry, and uu^* is a projection in B which is equivalent to 1 and hence infinite.

Conversely, let a be a nonzero element of A, and B the hereditary C^*-subalgebra generated by $f_\varepsilon(a^* a)$ for ε small enough that $f_\varepsilon(a^* a) \neq 0$, where f_ε is the continuous function that takes the value 0 on $(-\infty, \varepsilon/2]$ and the value 1 on $[\varepsilon, \infty)$, and is linear on $[\varepsilon/2, \varepsilon]$. Let u be an isometry in A with $p = uu^* \in B$. Then $a^* a \geq (\varepsilon/2)p$, so $z = u^* a^* au$ is invertible, and $z^{-1} u^* a^* au = 1$. $\qquad\square$

QUESTION 6.11.6. Is every infinite simple unital C^*-algebra purely infinite?

Call a projection in a unital C^*-algebra *very full* if it contains a subprojection equivalent to the identity. Any projection equivalent to a very full projection is very full. If A is simple, then a nontrivial projection in A is very full if and only if it is infinite (see the proof of 6.11.3).

THEOREM 6.11.7. *Let A be a properly infinite unital C^*-algebra, and $V_f(A)$ the subset of $V(A)$ consisting of the equivalence classes of very full projections of A. Then:*

(a) $V_f(A)$ *is a subsemigroup of $V(A)$.*
(b) $V_f(A)$ *is a group (with a different identity element than $V(A)$).*
(c) *The homomorphism from $V(A)$ to $K_0(A)$ induces an isomorphism from $V_f(A)$ onto $K_0(A)$.*

PROOF. (a) Since A contains a sequence of mutually orthogonal projections equivalent to the identity, it follows easily that every projection in $M_\infty(A)$ is equivalent to a projection in A. So if p and q are very full projections in A, then $p \oplus q$ is equivalent to a projection in A, which is very full since it contains a copy of p. Thus $V_f(A)$ is a subsemigroup.

(b) Notice first that if p and q are projections in A with p very full, then $q \sim q' \le p$ such that $p - q'$ is very full, since p contains two mutually orthogonal subprojections equivalent to 1, so q' can be taken to be a copy of q under one of them. In particular, if p is very full, then $p \sim p' \le p$ with $p - p'$ very full.

If p, q are very full and $p \sim p' \le p$ with $p - p'$ very full, we claim that $[q] = [p - p'] + [q]$. Let $q \sim q' \le q$ with $q - q'$ very full. We may assume $p \le q'$ by replacing it by an equivalent projection; then

$$[q] = [q'] = [q' - p] + [p] = [q' - p] + [p'] = [q' - p + p'].$$

So $[q] + [p - p'] = [q' - p + p'] + [p - p'] = [q'] = [q]$.

Thus $[p - p']$ is an identity for $V_f(A)$ for any very full p, and by the uniqueness of an identity in a semigroup $p - p' \sim q - q'$ for any other very full q.

To get inverses, let p be very full, and let p', p'' be orthogonal subprojections of p, equivalent to p, with $p - p' - p''$ very full. Then $[p] + [p - p' - p''] = [p - p' - p''] + [p''] = [p - p']$, so $[p - p' - p'']$ is the inverse of p.

(c) If q is any projection in A and p a very full projection with $[p]$ the identity of $V_f(A)$, then p and q have orthogonal representatives $p', q' \in A$; $p' + q'$ is very full and has the same class as q in $K_0(A)$. Thus the map from $V_f(A)$ to $K_0(A)$ is surjective, since the image is a subgroup containing $K_0(A)_+$. If p is very full and the class of p is 0 in $K_0(A)$, then there is a $p' \in A$ with $p' \sim p$, and a $q \in A$, $q \perp p'$, such that $q \sim p' + q$. But then q is very full, and by cancellation in $V_f(A)$ we get that $[p]$ is the identity of $V_f(A)$, i.e. the map is injective. □

COROLLARY 6.11.8. *If A is a properly infinite (unital) C^*-algebra, then*

$$K_0(A)_+ = \Sigma(A) = K_0(A).$$

COROLLARY 6.11.9. *Two infinite projections in a simple unital C^*-algebra are equivalent if and only if they have the same K_0-class. Two infinite projections with infinite complements which have the same K_0-class are unitarily equivalent. Two nontrivial projections with the same K_0-class in a purely infinite C^*-algebra are unitarily equivalent.*

7. Theory of AF Algebras

A particularly good illustration of the theory developed in this chapter is the dimension group approach to the theory of AF algebras. This was the first place that K-theory explicitly appeared in the theory of C^*-algebras. This classification has been greatly expanded in recent years to include many non-AF algebras; the methods used, while considerably more complicated and sophisticated technically, are basically similar to those described in this section. (See [Elliott 1996] for a survey.)

7.1. Basic Definitions

DEFINITION 7.1.1. A C^*-algebra A is an *AF algebra* if A is an inductive limit of a sequence of finite-dimensional C^*-algebras.

EXAMPLES 7.1.2. The CAR algebra; more generally, UHF and matroid C^*-algebras; $\mathbb{K}$; $C_0(X)$, when X is locally compact, totally disconnected, and second countable.

AF algebras are in some sense the "zero-dimensional" C^*-algebras, although real rank zero (6.5) is now regarded as the most natural noncommutative analog of zero-dimensionality.

AF algebras inherit many nice structural properties from their finite-dimensional subalgebras. For example, every AF algebra is stably finite and has real rank zero (HP); and a unital AF algebra has connected unitary group and stable rank 1. The relations $\sim$, $\sim_u$, and $\sim_h$ coincide for AF algebras. The class of AF algebras is closed under stable isomorphism (this is a nontrivial fact due to Elliott; see [Effros 1981, 9.4]) and under extensions (a theorem of Brown, proved using K-theory; see [Effros 1981, 9.9]).

AF algebras were first studied by Bratteli [1972], following analysis of special cases by Glimm [1960] (UHF algebras) and Dixmier [1967] (matroid C^*-algebras). They have a simple enough structure to analyze quite thoroughly, yet form a rich enough class to illustrate many general phenomena in C^*-algebra theory. For example, much of the general structure theory of stably finite simple C^*-algebras has been motivated by the study of simple AF algebras (see [Blackadar 1983a] for a survey).

We will not attempt to give a complete development of AF algebra theory here; see [Effros 1981] for an excellent and comprehensive study. Another good reference is [Goodearl 1982]. We will develop only the most basic aspects, which illustrate the theory of Sections 5 and 6.

7.2. Representation by Diagrams

A finite-dimensional C^*-algebra is a direct sum of matrix algebras. If A and B are finite-dimensional C^*-algebras with $A = A_1 \oplus \cdots \oplus A_r$, $B = B_1 \oplus \cdots \oplus B_s$, and $\phi : A \to B$, then ϕ can be described by a matrix of nonnegative integers $R_\phi = (m_{ij})$ ($1 \le i \le s$, $1 \le j \le r$): write $\phi = (\phi_1, \ldots, \phi_s)$, where $\phi_i : A \to B_i$ is the i-th coordinate function. Regard ϕ_i as a representation of $A_1 \oplus \ldots \oplus A_r$—it will break up as a direct sum of irreducible representations. There is exactly one irreducible representation of A for each summand A_j; let m_{ij} be the multiplicity of this representation in ϕ_i, called the multiplicity of the partial embedding of A_j into B_i.

Recall the following fact from linear algebra (see [Takesaki 1979, I.11.9], for example):

PROPOSITION 7.2.1. ϕ is characterized up to unitary equivalence by R_ϕ, i.e. if $\phi, \psi : A \to B$ with $R_\phi = R_\psi$, then there is a unitary $u \in B$ with $\psi(x) = u\phi(x)u^*$ for all $x \in A$.

Conversely, given any $s \times r$ matrix R with nonnegative integer entries, there is a *-homomorphism $\phi : A \to B$ with $R_\phi = R$ as long as the summands of B are of large enough dimension to absorb the embeddings. $\phi_i(x_1, \ldots, x_s) = \mathrm{diag}(x_1, \ldots, x_2, \ldots, \ldots, x_s, 0, \ldots, 0)$, with x_j repeated m_{ij} times. The map ϕ is injective if and only if no column of R_ϕ is zero.

Thus an AF algebra $A = \varinjlim A_n$ is completely determined up to isomorphism by the sizes of the central summands of the A_n and the multiplicities of the partial embeddings of A_n into A_{n+1}. This information can be conveniently displayed in a graph called a Bratteli diagram, with one row for each n, one node in the n-th row for each summand of A_n, one edge between each pair of nodes in successive rows for each partial embedding of nonzero multiplicity, and a positive integer for each node giving the size of the corresponding summand, and for each edge giving the multiplicity of the corresponding partial embedding. Conversely, any such Bratteli diagram defines a (unique) AF algebra.

Much structural information about A (for example, the ideal structure) can be read off from the Bratteli diagram. In fact, since the diagram determines the algebra, in principle all of the structure of A is contained within the diagram in some sense.

However, one major problem restricts the usefulness of the study of AF algebras by diagrams: many quite different diagrams yield isomorphic algebras, and there is no known reasonable algorithm for generating a diagram from an algebra or determining when two diagrams give isomorphic algebras.

7.3. Dimension Groups

K-theory gives an alternate approach. In the situation described above, we have $K_0(A) \cong \mathbb{Z}^r$ and $K_0(B) \cong \mathbb{Z}^s$ with the ordinary orderings. $\phi_* : \mathbb{Z}^r \to \mathbb{Z}^s$ is given by multiplication by R_ϕ. The results of 7.2 translate into:

PROPOSITION 7.3.1. *Let A and B be finite-dimensional C^*-algebras.*

(a) *If $\sigma : K_0(A) \to K_0(B)$ is a homomorphism of scaled ordered groups, then there is a $*$-homomorphism $\phi : A \to B$ with $\phi_* = \sigma$.*

(b) *If $\phi, \psi : A \to B$ with $\phi_* = \psi_*$, then ϕ and ψ are unitarily equivalent. ϕ is injective if and only if ϕ_* sends no generator of $K_0(A)$ to 0, if and only if $(\ker \phi_*) \cap K_0(A)_+ = 0$.*

If $A = \varinjlim A_n$, then $K_0(A) = \varinjlim K_0(A_n)$ as a scaled ordered group. This is called the *dimension group* of A. So $K_0(A) = \varinjlim(\mathbb{Z}^{r_n}, R_n)$ as scaled ordered groups.

We now prove the fundamental theorem of AF algebra theory, due to Elliott [1976], which says that the dimension group of A is a complete isomorphism invariant for A (among AF algebras).

THEOREM 7.3.2. *If A and B are AF algebras and $\sigma : K_0(A) \to K_0(B)$ is an isomorphism of scaled ordered groups, then there is an isomorphism $\phi : A \to B$ with $\phi_* = \sigma$.*

PROOF. Let $A = \varinjlim(A_n, \alpha_{nm})$ and $B = \varinjlim(B_n, \beta_{nm})$. We seek $*$-homomorphisms ϕ_i and ψ_i making the triangles in the following diagram commutative:

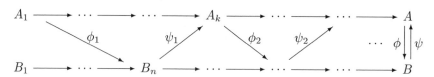

Then the maps $\phi : A \to B$ and $\psi : B \to A$ defined by the ϕ_i and ψ_i will be $*$-isomorphisms which are inverses of each other. We also require that $\phi_* = \sigma$ (so $\psi_* = \sigma^{-1}$), i.e. the induced diagram on K-theory should be:

$$
\begin{array}{ccccccc}
K_0(A_1) \to \cdots \to & \cdots \to & K_0(A_k) \to \cdots \to & \cdots \to & K_0(A) \\
\end{array}
$$

We will first construct the maps ϕ_{i*} and ψ_{i*} inductively, and then lift to the algebra level by 7.3.1. The key property of the ordered groups $K_0(A_n) \cong \mathbb{Z}^{r_n}$ is the fact that the algebraic and order structures are determined by a finite number of conditions.

We begin by constructing ϕ_{1*}. Regard $\sigma \circ \alpha_{1*}$ as a map from $K_0(A_1)$ to $K_0(B)$. This is a homomorphism of scaled ordered groups. Since $K_0(B) = \varinjlim K_0(B_n)$ as a scaled ordered group, for sufficiently large n there is a homomorphism $\phi_{1*} : K_0(A_1) \to K_0(B_n)$ of scaled ordered groups with $\beta_{n*} \circ \phi_{1*} = \sigma \circ \alpha_{1*}$ [let $x_1, \ldots, x_{r_1}$ be the images of the standard generators $\mathbf{e}_1, \ldots, \mathbf{e}_{r_1}$ of $K_0(A_1)$, and find positive preimages y_i for the x_i in some $K_0(B_n)$. If $c_1\mathbf{e}_1 + \cdots + c_{r_1}\mathbf{e}_{r_1}$ is the distinguished order unit of $K_0(A_1)$, then $c_1 x_1 + \cdots + c_{r_1} x_{r_1}$ is in $\Sigma(B)$, so by increasing n if necessary we may assume that $c_1 y_1 + \cdots + c_{r_1} y_{r_1}$ is in $\Sigma(B_n)$. ϕ_{1*} sends $\mathbf{e}_i$ to y_i.] Then by 7.3.1(a) there is a map $\phi_1 : A_1 \to B_n$ implementing ϕ_{1*}. Now regard $\sigma^{-1} \circ \beta_{n*}$ as a map from $K_0(B_n)$ to $K_0(A)$, and construct ψ_{1*} in the same manner. By construction we have $\alpha_{k*} \circ \psi_{1*} \circ \phi_{1*} = \alpha_{1*}$ as maps from $K_0(A_1)$ to $K_0(A)$, so (since $K_0(A_1)$ is finitely presented) by increasing k if necessary we may make $\psi_{1*} \circ \phi_{1*} = \alpha_{1k*}$.

Now let $\omega : B_n \to A_k$ be a lift of ψ_{1*}. Since $\omega \circ \phi_1$ and α_{1k} agree on K-theory, by 7.3.1(b) there is a unitary $u \in A_k$ with $\alpha_{1k}(x) = u(\omega \circ \phi_1)(x)u^*$ for all $x \in A_1$. Set $\psi_1(y) = u\omega(y)u^*$ for $y \in B_n$. Continue inductively. $\qquad\square$

Note that the proof of 7.3.2 actually shows the stronger result that ϕ may be chosen to map $\underrightarrow{\text{alglim}}\, A_n$ isomorphically onto $\underrightarrow{\text{alglim}}\, B_n$, i.e. any two dense locally finite *-subalgebras of an AF algebra A are isomorphic.

COROLLARY 7.3.3. *Two AF algebras are stably isomorphic if and only if their dimension groups are isomorphic as ordered groups.*

PROOF. The scale of a stable C^*-algebra is the whole positive cone. $\qquad\square$

There is also a version of 7.3.2 for homomorphisms, with virtually the same proof:

THEOREM 7.3.4. *If A and B are AF algebras and $\sigma : K_0(A) \to K_0(B)$ is a homomorphism of scaled ordered groups, then there is a *-homomorphism $\phi : A \to B$ with $\phi_* = \sigma$. If A and B are unital and $\sigma([1_A]) = [1_B]$, then ϕ may be chosen unital.*

One can read much of the structure of an AF algebra A from its dimension group. For example, the (closed two-sided) ideals of A are in one-one correspondence with the order ideals of $K_0(A)$. In particular, A is simple if and only if $K_0(A)$ is simple. Also, the (semifinite densely defined) traces on A are in one-one correspondence with the positive homomorphisms from $K_0(A)$ to $\mathbb{R}$. The norm of a trace τ is just $\sup\{\tau(x) : x \in \Sigma(A)\}$; so a trace is finite if and only if it is bounded on $\Sigma(A)$. If A is unital, the tracial states of A correspond exactly to the states on $(K_0(A), K_0(A)_+, [1_A])$ (6.9.2).

7.4. Classification of Dimension Groups

To complete the general theory, we need a characterization of which ordered groups can occur as dimension groups of AF algebras. The dimension groups are, of course, exactly the ordered groups which can be written as $\varinjlim \mathbb{Z}^{r_n}$ (with ordinary ordering), but this condition is frequently nearly impossible to check. An intrinsic characterization is needed.

There are three fundamental properties of dimension groups. First, any such group is clearly countable. Secondly, it is easy to see that a dimension group is unperforated (and in particular torsion-free), since $\mathbb{Z}^n$ is unperforated. And finally, a dimension group G has the *Riesz Interpolation Property*: if $x_1, x_2, y_1, y_2 \in G$ with $x_1, x_2 \leq y_1, y_2$, then there is a $z \in G$ with $x_1, x_2 \leq z \leq y_1, y_2$. [If $x_1, x_2 \leq y_1, y_2$ in $K_0(A)$, then the same relation must hold among suitable preimages in $K_0(A_n) \cong \mathbb{Z}^{r_n}$ for some n. But $\mathbb{Z}^r$ is a lattice, so interpolation obviously holds there.]

The theorem of Effros, Handelman, and Shen states that these three properties characterize dimension groups:

THEOREM 7.4.1. *An ordered group is a dimension group if and only if it is countable, unperforated, and has the Riesz Interpolation Property.*

See [Effros 1981, 3.1] for a proof.

COROLLARY 7.4.2. *Any countable totally ordered group is a dimension group.*

7.4.2 was obtained by Elliott [1979] before 7.4.1 was proved.

The description of dimension groups can be made more explicit in the case of simple AF algebras. We have the following result of Effros, Handelman, and Shen [Effros et al. 1980, 3.1], which is a fairly easy consequence of 7.4.1:

THEOREM 7.4.3. *Let G be a countable torsion-free abelian group, Δ a metrizable Choquet simplex, and $\rho : G \to \mathrm{Aff}(\Delta)$ a homomorphism with uniformly dense range. Then G becomes a simple dimension group with the strict ordering from ρ, i.e. with $G_+ = \{0\} \cup \{x \in G : \rho(x) > 0 \text{ everywhere}\}$. If G contains an element u with $\rho(u)$ the constant function 1, then (G, G_+, u) is isomorphic to the scaled dimension group of a (unique) simple unital AF algebra A with $T(A) \cong \Delta$.*

The converse is also true [Blackadar 1980b, 3.1]: if A is an infinite-dimensional simple unital AF algebra, then the state space of $K_0(A)$ is $T(A)$; the canonical map ρ from $K_0(A)$ to $\mathrm{Aff}(T(A))$ has uniformly dense range, and $K_0(A)$ has the strict ordering from ρ (6.9.2, 6.9.3).

Among other things, this gives an easy way of constructing simple unital AF algebras with arbitrary trace simplexes (cf. [Goodearl 1977/78; Blackadar 1980b]).

7.4.4. If Δ is an $(n\text{-}1)$-simplex, then $\mathrm{Aff}(\Delta) \cong \mathbb{R}^n$, so as a special case of 7.4.3 it follows that if G is countable and torsion-free, and $\rho : G \to \mathbb{R}^n$ is a homomorphism with dense range, then G is a simple dimension group with the strict ordering from ρ. In particular, a countable dense subgroup of $\mathbb{R}^n$ is a simple dimension group with the strict ordering.

7.5. UHF Algebras

An important special case is $n = 1$. Any countable dense subgroup of $\mathbb{R}$ is a simple dimension group. The dense subgroups of $\mathbb{Q}$ containing $\mathbb{Z}$ correspond to the UHF algebras (taking 1 as the order unit), and the classification in [Glimm 1960] corresponds exactly to the classification of these subgroups.

These are in one-one correspondence with the *generalized integers*, formal products $q = 2^{m_2} 3^{m_3} 5^{m_5} \ldots$, where an infinite number of primes and infinite exponents are allowed. The subgroup of $\mathbb{Q}$ corresponding to q is the group, denoted $\mathbb{Z}_{(q)}$, of all rational numbers whose denominators "divide" q. (If q is an ordinary integer, $\mathbb{Z}_{(q)}$ is not dense, and the corresponding algebra is $\mathbb{M}_q$.) The algebra corresponding to the dyadic rationals $\mathbb{D} = \mathbb{Z}_{(2^\infty)}$ is the CAR algebra. More generally, the classification of matroid C^*-algebras (ones stably isomorphic to UHF algebras) done in [Dixmier 1967] is a special case of the classification in 6.6. There are many other interesting countable dense subgroups of $\mathbb{R}$, for example $\mathbb{Z} + \mathbb{Z}\theta$, θ irrational; these groups are closely related to the structure of irrational rotation algebras (10.11.6) [Effros 1981, 10.1; Rieffel 1981; Pimsner and Voiculescu 1980b].

7.6. Other Simple AF Algebras

Non-totally-ordered simple dimension groups are particularly interesting, since they illustrate phenomena which cannot happen in factors. For example, consider the following four dimension groups:

$G_1 = \mathbb{Q}^2$ with strict ordering;
$G_2 = \mathbb{Q}^2$ with strict ordering from first coordinate;
$G_3 = \mathbb{D}^2$ with strict ordering from first coordinate;
$G_4 = \mathbb{D} \oplus \mathbb{Z}$ with strict ordering from first coordinate.

Choose order units $u_i \in G_i$ by $u_1 = (1,1)$, $u_2 = u_3 = u_4 = (1,0)$. Let A_i be the simple unital AF algebra with dimension group (G_i, u_i).

7.6.1. These groups can be pictured geometrically as subsets of $\mathbb{R}^2$. In G_1, the positive cone consists of the origin and the points of the group in the open first quadrant; in $G_2 - G_4$ the positive cone is the origin and the points in the open right half plane. In G_1, the hereditary generating subset $\Sigma(A_1)$ consists of $(0,0)$, $(1,1)$, and the points in the open square $\{(x,y) : 0 < x < 1, 0 < y < 1\}$. In $G_2 - G_4 \Sigma(A_i)$ consists of $(0,0)$, $(1,0)$, and the open vertical strip $\{(x,y) : 0 < x < 1\}$. G_1 has two extremal states τ_1 and τ_2, corresponding to the projections onto the coordinate axes; the others have a unique state corresponding to projection onto the x-axis (and they have the strict ordering from this projection). So A_1 has two extremal tracial states, also called τ_1 and τ_2, and the others a unique tracial state τ.

7.6.2. None of the algebras has comparability of projections. For example, in A_1 consider projections p and q with $[p] = (\frac{1}{2}, \frac{1}{4})$, $[q] = (\frac{1}{4}, \frac{1}{2})$. Then $\tau_1(p) = \frac{1}{2}$, $\tau_1(q) = \frac{1}{4}$, $\tau_2(p) = \frac{1}{4}$, $\tau_2(q) = \frac{1}{2}$, so they cannot be comparable. Even though $A_2 - A_4$ have unique trace, they also fail to have comparability of projections: take p and q with $[p] = (\frac{1}{2}, 0)$, $[q] = (\frac{1}{2}, 1)$. (Note that these two elements are incomparable.) Although p and q are incomparable, they are "almost comparable" in the sense that whenever $r \preceq p$, $r \neq p$, then $r \preceq q$ and vice versa, and similarly if $p \preceq s$, $p \neq s$.

7.6.3. The C^*-algebras stably isomorphic to A_i can easily be described as in 6.6; they are parametrized by the upward directed hereditary generating subsets Σ of G_{i+}.

In $G_2 - G_4$, there are three possibilities for Σ:

$\Sigma_{(a,b)} = \{(0,0), (a,b)\} \cup \{(x,y) : 0 < x < a\}$ for some $(a,b) \in G_{i+}$;

$\Sigma_a^0 = \{(0,0)\} \cup \{(x,y) : 0 < x < a\}$ for some $a \in \mathbb{R}_+$;

$\Sigma_\infty^0 = G_{i+}$.

The sets $\Sigma_{(a,b)}$ correspond to the unital algebras, Σ_a^0 to the nonunital algebras with finite trace, and Σ_∞^0 to the stable algebra. This parametrization is not one-to-one, since by 7.3.2 any two Σ conjugate under an order-automorphism of G_i will yield isomorphic algebras. In G_2, we have $\Sigma_a^0 \sim \Sigma_b^0$ if and only if $b = qa$ for some $q \in \mathbb{Q}$, under the automorphism $(x,y) \to (qx,y)$, so the isomorphism classes of nonunital finite C^*-algebras stably isomorphic to A_2 are parametrized by the group $\mathbb{R}/\mathbb{Q}$. This classification is virtually identical to the classification of matroid C^*-algebras with dimension group $\mathbb{Q}$. In G_3 or G_4, $\Sigma_a^0 \sim \Sigma_b^0$ if and only if $b = 2^n a$ for some $n \in \mathbb{Z}$, so the finite nonunital algebras are parametrized by $\mathbb{R}_+/\{2^n : n \in \mathbb{Z}\} \cong \mathbb{R}/\mathbb{Z}$.

There is also much collapsing among the $\Sigma_{(a,b)}$. In G_2, all the $\Sigma_{(a,b)}$ are conjugate, i.e. any unital C^*-algebra stably isomorphic to A_2 is actually isomorphic to A_2. (The UHF algebra with dimension group $\mathbb{Q}$ has the same property.) In G_3, the order-preserving automorphisms are all of the form $(x,y) \to (2^n x, rx + 2^m y)$, where $m, n \in \mathbb{Z}$ and $r \in \mathbb{D}$, so $\Sigma_{(a,b)} \sim \Sigma_{(c,d)}$ if and only if there are such numbers with $c = 2^n a$, $d = ra + 2^m b$. So there are many unital algebras in this stable isomorphism class, and they are not all formed as matrix algebras over a fixed "fundamental" algebra as is the case with the CAR algebra. The automorphisms of G_4 are a bit harder to characterize.

7.6.4. The situation with G_1 is similar, but there is one interesting new phenomenon. We have these possibilities for Σ:

$\Sigma_{(a,b)} = \{(0,0), (a,b)\} \cup \{(x,y) : 0 < x < a, 0 < y < b\}$ for some $(a,b) \in G_{1+}$;

$\Sigma_{a,b}^0 = \{(0,0)\} \cup \{(x,y) : 0 < x < a, 0 < y < b\}$ for some $(a,b) \in \mathbb{R}_+^2$;

$\Sigma_{a,\infty}^0 = \{(0,0)\} \cup \{(x,y) : 0 < x < a, y > 0\}$ for some $a \in \mathbb{R}_+$;

$\Sigma_{\infty,b}^0 = (0,0) \cup \{(x,y) : 0 < y < b, x > 0\}$ for some $b \in \mathbb{R}_+$;

$\Sigma_{\infty,\infty}^0 = G_{1+}$.

As before, the $\Sigma_{(a,b)}$ correspond to unital algebras, $\Sigma_{\infty,\infty}^0$ to the stable algebra, and $\Sigma_{a,b}^0$ to the nonunital algebras with both extremal traces finite. But in this case we also have sets of the form $\Sigma_{a,\infty}^0$ and $\Sigma_{\infty,b}^0$, which correspond to algebras with one finite extremal trace and one infinite one. These algebras are simple and have a finite trace, but are not algebraically simple. A further discussion of examples of this kind is found in [Blackadar 1980b, 4.4-4.6].

7.6.5. By choosing other countable dense subgroups of $\mathbb{R}^2$, one can construct other interesting examples similar to A_1 which have various pathological automorphism properties.

A dimension group somewhat similar to G_4 was the essential tool used in the construction of the first simple unital projectionless C^*-algebra [Blackadar 1981].

7.7. EXERCISES AND PROBLEMS

7.7.1. Construct a Bratteli diagram for A_3 as follows.

(a) Let $v_n = (2^{-2n+1}, 2^{-n})$, $w_n = (2^{-2n+1}, -2^{-n})$ in $\mathbb{R}^2$; let Γ_n be the subgroup of $\mathbb{R}^2$ generated by v_n and w_n, with $(\Gamma_n)_+$ the semigroup generated by v_n, w_n, 0. $v_n = 3v_{n+1} + w_{n+1}$, $w_n = v_{n+1} + 3w_{n+1}$, so $\Gamma_n \subseteq \Gamma_{n+1}$, $(\Gamma_n)_+ \subseteq (\Gamma_{n+1})_+$; $\bigcup \Gamma_n = \mathbb{D}^2 = G_3$, $\bigcup (\Gamma_n)_+ = G_{3+}$.

(b) $(\Gamma_n, (\Gamma_n)_+, (1,0))$ is isomorphic to $\mathbb{Z}^2$ with ordinary ordering and order unit $(2^{2n-2}, 2^{2n-2})$; the embedding $\Gamma_n \to \Gamma_{n+1}$ corresponds to the homomorphism from $\mathbb{Z}^2$ to $\mathbb{Z}^2$ with matrix $R = \left[\begin{smallmatrix} 3 & 1 \\ 1 & 3 \end{smallmatrix}\right]$. So one obtains a *stationary* inductive system [Effros 1981, §6] for G_3, i.e. one of the form $\varinjlim (\mathbb{Z}^{r_n}, R_n)$ with constant r_n and R_n.

(c) Write down the corresponding Bratteli diagram. The n-th algebra is a direct sum of two $4^{n-1} \times 4^{n-1}$ matrix algebras.

(d) Modify the construction to give a Bratteli diagram for A_2. Does G_2 have a stationary inductive system?

7.7.2. (a) Show that a procedure such as in 7.7.1 does not work to give a diagram for A_4. G_4 is not *ultrasimplicially ordered*: it cannot be written as an inductive limit $(\mathbb{Z}^{r_n}, R_n)$ (in this case necessarily $r_n = 2$) with R_n injective.

(b) A diagram for A_4 can be constructed as follows. Let $u_n = (2^{-2n}, 1)$, $v_n = (2^{-2n}, -1)$, $w_n = (2^{-2n}, 0)$. Then G_4 is the group generated by $\{u_n, v_n, w_n\}$, with G_{4+} the semigroup generated by $\{u_n, v_n, w_n, 0\}$. Then $u_n = 2u_{n+1} + v_{n+1} + w_{n+1}$, $v_n = v_{n+1} + 3w_{n+1}$, $w_n = u_{n+1} + v_{n+1} + 2w_{n+1}$. Thus G_4 is isomorphic to the stationary inductive limit of $\mathbb{Z}^3$ with matrix

$$\begin{bmatrix} 2 & 0 & 1 \\ 1 & 1 & 1 \\ 1 & 3 & 2 \end{bmatrix}.$$

(c) Write down the corresponding Bratteli diagram.

7.7.3. Find a diagram for A_1 as follows.

(a) Let $v_1 = (\frac{2}{3}, \frac{1}{3})$, $w_1 = (\frac{1}{3}, \frac{2}{3})$, and choose v_n and w_n inductively by the following procedure: if $v_n = (a_n, b_n)$, $w_n = (b_n, a_n)$ with $a_n > b_n$, set $m_n = a_n/b_n$, and let m_{n+1} be a rational number greater than m_n. Let x be the point of intersection of the line through v_n with slope m_{n+1} and the line through the origin with slope $1/m_{n+1}$; let $y = (b, a)$, where $(a, b) = v_n - x$. Then y is parallel to x. Choose v_{n+1} to be a point with positive rational coordinates such that both $\boldsymbol{x}$ and $\boldsymbol{y}$ are integral multiples of $(n!)\boldsymbol{v}_{n+1}$. Let $\boldsymbol{w}_{n+1}$ be $\boldsymbol{v}_{n+1}$ with coordinates reversed.

(b) Show that $\boldsymbol{v}_n = k_n\boldsymbol{v}_{n+1}+l_n\boldsymbol{w}_{n+1}$, $\boldsymbol{w}_n = l_n\boldsymbol{v}_{n+1}+k_n\boldsymbol{w}_{n+1}$ for some positive integers k_n, l_n. If

$$R_n = \begin{bmatrix} k_n & l_n \\ l_n & k_n \end{bmatrix},$$

then $\varinjlim(\mathbb{Z}^2, R_n) \cong (G_1, G_{1+})$.

(c) Write down the corresponding Bratteli diagram. The n-th algebra is a direct sum of two copies of the $p_n \times p_n$ matrices, where $p_n = \prod_{i=1}^{n-1}(k_i + l_i)(p_1 = 1)$.

(d) G_1 cannot have a stationary inductive system, since any stationary dimension group has a unique state [Effros 1981, 6.1].

7.7.4. A dimension group is 2-*symmetric* if it is an inductive limit of the form $(\mathbb{Z}^2, R_n)$, with

$$R_n = \begin{bmatrix} k_n & l_n \\ l_n & k_n \end{bmatrix}.$$

G_1, G_2, G_3 are 2-symmetric. A 2-symmetric group is ultrasimplicially ordered.

(a) The 2-symmetric dimension groups are classified as follows [Fack and Maréchal 1979]. There are four types, the first two being degenerate. We say that two generalized integers are *compatible* if they are either both "odd" or if both have a factor of 2^∞.

(1) If all but finitely many l_n are zero, then the limit group is $\mathbb{Z}_{(q)}\oplus\mathbb{Z}_{(q)}$ with the ordinary ordering, for some generalized integer q. The corresponding algebra is a direct sum of two isomorphic UHF algebras (or matrix algebras).

(2) If R_n is not invertible (i.e. $k_n = l_n$) for infinitely many n, then the limit group is $\mathbb{Z}_{(q)}$ for some generalized integer q with a factor of 2^∞. The corresponding algebras are UHF algebras.

(3) The limit group can be the subgroup $G_{p,q,\infty}$ of $\mathbb{Q}^2$ generated by $\mathbb{Z}_{(p)} \oplus \mathbb{Z}_{(q)}$ and $(1/2, 1/2)$ for any compatible generalized integers p and q (q can be an odd ordinary integer if p is "odd"), with strict ordering from the first coordinate. The corresponding algebras are simple with unique trace.

(4) The limit group can be the subgroup $G_{p,q,\alpha}$ of $\mathbb{Q}^2$ generated by $\mathbb{Z}_{(p)} \oplus \mathbb{Z}_{(q)}$ and $(1/2, 1/2)$ for any compatible generalized (not ordinary) integers p and q which are not both (ordinary) integer multiples of r^∞ for the same prime r, with the positive cone the open wedge between the rays of slope $\pm\alpha$, for any $\alpha > 1$. The corresponding algebras are simple with two extremal traces.

In each case, all such groups occur. In cases (3) and (4), $G_{p,q,\alpha} \cong G_{p',q',\alpha'}$ if and only if there are positive rational numbers a and b with $p' = ap$, $q' = bq$, $\alpha' = (b/a)\alpha$.

(b) To generate symmetric diagrams in cases (3) and (4), choose integers p_n, q_n of the same parity with $0 < q_n < p_n$, $\prod p_n = p$, $\prod q_n = q$, and such that the infinite product $\prod \frac{p_n}{q_n}$ diverges to $+\infty$ in case (3) or converges to α in case (4). (Show that it is possible to find such p_n and q_n if and only if the conditions of (3) or (4) are satisfied.) Then set $k_n = (p_n + q_n)/2$, $l_n = (p_n - q_n)/2 = p_n - k_n$. Show that the right group is obtained in this way.

(c) Let $G = \mathbb{Z}_{(3^\infty)} \oplus \mathbb{Z}$ with strict ordering from the first coordinate, and let H be the subgroup of G of all pairs $(m/3^k, n)$ with m and n of the same parity. Then $H \cong G_{3^\infty,1,\infty}$ is 2-symmetric. Show that G is not even ultrasimplicially ordered [Elliott 1979, 2.7].

(d) Show that $\mathbb{D}^2$ with strict ordering is not 2-symmetric. Show why a procedure such as in 7.7.3 does not work to give a symmetric diagram (an insoluble system of Diophantine equations arises). Is this group ultrasimplicially ordered?

(e) Show that the 2-symmetric dimension groups are exactly the dimension groups of fixed-point subalgebras of UHF algebras under product-type symmetries [Fack and Maréchal 1979].

7.7.5. Prove the following theorem [Blackadar 1981, 3.1; Herman and Rosenberg 1981, 3.4]:

THEOREM. *Let A be an AF algebra. Give the automorphism group* $\mathrm{Aut}(A)$ *the topology of pointwise (norm-)convergence. Then the following subgroups of* $\mathrm{Aut}(A)$ *coincide:*

$\overline{\mathrm{In}}(A)$, *the closure of the inner automorphisms* $\mathrm{In}(A)$ *of the form* ad u, $u \in U_1(\tilde{A})$.
$\overline{\mathrm{Inn}}(A)$, *the closure of the inner automorphisms* $\mathrm{Inn}(A)$ *of the form* ad u, $u \in U_1(M(A))$.
$\mathrm{Aut}(A)_0$, *the connected component of the identity.*
$\mathrm{Aut}(A)_p$, *the path component of the identity.*
$\mathrm{Id}(A) = \{\alpha \in \mathrm{Aut}(A) \mid \alpha_* = 1 \text{ on } K_0(A)\}$.

This group of automorphisms is called the group of *approximately inner automorphisms*. In fact, if α is approximately inner, then there is a norm-continuous path $\langle u_t \rangle$ of unitaries in $\tilde{A}$ such that ad $u_t \to \alpha$ as $t \to \infty$.

(a) Show that $\mathrm{Id}(A) = \{\alpha \in \mathrm{Aut}(A) \mid \alpha(p) \sim p \text{ for all projections } p \in A\}$, using cancellation and the fact that the scale Σ generates $K_0(A)$ if A is AF.

(b) $\overline{\mathrm{In}}(A)$, $\overline{\mathrm{Inn}}(A)$, and $\mathrm{Aut}(A)_0$ are obviously closed subgroups. $\mathrm{Id}(A)$ is also closed by 4.6.6.

(c) Clearly $\overline{\mathrm{In}}(A) \subseteq \overline{\mathrm{Inn}}(A)$, $\mathrm{Aut}(A)_p \subseteq \mathrm{Aut}(A)_0$, and $\mathrm{Inn}(A) \subseteq \mathrm{Id}(A)$, so $\overline{\mathrm{Inn}}(A) \subseteq \mathrm{Id}(A)$. (Of course, if A is unital, then $\mathrm{In}(A) = \mathrm{Inn}(A)$.)

(d) If $\alpha \notin \mathrm{Id}(A)$, let p be a projection in A with $\alpha(p) \nsim p$. Then $H = \{\beta \in \mathrm{Aut}(A) \mid \beta(p) \sim p\}$ is a subgroup of $\mathrm{Aut}(A)$ containing $\mathrm{Id}(A)$ but not α. H is open by 4.6.6. So $\mathrm{Aut}(A)_0 \subseteq \mathrm{Id}(A)$.

(e) Show that $\mathrm{Id}(A) \subseteq \overline{\mathrm{In}}(A) \cap \mathrm{Aut}(A)_p$ as follows, completing the proof of the theorem.

(1) If $A = \overline{\bigcup A_n}$ with A_n finite-dimensional, then the commutant of A_n in A is $\overline{\bigcup R_k}$, where $R_k (k \geq n)$ is the commutant of A_n in A_k. Thus the commutant of A_n in A is an AF algebra [Blackadar 1980a, 2.1].

(2) If $A = \overline{\bigcup A_n}$ with A_n finite-dimensional, and $\alpha \in \mathrm{Id}(A)$, write $A_n + \mathbb{C}1 \subseteq \tilde{A}$ as $A_n^{(1)} \oplus \ldots \oplus A_n^{(r)}$ with $A_n^{(k)}$ a matrix algebra. Let $(e_{ij}^{(k)})$ be a set of matrix units for $A_n^{(k)}$, and let $f_{ij}^{(k)} = \alpha(e_{ij}^{(k)})$. Since $\alpha \in \mathrm{Id}(A)$, there is a partial isometry $w_k \in \tilde{A}$ with $w_k^* w_k = e_{11}^{(k)}$, $w_k w_k^* = f_{11}^{(k)}$. Let $u_n = \sum_k \sum_i f_{i1}^{(k)} w_k e_{1i}^{(k)}$. Then u_n is a unitary in $\tilde{A}$ with $\alpha|_{A_n} = (\mathrm{ad}\, u_n)|_{A_n}$ [Bratteli 1972, 3.8].

(3) $u_n^* u_{n+1}$ is in the commutant of A_n in $\tilde{A}$. Let $v_t (n \leq t \leq n+1)$ be a path of unitaries in the commutant of A_n from 1 to $u_n^* u_{n+1}$, and set $u_t = u_n v_t$ for $n \leq t \leq n+1$. $\mathrm{ad}\, u_t = \alpha$ on A_n for $t \geq n$, so $\mathrm{ad}\, u_t \to \alpha$ as $t \to \infty$.

K-Theory for Operator Algebras
MSRI Publications
Volume 5, second edition, 1998

CHAPTER IV

K_1-THEORY AND BOTT PERIODICITY

In this chapter, we will define the higher K-groups of a Banach algebra and relate them to suspensions in Section 8, and then prove the Bott Periodicity Theorem and establish the fundamental K-theory exact sequence in Section 9.

8. Higher K-Groups

8.1. Definition of $K_1(A)$

Let A be a local Banach algebra. Recall that $\mathrm{GL}_n(A) = \{x \in \mathrm{GL}_n(A^+) : x \equiv 1_n \bmod M_n(A)\}$. $\mathrm{GL}_n(A)$ is a closed normal subgroup of $\mathrm{GL}_n(A^+)$. We embed $\mathrm{GL}_n(A)$ into $\mathrm{GL}_{n+1}(A)$ by $x \to \mathrm{diag}(x, 1)$. [This embedding is the "exponential" of the embedding of $M_n(A)$ into $M_{n+1}(A)$ considered in Chapter III. This is the appropriate analog, since the connection between K_0 and K_1 is given by exponentiation.]

Let $\mathrm{GL}_\infty(A) = \varinjlim \mathrm{GL}_n(A)$. $\mathrm{GL}_\infty(A)$ is a topological group with the inductive limit topology. $\mathrm{GL}_\infty(A)$ can be thought of as the group of invertible infinite matrices which have diagonal elements in $1_{A^+} + A$, off-diagonal elements in A, and only finitely many entries different from 0 or 1. We will identify elements of $\mathrm{GL}_n(A)$ with their images in $\mathrm{GL}_\infty(A)$.

The embedding of $\mathrm{GL}_n(A)$ into $\mathrm{GL}_{n+1}(A)$ maps $\mathrm{GL}_n(A)_0$ into $\mathrm{GL}_{n+1}(A)_0$, and $\mathrm{GL}_\infty(A)_0 = \varinjlim \mathrm{GL}_n(A)_0$.

DEFINITION 8.1.1. $K_1(A) = \mathrm{GL}_\infty(A)/\mathrm{GL}_\infty(A)_0 = \varinjlim[\mathrm{GL}_n(A)/\mathrm{GL}_n(A)_0]$.

This is related to, but not the same as, the group K_1^{alg} of algebraic K-theory: $K_1^{alg}(A)$ is the quotient of $\mathrm{GL}_\infty(A)$ by its commutator subgroup. See [Karoubi 1978, II.6.13] for the relationship.

$K_1(A)$ is countable if A is separable, since nearby invertible elements are in the same component. By the same reasoning (using 3.1.4) we have $K_1(A) \cong K_1(\bar{A})$. If A is a local C^*-algebra, then $U_n(A)/U_n(A)_0 \cong \mathrm{GL}_n(A)/\mathrm{GL}_n(A)_0$, and so $K_1(A)$ is isomorphic to $U_\infty(A)/U_\infty(A)_0$, since there is a deformation retraction of $\mathrm{GL}_n(A)$ onto $U_n(A)$ given by polar decomposition. If A is a C^*-algebra, $K_1(A)$ is also isomorphic to $U_1(A \otimes \mathbb{K})/U_1(A \otimes \mathbb{K})_0$.

EXAMPLES 8.1.2. (a) $K_1(\mathbb{C}) = 0$, and more generally K_1 of any von Neumann algebra (or AW*-algebra) is 0 (the unitary group of a von Neumann algebra is connected by spectral theory). K_1 of any AF algebra is also 0 for the same reason.

(b) If $A = C(S^1)$, then $U_1(A)/U_1(A)_0 \cong \mathbb{Z}$ by sending a function to its winding number around 0. We will show that the map from $U_1(A)/U_1(A)_0$ to $K_1(A)$ is an isomorphism in this case, so $K_1(C(S^1)) \cong \mathbb{Z}$.

(c) The map from $\mathrm{GL}_n(A)/\mathrm{GL}_n(A)_0$ to $\mathrm{GL}_{n+1}(A)/\mathrm{GL}_{n+1}(A)_0$ need not be an isomorphism, and hence the map from $\mathrm{GL}_n(A)/\mathrm{GL}_n(A)_0$ to $K_1(A)$ need not be an isomorphism. For example, let $A = C(S^3)$. Then $U_1(A)/U_1(A)_0$ is trivial since every map from S^3 to S^1 is homotopic to a constant [Spanier 1966, 7.2.12]; but the homeomorphism $S^3 \cong SU(2)$ gives a unitary in $C(S^3, \mathbb{M}_2) \cong M_2(A)$ which is not in the connected component of the identity. See problem 6.10.1.

(d) The group $\mathrm{GL}_1(A)/\mathrm{GL}_1(A)_0$ need not be abelian in general. For example, let $A = C(U(2) \times U(2), \mathbb{M}_2)$ (where $U(2) = U_2(\mathbb{C})$). Define unitaries u and v by $u(x,y) = x$, $v(x,y) = y$. Then uv and vu are in different connected components of $U_1(A)$ [Araki et al. 1960]. However, $\mathrm{diag}(uv, 1)$ and $\mathrm{diag}(vu, 1)$ are in the same component of $U_2(A)$ (3.4.1), so the map from $U_1(A)/U_1(A)_0$ to $U_2(A)/U_2(A)_0$ is not injective in this case.

If $u \in \mathrm{GL}_n(A)$, we write $[u]$ for its image in $K_1(A)$.

PROPOSITION 8.1.3. $K_1(A)$ is an abelian group; in fact, $[u][v] = [\mathrm{diag}(u, v)]$.

PROOF. Follows immediately from 3.4.1.   □

If C is a commutative local Banach algebra, then the map from $\mathrm{GL}_1(C)/\mathrm{GL}_1(C)_0$ to $K_1(C)$ is always injective, since the determinant gives a continuous cross section for the embedding $\mathrm{GL}_1(C) \to \mathrm{GL}_\infty(C)$. Even in the commutative C^*-algebra case, the map $U_1(C)/U_1(C)_0 \to K_1(C)$ need not be surjective (8.1.2(c)), and the map $U_n(C)/U_n(C)_0 \to K_1(C)$ need not be injective (or surjective) for $n > 1$ (8.1.2(d)).

There is evidence that the map $U_1(A)/U_1(A)_0 \to K_1(A)$ is always an isomorphism when A is a simple unital C^*-algebra, although this remains an open problem.

As a simple example of this, we have:

PROPOSITION 8.1.4. Let A be a purely infinite C^*-algebra (6.11.4). Then the standard map from $U_1(A)/U_1(A)_0$ to $K_1(A)$ is an isomorphism.

PROOF. If u is a unitary in $A \otimes \mathbb{K}$, then u is close to a unitary in $M_n(A) + \mathbb{C}1$ for some n. If p is a nontrivial projection in A equivalent to 1, then $1_{M_n(A)}$ is unitarily equivalent to a subprojection of p within $M_{2n}(A)$ (4.3.1), hence homotopic within $M_{4n}(A)$ (4.4.1), i.e. there is a path of unital endomorphisms of $M_{4n}(A)$ from the identity to an endomorphism sending $1_{M_n(A)}$ to a subprojection of p. Thus u is connected to a unitary of the form $v + (1 - p)$, where v is a unitary in pAp, i.e. the map from $U_1(A)/U_1(A)_0$ to $K_1(A)$ is surjective. Also, since $1_A - p$ contains a sequence of mutually orthogonal projections equivalent to p (6.11.2), there is an isomorphism from $pAp \otimes \mathbb{K}$ into A which sends $x \otimes e_{11}$ to x for $x \in pAp$. So if v is in $U_1(pAp \otimes \mathbb{K})_0$, it is in $U_1(A)_0$, i.e. the map is injective. $\qquad \square$

The theory of stable rank [Rieffel 1983b] asserts that the map

$$\mathrm{GL}_n(A)/\mathrm{GL}_n(A)_0 \to K_1(A)$$

is an isomorphism for sufficiently large n if $\mathrm{sr}(A)$ is finite, and gives some information on the smallest n for which this is true. The question of when this map is injective or surjective is a typical question of nonstable K-theory (6.5).

8.1.5. If $\phi : A \to B$, then ϕ extends uniquely to a unital map from A^+ to B^+, and hence defines a homomorphism $\phi_* : K_1(A) \to K_1(B)$. Also, if $A = \varinjlim(A_i, \phi_{ij})$, then $K_1(A) \cong \varinjlim(K_1(A_i), \phi_{ij*})$ (see 3.3.3). So K_1 is a functor from local Banach algebras to abelian groups which commutes with inductive limits.

8.2. Suspensions

DEFINITION 8.2.1. The *suspension* of A, denoted SA, is

$$\{f : \mathbb{R} \to A \mid f \text{ continuous}, \lim_{x \to \infty} \|f(x)\| = 0\}.$$

With pointwise operations and sup norm, SA is a local Banach algebra which is complete if A is; if A is a C^*-algebra, then so is $SA \cong C_0(\mathbb{R}) \otimes A$.

We have $S(M_n(A)) \cong M_n(SA)$, and the map $\phi : A \to B$ induces $S\phi : SA \to SB$. Moreover,

$$(SA)^+ \cong \{f : [0,1] \to A^+ \mid f \text{ continuous}, f(0) = f(1) = \lambda 1, f(t) = \lambda 1 + x_t \text{ for } x_t \in A\}$$
$$\cong \{f : S^1 \to A \mid f \text{ continuous}, f(z) = \lambda 1 + x_z, x_z \in A, x_1 = 0\}.$$

THEOREM 8.2.2. $K_1(A)$ *is naturally isomorphic to* $K_0(SA)$, *i.e. there is an isomorphism* $\theta_A : K_1(A) \to K_0(SA)$ *such that, whenever* $\phi : A \to B$, *the following diagram commutes:*

$$
\begin{array}{ccc}
K_1(A) & \xrightarrow{\phi_*} & K_1(B) \\
\theta_A \downarrow & & \downarrow \theta_B \\
K_0(SA) & \xrightarrow{S\phi_*} & K_0(SB)
\end{array}
$$

(*In the language of category theory,* θ *gives an invertible natural transformation from* K_1 *to* $K_0 \circ S$.)

PROOF. Let $u \in \mathrm{GL}_n(A)$. Take a path z_t from 1_{2n} to $\mathrm{diag}(u, u^{-1})$ in $\mathrm{GL}_{2n}(A)$ (note that the path given in the proof of 3.4.1 lies in $\mathrm{GL}_{2n}(A)$). Set $e_t = z_t p_n z_t^{-1}$. Then $e = (e_t)$ is an idempotent in $M_{2n}((SA)^+)$. Set $\theta_A([u]) = [e] - [p_n]$ (where p_n also denotes the corresponding element of $M_\infty((SA)^+)$, i.e. the constant function p_n).

(1) Show θ_A is well defined. If $[u] = [v]$, we may assume that n is large enough that there are paths a_t from 1 to $v^{-1}u$ and b_t from 1 to vu^{-1} in $\mathrm{GL}_n(A)$. Let z_t and w_t be paths for $\mathrm{diag}(u, u^{-1})$ and $\mathrm{diag}(v, v^{-1})$, e_t and f_t the corresponding projections. Let $x_t = w_t \, \mathrm{diag}(a_t, b_t) z_t^{-1}$. Then x is an invertible element of $M_{2n}((SA)^+)$ which conjugates e to f, so $[e] - [p_n] = [f] - [p_n]$ in $K_0(SA)$.

(2) Conversely, if $\theta_A([u]) = \theta_A([v])$, let z, w, e, f be as above. For sufficiently large $k > n$, there is an invertible $x \in (M_k(SA))^+$ which conjugates e to f. Then $w_t^{-1} x_t z_t$ commutes with p_n in $M_k(A^+)$ (identifying z, w, e, f, p_n with their images in $M_k(A^+)$), so $w_t^{-1} x_t z_t$ must be of the form $\mathrm{diag}(c_t, d_t)$ for each t, where $c_t \in \mathrm{GL}_n(A)$ and $d_t \in \mathrm{GL}_{k-n}(A)$. We have $c_0 = 1$, $c_1 = v^{-1}u$.

(3) To show θ_A is onto, let $[f] - [p_n] \in K_0(SA)$, i.e. f is an idempotent in $M_k((SA)^+)$ with $f \equiv p_n \bmod M_k(SA)$. So f is a path $f_t \in M_k(A^+)$ with $f_0 = f_1 = p_n$ and $f_t \equiv p_n \bmod M_k(A)$ for all t. We may assume $k \geq 2n$. Let $w_t \in \mathrm{GL}_k(A)$ with $w_t p_n w_t^{-1} = f_t$, $w_0 = 1_k$, $w_t \equiv 1_k$ for all t, as in 3.4.1. (Check that the formula given there yields $w_t \in \mathrm{GL}_2(A)$ if the path is constant mod A.) Then $w_1 p_n w_1^{-1} = p_n$, so w_1 must be of the form $\mathrm{diag}(u, v)$ for some $u \in \mathrm{GL}_n(A)$, $v \in \mathrm{GL}_{k-n}(A)$. By expanding k if necessary we may assume v is connected by a path b_t to $\mathrm{diag}(u^{-1}, 1_{k-2n})$ (add on $\mathrm{diag}(v^{-1}, u^{-1})$ to everything if necessary). Then $a_t = v^{-1} b_t$ connects 1_{k-n} to $v^{-1} \mathrm{diag}(u^{-1}, 1_{k-2n})$. Let z_t be a path from 1_k to $\mathrm{diag}(u, u^{-1}, 1_{k-2n})$ in $\mathrm{GL}_k(A)$, and $e_t = z_t p_n z_t^{-1}$. Then $x_t = w_t \, \mathrm{diag}(1, a_t) z_t^{-1}$ has $x_0 = x_1 = 1$, so $x \in M_k((SA)^+)$; x is invertible and conjugates e to f. So $[f] - [p_n] = [e] - [p_n] = \theta_A([u])$. The fact that θ_A is a homomorphism and the functoriality of θ are clear, since all of the group operations are diagonal sums, which commute with θ and S. □

COROLLARY 8.2.3. *If* $0 \to J \xrightarrow{\iota} A \xrightarrow{\pi} A/J \to 0$ *is an exact sequence of local Banach algebras, then the induced sequence* $K_1(J) \xrightarrow{\iota_*} K_1(A) \xrightarrow{\pi_*} K_1(A/J)$ *is exact in the middle.*

8.3. Long Exact Sequence of K-Theory

Just as with K_0, we cannot make the sequence exact at the ends by adding 0's. For example, if $A = C([0, 1])$, $J = C_0((0, 1))$, then the unitary $u(t) = e^{2\pi i t}$ in J^+ gives a nontrivial element of $K_1(J)$ which becomes trivial in $K_1(A)$, so the map from $K_1(J)$ to $K_1(A)$ is not injective. Similarly, if $A = C(\bar{D}^2)$, $J = C_0(D^2)$, where D^2 is the open unit disk, then $A/J \cong C(S^1)$. $K_1(A)$ is trivial, but $K_1(A/J) \cong \mathbb{Z}$, so the map from $K_1(A)$ to $K_1(A/J)$ is not surjective.

Instead, we can define a connecting map $\partial : K_1(A/J) \to K_0(A)$ which makes a long exact sequence

$$K_1(J) \xrightarrow{\iota_*} K_1(A) \xrightarrow{\pi_*} K_1(A/J) \xrightarrow{\partial} K_0(J) \xrightarrow{\iota_*} K_0(A) \xrightarrow{\pi_*} K_0(A/J).$$

DEFINITION 8.3.1. Let $u \in \mathrm{GL}_n(A/J)$, and let $w \in \mathrm{GL}_{2n}(A)$ be a lift of $\mathrm{diag}(u, u^{-1})$. Define $\partial([u]) = [w p_n w^{-1}] - [p_n] \in K_0(J)$. (We have $\partial([u]) \in K_0(J)$ because $\mathrm{diag}(u, u^{-1})$ commutes with p_n, so $w p_n w^{-1} \in M_{2n}(J^+)$, and its image mod $M_{2n}(J)$ is p_n.)

To show that ∂ is well defined, suppose we lift $\mathrm{diag}(u, u^{-1})$ to w', and set $z = w'w^{-1}$. Then $z \in \mathrm{GL}_{2n}(J)$, and $[zwp_n w^{-1}z^{-1}] - [p_n] = [wp_n w^{-1}] - [p_n]$, so the definition is independent of the lift. If we replace u by u' with $[u'] = [u]$, then $v = u^{-1}u' \in \mathrm{GL}_n(A/J)_0$, so we can lift v to a and $uv^{-1}u^{-1}$ to b in $\mathrm{GL}_n(A)$ by 3.4.4. Then $\mathrm{diag}(u', u'^{-1}) = \mathrm{diag}(uv, v^{-1}u^{-1}) = \mathrm{diag}(u, u^{-1})\,\mathrm{diag}(v, uv^{-1}u^{-1})$ which lifts to $w\,\mathrm{diag}(a, b)$. $\mathrm{diag}(a, b)$ commutes with p_n.

∂ is obviously a homomorphism since both operations are diagonal sum.

8.3.2. The map ∂ is called the *index map*. The reason is the following. Suppose A is a unital C^*-algebra and u is a unitary in $M_n(A/J)$. If u lifts to a partial isometry $v \in M_n(A)$, then $\mathrm{diag}(u, u^{-1})$ lifts to the unitary

$$w = \begin{bmatrix} v & 1 - vv^* \\ 1 - v^*v & v^* \end{bmatrix}$$

so $\partial([u]) = [wp_n w^{-1}] - [p_n] = [\mathrm{diag}(vv^*, 1 - v^*v)] - [p_n] = [1 - v^*v] - [1 - vv^*]$. In the special case $A = \mathbb{B}(\mathbb{H})$, $J = \mathbb{K}(\mathbb{H})$, and $K_0(\mathbb{K})$ is identified with $\mathbb{Z}$ in the standard way, the map ∂ is exactly the map which sends a unitary in the Calkin algebra to its Fredholm index.

Unitaries in a quotient do not lift to partial isometries in general, so the definition of ∂ must be stated in the more complicated way given above.

PROPOSITION 8.3.3. ∂ *makes the sequence exact at* $K_1(A/J)$.

PROOF. If $v \in \mathrm{GL}_n(A)$, then $\mathrm{diag}(\pi(v), \pi(v)^{-1})$ has a lifting $\mathrm{diag}(v, v^{-1})$ which commutes with p_n, so $\partial([\pi(v)]) = 0$. Conversely, if $u \in \mathrm{GL}_n(A/J)$ and $\partial([u]) = 0$, let w be a lift of $\mathrm{diag}(u, u^{-1})$. Then $[wp_n w^{-1}] - [p_n] = [xp_n x^{-1}] - [p_n]$ for some $x \in \mathrm{GL}_{2n}(J^+)$, with $x \equiv a \bmod J$, where $a \in \mathrm{GL}_{2n}(\mathbb{C})$ commutes with p_n (regarding a as a member of $\mathrm{GL}_{2n}(J^+)$). By increasing n we may assume $wp_n w^{-1} = xp_n x^{-1}$. Write $y = xa^{-1} \in \mathrm{GL}_{2n}(J)$; then $y^{-1}w$ is a lift of $\mathrm{diag}(u, u^{-1})$, and commutes with p_n, so $y^{-1}w = \mathrm{diag}(c, d)$ for some c, d. c is a lift of u. $\square$

PROPOSITION 8.3.4. ∂ *makes the sequence exact at* $K_0(J)$.

PROOF. It is trivial that $\partial([u]) \to 0$ in $K_0(A)$. Conversely, if $[e] - [p_n] \to 0$ in $K_0(A)$, then by increasing n we may suppose $e = wp_n w^{-1}$ for some $w \in \mathrm{GL}_{4n}(A)_0$. In $\mathrm{GL}_{4n}(A/J)$, $\pi(w) = \mathrm{diag}(u_1, u_2)$, since $e \equiv p_n \bmod J$, and $\pi(w) \in \mathrm{GL}_{4n}(A/J)_0$. By increasing n we may assume that u_2 and $\mathrm{diag}(u_1^{-1}, 1)$ are in the same component, and so $\mathrm{diag}(u_1^{-1}, 1)u_2^{-1}$ has a lift v. Then, if $z = \mathrm{diag}(1, v)w$, we have $\pi(z) = \mathrm{diag}(u_1, u_1^{-1}, 1)$ and $e = zp_n z^{-1}$, so $[e] - [p_n] = \partial([u_1])$. $\square$

8.3.5. We can also define higher K-groups by

$$K_2(A) = K_1(SA) = K_0(S^2 A), \ldots, K_n(A) = K_0(S^n A).$$

We then have connecting maps from $K_{n+1}(A/J)$ to $K_n(J)$ for each n by suspension, and an infinite long exact sequence

$$\cdots \xrightarrow{\partial} K_n(J) \xrightarrow{\iota_*} K_n(A) \xrightarrow{\pi_*} K_n(A/J) \xrightarrow{\partial} K_{n-1}(J) \xrightarrow{\iota_*} \cdots \xrightarrow{\pi_*} K_0(A/J).$$

PROPOSITION 8.3.6. *If* $0 \to J \to A \to A/J \to 0$ *is a split exact sequence of local Banach algebras, then* $0 \to K_n(J) \to K_n(A) \to K_n(A/J) \to 0$ *is a split exact sequence for all* n.

PROOF. All the connecting maps are 0 since everything in $K_n(A/J)$ lifts.   □

COROLLARY 8.3.7. $K_1(A) \cong K_1(A^+)$ *for all* A. *More precisely, the inclusion of* A *into* A^+ *induces an isomorphism.*

PROOF. Use 8.3.6 plus the fact that $K_1(\mathbb{C}) = 0$.   □

9. Bott Periodicity

In this section, we will show that $K_0(A)$ is naturally isomorphic to $K_1(SA)$, and hence to $K_2(A)$. As a consequence, the long exact sequence of 8.3.5 becomes a cyclic 6-term exact sequence. Throughout, A will be a local Banach algebra.

9.1. Basic Definitions

We have a split exact sequence $0 \to SA \to \Omega A \to A \to 0$, where $\Omega A = C(S^1, A)$, which induces a split exact sequence $0 \to K_1(SA) \to K_1(\Omega A) \to K_1(A) \to 0$, so $K_1(SA) = \ker \eta_*$, where $\eta : \Omega A \to A$ is evaluation at 1. This will be our *standard picture* of $K_1(SA)$. So $K_1(SA)$ may be viewed as the group of homotopy equivalence classes of loops in $\mathrm{GL}_\infty(A)$ with base point 1. The group operation is pointwise multiplication, but may alternately be taken as the ordinary concatenation multiplication of loops [Spanier 1966, 1.6.10], i.e. $K_1(SA) \cong \pi_1(\mathrm{GL}_\infty(A))$.

If e is an idempotent in $M_n(A^+)$, write $f_e(z) = ze + (1 - e) \in \mathrm{GL}_n(\Omega(A^+)) \cong C(S^1, \mathrm{GL}_n(A^+))$. A loop of this sort is called an *idempotent loop*. If $e_1 \equiv e_2$ mod $M_n(A)$, then $f_{e_1} f_{e_2}^{-1} \in \mathrm{GL}_n(\Omega A)$, taking the value 1 at $z = 1$. If $e_1 \sim_h e_2$, then f_{e_1} is homotopic to f_{e_2} as elements of $\mathrm{GL}_n(\Omega(A^+))$ taking the value 1 at 1, i.e. as loops in $\mathrm{GL}_n(A^+)$ with base point 1. We will write $f_{e_1} \sim_h f_{e_2}$ to denote this type of homotopy of elements.

DEFINITION 9.1.1. The homomorphism $\beta_A : K_0(A) \to K_1(SA)$ defined by $\beta_A([e] - [p_n]) = [f_e f_{p_n}^{-1}]$ is called the *Bott map* for A.

β_A is well defined by the discussion in 9.1. The Bott map construction is clearly functorial (natural), i.e. if $\phi : A \to B$, then the following diagram is commutative:

$$
\begin{array}{ccc}
K_0(A) & \xrightarrow{\phi_*} & K_0(B) \\
\beta_A \downarrow & & \downarrow \beta_B \\
K_1(SA) & \xrightarrow{\phi_*} & K_1(SB)
\end{array}
$$

9.2. Proof of the Theorem

THEOREM 9.2.1 (BOTT PERIODICITY). β_A *is an isomorphism.*

This theorem is probably the central result of K-theory. The proof will proceed in a number of steps.

LEMMA 9.2.2. *If in the following diagram β_{A^+} and $\beta_{\mathbb{C}}$ are isomorphisms, then so is β_A:*

$$
\begin{array}{ccccccccc}
0 & \longrightarrow & K_0(A) & \longrightarrow & K_0(A^+) & \longrightarrow & K_0(\mathbb{C}) & \longrightarrow & 0 \\
& & \downarrow{\scriptstyle \beta_A} & & \downarrow{\scriptstyle \beta_{A^+}} & & \downarrow{\scriptstyle \beta_{\mathbb{C}}} & & \\
0 & \longrightarrow & K_1(SA) & \longrightarrow & K_1(S(A^+)) & \longrightarrow & K_1(S\mathbb{C}) & \longrightarrow & 0
\end{array}
$$

So it suffices to prove 9.2.1 for A unital.

This is a special case of the Five Lemma of homological algebra. The proof is a straightforward diagram chase.

So for the rest of the proof we will assume A is unital, and think of $\beta : [e] \to [f_e]$.

9.2.3. $\mathrm{GL}_n(C(S^1, A))$ has a local affine structure: loops which are sufficiently uniformly close together are "linearly homotopic" via the line segment between. Any homotopy in $\mathrm{GL}_\infty(C(S^1, A))$ can be approximated arbitrarily closely in the topology of uniform convergence by a "polygonal" homotopy with the same endpoints. Also, if f is a loop in $\mathrm{GL}_n(A)$ and g is a loop in $M_n(A)$ which is sufficiently uniformly close to f, then g is a loop in $\mathrm{GL}_n(A)$.

DEFINITION 9.2.4. A *polynomial loop* is a loop of the form $f(z) = a_0 + a_1 z + \cdots + a_m z^m$ with $a_i \in M_\infty(A)$ and $f(z) \in \mathrm{GL}_\infty(A)_0$ for all z. A *linear loop* is a polynomial loop of degree 1, i.e. $f(z) = a_0 + a_1 z$.

The range of β_A is contained in the group generated by the linear loops.

LEMMA 9.2.5. *Every loop can be uniformly approximated arbitrarily closely by a quotient of polynomial loops. In fact, every loop can be approximated by a trigonometric polynomial loop of the form $\sum_{k=-N}^{N} a_k z^k$.*

PROOF. Let L be the set of loops which can be so approximated. L contains all scalar-valued loops by the Stone–Weierstrass Theorem; L also clearly contains all constant loops. Then L contains all polygonal loops. □

LEMMA 9.2.6. *If f and g are polynomial loops, and $f \sim_h g$ as loops, then $z^m f \sim_h z^m g$ as polynomial loops of degree $\leq 2m$ for some sufficiently large m, via a polygonal homotopy.*

PROOF. Approximate the homotopy by a polygonal homotopy, and approximate the vertices by trigonometric polynomial loops. The polygonal homotopy with these vertices gives the desired homotopy. □

LEMMA 9.2.7. *There is a continuous function μ_m from the set of polynomial loops of degree $\leq m$ to the set of linear loops. $f \sim_h \mu_m(f)$.*

PROOF. If $f(z) = a_0 + \cdots + a_m z^m$, identify f with

$$\mathrm{diag}(f, 1, \ldots, 1) \in M_{(m+1)n}(A) \cong M_{m+1}(M_n(A)).$$

Set

$$\mu_m(f)(z) = \begin{bmatrix} a_0 & a_1 & \cdots & a_{m-1} & a_m \\ -z & 1 & \cdots & 0 & 0 \\ & & \cdot & & \cdot \\ & & \cdot & & \cdot \\ & & \cdot & & \cdot \\ 0 & 0 & \cdots & -z & 1 \end{bmatrix}$$

$\mu_m(f)$ is clearly a linear loop, and $f \to \mu_m(f)$ is continuous. We can recover f from $\mu_m(f)$ by the following process. We have

$$\begin{bmatrix} 1 & 0 & \cdots & 0 & -a_m \\ 0 & 1 & \cdots & 0 & 0 \\ & & & & \cdot \\ & & & 1 & 0 \\ 0 & 0 & \cdots & 0 & 1 \end{bmatrix} \mu_m(f) \begin{bmatrix} 1 & 0 & \cdots & 0 & 0 \\ 0 & 1 & \cdots & 0 & 0 \\ & & & & \cdot \\ & & & 1 & 0 \\ 0 & 0 & \cdots & z & 1 \end{bmatrix} = \begin{bmatrix} a_0 & a_1 & \cdots & (a_{m-1}+a_m z) & 0 \\ -z & 1 & \cdots & & 0 \\ & & \cdot & & \cdot \\ & & z & 1 & 0 \\ 0 & 0 & \cdots & 0 & 1 \end{bmatrix}$$

Each of these factors can be shrunk to the identity within invertible elements just by shrinking the off-diagonal term to 0. By repeating the process we obtain $f \sim_h \mu_m(f)$. □

So every equivalence class in $K_1(SA)$ is represented by a quotient of linear loops with denominator z^m. Two homotopic loops can be represented in such a way, for the same m, with the numerators homotopic within the set of linear loops.

LEMMA 9.2.8. *There is a continuous map γ from the set of linear loops to the set of idempotent loops. $f \sim_h \gamma(f)$.*

PROOF. Let $f(z) = a + bz$. $a + b \in \mathrm{GL}_n(A)_0$, so $f \sim_h (1/(a+b))(a+bz)$, so we may assume $a + b = 1$, $a = 1 - b$, i.e. $f(z) = bz + (1-b) = 1 + (z-1)b$. If $z \neq 1$, then $f(z) = (1-z)[(1-z)^{-1} - b] \in \mathrm{GL}_n(A)$, so $(1-z)^{-1}$ is not in $\sigma(b)$. So $\sigma(z) \cap \{\lambda \mid \mathrm{Re}\,\lambda = \frac{1}{2}\} = \varnothing$. Make e from b by holomorphic functional calculus. $\gamma(f)(z) = ez + (1-e) \sim_h f(z)$ by a linear homotopy. □

COROLLARY 9.2.9. *Every loop is homotopic to a quotient of an idempotent loop for a unique homotopy class of idempotents in $M_\infty(A)$ by z^m for some m. So β_A is an isomorphism.*

This completes the proof of Bott Periodicity. □

It is interesting to note that even if A is a C^*-algebra, one cannot work exclusively with unitaries in this proof (at least without a lot of additional work).

Note that there is no control over how large a matrix algebra one must deal with in order to handle a given loop in $\mathrm{GL}_n(A)$. Thus Bott Periodicity is purely a stable result in general. Some results on stable rank and its relation to K-theory can shed some light on the necessary size of expansion of the matrix algebras in certain instances. See [Rieffel 1983b] for a more detailed discussion of this and related matters.

Note also that even if A is stably finite and unital, so that there is a good order structure on $K_0(A)$, there is no natural order structure on $K_0(S^2A)(V(S^2A)$ is trivial, so $K_0(S^2A)_+ = 0)$, so the order structure of $K_0(A)$ is apparently completely lost under Bott Periodicity.

There is, however, some vestige of the order structure which remains. If τ is a trace on A, and $\tau_* : K_0(A) \to \mathbb{R}$ the corresponding state, then τ_* can be recovered from $K_1(SA)$ by integration: if $u \in U_n(SA)$ is a smooth loop (any class can be so represented), then

$$\tau_*([u]) = \frac{1}{2\pi i} \int_0^{2\pi} \tau(u'(t)u(t)^{-1})dt$$

We will study this and related matters in 10.10.

For C^*-algebras there is an elegant alternate proof of Bott Periodicity due to Cuntz (9.4.2).

9.2.10. In the case $A = \mathbb{C}$, the Bott map provides an isomorphism from $K_0(\mathbb{C})$ to $K_0(C_0(\mathbb{R}^2))$. This map can be described as the map which sends $[1]$ to $[q]-[p]$, where p and q are the projections in $M_2(C_0(\mathbb{R}^2)^+)$ defined by

$$p(z) = \begin{bmatrix} 1 & 0 \\ 0 & 0 \end{bmatrix}, \qquad q(z) = \frac{1}{1+z\bar{z}}\begin{bmatrix} z\bar{z} & z \\ \bar{z} & 1 \end{bmatrix}$$

($\mathbb{R}^2$ is identified with $\mathbb{C}$). q corresponds to the nontrivial line bundle on S^2 given by σ_1 of 1.1.2(c), or from identifying S^2 with $\mathbb{CP}^1$ and thus with the space $\mathrm{Pr}_1(\mathbb{M}_2)$ of one-dimensional projections in $\mathbb{M}_2$, and taking the line bundle

$$\{(p,x) \mid x \in \mathrm{Range}(p)\} \subseteq \mathrm{Pr}_1(\mathbb{M}_2) \times \mathbb{C}^2.$$

This q is often called the *Bott projection*. $[q] - [p]$ is called the *Bott element* in $K_0(C_0(\mathbb{R}^2))$.

9.3. Six-Term Exact Sequence

THEOREM 9.3.1 (STANDARD EXACT SEQUENCE). *Let* $0 \longrightarrow J \overset{\iota}{\longrightarrow} A \overset{\pi}{\longrightarrow} A/J \longrightarrow 0$ *be an exact sequence of local Banach algebras. Then the following six-term cyclic sequence is exact:*

$$
\begin{array}{ccccc}
K_0(J) & \overset{\iota_*}{\longrightarrow} & K_0(A) & \overset{\pi_*}{\longrightarrow} & K_0(A/J) \\
{\scriptstyle\partial}\uparrow & & & & \downarrow{\scriptstyle\partial} \\
K_1(A/J) & \overset{\pi_*}{\longleftarrow} & K_1(A) & \overset{\iota_*}{\longleftarrow} & K_1(J)
\end{array}
$$

The map $\partial : K_0(A/J) \to K_1(J)$ is the composition of the suspended index map $\partial : K_2(A/J) \to K_1(J)$ with the Bott map.

This follows immediately from 9.2.1 and 8.3.

9.3.2. The connecting map $\partial{:}K_0(A/J) \to K_1(J)$ is called the *exponential map*. An explicit formula for this map is given by $\partial([e] - [p_n]) = [exp(2\pi ix)]$, where e is an idempotent in $M_\infty((A/J)^+)$ with $e \equiv p_n \bmod M_\infty(A/J)$ and $x \in M_\infty(A^+)$ with $\pi(x) = e$. The derivation of this formula is an easy exercise.

If e lifts to an idempotent in $M_\infty(A^+)$, then $\partial([e] - [p_n]) = 0$. The exponential map is the obstruction to (stably) lifting idempotents [projections] from quotients, just as the index map is the obstruction to (stably) lifting invertibles [unitaries].

If A is a C^*-algebra, there is an alternate way of viewing the groups $K_0(A)$ and $K_1(A)$ using the outer multiplier algebra of A: $K_i(A)$ is isomorphic to K_{1-i} of the stable outer multiplier algebra of A (12.2.3). The proof uses the six-term exact sequence and the fact that the K-theory of a stable multiplier algebra is trivial.

The reader is urged to look ahead to Section 12 and to try problem 12.5.1. The point of view taken in that section is good motivation for the connections between K-theory, Ext-theory, and Kasparov theory.

9.4. EXERCISES AND PROBLEMS

9.4.1. Let A be a C^*-algebra. Then there is a split exact sequence

$$0 \to SA \to A \otimes C(S^1) \to A \to 0.$$

Use 8.3.6 and Bott periodicity to compute that

$$K_0(A \otimes C(S^1)) \cong K_1(A \otimes C(S^1)) \cong K_0(A) \oplus K_1(A).$$

This is a special case of the Pimsner–Voiculescu exact sequence (10.2.1), which tells how to compute the K-theory of a crossed product by $\mathbb{Z}$, and of the Künneth Theorem for tensor products (23.1.3), which tells how to compute the K-theory of a general tensor product.

9.4.2. Let u be the unilateral shift on a Hilbert space, and $T = C^*(u)$. T is called the *Toeplitz algebra* (it is isomorphic to the C^*-algebra generated by all Toeplitz operators on the unit disk with continuous symbol).

(a) $e = 1 - uu^*$ is a one-dimensional projection. T contains $\mathbb{K}$ as an essential ideal, and $T/\mathbb{K} \cong C(S^1)$. So there is a short exact sequence

$$0 \to \mathbb{K} \to T \to C(S^1) \to 0$$

This sequence does not split; in fact, this is the "standard" nonsplit extension of $C(S^1)$ by $\mathbb{K}$.

(b) T is the universal C^*-algebra generated by an isometry, i.e. if v is an isometry in any C^*-algebra, there is a canonical homomorphism from T onto $C^*(v)$ sending u to v.

(c) Let $q : T \to \mathbb{C}$ be the composition of the quotient map from T to $C(S^1)$ followed by evaluation at 1. If $j : \mathbb{C} \to T$ is the unital embedding, we have $q \circ j = 1_{\mathbb{C}}$, so $q_* \circ j_* : K_i(\mathbb{C}) \to K_i(T) \to K_i(\mathbb{C})$ is the identity.

(d) Show that $j_* \circ q_* : K_i(T) \to K_i(T)$ is also the identity, so that j_* and q_* are mutually inverse isomorphisms, as follows.

(1) Let $\hat{T}$ be the C^*-subalgebra of $T \otimes T$ generated by $\mathbb{K} \otimes T$ and $T \otimes 1$, and set $v = u \otimes 1$, $f = e \otimes 1 = 1 - vv^*$, $w = e \otimes u$, $g = e \otimes e$. Set

$$z_0 = v(1 - f)v^* + wv^* + vw^* + g,$$
$$z_1 = v(1 - f)v^* + fv^* + vf.$$

Then z_0 and z_1 are self-adjoint unitaries in $\hat{T}$, so there is a path (z_t) of unitaries in $\hat{T}$ from z_0 to z_1. Set $v_t = z_t v$, and let ϕ_t be the canonical homomorphism from T to $\hat{T}$ sending u to v_t. Then $\phi_0(u) = v(1 - f) + w$, $\phi_1(u) = v(1 - f) + g$.

(2) Let $\bar{T}$ be the subalgebra $\{(x, y) \mid \pi(x) = y\}$ of $\hat{T} \oplus T$, where $\pi : \hat{T} \to \hat{T}/\mathbb{K} \otimes T \cong T$ is the quotient map. $\bar{T}$ is a split extension of T by $\mathbb{K} \otimes T$. Define ψ_0, ψ_1, ψ, $\omega : T \to \bar{T}$ by $\psi_0(u) = (\phi_0(u), u)$, $\psi_1(u) = (\phi_1(u), u)$, $\psi(u) = (v(1-f), u)$, $\omega(u) = (w, 0)$. Then ψ_0 and ψ_1 are homotopic, $\psi_0 = \psi + \omega$ (note that ψ and ω have orthogonal ranges), and $\psi_1 = \psi + \omega \circ j \circ q$. Thus $\psi_* + \omega_* = \psi_* + \omega_* \circ j_* \circ q_* : K_i(T) \to K_i(T)$.

(3) Since $\bar{T}$ is a split extension of $\mathbb{K} \otimes T$, ω_* is injective by 8.3.6, i.e. $j_* \circ q_* : K_i(T) \to K_i(T)$ is the identity.

(e) Let $T_0 = \ker q = C^*(u - 1)$. Then there is a split exact sequence

$$0 \longrightarrow T_0 \longrightarrow T \xrightarrow{\ q\ } \mathbb{C} \longrightarrow 0.$$

Therefore $K_*(T_0) = 0$.

(f) If B is any C^*-algebra, define $j_B = 1_B \otimes j : B \to B \otimes T$, $q_B = 1_B \otimes q : B \otimes T \to B$. Then just as in (c), (d), (e), we have $j_{B*} : K_i(B) \to K_i(B \otimes T)$ is an isomorphism with inverse q_{B*}, and $K_*(B \otimes T_0) = 0$. (Use the fact that T is nuclear to show that tensoring with B preserves the exact sequences.)

(g) Apply the long exact sequence of K-theory to the extension

$$0 \to B \otimes \mathbb{K} \to B \otimes T_0 \to B \otimes C_0(\mathbb{R}) \to 0$$

to obtain an alternate proof of Bott periodicity [Cuntz 1984, § 4].

The Toeplitz algebra is interesting in its own right and is a valuable technical tool in many K-theory arguments.

9.4.3. Let A be a commutative Banach algebra with maximal ideal space X. Then there is a natural homomorphism $\gamma : A \to C_0(X)$, the Gelfand transform. γ induces homomorphisms $\gamma_* : K_i(A) \to K_i(C_0(X)) \cong K_c^{-i}(X)$. These homomorphisms are actually isomorphisms.

That $\gamma_* : K_1(A) \cong K_c^{-1}(X)$ is an isomorphism was proved by Arens using the theory of several complex variables. The other isomorphism can then be obtained using Bott periodicity (there is also a direct proof due to Grauert). There are two earlier related theorems, the Shilov Idempotent Theorem, which states that every idempotent in $C_0(X)$ is in the image of γ, and hence that the group of integer linear combinations of idempotents in A is isomorphic to $H_c^0(X;\mathbb{Z}) \cong C_c(X,\mathbb{Z})$, and the Arens–Royden Theorem, which states that γ induces an isomorphism of $\mathrm{GL}_1(A)/\mathrm{GL}_1(A)_0$ with $\mathrm{GL}_1(C_0(X))/\mathrm{GL}_1(C_0(X))_0 \cong H_c^1(X;\mathbb{Z})$.

A description of these results and some applications, along with a further discussion of K-theory for commutative Banach algebras, may be found in [Taylor 1975].

9.4.4. If A and B are C^*-algebras, denote by $[A, B]$ the set of homotopy classes of $*$-homomorphisms from A to B. Composition gives a well-defined map from $[A, B] \times [B, C]$ to $[A, C]$.

(a) For any A and B, there is a notion of "orthogonal direct sum" on $[A, B \otimes \mathbb{K}]$ making it into an abelian semigroup [use an isomorphism $M_2(\mathbb{K}) = \mathbb{K}$]. (Compare 15.6.)

(b) A homomorphism from $\mathbb{C}$ into a C^*-algebra B is just a choice of projection in B (the image of 1). So for any B, we may identify $[\mathbb{C}, B \otimes \mathbb{K}]$ with $V(B)$.

(c) A homomorphism ϕ from $S = C_0(\mathbb{R}) = C_0((0,1))$ into a C^*-algebra B is just a choice of unitary in B^+ which is 1 mod B (the image of $f(t) = e^{2\pi i t}$ under the extension of ϕ to $\tilde{\phi} : C(\mathbb{T}) \to B^+$). Thus $[S, B \otimes \mathbb{K}]$ may be identified with $K_1(B)$.

(d) Using Bott periodicity, $[S, SB \otimes \mathbb{K}]$ may be identified with $K_0(B)$. If B is unital, this identification agrees with the suspension of the identification induced by the identification of (b).

Notes for Chapter IV

Some of the exposition of this section is adapted from unpublished lecture notes of Larry Brown, which were in turn based on Taylor's article [1975]. The proof of Bott Periodicity given here is originally due to Atiyah [1968].

K-Theory for Operator Algebras
MSRI Publications
Volume **5**, second edition, 1998

CHAPTER V

K-THEORY OF CROSSED PRODUCTS

10. The Pimsner–Voiculescu Exact Sequence and Connes' Thom Isomorphism

In this section, we will develop exact sequences which allow computation of the K-groups of crossed products of C^*-algebras by $\mathbb{R}$ or cyclic groups.

10.1. Crossed Products

We begin with a very brief review of the general theory of crossed products, in order to establish notation. A complete discussion of the theory can be found in [Pedersen 1979, Chapter 7].

Let A be a C^*-algebra, G a locally compact group, and α a continuous homomorphism from G into $\mathrm{Aut}(A)$, the group of *-automorphisms of A with the topology of pointwise norm-convergence. A *covariant representation* of the covariant system (A, G, α) is a pair of representations (π, ρ) of A and G on the same Hilbert space such that $\rho(g)\pi(a)\rho(g)^* = \pi(\alpha_g(a))$ for all $a \in A$, $g \in G$. Each covariant representation of (A, G, α) gives a representation of the twisted convolution algebra $C_c(G, A)$ by integration, and hence a pre-C^*-norm on this *-algebra. The supremum of all these norms is a C^*-norm, and the completion of $C_c(G, A)$ with respect to this norm is called the *crossed product* of A by G under the action α, denoted $A \times_\alpha G$, or sometimes $C^*(G, A)$ or $C^*(G, A, \alpha)$. The *-representations of $A \times_\alpha G$ are in natural one-one correspondence with the covariant representations of the system (A, G, α).

If α is a single automorphism of A, we may regard α as giving an action of $\mathbb{Z}$ on A; the crossed product will be called the crossed product of A by α.

Roughly speaking, the idea of the crossed product construction is to embed A into a larger C^*-algebra in which the automorphisms become inner (where "inner" means "determined by a multiplier" in the nonunital case). This description is strictly correct only when G is discrete; in the general case both A and G are naturally embedded in the multiplier algebra of $A \times_\alpha G$.

One can also form the reduced crossed product of A by G, denoted $C_r^*(G, A)$ or $C_r^*(G, A, \alpha)$, which is a quotient of the (full) crossed product in general. If G is amenable, the two crossed products coincide. In this section we will be almost exclusively concerned with amenable groups (almost always with abelian groups).

One can also define twisted crossed products using cocycles.

EXAMPLES 10.1.1. (a) If the action of G is trivial, then $A \times_\alpha G \cong A \otimes_{\max} C^*(G)$; the reduced crossed product is $A \otimes_{\min} C^*_r(G)$. If G is amenable, then $C^*(G)$ and $C^*_r(G)$ coincide and are nuclear, so the maximal and minimal cross norms also coincide. As a special case, if $A = \mathbb{C}$, the crossed product is just $C^*(G)$.

(b) If A is commutative, i.e. $A = C_0(X)$ for a locally compact Hausdorff space X, then α corresponds to a group of homeomorphisms of X. The system (X, G, α) is called a dynamical system, especially in the case where G is $\mathbb{Z}$ or $\mathbb{R}$. A general covariant system (A, G, α) is sometimes called a C^*-dynamical system.

(c) If G is a semidirect product of groups $G = N \times_\alpha H$, then $C^*(G) \cong C^*(N) \times_\alpha H$. The C^*-algebra of a general group extension is given by a twisted crossed product. If $H = \mathbb{R}$, then every extension is untwisted; so the C^*-algebra of any simply connected solvable Lie group can be constructed by successive crossed products by $\mathbb{R}$. More generally, if G is a simply connected solvable Lie group, any crossed product $A \times_\alpha G$ can be written as an iterated crossed product by $\mathbb{R}$.

The most important result about crossed products for our purposes is the Takai Duality Theorem, a C^*-analog of a theorem of Takesaki about W*-crossed products. If G is a locally compact abelian group with dual group $\hat{G}$, then there is a natural action $\hat{\alpha}$ of $\hat{G}$ on $A \times_\alpha G$, called the *dual action*, with $\hat{\alpha}$ trivial on the image of A and $\hat{\alpha}_\gamma(g) = \langle g, \gamma \rangle g$ for $g \in G$ (identifying G with its image in the multiplier algebra).

THEOREM 10.1.2 (TAKAI DUALITY). *Let (A, G, α) be a covariant system with G abelian. Then $(A \times_\alpha G) \times_{\hat{\alpha}} \hat{G} \cong A \otimes \mathbb{K}(L^2(G))$. So the second crossed product is stably isomorphic to A, and is stable if G is infinite (if G has n elements, it is $M_n(A)$). Under this isomorphism, the second dual action $\hat{\hat{\alpha}}$ of G on the second crossed product is $\alpha \otimes \lambda$, where λ is the action of G on $\mathbb{K}(L^2(G))$ coming from the regular representation of G.*

See [Pedersen 1979, 7.9.3] for a discussion and proof.

The reader who is unfamiliar with Takai duality is urged to work out the details explicitly in the case $G = \mathbb{Z}_2$. This simple special case sheds light on the general situation. Also, this case will be especially important in the development of KK-theory; for example, Bott Periodicity may be regarded as a special case of duality for $\mathbb{Z}_2$.

10.2. Crossed Products by $\mathbb{Z}$ or $\mathbb{R}$

Crossed products have long been used to construct interesting C^*- and W*-algebras. Unfortunately, it has been rather difficult to obtain any good information about the internal structure of crossed products. There are rather extensive (although still incomplete) results about the ideal structure of crossed products, and in particular some sufficient conditions for a crossed product to be simple; but until the development of K-theory little could be said about the structure of projections in the crossed product, for example. The attempt to describe all equivalence classes of projections in the irrational rotation algebras (10.11.6) was one of the primary motivations for the work of Pimsner and Voiculescu. The solution of this problem was one of the first nontrivial applications of K-theory to the structure of C^*-algebras. (A few months earlier, Cuntz had used somewhat similar arguments to calculate the K-groups of the O_n (10.11.8).) The Pimsner–Voiculescu work is regarded as a milestone, and as a catalyst for the subsequent rapid development of K-theory for C^*-algebras.

The two fundamental results which allow computation of the K-theory of crossed products by $\mathbb{Z}$ or $\mathbb{R}$ are:

THEOREM 10.2.1 (PIMSNER–VOICULESCU EXACT SEQUENCE). *Let A be a C^*-algebra and $\alpha \in \operatorname{Aut}(A)$. Then there is a cyclic six-term exact sequence*

$$
\begin{array}{ccccc}
K_0(A) & \xrightarrow{\;1-\alpha_*\;} & K_0(A) & \xrightarrow{\;\iota_*\;} & K_0(A \times_\alpha \mathbb{Z}) \\
\uparrow & & & & \downarrow \\
K_1(A \times_\alpha \mathbb{Z}) & \xleftarrow{\;\iota_*\;} & K_1(A) & \xleftarrow{\;1-\alpha_*\;} & K_1(A)
\end{array}
$$

THEOREM 10.2.2 (CONNES' THOM ISOMORPHISM). *If $\alpha : \mathbb{R} \to \operatorname{Aut}(A)$, then $K_i(A \times_\alpha \mathbb{R}) \cong K_{1-i}(A)$ $(i = 0, 1)$.*

Connes' Thom Isomorphism is a generalization of Bott Periodicity (the case of trivial action), and is an analog (though not a generalization) of the ordinary Thom isomorphism, which says that if E is a K-oriented n-dimensional vector bundle over X, and E is itself regarded as a locally compact Hausdorff space, then $K^i(E) \cong K^{i+n \bmod 2}(X)$. The result is a bit surprising at first glance, since it says that the K-theory of a crossed product by $\mathbb{R}$ is independent of the action. An intuitive argument for this fact is that any action of $\mathbb{R}$ can be continuously deformed to a trivial action, and K-theory is insensitive to continuous deformations. This rough argument can be used as the basis of a proof of 10.2.2, using KK-theory [Fack and Skandalis 1981]. We will take an alternate approach in this chapter, and will return to the KK proof in Section 19.

The Pimsner–Voiculescu (P-V) exact sequence, which predates Connes' result, shows that the K-theory of a crossed product by $\mathbb{Z}$ is not independent of the action; but (roughly speaking) it depends only on the induced action on the K-theory of A. The P-V exact sequence is not in general enough to completely determine the K-theory of the crossed product, except in special cases (e.g. when the K-groups of A are free abelian groups); but it is nonetheless a powerful tool in determining the possibilities.

The P-V exact sequence is a C^*-algebra analog of an exact sequence for the algebraic K-theory of an algebraic crossed product by $\mathbb{Z}$, due to Farrell and Hsiang [1970]. The techniques of proof are, however, very different.

There are a number of proofs known of both 10.2.1 and 10.2.2. The original proof of Pimsner and Voiculescu [1980a] involved forming a "Toeplitz extension" by setting $B = C_0(X)$, where $X = \mathbb{Z} \cup \{+\infty\}$, $\beta \in \text{Aut}(B)$ leaving $+\infty$ fixed and left translating on $\mathbb{Z}$. Set $C = A \otimes B$, $\gamma = \alpha \otimes \beta$. If J is the ideal $C_0(\mathbb{Z})$ of B, then $I = A \otimes J$ is an ideal of C invariant under γ, so we have an extension

$$0 \longrightarrow I \times_\gamma \mathbb{Z} \longrightarrow C \times_\gamma \mathbb{Z} \longrightarrow Q \longrightarrow 0$$

It is easy to see that $Q \cong A \times_\alpha \mathbb{Z}$, and it follows easily from Takai duality that $I \times_\gamma \mathbb{Z} \cong A \otimes \mathbb{K}$. The major part of the work is to show that the K-theory of $C \times_\gamma \mathbb{Z}$ is naturally isomorphic to the K-theory of A. When these identifications are made, the P-V exact sequence is just the six-term exact sequence of this extension. See 19.9.2 for a description of the argument.

Rieffel [1982] has shown how to adapt this proof to give Connes' result: replace $\mathbb{Z}$ by $\mathbb{R}$ in the construction, and then show that the K-theory of $C \times_\gamma \mathbb{R}$ is trivial so that the connecting maps in the six-term sequence must be isomorphisms.

We will take a somewhat different approach. The P-V exact sequence can be obtained rather easily from the Thom isomorphism. We will first do this, develop some variations and consequences of the P-V sequence, and then give a proof of the Thom isomorphism which is a combination of the arguments of Rieffel [1982] and unpublished work of Pimsner and Voiculescu.

10.3. The Mapping Torus

The P-V exact sequence can be obtained almost immediately from the Thom isomorphism by invoking a theorem of Green [1977] on Morita equivalence. We will give an alternate argument, based on the elementary structure of crossed products. The concept we need is that of the mapping torus:

DEFINITION 10.3.1. Let $\alpha \in \text{Aut}(A)$. The *mapping torus* of α is $M_\alpha = \{f : \mathbb{R} \to A | f(x+1) = \alpha(f(x)) \text{ for all } x \in \mathbb{R}\} \cong \{f : [0,1] \to A | f(1) = \alpha(f(0))\}$.

There is an exact sequence $0 \to SA \to M_\alpha \to A \to 0$.

We need the following structure theorem for crossed products. If β is an action of $\mathbb{R}$ on a C^*-algebra B, and β is trivial on $\mathbb{Z}$, then β drops to an action of $\mathbb{T} = \mathbb{R}/\mathbb{Z}$, also denoted β. β induces an action $\hat{\beta}$ of $\hat{\mathbb{T}} = \mathbb{Z}$ on $B \times_\beta \mathbb{T}$, i.e. an automorphism of $B \times_\beta \mathbb{T}$.

PROPOSITION 10.3.2. $B \times_\beta \mathbb{R}$ *is isomorphic to the mapping torus of* $\hat{\beta}$ *on* $B \times_\beta \mathbb{T}$.

PROOF. Represent B faithfully on a Hilbert space $\mathbb{H}$. The regular representation of $B \times_\beta \mathbb{R}$ on $\mathbb{H} \otimes L^2(\mathbb{R})$ is faithful, and is obtained by inducing the regular representation ρ of $B \times_\beta \mathbb{Z}$ on $\mathbb{H} \otimes L^2(\mathbb{Z})$ to $B \times_\beta \mathbb{R}$. But β is trivial on $\mathbb{Z}$, so ρ is equivalent to the representation σ of $B \otimes C^*(\mathbb{Z}) \cong B \otimes C(\mathbb{T})$ on $\mathbb{H} \otimes L^2(\mathbb{T})$ given by $\sigma(b \otimes 1) = b \otimes 1$, $\sigma(1 \otimes u) = 1 \otimes M$, where M is multiplication by z and

u is the generator of $\mathbb{Z}$. The induced representation λ acts on

$$\{f : \mathbb{R} \to \mathbb{H} \otimes L^2(\mathbb{T}) \mid f(t+1) = (1 \otimes M)f(t), \int_0^1 \|f\|^2 < \infty\}$$

with action $[\lambda(b)f](t) = (\beta_t(b) \otimes 1)f(t)$, $[\lambda(r)f](t) = f(t-r)$.

We may identify this Hilbert space with

$$L^2([0,1]) \otimes (L^2([0,1]) \otimes \mathbb{H}) \cong L^2([0,1]^2) \otimes \mathbb{H},$$

interchange the order in which the variables are written, and conjugate by the unitary U defined by $Uf(s,t) = e^{2\pi i st}f(s,t)$, to convert λ into the representation μ on $L^2([0,1]^2) \otimes \mathbb{H}$ with action $[\mu(b)f](s,t) = \beta_t(b)f(s,t)$, $[\mu(r)f](s,t) = e^{2\pi i rs}f(s, t-r \text{ [mod 1]})$.

For fixed s, the operators $\mu(b)$ and $\mu(r)$ give operators on $L^2([0,1]) \otimes \mathbb{H}$; for fixed b or r these are continuous in s; thus $B \times_\beta \mathbb{R}$ may be regarded as a subalgebra of $C([0,1], \mathbb{B}(L^2([0,1]) \otimes \mathbb{H}))$. For fixed s, the operators $\{\mu(b)\mu(\phi) \mid b \in B, \phi \in C_c(\mathbb{R})\}$ generate the canonical copy of $B \times_\beta \mathbb{Z}$ in $L^2([0,1]) \otimes \mathbb{H}$ (if ϕ has small support, then there is a continuous function ψ on $\mathbb{T}$ such that $\mu(b)\mu(\phi)$ is the canonical action of $b\psi$), so $B \times_\beta \mathbb{R} \subseteq C([0,1], B \times_\beta \mathbb{Z})$, and for each $a \in B \times_\beta \mathbb{Z}$ and each s there is an $f \in B \times_\beta \mathbb{R}$ with $f(s) = a$. The values of any such f at 0 and 1 match up via $\hat{\beta}$, so $B \times_\beta \mathbb{R}$ may be identified with a subalgebra of the mapping torus of $\hat{\beta}$. Furthermore, the unitary $\mu(1)$ is a multiplier of $B \times_\beta \mathbb{R}$; this unitary corresponds to the continuous function $f(s) = e^{2\pi i s}1$. Thus any continuous scalar-valued function f with $f(0) = f(1)$ is a multiplier. A simple partition of unity argument shows that $B \times_\beta \mathbb{R}$ contains the whole mapping torus. □

The identical result with virtually the same proof is valid in the case of an action β of $\mathbb{Z}$ which is trivial on $n\mathbb{Z}$. The action drops to an action of $\mathbb{Z}_n$; the dual action $\hat{\beta}$ is an action of $\hat{\mathbb{Z}}_n = \mathbb{Z}_n$, i.e. an automorphism $\hat{\beta}$ of $B \times_\beta \mathbb{Z}_n$ of order n.

PROPOSITION 10.3.3. $B \times_\beta \mathbb{Z}$ is isomorphic to the mapping torus of $\hat{\beta}$ on $B \times_\beta \mathbb{Z}_n$.

The details of the proof may be found in [Blackadar 1983a, 7.8.1].

10.3.2 and 10.3.3 are special cases of a general result: if G is a locally compact abelian group, H a closed subgroup, B a C^*-algebra, and $\beta : G \to \text{Aut}(B)$ a homomorphism with $H \subseteq \ker(\beta)$ (or just with $\beta|_H$ inner), then $B \times_\beta G$ can be written as a fibered algebra over $\hat{H}$ with fiber $B \times_\beta (G/H)$ (cf. [Olesen and Pedersen 1986]). Presumably a similar result holds in the nonabelian case.

10.4. Proof of the P-V Sequence

We now derive the P-V exact sequence. Suppose $\alpha \in \text{Aut}(A)$, and let $B = A \times_\alpha \mathbb{Z}$. There is an action $\beta = \hat{\alpha}$ of $\mathbb{T}$ on B; we may regard β as an action of

$\mathbb{R}$ on B with $\mathbb{Z}$ acting trivially. We have $B \times_\beta \mathbb{T} \cong A \otimes \mathbb{K}$ by Takai duality, and $B \times_\beta \mathbb{R} \cong M_{\hat{\beta}}$. Thus there is a short exact sequence

$$0 \longrightarrow S(A \otimes \mathbb{K}) \longrightarrow B \times_\beta \mathbb{R} \longrightarrow A \otimes \mathbb{K} \longrightarrow 0$$

From the Thom isomorphism, we have $K_i(B \times_\beta \mathbb{R}) \cong K_{1-i}(B)$; thus the P-V exact sequence is just the six-term exact sequence associated to this extension.

It only remains to show that the connecting maps in this exact sequence are of the form $1 - \alpha_*$, which comes from the following two propositions.

PROPOSITION 10.4.1. *Let* $\alpha \in \text{Aut}(A)$, *and let* M_α *be the mapping torus. If* $K_i(SA)$ *is identified with* $K_{1-i}(A)$ *via the Bott map, then the connecting maps in the six-term exact sequence*

$$
\begin{array}{ccc}
K_1(A) & \longrightarrow K_0(M_\alpha) \longrightarrow & K_0(A) \\
\partial \uparrow & & \downarrow \partial \\
K_1(A) & \longleftarrow K_1(M_\alpha) \longleftarrow & K_0(A)
\end{array}
$$

are of the form $1 - \alpha_*$.

PROOF. α extends uniquely to an automorphism of A^+ and hence to $M_n(A^+)$. Let p be a projection in $M_n(A^+)$, congruent to 1_k. Then the image of $[p] - [1_k]$ in $K_1(SA)$ under the connecting map is $[exp\{2\pi i(tp + (1-t)\alpha(p) - 1_k)\}] = [exp\{2\pi i(tp - t\alpha(p))\}]$, which is exactly the image of $[p] - [\alpha(p)]$ under the Bott map. The argument for the other connecting map is similar, or may be obtained immediately by suspension. $\square$

PROPOSITION 10.4.2. *Let* $\alpha \in \text{Aut}(A)$, $\gamma \in \text{Aut}(\mathbb{K})$. *If* $K_i(A)$ *is identified with* $K_i(A \otimes \mathbb{K})$ *in the standard way, then* $\alpha_* = (\alpha \otimes \gamma)_*$.

PROOF. Since γ is homotopic to 1 (every automorphism of $\mathbb{K}$ is induced by a unitary in $\mathbb{B}$, and $U_1(\mathbb{B})$ is connected), $\alpha \otimes \gamma$ is homotopic to $\alpha \otimes 1$. $\square$

It is worth explicitly recording the following fact from the discussion above:

PROPOSITION 10.4.3. *Let* α *be an automorphism of a* C^*-*algebra* A. *Then* $K_i(A \times_\alpha \mathbb{Z}) \cong K_{1-i}(M_\alpha)$ *for* $i = 0, 1$. *The isomorphism is natural with respect to covariant homomorphisms.*

10.5. Homotopy Invariance

The next simple proposition shows that the mapping torus of α depends only on the homotopy class of α in $\text{Aut}(A)$.

PROPOSITION 10.5.1. *Let* $\alpha, \beta \in \text{Aut}(A)$. *If* α *and* β *are homotopic, then* $M_\alpha \cong M_\beta$. *More generally, if there is a* $\gamma \in \text{Aut}(A)$ *with* α *homotopic to* $\gamma^{-1} \circ \beta \circ \gamma$, *then* $M_\alpha \cong M_\beta$.

PROOF. Under the hypotheses, $\gamma_0 = \beta^{-1} \circ \gamma \circ \alpha$ is homotopic to $\gamma_1 = \gamma$ via a path $\{\gamma_t\}$. If $f \in M_\alpha$, define $\phi(f)$ by $[\phi(f)](t) = \gamma_t(f(t))$. Then $\phi(f)$ is a continuous function, and $\beta([\phi(f)](0)) = (\beta \circ \gamma_0)(f(0)) = (\gamma \circ \alpha)(f(0)) = \gamma(f(1)) = [\phi(f)](1)$, so $\phi(f) \in M_\beta$. ϕ is clearly an isomorphism.   □

10.5.1 gives a somewhat sharper result than the P-V sequence on the extent to which the K-groups of $A \times_\alpha \mathbb{Z}$ depend on α:

COROLLARY 10.5.2. *The K-groups of $A \times_\alpha \mathbb{Z}$ depend up to isomorphism only on the homotopy class of α in* $\mathrm{Aut}(A)$.

It is not true in general that the K-groups of $A \times_\alpha \mathbb{Z}$ are determined by α_*; see [Rosenberg and Schochet 1986, 10.6].

10.6. Exact Sequence for Crossed Products by $\mathbb{T}$

A byproduct of our approach to the P-V sequence is an exact sequence for crossed products by $\mathbb{T}$. If α is an action of $\mathbb{T}$ on A, then by regarding α as an action of $\mathbb{R}$ and identifying the K-groups of $A \times_\alpha \mathbb{R}$ with those of A by the Thom isomorphism, we obtain an exact sequence

$$
\begin{array}{ccc}
K_0(A \times_\alpha \mathbb{T}) \xrightarrow{\ 1 - \hat{\alpha}_* \ } K_0(A \times_\alpha \mathbb{T}) \longrightarrow K_0(A) \\
\Big\uparrow q_* \qquad\qquad\qquad\qquad\qquad\qquad \Big\downarrow q_* \\
K_1(A) \longleftarrow K_1(A \times_\alpha \mathbb{T}) \xleftarrow{\ 1 - \hat{\alpha}_* \ } K_1(A \times_\alpha \mathbb{T})
\end{array}
$$

where $q : A \times_\alpha \mathbb{R} \to A \times_\alpha \mathbb{T}$ is the quotient map. (This sequence is exactly the P-V sequence for the action $\hat{\alpha}$, when $(A \times_\alpha \mathbb{T}) \times_{\hat{\alpha}} \mathbb{Z}$ is identified with $A \otimes \mathbb{K}$ by Takai duality.)

This sequence, however, is of limited use in actually calculating $K_*(A \times_\alpha \mathbb{T})$, since each of the K-groups appears in two places in the sequence.

10.7. Exact Sequence for Crossed Products by Finite Cyclic Groups

In the same manner, we can obtain an exact sequence relating the K-theory of a crossed product by $\mathbb{Z}_n$ to the K-theory of the corresponding crossed product by $\mathbb{Z}$.

THEOREM 10.7.1. *Let $\alpha \in \mathrm{Aut}(A)$ with $\alpha^n = 1$. Then we have an exact sequence*

$$
\begin{array}{ccc}
K_0(A \times_\alpha \mathbb{Z}_n) \xrightarrow{\ 1 - \hat{\alpha}_* \ } K_0(A \times_\alpha \mathbb{Z}_n) \longrightarrow K_1(A \times_\alpha \mathbb{Z}) \\
\Big\uparrow q_* \qquad\qquad\qquad\qquad\qquad\qquad \Big\downarrow q_* \\
K_0(A \times_\alpha \mathbb{Z}) \longleftarrow K_1(A \times_\alpha \mathbb{Z}_n) \xleftarrow{\ 1 - \hat{\alpha}_* \ } K_1(A \times_\alpha \mathbb{Z}_n)
\end{array}
$$

This sequence is not really too useful for computing the K-theory of $A \times_\alpha \mathbb{Z}_n$, since these groups appear in four places in the sequence. However, the sequence can be used to gain some information about the groups.

Unlike the case of crossed products by $\mathbb{Z}$, the K-theory of $A \times_\alpha \mathbb{Z}_n$ does not depend only on the homotopy class of α in $\mathrm{Aut}(A)$: if A is the UHF algebra of type 2^∞ (the CAR algebra), then all automorphisms are homotopic (7.7.5), but there are many possibilities for $K_0(A \times_\alpha \mathbb{Z}_2)$ for various α of order 2 (7.7.4). It is not even true that the K-theory of $A \times_\alpha \mathbb{Z}_n$ depends only on the homotopy class of α within the automorphisms of A of order dividing n (G. Segal, unpublished; communicated by N. C. Phillips).

10.8. Crossed Products by Amalgamated Free Products

Pimsner and Voiculescu [1982] were able to show, by a generalization of the Toeplitz extension construction, that the following result holds for *reduced* crossed products by free groups.

THEOREM 10.8.1. *Let* $\alpha_j (j = 1, \ldots, n) \in \mathrm{Aut}(A)$, *so that the* α_j *define an action* α *of* $\mathbb{F}_n$, *the free group on* n *generators, on* A. *Then the following sequence is exact, where all crossed products are reduced crossed products:*

$$
\begin{array}{ccccc}
\displaystyle\bigoplus_{j=1}^{n} K_0(A) & \xrightarrow{\ \sigma\ } & K_0(A) & \longrightarrow & K_0(C_r^*(\mathbb{F}_n, A)) \\
\Big\uparrow & & & & \Big\downarrow \\
K_1(C_r^*(\mathbb{F}_n, A)) & \longleftarrow & K_1(A) & \xleftarrow{\ \sigma\ } & \displaystyle\bigoplus_{j=1}^{n} K_1(A)
\end{array}
$$

where $\sigma = \sum_{j=1}^{n}(1 - \alpha_{j*})$.

COROLLARY 10.8.2. *The simple unital* C^*-*algebra* $C_r^*(\mathbb{F}_n)$ $(n \geq 2)$ *contains no nontrivial projections.*

This corollary verified a conjecture of Kadison. Crossed products were considered by several authors in an attempt to solve the Kaplansky problem of whether there exists a simple C^*-algebra with no nontrivial projections. Effros and Hahn [1967] conjectured that the irrational rotation algebras (10.11.6) were projectionless, subsequently disproved by the Powers–Rieffel construction. The principal difficulty faced by these earlier authors was the total lack of tools for studying projections in crossed products until the P-V work. Kaplansky's problem was solved in [Blackadar 1980a; 1981], using a mapping torus construction and properties of AF algebras. Subsequent examples have been constructed by several authors (e.g. 10.11.7). Cuntz has also found a much simpler direct proof of 10.8.2 using Fredholm modules [Cuntz 1983b; Connes 1994, IV.5] (cf. 10.11.11).

As a result of the P-V result on crossed products by free groups, there is a natural conjecture for crossed products by amalgamated free products:

CONJECTURE 10.8.3. *Let* $G = H *_L K$ *be the amalgamated free product of (discrete) groups* H *and* K *over* L. *Let* α *be an action of* G *on* A. *Then for* *either the full or reduced crossed products there is a six-term exact sequence*

$$K_0(A \times_\alpha L) \longrightarrow K_0(A \times_\alpha H) \oplus K_0(A \times_\alpha K) \longrightarrow K_0(A \times_\alpha G)$$

$$K_1(A \times_\alpha G) \longleftarrow K_1(A \times_\alpha H) \oplus K_1(A \times_\alpha K) \longleftarrow K_1(A \times_\alpha L)$$

(where the restriction of α to a subgroup of G is also denoted α). The horizontal maps are the K-theory maps induced by inclusions and differences of inclusions respectively.

Special cases of this sequence have been proved. Lance [1983] has proved the conjecture for free products of "nice" groups if $A = \mathbb{C}$; Lance's results apply for free products of cyclic groups. Natsume [1985] extended Lance's results to amalgamated free products, including such groups as $SL(2, \mathbb{Z}) \cong \mathbb{Z}_4 *_{\mathbb{Z}_2} \mathbb{Z}_6$. Some other cases have been proved by Kasparov [1984b]. Cuntz [1982b] showed that similar results hold for full crossed products (10.11.11); he also introduced the concept of K-amenability (20.9) to compare the K-theory of full and reduced crossed products. Waldhausen [1978] has obtained related results in the purely algebraic setting.

10.9. Proof of the Thom Isomorphism

We prove the Thom isomorphism by means of a series of reductions. As in [Rieffel 1982], if α is an action of $\mathbb{R}$ on a C^*-algebra A, we form the *Wiener–Hopf* extension as follows. Let $C = C_0(\mathbb{R} \cup \{+\infty\})$, and let τ be the action of $\mathbb{R}$ on C obtained by fixing $+\infty$ and translating $\mathbb{R}$. Let $CA = C \otimes A$, γ the diagonal action $\tau \otimes \alpha$. CA has an invariant ideal $SA = C_0(\mathbb{R}) \otimes A$; we also denote the action of R on SA by γ.

LEMMA 10.9.1. $SA \times_\gamma \mathbb{R} \cong SA \times_{\tau \otimes 1} \mathbb{R} \cong A \otimes \mathbb{K}$.

PROOF. The first isomorphism is given by sending $g \in C_c(\mathbb{R}, C_c(\mathbb{R}, A)) \subseteq SA \times_\gamma \mathbb{R}$ to $\tilde{g}$, where $[\tilde{g}(s)](t) = \alpha_{-t}([g(s)](t))$. The second isomorphism comes from Takai duality. $\qquad \square$

Thus the six-term exact sequence for the extension

$$0 \longrightarrow SA \times_\gamma \mathbb{R} \longrightarrow CA \times_\gamma \mathbb{R} \longrightarrow A \times_\alpha \mathbb{R} \longrightarrow 0$$

becomes

$$K_1(A) \longrightarrow K_1(CA \times_\gamma \mathbb{R}) \longrightarrow K_1(A \times_\alpha \mathbb{R})$$

$$exp \uparrow \qquad\qquad\qquad\qquad\qquad\qquad \downarrow index$$

$$K_0(A \times_\alpha \mathbb{R}) \longleftarrow K_0(CA \times_\gamma \mathbb{R}) \longleftarrow K_0(A)$$

The Thom isomorphism amounts to the statement that *exp* and *index* are isomorphisms.

LEMMA 10.9.2. *If, for every A and α, the map index in the above exact sequence is surjective, then for every A and α the maps exp and index are isomorphisms.*

PROOF. Suppose *index* is always surjective. Then $K_1(A \times_\alpha \mathbb{R}) = 0$ implies $K_0(A) = 0$. Applying this fact with A replaced by $A \times_\alpha \mathbb{R}$ and α replaced by $\hat\alpha$, and applying Takai duality, we have that $K_1(A) = 0$ implies that $K_0(A \times_\alpha \mathbb{R}) = 0$ for any A and α. Since suspension (with trivial action) commutes with crossed products, we also have that $K_0(A) = 0$ implies $K_1(A \times_\alpha \mathbb{R}) = 0$ for any A and α. Apply these implications with A replaced by CA and α replaced by γ (recall that $K_*(CA) = 0$ since CA is contractible) to conclude that $K_*(CA \times_\gamma \mathbb{R}) = 0$ for any A and α. The result now follows from exactness. $\square$

So we have reduced the problem down to showing that *index* is always surjective; by exactness this is the same as showing that the map $\phi : A \to CA \times_\gamma \mathbb{R}$ defined in 10.9.1 induces the zero map on K-theory. The argument which follows is due to Pimsner and Voiculescu (unpublished); I am grateful to M. Rieffel for calling this proof to my attention and for developing the exposition given here.

We may describe ϕ concretely as follows. Set $E(s,t) = e^{s/2}e^{-t}\chi(t)\chi(t-s) \in L^1(\mathbb{R}^2)$, where χ is the characteristic function of $(0,\infty)$; then E defines an element of $C_0(\mathbb{R}) \times_\tau \mathbb{R}$ (even though E is not continuous in t, E can be approximated in norm by elements of $L^1(\mathbb{R}, C_c(\mathbb{R})) \subseteq C_0(\mathbb{R}) \times_\tau \mathbb{R}$), and in fact E is a "standard" one-dimensional projection in $C_0(\mathbb{R}) \times_\tau \mathbb{R} \cong \mathbb{K}$. Then ϕ may be taken to be given by the formula $[\phi(a)](s,t) = \alpha_t(a)E(s,t)$. ϕ maps A into the corner $E(CA \times_\gamma \mathbb{R})E$ of $CA \times_\gamma \mathbb{R}$.

There is an embedding μ of A into the multiplier algebra $M(CA \times_\gamma \mathbb{R})$ as "constant functions", i.e. if $g \in C_c(\mathbb{R} \times (\mathbb{R} \cup \{+\infty\}), A) \subseteq CA \times_\gamma \mathbb{R}$, then $[\mu(a)g](s,t) = ag(s,t)$, $[g\mu(a)](s,t) = g(s,t)\alpha_s(a)$.

Let $B = CA \times_\gamma \mathbb{R} + \mu(A) + \mathbb{C}1 \subseteq M(CA \times_\gamma \mathbb{R})$. Then B is a split extension

$$0 \longrightarrow CA \times_\gamma \mathbb{R} \longrightarrow B \mathrel{\substack{\longrightarrow \\ \longleftarrow}} \tilde{A} \longrightarrow 0$$

and therefore the map $i_* : K_0(CA \times_\gamma \mathbb{R}) \to K_0(B)$ is injective. Thus we need only show that $\psi_* : K_0(A) \to K_0(B)$ is the zero map, where $\psi = i \circ \phi$. What we will actually show is that $\mu_* = \mu_* + \psi_*$.

We define a function F by $F(s,t) = e^{-s/2}\chi(s)\chi(t-s)$.

LEMMA 10.9.3. *F defines an element of $C_0(\mathbb{R} \cup \{+\infty\}) \times_\tau \mathbb{R}$.*

PROOF. The difficulty is similar to the problem with E, namely that F is not continuous in t. But if $G_\varepsilon(s,t) = e^{-s/2}g_\varepsilon(s)g_\varepsilon(t-s)$, where g_ε agrees with χ except on $[0,\varepsilon]$ where it is linear, then $G_\varepsilon \to F$ as $\varepsilon \to 0$. $\square$

We have $F^*(s,t) = e^{s/2}\chi(-s)\chi(t)$. Now define

$$S = 1 - F \in (C_0(\mathbb{R} \cup \{+\infty\}) \times_\tau \mathbb{R})^\sim.$$

LEMMA 10.9.4. *$S^*S = 1$, $SS^* = 1 - E$.*

The proof is a simple computation.

We define $E_\varepsilon = \frac{1}{\varepsilon}E\left(\frac{s}{\varepsilon}, \frac{t}{\varepsilon}\right)$, $F_\varepsilon = \frac{1}{\varepsilon}F\left(\frac{s}{\varepsilon}, \frac{t}{\varepsilon}\right)$, $S_\varepsilon = 1 - F_\varepsilon$. Then, as above, $S_\varepsilon^* S_\varepsilon = 1$, $S_\varepsilon S_\varepsilon^* = 1 - E_\varepsilon$.

Now suppose A is *unital*. We may then regard $C_0(\mathbb{R}) \times_\tau \mathbb{R}$ in a natural way as a subalgebra of $CA \times_\gamma \mathbb{R}$, and hence we may regard E_ε, F_ε, S_ε as elements of B. Set $\omega_\varepsilon(a) = S_\varepsilon \mu(a) S_\varepsilon^*$, $[\phi_\varepsilon(a)](s,t) = \alpha_t(a)E_\varepsilon(s,t)$, $\psi_\varepsilon = i \circ \phi_\varepsilon$.

LEMMA 10.9.5. *ω_ε is a *-homomorphism from A into $(1 - E_\varepsilon)B(1 - E_\varepsilon)$, ψ_ε a *-homomorphism from A into $E_\varepsilon B E_\varepsilon$, and ω_ε and ψ_ε are point-norm continuous in ε (for $\varepsilon > 0$).*

PROOF. Obvious. □

LEMMA 10.9.6. *$\omega_* = \mu_*$ as maps from $K_0(A)$ to $K_0(B)$.*

PROOF. If p is a projection in $M_n(A) \cong A \otimes \mathbb{M}_n$, then $(S \otimes 1)(\mu \otimes 1)(p)$ is a partial isometry from $(\mu \otimes 1)(p)$ to $(\omega \otimes 1)(p)$, so $\mu_*([p]) = \omega_*([p])$. □

If we set $\mu_\varepsilon = \omega_\varepsilon + \psi_\varepsilon$, then μ_ε is a point-norm continuous path of *-homomorphisms from A to B (note that ω_ε and ψ_ε have orthogonal ranges), and we have $\mu_{\varepsilon*} = \omega_{\varepsilon*} + \psi_{\varepsilon*} = \mu_* + \psi_*$. The next lemma shows that $\mu_{\varepsilon*} = \mu_*$, so that $\psi_* = 0$, completing the proof that *index* is surjective in the unital case.

LEMMA 10.9.7. *As $\varepsilon \to 0$, $\mu_\varepsilon \to \mu$ in the point-norm topology.*

PROOF. We need to show that $S_\varepsilon \mu(a) S_\varepsilon^* + \psi_\varepsilon(a) \to \mu(a)$ for each $a \in A$. It suffices to show that for every $a \in A$,

$$\|S_\varepsilon \mu(a) S_\varepsilon^* - S_\varepsilon S_\varepsilon^* \mu(a)\| \to 0$$

$$\|\psi_\varepsilon(a) - E_\varepsilon \mu(a)\| \to 0$$

We need a sublemma for both parts:

LEMMA 10.9.8. *For any $a \in A$, $\lim_{\varepsilon \to 0} \frac{1}{\varepsilon} \int_0^\infty \|a - \alpha_{-s}(a)\| e^{-s/2\varepsilon} ds = 0$.*

PROOF.

$$\frac{1}{\varepsilon} \int_0^\infty \|a - \alpha_{-s}(a)\| e^{-s/2\varepsilon} ds$$

$$= \frac{1}{\varepsilon} \int_0^{\sqrt{\varepsilon}} \|a - \alpha_{-s}(a)\| e^{-s/2\varepsilon} ds + \frac{1}{\varepsilon} \int_{\sqrt{\varepsilon}}^\infty \|a - \alpha_{-s}(a)\| e^{-s/2\varepsilon} ds$$

$$\leq \sup_{0 < s < \sqrt{\varepsilon}} \|a - \alpha_{-s}(a)\| \frac{1}{\varepsilon}(-2\varepsilon)e^{-s/2\varepsilon}\Big|_0^{\sqrt{\varepsilon}} + \frac{2}{\varepsilon}\|a\|(-2\varepsilon)e^{-s/2\varepsilon}\Big|_{\sqrt{\varepsilon}}^\infty$$

$$= \sup_{0 < s < \sqrt{\varepsilon}} \|a - \alpha_{-s}(a)\| 2(1 - e^{-1/2\sqrt{\varepsilon}}) + 4\|a\|e^{-1/2\sqrt{\varepsilon}}$$

As $\varepsilon \to 0$, $\sup_{0 < s < \sqrt{\varepsilon}} \|a - \alpha_{-s}(a)\| \to 0$, so both terms approach 0. □

PROOF OF 10.9.7 (CONT.). Suppose $a \in A$. Then

$$\|S_\varepsilon \mu(a) S_\varepsilon^* - S_\varepsilon S_\varepsilon^* \mu(a)\| \leq \|\mu(a) S_\varepsilon^* - S_\varepsilon^* \mu(a)\|$$

$$\|\mu(a)(1 - F_\varepsilon^*) - (1 - F_\varepsilon^*)\mu(a)\| = \|\mu(a) F_\varepsilon^* - F_\varepsilon^* \mu(a)\|$$

Using the formula for F_ε^* and the fact that the norm in $CA \times_\gamma \mathbb{R}$ is dominated by the norm in $L^1(\mathbb{R}, CA)$, we have

$$\|\mu(a) F_\varepsilon^* - F_\varepsilon^* \mu(a)\| \leq \int_{-\infty}^{\infty} \left[\sup_t \|(a - \alpha_s(a)) \tfrac{1}{\varepsilon} e^{s/2\varepsilon} \chi(-s)\chi(t)\| \right] ds$$

$$= \frac{1}{\varepsilon} \int_{-\infty}^{\infty} \|a - \alpha_s(a)\| \, \chi(-s) e^{s/2\varepsilon} ds$$

$$= \frac{1}{\varepsilon} \int_0^{\infty} \|a - \alpha_{-s}(a)\| e^{-s/2\varepsilon} ds$$

This integral goes to 0 by 10.9.8.

We now consider the other expression. First note that

$$\lim_{\varepsilon \to 0} \left[\sup_{t \geq 0} \{\|\alpha_t(a) - a\| e^{-t/\varepsilon}\} \right] = 0$$

so for any $\delta > 0$ there is an ε_0 such that for all $\varepsilon < \varepsilon_0$, $\sup_{t \geq 0}\{\|\alpha_t(a) - a\| e^{-t/\varepsilon}\} < \delta$.

The second expression can be written

$$\|\alpha_t(a) E_\varepsilon(s, t) - E_\varepsilon(s, t) \alpha_s(a)\|$$

In a similar way as above, this expression is dominated by

$$\int_{-\infty}^{\infty} \left[\sup_t \|(\alpha_t(a) - \alpha_s(a)) E_\varepsilon(s, t)\| \right] ds$$

$$= \int_{-\infty}^{\infty} \left[\sup_t \|(\alpha_t(a) - \alpha_s(a)) \tfrac{1}{\varepsilon} \chi(t) e^{s/2\varepsilon} e^{-t/\varepsilon} \chi(t - s)\| \right] ds.$$

Consider the integrals $\int_0^{\infty}$ and $\int_{-\infty}^0$ separately.

For $\int_0^{\infty}$, we need only consider $t \geq s \geq 0$, so the integrand becomes

$$\sup_{t \geq s}\{\|\alpha_t(a) - \alpha_s(a)\| \tfrac{1}{\varepsilon} e^{s/2\varepsilon} e^{-t/\varepsilon}\} = \sup_{r \geq 0}\{\|\alpha_{r+s}(a) - \alpha_s(a)\| \tfrac{1}{\varepsilon} e^{s/2\varepsilon} e^{-(r+s)/\varepsilon}\}$$

$$= \sup_{r \geq 0}\{\|\alpha_r(a) - a\| e^{r/\varepsilon}\} \tfrac{1}{\varepsilon} e^{-s/2\varepsilon}$$

$$\leq \delta \tfrac{1}{\varepsilon} e^{-s/2\varepsilon} \quad \text{for } \varepsilon < \varepsilon_0.$$

So $\int_0^{\infty} \leq \int_0^{\infty} \delta \tfrac{1}{\varepsilon} e^{-s/2\varepsilon} ds = 2\delta$ for $\varepsilon < \varepsilon_0$.

For $\int_{-\infty}^{0}$, the $\chi(t-s)$ is irrelevant since we must only consider $t \geq 0$ because of the $\chi(t)$ term. We have

$$\|\alpha_t(a) - \alpha_s(a)\| \, e^{-t/\varepsilon} \leq \{\|\alpha_t(a) - a\| + \|a - \alpha_s(a)\|\} e^{-t/\varepsilon}$$
$$\leq \|\alpha_t(a) - a\| \, e^{-t/\varepsilon} + \|a - \alpha_s(a)\|$$
$$\leq \delta + \|a - \alpha_s(a)\| \quad \text{for } \varepsilon < \varepsilon_0.$$

So $\int_{-\infty}^{0} \leq \int_{-\infty}^{0}(\|a - \alpha_s(a)\| + \delta)\frac{1}{\varepsilon}e^{s/2\varepsilon}ds = \int_{-\infty}^{0}\|a - \alpha_s(a)\|\frac{1}{\varepsilon}e^{s/2\varepsilon}ds + 2\delta$ for $\varepsilon < \varepsilon_0$. This integral goes to 0 by 10.9.8. $\qquad\square$

We have now proved that *index* is surjective whenever A is unital. For general A, we have a commutative diagram

$$
\begin{array}{ccccc}
K_1(A \times_\alpha \mathbb{R}) & \longrightarrow & K_1(A^+ \times_\alpha \mathbb{R}) & \longrightarrow & K_1(C^*(\mathbb{R})) \\
\downarrow & & \downarrow & & \downarrow \\
K_1(\mathbb{K}) & \longrightarrow & K_0(A) & \longrightarrow K_0(A^+) & \longrightarrow & K_0(\mathbb{K})
\end{array}
$$

with exact rows, where the vertical maps are given by *index*. We know that the last two *index* maps are surjective, and in fact the last one must be an isomorphism since both groups in the last column are isomorphic to $\mathbb{Z}$. Furthermore, the map from $K_0(A)$ to $K_0(A^+)$ is injective since $K_1(\mathbb{K}) = 0$. A simple diagram chase then shows that the first *index* map must also be surjective.

This completes the proof of the Thom isomorphism. $\qquad\square$

It is interesting to note that we can obtain an alternate proof of Bott Periodicity from this argument. The only place we used Bott Periodicity in the proof of the Thom isomorphism was in 10.9.2. We could have instead used the long exact sequence of K-theory in the argument of 10.9.2 to conclude from the surjectivity of *index* that $K_{i+1}(A \times_\alpha \mathbb{R}) \cong K_i(A)$ for all A, α, i. Applying this result to a trivial action (with $i = 0$) yields Bott Periodicity.

10.10. Order Structure and Traces on $A \times_\alpha \mathbb{Z}$

It is a highly nontrivial matter to determine the order structure on $K_0(A \times_\alpha \mathbb{Z})$. Even if the action is trivial, so that $A \times_\alpha \mathbb{Z} \cong C(S^1, A)$, there is no known way to calculate the order structure. For example, if $A = C(\mathbb{T}^3)$, then $K_0(A) \cong \mathbb{Z}^4$ with strict ordering from the first coordinate; but $K_0(C(\mathbb{T}^4)) \cong \mathbb{Z}^8$ is perforated (6.10.2).

One can, however, obtain quite a bit of partial information: it is (at least theoretically) possible to calculate the range of any state on $K_0(A \times_\alpha \mathbb{Z})$ which comes from a tracial state on the crossed product (by [Blackadar and Rørdam 1992], every state on K_0 arises this way). In good cases this calculation can be done rather easily and gives complete information on the order structure.

The results of this section are due to Pimsner [1985], based on earlier work of Connes [1981] (see also [Exel 1985] and [Kishimoto and Kumjian 1998]). Special cases of these calculations were done by Pimsner–Voiculescu [1980a] and Elliott

[1984]. Pimsner's results are actually more general than those described here; the most general results are for reduced crossed products by free groups, and the results require use of Connes' cyclic cohomology. For simplicity, we will only consider crossed products by $\mathbb{Z}$, and we will not get into cyclic cohomology at all.

Since order structures and states on $K_0(A)$ only make good sense when A is unital, we will only consider unital C^*-algebras, although some of the results carry through to the nonunital case.

10.10.1. We begin with a brief review of the de la Harpe–Skandalis determinant associated with a trace [Harpe and Skandalis 1984]. If τ is a tracial state on A which is invariant under α, then there is a canonical extension of τ to a tracial state on $A \times_\alpha \mathbb{Z}$ (the extension is defined by $\tau(\sum a_n U^n) = \tau(a_0)$). While this may not be the only extension, it turns out that any other extension defines the same state on $K_0(A \times_\alpha \mathbb{Z})$. Conversely, of course, any tracial state on the crossed product restricts to an invariant tracial state on A. Any trace on A extends uniquely to $M_n(A)$ (the extended trace has norm n). We will use the same symbol τ to denote the original trace and its canonical extensions, and also to denote the corresponding state of K_0.

If τ is an invariant trace on A, we define a function $\Lambda_\tau : K_0(A \times_\alpha \mathbb{Z}) \cong K_1(M_\alpha) \to \mathbb{R}$ by

$$\Lambda_\tau([u]) = \frac{1}{2\pi i} \int_0^1 \tau(u'(t)u(t)^{-1}) \, dt$$

where u is a piecewise-smooth path in $U_\infty(M_\alpha) = \{f : [0,1] \to U_\infty(A) \mid f(1) = \alpha(f(0))\}$ (any unitary in $U_\infty(M_\alpha)$ is homotopic to one of these). It is easy to check that Λ_τ is a homomorphism which extends the map $\tau_* : K_0(A) \to \mathbb{R}$.

Similarly, we obtain

$$\Delta_\tau : U_\infty(A)_0 \to \mathbb{R}/\tau(K_0(A))$$

defined by

$$\Delta_\tau(u) = \rho\left[\frac{1}{2\pi i} \int_0^1 \tau(\xi'(t)\xi(t)^{-1}) \, dt\right]$$

where $\xi : [0,1] \to U_\infty(A)$ is any piecewise smooth path of unitaries from 1 to u and $\rho : \mathbb{R} \to \mathbb{R}/\tau(K_0(A))$ is the quotient map.

An alternate formula for Δ_τ is

$$\Delta_\tau\left(\prod e^{2\pi i x_k}\right) = \sum \rho(\tau(x_k)) \quad \text{for } x_k = x_k^*$$

It is clear that Δ_τ is a group homomorphism. It is also easy to check that if A is commutative, so that the determinant $\mathrm{Det} : U_\infty(A) \to U_1(A)$ is defined, then $\Delta_\tau = \Delta_\tau \circ \mathrm{Det}$.

10.10.2. Now suppose $0 \longrightarrow J \overset{j}{\longrightarrow} B \overset{q}{\longrightarrow} A \longrightarrow 0$ is an exact sequence of C^*-algebras. If τ is a trace on A, and $u \in U_\infty(J)$ with $j(u) \in U_\infty(B)_0$ (i.e.

$[u] \in \ker j_* : K_1(J) \to K_1(B))$, then we may define $\underline{\Delta}_\tau(u) = \underline{\Delta}_{\tau \circ q}(j(u))$. If $u \in U_\infty(J)_0$, one may choose the map ξ of 10.10.1 to lie within $U_\infty(J)$ and so $\underline{\Delta}_\tau(u) = 0$; thus $\underline{\Delta}_\tau$ drops to a well defined group homomorphism from $\ker j_*$ to $\mathbb{R}/\tau \circ q(K_0(B))$.

PROPOSITION 10.10.3. *The following sequence is exact:*

$$0 \longrightarrow \tau \circ q(K_0(B)) \longrightarrow \tau(K_0(A)) \overset{\rho}{\longrightarrow} \underline{\Delta}_\tau(\ker j_*) \longrightarrow 0$$

(where the first map is the inclusion of two subgroups of $\mathbb{R}$, *and* ρ *is the restriction of the quotient map from* $\mathbb{R}$ *to* $\mathbb{R}/\tau \circ q(K_0(B))$*).*

PROOF. From the six-term exact sequence of K-theory, we have $\ker j_* = \partial(K_0(A))$. If p is a projection in $M_\infty(A)$, choose $x = x^* \in M_\infty(B)$ with $q(x) = p$. Then $u = e^{2\pi ix} \in \ker j_*$, and $\underline{\Delta}_\tau(u) = \rho(\tau(p))$, so $\rho(\tau(K_0(A)) \subseteq \underline{\Delta}_\tau(\ker j_*)$. Differences of such elements fill up all of $\underline{\Delta}_\tau(\ker j_*)$.   □

We now come to the main result:

THEOREM 10.10.4. *Let* $\alpha \in \mathrm{Aut}(A)$, *and let* τ *be an invariant tracial state of* A. *Then the map* $\underline{\Delta}_\tau^\alpha$ *from* $\ker(1 - \alpha_*) \subseteq K_1(A)$ *to* $\mathbb{R}/\tau(K_0(A))$ *defined by*

$$\underline{\Delta}_\tau^\alpha([u]) = \underline{\Delta}_\tau(u\alpha(u^{-1}))$$

$(u \in U_\infty(A),\ u\alpha(u^{-1}) \in U_\infty(A)_0)$ *is a well defined group homomorphism, and the following sequence is exact:*

$$0 \longrightarrow \tau(K_0(A)) \longrightarrow \tau(K_0(A \times_\alpha \mathbb{Z})) \overset{\rho}{\longrightarrow} \underline{\Delta}_\tau^\alpha(\ker(1 - \alpha_*)) \longrightarrow 0$$

PROOF. Apply 10.10.3 to the extension in 10.4 corresponding to the P-V exact sequence.   □

If A is commutative, we can be more precise:

THEOREM 10.10.5. *Under the conditions of 10.10.4, suppose* $A = C(X)$ *is commutative. Then* $\rho(\tau(K_0(A \times_\alpha \mathbb{Z}))$ *is equal to the image of* $U_1(A)$ *in* $\ker(1 - \alpha_*)$ *under* $\underline{\Delta}_\tau^\alpha$. *So to compute the range of* τ *on* $K_0(A \times_\alpha \mathbb{Z})$, *we need only compute* $\tau(K_0(A))$ *and* $\underline{\Delta}_\tau(u\alpha(u^{-1}))$ *for one* u *in each component of* $U_1(A)$ *which is left fixed by* α, *i.e. for each element of* $H^1(X; \mathbb{Z})$ *left fixed by* α^*.

PROOF. This follows immediately from the fact that $\Delta_\tau = \Delta_\tau \circ \mathrm{Det}$.   □

COROLLARY 10.10.6. *Let* X *be a compact space with* $H^1(X; \mathbb{Z}) = 0$, α *a homeomorphism of* X, *and* τ *the trace on* $C(X)$ *corresponding to an invariant probability measure (such a measure always exists). Then the range of the dual trace on* $K_0(C(X) \times_\alpha \mathbb{Z})$ *is the same as the range of* τ *on* $K_0(C(X))$. *If* X *is connected and* α *is minimal, then* $C(X) \times_\alpha \mathbb{Z}$ *is a simple unital* C^*-*algebra with no nontrivial projections.*

PROOF. The first part follows immediately from 10.10.5. For the second part, if X is connected the range of τ on $K_0(C(X))$ is $\mathbb{Z}$ (the value on any projection is the dimension of the corresponding vector bundle). If α is minimal, then $C(X) \times_\alpha \mathbb{Z}$ is a simple unital C^*-algebra by [Pedersen 1979, 8.11.2]. If τ is the dual trace to any invariant measure, then the range of τ on $K_0(A \times_\alpha \mathbb{Z})$ is $\mathbb{Z}$; since τ must be faithful it follows that 1 is a minimal projection.   □

10.11. EXERCISES AND PROBLEMS

10.11.1. Let A be a stably finite unital C^*-algebra, and $\alpha \in \mathrm{Aut}(A)$. Show using the P-V sequence that $K_1(A \times_\alpha \mathbb{Z})$ is nontrivial. Conclude that a crossed product of a unital AF algebra by $\mathbb{Z}$ is never AF. (A crossed product of a nonunital AF algebra by $\mathbb{Z}$ can be AF: if $\mathbb{Z}$ acts on $C_0(\mathbb{Z})$ by translation, the crossed product is isomorphic to $\mathbb{K}$ by Takai duality.)

10.11.2. Let A be the simple unital AF algebra A_3 defined in 7.6, and define $\sigma : \mathbb{D}^2 \to \mathbb{D}^2$ by $\sigma(a, b) = (a, -2b)$. Then σ is an automorphism of the scaled dimension group of A, so there is an automorphism α of A with $\alpha_* = \sigma$ by 7.3.2. Set $B = A \times_\alpha \mathbb{Z}$.

(a) No (nonzero) power of α_* is the identity, so no power of α is inner; thus by [Olesen and Pedersen 1978] (cf. [Pedersen 1979, 8.11.12]) B is simple.

(b) B is stably finite since the trace on A is invariant under α and thus extends to a trace on B.

(c) By the P-V sequence we have $K_0(B) \cong \mathbb{D}^2/(0 \oplus 3\mathbb{D}) \cong \mathbb{D} \oplus \mathbb{Z}_3$. Thus a stably finite simple unital C^*-algebra can have torsion in its K_0-group. (See [Elliott 1996] and its references for many more examples of this type.)

(d) Show that the ordering on $K_0(B) \cong \mathbb{D} \oplus \mathbb{Z}_3$ is strict ordering from the first coordinate; the order unit is $(1, 0)$.

Using a similar construction one can obtain a stably finite simple unital C^*-algebra with n-torsion in its K_0-group, for any $n > 2$ (hence also for $n = 2$).

10.11.3. The following question about AF algebras is open:

Question. Let A be an AF algebra, σ an automorphism of the scaled ordered group $K_0(A)$ with $\sigma^n = 1$. Is there an automorphism α of A with $\alpha_* = \sigma$ and $\alpha^n = 1$?

This question is important in studying the algebraic structure of $\mathrm{Aut}(A)$.

10.11.4. (a) Regard $M_k(C(S^1)) \cong C(S^1, \mathbb{M}_k)$ as

$$\{f : [0, 1] \to \mathbb{M}_k \mid f \text{ continuous, } f(0) = f(1)\}.$$

Define $\phi_k : M_k(C(S^1)) \to M_{2k}(C(S^1))$ by

$$[\phi_k(f)](t) = u_t \cdot \mathrm{diag}\left(f\left(\frac{t}{2}\right), f\left(\frac{t+1}{2}\right)\right) \cdot u_t^*$$

where u_t is as in 3.4.1. ϕ_k is called a *standard twice-around embedding*. (More generally, one could define an r-times-around embedding.)

(b) Let $B = \varinjlim (M_{2^n}(C(S^1)), \phi_{2^n})$. B is called the *Bunce–Deddens algebra of type* 2^∞. Show that B is simple, unital, and stably finite, and that $K_0(B) \cong \mathbb{D}$, $K_1(B) \cong \mathbb{Z}$. (Given any generalized integer q, one can similarly define a Bunce–Deddens algebra of type q, whose K_0-group is $\mathbb{Z}_{(q)}$ and whose K_1 is $\mathbb{Z}$. The classification and structure of Bunce–Deddens algebras resembles that of UHF algebras.)

(c) Show that B is isomorphic to $C(S^1) \times_\rho G$, where G is the group of all 2^n-th roots of unity (for all n) acting on S^1 by translation (rotation). The other Bunce–Deddens algebras can similarly be obtained as crossed products of $C(S^1)$ by dense torsion subgroups of $\mathbb{T}$ acting by translation [Green 1977]. Dually, B is isomorphic to $C(X) \times_\tau \mathbb{Z}$, where X is the 2-adic integers and $\mathbb{Z} \subseteq X$ acts by translation. The Bunce–Deddens algebras can also be obtained as C^*-algebras generated by weighted shifts [Bunce and Deddens 1975].

10.11.5. (a) Let $\alpha \in \mathrm{Aut}(C(S^1))$ be defined by $[\alpha(f)](z) = f(\bar{z})$, i.e. α corresponds to reflection across the x-axis. Show that $D = C(S^1) \times_\alpha \mathbb{Z}_2$ is isomorphic to $C^*(\mathbb{Z}_2 * \mathbb{Z}_2)$ (6.10.4), and that $K_1(D)$ is trivial. However, $D \times_{\hat{\alpha}} \mathbb{Z}_2 \cong M_2(C(S^1))$, so $K_1(D \times_{\hat{\alpha}} \mathbb{Z}_2) \cong \mathbb{Z}$. Thus a crossed product of a unital C^*-algebra with trivial K_1 by a finite group can have nontrivial K_1.

(b) The embedding $\phi_k : \mathbb{M}_k \otimes C(S^1) \to \mathbb{M}_{2k} \otimes C(S^1)$ commutes with $1 \otimes \alpha$; thus there is an induced symmetry β of the Bunce–Deddens algebra B of type 2^∞ such that $A = B \times_\beta \mathbb{Z}_2$ (which is simple) has trivial K_1. $A \times_{\hat{\beta}} \mathbb{Z}_2 \cong M_2(B)$ ($\cong B$) by Takai duality; so even the crossed product of a very nice stably finite simple unital C^*-algebra with trivial K_1 (connected unitary group) by $\mathbb{Z}_2$ can have nontrivial K_1 (disconnected unitary group).

(c) Show that $K_0(A) \cong \mathbb{D} \oplus \mathbb{Z}$ with strict ordering from the first coordinate, by writing A as an inductive limit of copies of $\mathbb{M}_{2^n} \otimes C^*(\mathbb{Z}_2 * \mathbb{Z}_2)$ under "twice-around" embeddings. Thus A has the same K-theory as the AF algebra A_4 of 7.6, and $\hat{\beta}_*$ is the automorphism of $\mathbb{D} \oplus \mathbb{Z}$ defined by $\hat{\beta}_*(a, b) = (a, -b)$.

(d) A is an AF algebra [Blackadar 1990] (and hence is isomorphic to A_4). Since $B = A \times_{\hat{\beta}} \mathbb{Z}_2$, the crossed product of an AF algebra by a symmetry need not be AF (alternately, the fixed-point subalgebra of a symmetry of an AF algebra need not be AF).

This example in part inspired the extensive classification program of G. Elliott, the subject of a great deal of attention in recent years. See [Elliott 1996] for a survey.

10.11.6. Let θ be an irrational number between 0 and 1. Consider the homeomorphism α_θ of S^1 given by $\alpha_\theta(z) = e^{2\pi i \theta} z$, i.e. rotation by $2\pi\theta$. Set $A_\theta = C(S^1) \times_{\alpha_\theta} \mathbb{Z}$. A_θ is called the *irrational rotation algebra* with angle $2\pi\theta$.

(a) A_θ is simple since α_θ is minimal. Show that any two unitaries u, v in a C^*-algebra satisfying $uv = e^{2\pi i \theta} vu$ generate a C^*-subalgebra canonically isomorphic to A_θ.

(b) Use the P-V exact sequence to calculate $K_0(A_\theta) \cong K_1(A_\theta) \cong \mathbb{Z}^2$.

(c) Show that A_θ has a unique trace τ, the dual trace to Lebesgue measure on S^1 (the only measure invariant under α_θ).

(d) Let $u(z) = z$. Then $[u]$ generates $K_1(C(S^1)) \cong \mathbb{Z}$. $u\alpha_\theta(u^{-1}) = e^{2\pi i\theta}1$, so $\underline{\Delta}^{\alpha_\theta}_\tau([u]) = \theta$. Conclude from 10.10.5 that $\tau(K_0(A_\theta)) = \mathbb{Z} + \mathbb{Z}\theta$. Hence from (b), τ maps $K_0(A)$ isomorphically onto $\mathbb{Z} + \mathbb{Z}\theta$.

(e) Conclude that $A_{\theta_1} \cong A_{\theta_2}$ if and only if $\theta_1 = \theta_2$ or $\theta_1 = 1 - \theta_2$.

The Powers–Rieffel construction [Rieffel 1981] shows that if $\lambda \in (\mathbb{Z} + \mathbb{Z}\theta) \cap [0, 1]$, then there is a projection in A_θ of trace λ, so that $K_0(A_\theta)$ is actually isomorphic to $\mathbb{Z} + \mathbb{Z}\theta$ as an ordered group. Rieffel [1983a] also showed that the A_θ have cancellation (6.5), so that two projections in a matrix algebra are unitarily equivalent if and only if they have the same trace.

The irrational rotation algebras come up naturally in several interesting contexts. For example, the foliation C^*-algebra of a Kronecker foliation of $\mathbb{T}^2$ is the stable algebra of an irrational rotation algebra [Connes 1982]. Irrational rotation algebras are a natural setting for Connes' noncommutative differential geometry [Connes 1994]. Irrational rotation algebras are AT algebras, inductive limits of direct sums of matrix algebras over $C(\mathbb{T})$ [Elliott and Evans 1993]. There is much fascinating structure yet to be explored in these algebras.

It is also interesting to study higher-dimensional analogs of the A_θ, sometimes called "noncommutative tori" (cf. [Elliott 1984; Rieffel 1988]).

10.11.7. There exists a minimal homeomorphism α of the 3-sphere [Fathi and Herman 1977]. By 10.10.6, $A = C(S^3) \times_\alpha \mathbb{Z}$ is a simple unital C^*-algebra with no nontrivial projections.

A similar example, also due to Connes [1982, §12], is obtained by taking Γ to be a discrete cocompact subgroup of $SL(2, \mathbb{R})$ such that the quotient $V = SL(2, \mathbb{R})/\Gamma$ is a homology 3-sphere (e.g. the group of products of an even number of hyperbolic reflections across the sides of a regular triangle with corner angles $\pi/4$ in the Poincaré disk). Let α be the homeomorphism of V given by left translation by $\left[\begin{smallmatrix} 1 & 0 \\ 1 & 1 \end{smallmatrix}\right]$.

Calculate the K-theory of these C^*-algebras. It is not true that $K_0(A) \cong \mathbb{Z}$, i.e. there are projections in matrix algebras over A which are not (stably) equivalent to "diagonal" projections. Show that a simple unital projectionless C^*-algebra A constructed by the method of 10.10.6 never has $K_0(A) \cong \mathbb{Z}$.

10.11.8. Fix n with $1 < n < \infty$; let O_n be as in 6.3.2, and let A be the UHF algebra with dimension group $\mathbb{Z}_{(n^\infty)}$ (7.5).

(a) There is an automorphism α of $A \otimes \mathbb{K}$ with $(A \otimes \mathbb{K}) \times_\alpha \mathbb{Z} \cong O_n \otimes \mathbb{K}$: choose a one-dimensional projection e in $\mathbb{M}_n$, and let B be the completion of the *-algebra of sums of formal tensors $\bigotimes_{k \in \mathbb{Z}} x_k$, where $x_k \in \mathbb{M}_n$, $x_k = e$ for sufficiently large negative k, and $x_k = 1$ for sufficiently large positive k. Then $B \cong A \otimes \mathbb{K}$ (B can be alternately described as an inductive limit of algebras of the form $B_i = \otimes_{k \geq -i} \mathbb{M}_n)$. If α is the tensor product shift automorphism on B, then $B \times_\alpha \mathbb{Z} \cong O_n \otimes \mathbb{K}$.

(b) α_* is multiplication by n on $\mathbb{Z}_{(n^\infty)}$; thus from the P-V exact sequence $K_0(O_n) \cong \mathbb{Z}_{(n^\infty)}/(1-n)\mathbb{Z}_{(n^\infty)} \cong \mathbb{Z}_{n-1}$. Show that this isomorphism sends $[1]$ to a generator of $\mathbb{Z}_{n-1}$.

(c) Conclude that O_n is not isomorphic to $M_k(O_m)$ for any k if $n \neq m$, and that $M_k(O_n) \not\cong O_n$ unless k and $n-1$ are relatively prime.

(d) If p and q are nonzero projections in $M_k(O_n)$, then $p \sim q$ if and only if $[p] = [q]$ in $K_0(O_n)$ (6.11.9). Thus $M_k(O_n) \cong O_n$ if $k \equiv 1 \bmod n-1$.

(e) The P-V exact sequence also yields that $K_1(O_n) = 0$.

(f) $M_k(O_n) \cong O_n$ if k and $n-1$ are relatively prime. $M_{(n^\infty)} \otimes O_n \cong O_n$. In fact, if A is a purely infinite separable nuclear C^*-algebra in the bootstrap class N (22.3.4), with $(K_0(A), [1_A]) \cong (\mathbb{Z}_{n-1}, 1)$ and $K_1(A) = 0$, then $A \cong O_n$ (this is a special case of the remarkable classification in [Kirchberg 1998]).

These results were first proved by Cuntz [1981b] before development of the P-V sequence; the arguments were somewhat similar to those necessary to obtain the P-V sequence in the above situation. Cuntz also studied the simple C^*-algebra O_∞ generated by a sequence of isometries with mutually orthogonal ranges. $K_0(O_\infty) \cong \mathbb{Z}$, $K_1(O_\infty) = 0$.

10.11.9. Let $A = (a_{ij})$ be an $n \times n$ matrix of 0's and 1's. O_A is defined to be the universal C^*-algebra generated by n partial isometries $s_1, \ldots, s_n$ with the relations $s_i^* s_i = \sum_{j=1}^n a_{ij} s_j s_j^*$ [Cuntz and Krieger 1980; Cuntz 1981a]. If A consists entirely of 1's, then $O_A = O_n$.

(a) O_A is simple if A is irreducible, except in degenerate cases.

(b) As in 10.11.8, $O_A \otimes \mathbb{K}$ can be written as a crossed product $B \times_\alpha \mathbb{Z}$ for a stable AF algebra B. The dimension group of B is the stationary direct limit $(\mathbb{Z}^n, \phi_A)$, where ϕ_A is multiplication by A^t. α_* corresponds to the shift on this direct limit.

(c) From the P-V exact sequence, we obtain that $K_0(O_A) \cong \mathbb{Z}^n/(1-A^t)\mathbb{Z}^n$ and $K_1(O_A) \cong \ker(1-A^t)$ (regarded as an endomorphism of $\mathbb{Z}^n$). So if $\operatorname{Det}(1-A^t) = 0$, $K_0(O_A)$ and $K_1(O_A)$ are infinite groups.

The O_A arise naturally in the study of topological Markov chains [Cuntz and Krieger 1980].

10.11.10. There exist many infinite-dimensional stably finite simple unital C^*-algebras A with $K_*(A) \cong K_*(\mathbb{C})$, i.e. $K_0(A) \cong \mathbb{Z}$, $K_1(A) = 0$. (Such a C^*-algebra necessarily contains a minimal projection.) Here is perhaps the simplest example.

(a) Let $\mathbb{F}_\infty$ be the free group on a countable number of generators. Then the commutator subgroup of $\mathbb{F}_\infty$ is isomorphic to $\mathbb{F}_\infty$. Let $G_n = \mathbb{F}_\infty$ and $\phi_{n,n+1} : G_n \to G_{n+1}$ be an isomorphism onto the commutator subgroup, and let $G = \varinjlim (G_n, \phi_{n,n+1})$.

(b) Show that $K_0(C_r^*(\mathbb{F}_\infty)) \cong (\mathbb{Z}, 1)$ as a scaled ordered group, and

$$K_1(C_r^*(\mathbb{F}_\infty)) \cong \mathbb{Z}^\infty,$$

with the generators of $\mathbb{F}_\infty$ (regarded as unitaries in the C^*-algebra) as generators of K_1. (See 10.11.11(h).)

(c) $C_r^*(G) \cong \varinjlim(C_r^*(G_n), \phi_{n,n+1})$. Show that

$$\phi_{n,n+1*} : K_0(C_r^*(G_n)) \to K_0(C_r^*(G_{n+1}))$$

is the identity map on $\mathbb{Z}$, and that $\phi_{n,n+1*} : K_1(C_r^*(G_n)) \to K_1(C_r^*(G_{n+1}))$ is the zero map (use 3.4.1). Thus $K_0(C_r^*(G)) \cong \mathbb{Z}$ and $K_1(C_r^*(G)) = 0$.

This example is not nuclear. A more interesting nuclear example is given in [Jiang and Su 1997].

10.11.11. K-THEORY OF AMALGAMATED FREE PRODUCTS.

(a) Let A, B, D be C^*-algebras, and $\phi_1 : D \to A$, $\phi_2 : D \to B$ embeddings. Regard D as a C^*-subalgebra of A and B via ϕ_i. The *amalgamated free product* of A and B over D, denoted $A *_D B$, is the universal C^*-algebra generated by homomorphic images of A and B which agree on D. (It can be formally defined as the completion of the algebraic free product of A and B with respect to an evident pre-C^*-norm.) If $D = 0$, then $A * B = A *_D B$ is the *free product* of A and B; if A and B are unital and $D = \mathbb{C}1$, then $A *_\mathbb{C} B$ is the *unital free product* of A and B. Show that the natural homomorphisms $\iota_1 : A \to A *_D B$ and $\iota_2 : B \to A *_D B$ are injective [Blackadar 1978, 3.1], so we may identify A and B with C^*-subalgebras of $A *_D B$. (See [Blackadar 1985b; Loring 1997] for a general discussion of universal C^*-algebras defined by generators and relations.)

(b) Suppose there are retractions $r_1 : A \to D$ and $r_2 : B \to D$. (This will be automatic in the case of free products, taking $r_i = 0$.) Then there is an induced retraction $r : A *_D B \to D$. Let $P = \{(a,b) \mid r_1(a) = r_2(b)\} \subseteq A \oplus B$ be the pullback (15.3); let $i : D \to P$ be the diagonal inclusion, and $g = i \circ r :$ $A *_D B \to P$. Define another map $k : A *_D B \to P$ by setting $k(a) = (a, r_1(a))$, $k(b) = (r_2(b), b)$, and using the universal property of $A *_D B$. Finally, define $f : P \to M_2(A *_D B)$ by $f((a,b)) = \text{diag}(a,b)$.

(c) $(1 \otimes k) \circ f : P \to M_2(P)$ sends (a,b) to $\text{diag}((a, r_2(b)), r_1(a), b))$; this homomorphism is homotopic to $1_P \oplus (k \circ g)$ via conjugation by the unitaries $(1, u_t)$, where u_t is as in 3.4.1. So $k_* \circ f_* - k_* \circ g_*$ is the identity map on $K_i(P)$.

(d) $h_1 = f \circ k : A *_D B \to M_2(A *_D B)$ is homotopic to $h_0 = 1_{A*_D B} \oplus (g \circ k)$ via the path of homomorphisms h_t defined by $h_t(a) = \text{diag}(a, r_1(a))$, $h_t(b) = u_t \cdot \text{diag}(b, r_2(b)) \cdot u_t^*$, using the universal property of $A *_D B$. So $f_* \circ k_* - g_* \circ k_*$ is the identity map on $K_i(A *_D B)$.

(e) Conclude that $k_* : K_i(A *_D B) \to K_i(P)$ is an isomorphism with inverse $(f_* - g_*)$, so there is an exact sequence

$$0 \longrightarrow K_i(D) \xrightarrow{(\phi_{i*}, -\phi_{2*})} K_i(A) \oplus K_i(B) \xrightarrow{\iota_{1*} + \iota_{2*}} K_i(A *_D B) \longrightarrow 0$$

which splits naturally. If $D = 0$, then $P = A \oplus B$, so $K_i(A * B) \cong K_i(A) \oplus K_i(B)$ for $i = 0, 1$, If A and B are unital and $D = \mathbb{C}1$, then $K_0(A *_{\mathbb{C}} B) \cong (K_0(A) \oplus K_0(B))/\langle([1_A], -[1_B])\rangle$ and $K_1(A *_{\mathbb{C}} B) \cong K_1(A) \oplus K_1(B)$.

(f) If G_1, G_2, and H are discrete groups and $\phi_i : H \to G_i$ is an embedding, then $C^*(G_1 *_H G_2) \cong C^*(G_1) *_{C^*(H)} C^*(G_2)$. In particular, $C^*(G_1 * G_2) \cong C^*(G_1) *_{\mathbb{C}} C^*(G_2)$. In this case, there are retractions r_1 and r_2 given by the trivial representation.

(g) Conclude that if G_1 and G_2 are discrete groups, then $K_0(C^*(G_1 * G_2)) \cong (K_0(C^*(G_1)) \oplus K_0(C^*(G_2)))/\langle([1_{C^*(G_1)}], -[1_{C^*(G_2)}])\rangle$ and $K_1(C^*(G_1 * G_2)) \cong K_1(C^*(G_1)) \oplus K_1(C^*(G_2))$. In particular, $K_0(C^*(\mathbb{F}_n)) \cong \mathbb{Z}$, $K_1(C^*(\mathbb{F}_n)) \cong \mathbb{Z}^n$.

(h) If G_1 and G_2 are amenable, then the quotient map $\lambda : C^*(G_1 * G_2) \to C_r^*(G_1 * G_2)$ induces an isomorphism on K-theory (20.9), so the results of (g) are valid also for $C_r^*(\mathbb{F}_n)$.

The results of this problem are due to Cuntz [1982b; 1983b]. Related results were also obtained by Brown [1981]. The existence of retractions hypothesis can be relaxed [Germain 1997].

10.11.12. Develop exact sequences for the K-theory of twisted crossed products. Very little is known about this situation.

10.11.13. [Cuntz 1987] (a) Let A be a C^*-algebra, and let $QA = A * A$ be the free product (10.11.11) of A with itself. QA has the universal property that any pair (ϕ, ψ) of *-homomorphisms from A to a C^*-algebra B define a unique homomorphism $q(\phi, \psi)$ from QA to B. In particular, there is a *-homomorphism $q(id_A, id_A)$ from QA to A. Let qA be the kernel of $q(id_A, id_A)$. Q and q are functors: if $\phi : A \to B$ is a *-homomorphism, there is a natural induced *-homomorphism $q\phi : QA \to QB$ which takes qA into qB.

(b) There are natural *-homomorphisms $\iota = \iota_1$ and $\bar{\iota} = \iota_2$ from A to QA as in 10.11.11(a). For $x \in A$, define $q(x) = \iota(x) - \bar{\iota}(x)$. Then $q(x) \in qA$, and qA is the ideal of QA generated by $\{q(x) : x \in A\}$. q is not a homomorphism; in fact, it is more like a derivation: $q(xy) = \iota(x)q(y) + q(x)\iota(y) - q(x)q(y)$.

(c) There are natural *-homomorphisms $\pi_0 = q(id_A, 0)$ and $\pi_1 = q(0, id_A)$ from QA to A, and by restriction from qA to A. $\pi_0(q(x)) = x$, $\pi_1(q(x)) = -x$. There is also a symmetry τ of QA, taking qA to itself, obtained by interchanging the two copies of A. $\bar{\iota} = \tau \circ \iota$ and $\pi_1 = \pi_0 \circ \tau$.

(d) By 10.11.11(e) the maps $\pi_0 \oplus \pi_1 : QA \to A \oplus A$ and $\psi : A \oplus A \to M_2(QA)$ defined by $\psi(x, y) = \mathrm{diag}(\iota(x), \bar{\iota}(y))$ are homotopy inverses, and hence give isomorphisms between $K_*(QA)$ and $K_*(A \oplus A)$.

(e) $K_*(QA)$ is naturally isomorphic to $K_*(qA) \oplus K_0(A)$ by applying the 6-term K-theory exact sequence to the split exact sequence

$$0 \longrightarrow qA \longrightarrow QA \overset{\iota}{\underset{}{\longleftrightarrow}} \tilde{A} \longrightarrow 0.$$

The second summand corresponds to ι_*. Thus $\pi_{0*} : K_*(qA) \to K_*(A)$ is an isomorphism, with inverse $\iota_* - \bar{\iota}_*$.

(f) By (e), $K_0(q\mathbb{C}) \cong K_0(\mathbb{C}) \cong \mathbb{Z}$, and there is a canonical generator $e \in K_0(q\mathbb{C})$ corresponding to $1_{\mathbb{C}}$. The map that takes $[\phi] \in [q\mathbb{C}, B \otimes \mathbb{K}]$ (9.4.4) to $\phi_*(e) \in K_0(B)$ is an isomorphism from $[q\mathbb{C}, B \otimes \mathbb{K}]$ onto $K_0(B)$.

(g) $Q\mathbb{C} = \{f \in C([0,1], \mathbb{M}_2) \mid f(0) = \mathrm{diag}(\alpha, 0), f(1) = \mathrm{diag}(\beta, \gamma)\}$ is the universal C^*-algebra generated by two projections. $(Q\mathbb{C})^+$ is the algebra of 6.10.4. $q\mathbb{C} = \{f \in Q\mathbb{C} \mid \alpha = 0\} = \{f \in C_0((0,1], \mathbb{M}_2) \mid f(1) \text{ diagonal}\} \cong C_0(\mathbb{R}) \times_\sigma \mathbb{Z}_2$, where σ is the "flip" automorphism $(\sigma(f))(t) = f(-t)$. $Q\mathbb{C}$, $(Q\mathbb{C})^+$, and $q\mathbb{C}$ are all semiprojective (4.7.1) [Loring 1997].

11. Equivariant $\boldsymbol{K}$-Theory

In this section, we develop the basics of equivariant K-theory for compact groups. Readers interested in a more extensive development of the theory may consult [Phillips 1987] (on which this section is largely based), [Segal 1968], and [Rosenberg and Schochet 1986]. Some further developments can also be found in Section 20, including a description of an equivariant theory for noncompact groups.

The first notions of equivariant K-theory were due to Atiyah and Segal; see [Segal 1968].

Throughout this section (except in 11.10) we will assume that all groups are *compact*.

11.1. Group Algebras

We begin by reviewing the structure of $L^1(G)$ and $C^*(G)$ for G compact. Recall that every (unitary) representation of G is a direct sum of irreducible representations, and that every irreducible representation is on a finite-dimensional space. If χ is the normalized character of an irreducible representation of G, then χ is a self-adjoint central idempotent in $L^1(G)$, and $\chi L^1(G)\chi$ is a direct summand of $L^1(G)$ which is isomorphic to $\mathbb{M}_n$, where n is the dimension of the representation. The summands corresponding to different irreducible representations are orthogonal, and the span of all the summands is dense in $L^1(G)$. We may summarize the situation in the following proposition.

PROPOSITION 11.1.1. *Let G be a compact group. Then there is an increasing net $\langle p_i \rangle$ of central projections of $L^1(G)$ such that the following conditions are satisfied.*

(a) *$p_i L^1(G) p_i = p_i C^*(G) p_i$ is finite-dimensional.*

(b) *(p_i) forms an approximate identity for $C^*(G)$, and more generally for $A \times_\alpha G$ for any covariant system (A, G, α) with A unital (where $L^1(G)$ is embedded in the standard way in the crossed product).*

(c) $\bigcup p_i L^1(G) p_i$ is dense in $L^1(G)$, and more generally $\bigcup p_i L^1(G, A, \alpha) p_i$ is dense in $L^1(G, A, \alpha)$ for any covariant system, although (p_i) is not an approximate identity for $L^1(G)$ in general.

If G is second countable, then $\langle p_i \rangle$ can be chosen to be an increasing sequence.

COROLLARY 11.1.2. *$C^*(G)$ is an AF algebra, a direct sum of matrix algebras. $\cup p_i L^1(G) p_i$ is a local Banach algebra under either the L^1-norm or the C^*-norm. Thus the standard inclusion $L^1(G) \to C^*(G)$ induces an isomorphism on K-theory. If A is $L^1(G)$, $C^*(G)$, or the dense local subalgebra, then $K_0(A) \cong \bigoplus_{\chi \in G} \mathbb{Z}$ and $K_1(A) = 0$.*

DEFINITION 11.1.3. The *representation ring $R(G)$* of G is the ring whose elements are formal differences of equivalence classes of finite-dimensional representations of G, with direct sum and tensor product as the ring operations. The trivial one-dimensional representation is the multiplicative identity.

By decomposing a representation into a direct sum of irreducibles, the additive group of $R(G)$ can be identified with $K_0(A)$ as in 11.1.2.

11.2. Projective Modules

There are three standard ways of viewing elements of $K_0(A)$, as (differences of equivalence classes of) idempotents, projective modules, or (if A is commutative) vector bundles. All three have natural analogs in the equivariant setting.

Suppose A is a unital C^*-algebra. If E is a finitely generated projective A-module, i.e. a direct summand of A^n for some n (we consider only *right* modules), then E has a uniquely determined topology independent of how it is realized as a direct summand. (E is a Banach space under many equivalent natural norms.) Write $\mathbb{L}(E)$ for the set of all bounded linear operators on E and $\mathbb{B}(E)$ for the subalgebra of $\mathbb{L}(E)$ consisting of module maps. If E is realized as a summand of A^n, then $E = pA^n$ for some projection p in $M_n(A)$; thus $\mathbb{B}(E)$ is isomorphic to the C^*-algebra $pM_n(A)p$. $\mathbb{L}(E)$ also, of course, has a strong operator topology which is independent of how the norm on E is chosen.

DEFINITION 11.2.1. Let (A, G, α) be a C^*-covariant system, with G compact and A unital. A (finitely generated) *projective (A, G, α)-module* is a pair (E, λ), where E is a (finitely generated) projective A-module and λ is a strongly continuous homomorphism from G into the invertible elements in $\mathbb{L}(E)$, such that

$$\lambda_g(ea) = \lambda_g(e)\alpha_g(a) \quad \text{for } g \in G \text{ and } a \in A.$$

(Note that λ_g is not in general a module homomorphism.)

We sometimes suppress the λ and simply refer to E as a projective (A, G, α)-module. The words "finitely generated" will also always be understood for projective modules.

PROPOSITION 11.2.2. *If (E, λ) is a projective (A, G, α)-module, then E becomes an $L^1(G, A, \alpha)$-module under*

$$e\phi = \int_G \lambda_g^{-1}(e)\alpha_g^{-1}(\phi(g))\, dg \quad \text{for } \phi \in L^1(A, G, \alpha) \text{ and } e \in E.$$

PROOF. It is clear that the formula makes E into a $C_c(G, A)$-module. For any fixed e, $\{\lambda_g(e) \mid g \in G\}$ is bounded, so that $\|\lambda\|_\infty = \sup \|\lambda_g\| < \infty$ by the Banach–Steinhaus Theorem. If $\phi \in C_c(G, A)$, then we have

$$\|e\phi\| \leq \int_G \|\lambda_{g^{-1}}(e)\| \, \|\phi(g)\| \, dg \leq \|\lambda\|_\infty \|e\| \, \|\phi\|_1.$$

So the action of C_c extends to L^1. $\square$

The standard way to form projective (A, G, α)-modules is as follows. Let π be a representation of G on a finite-dimensional vector space V. Then $E = V \otimes A$ (tensor product over $\mathbb{C}$) is a right A-module, and becomes a projective (A, G, α)-module if given the diagonal action of G. The modules constructed in this way may be regarded as "free modules."

We will now show that every projective (A, G, α)-module is a direct summand of one of these "free modules."

PROPOSITION 11.2.3. *If (E, λ) is a projective (A, G, α)-module, then there is a finite-dimensional representation space V of G such that E is equivariantly isomorphic to a direct summand of $V \otimes A$.*

PROOF. Let $\{e_1, \ldots, e_k\}$ be a set of generators for E as an A-module, and let $\{p_n\}$ be a set of projections as in 11.1.1. We first claim that if n is sufficiently large, then $\{e_1 p_n, \ldots, e_k p_n\}$ generate E. This is a consequence of the projectivity of E: since the map T from A^k to E sending $(a_1, \ldots, a_k)$ to $\sum e_i a_i$ has a bounded right inverse, any sufficiently small perturbation of T also has a right inverse and is therefore surjective.

For such an n, let V be the subspace of E spanned by $\{e_i p_n L^1(G) p_n\}$ (where $L^1(G)$ acts on E as in 11.2.2). Then V is finite-dimensional, and is invariant for the action of G (G multiplies $L^1(G)$ into itself, and p_n is central).

Give $V \otimes A$ the diagonal action of G to make it into a "free" (A, G, α)-module, and define $S : V \otimes A \to E$ by $S(e \otimes a) = ea$. Then S is an equivariant surjective A-module homomorphism. S has a right inverse by projectivity, and by averaging an arbitrary right inverse over G one obtains an equivariant right inverse, which exhibits E as a direct summand of $V \otimes A$. $\square$

11.3. Projections

As in the non-equivariant case, projective (A, G, α)-modules can be identified with projections. If p is a G-invariant projection in $\mathbb{B}(V \otimes A) \cong \mathbb{L}(V) \otimes A$, then $p(V \otimes A)$ is a projective (A, G, α)-module; 11.2.3 shows that every projective (A, G, α)-module arises in this way.

If W is another finite-dimensional representation space of G, we may identify the set of A-module homomorphisms from $V \otimes A$ to $W \otimes A$ with $\mathbb{L}(V, W) \otimes A$. There is a natural action of G on $\mathbb{L}(V, W) \otimes A$, given by

$$g \cdot t = (\lambda_g \otimes \alpha_g) \circ t \circ (\mu_g \otimes \alpha_g)^{-1}.$$

With this action, the modules $p(V \otimes A)$ and $q(W \otimes A)$ are isomorphic as (A, G, α)-modules if and only if there are G-invariant elements $u \in \mathbb{L}(V, W) \otimes A$ and $v \in \mathbb{L}(W, V) \otimes A$ with $uv = p$, $vu = q$. If there exist such u, v, then p and q are said to be *Murray–von Neumann equivalent*.

11.4. G-Vector Bundles

Suppose $A = C(X)$ is commutative. The actions of G on $C(X)$ exactly correspond to continuous actions of G on X, continuous meaning that the function $(g, x) \rightarrow g \cdot x$ from $G \times X$ to X is continuous. A *G-action* on X is always understood to be continuous.

DEFINITION 11.4.1. Let X be a space with a G-action. A *G-vector bundle* is a vector bundle E over X with a G-action by vector bundle automorphisms such that the projection map $p : E \rightarrow X$ is equivariant.

There is an obvious notion of morphism and isomorphism of G-vector bundles.

To each vector bundle E over X there corresponds naturally a finitely generated projective $C(X)$-module, the set $\Gamma(E)$ of continuous sections. If E is a G-vector bundle, then $\Gamma(E)$ has a natural induced G-action making $\Gamma(E)$ a projective $(C(X), G, \alpha)$-module.

There is an equivariant version of Swan's Theorem:

THEOREM 11.4.2. *Let X be a compact space with G-action α. Then the correspondence $[E] \rightarrow [\Gamma(E)]$ is a one-one correspondence between the isomorphism classes of G-vector bundles over X and the isomorphism classes of projective $(C(X), G, \alpha)$-modules.*

See [Phillips 1987, 2.3.1] for a proof; the proof is basically the same as Swan's with the G-actions woven in. This result is due to Atiyah, at least in the case where G is finite [Atiyah 1967, p. 41].

11.5. Definition of Equivariant K_0

Let (A, G, α) be a covariant system, with A unital. There is an obvious binary operation on the set of projective (A, G, α)-modules, direct sum. In the idempotent picture, one defines the direct sum of $p \in \mathbb{L}(V) \otimes A$ and $q \in \mathbb{L}(W) \otimes A$ to be $\mathrm{diag}(p, q) \in \mathbb{L}(V \oplus W) \otimes A$. The corresponding notion for G-vector bundles is (ordinary) Whitney sum. It is clear that this binary operation respects equivalence classes, and drops to an associative binary operation on the set $V^G(A)$ of equivalence classes, making $V^G(A)$ into a commutative semigroup with identity (the 0-module).

DEFINITION 11.5.1. $K_0^G(A)$ is the Grothendieck group of $V^G(A)$. If $A = C(X)$, we sometimes write $K_G^0(X)$ for $K_0^G(C(X))$. If it is necessary to explicitly keep track of the action, we will write $K_0^G(A, \alpha)$.

$K_0^G(A)$ is not only an abelian group; it has a natural structure as an $R(G)$-module, with the action of $[W] \in R(G)$ sending $p(V \otimes A)$ to $(1 \otimes p)(W \otimes V \otimes A)$. Note that if $\mathbb{C}$ is given the trivial G-action, then $K_0^G(\mathbb{C}) \cong R(G)$.

PROPOSITION 11.5.2. K_0^G is a functor from unital covariant G-systems to $R(G)$-modules.

PROOF. If $\phi : A \to B$ is an equivariant *-homomorphism, we define

$$\phi_*[p(V \otimes A)] = [p((V \otimes A) \otimes_A B)] = [(1_{\mathbb{L}(V)} \otimes \phi)(p)(V \otimes B)].$$

It is easy to check that this map has all desired properties.   □

Note that for $V \otimes A$ and $W \otimes A$ to represent the same element of $K_0^G(A)$, it is not enough that there exists an equivariant isomorphism from $\mathbb{L}(V) \otimes A$ to $\mathbb{L}(W) \otimes A$. For example, let $G = \mathbb{T}^1$ and $A = \mathbb{C}$ with trivial action. If V and W are one-dimensional representation spaces corresponding to different representations, then $\mathbb{L}(V) \cong \mathbb{L}(W) \cong \mathbb{C}$ with trivial action, but $[V] \neq [W]$ in $K_0^G(\mathbb{C}) = R(G)$.

We now define $K_0^G(A)$ for nonunital A just as in the nonequivariant case:

DEFINITION 11.5.3. $K_0^G(A)$ is the kernel of $\pi_* : K_0^G(A^+) \to K_0^G(\mathbb{C})$, where A^+ has the induced action from α and $\mathbb{C}$ has the trivial action.

It is easy to see that this definition agrees with the previous one if A is unital, and one obtains a functor from covariant G-systems to $R(G)$-modules.

11.6. Homotopy Invariance

Almost all of the results of Chapter II have straightforward analogs in the equivariant case. The only one which will be relevant for our purposes is the analog of 4.3.2:

PROPOSITION 11.6.1. Let A be a (local) Banach algebra, and H a subgroup of $\mathrm{Aut}(A)$. If e and f are H-invariant idempotents in A and $\|e - f\| < 1/\|2e - 1\|$, then there is an H-invariant invertible element $u \in \tilde{A}$ with $ueu^{-1} = f$.

PROOF. The proof is identical to the proof of 4.3.2; note that the v defined there is H-invariant.   □

COROLLARY 11.6.2. Let (A, G, α) and (B, G, β) be covariant systems, and ϕ_t $(0 \leq t \leq 1)$ a pointwise-continuous path of equivariant *-homomorphisms from A to B. Then $\phi_{0*} = \phi_{1*}$.

11.7. Relation with Crossed Products

In this paragraph we prove the main result of this section, that equivariant K-theory can be identified with the ordinary K-theory of a crossed product. This theorem is due to Atiyah if A is commutative and G is finite, P. Green [1982] in the case A is commutative and G arbitrary, and to Green and J. Rosenberg (unpublished, but see [Rosenberg 1984]) and independently to P. Julg [1981] in the general case.

THEOREM 11.7.1. *Let (A, G, α) be a covariant system (with G compact). There is a natural isomorphism $K_0^G(A) \cong K_0(A \times_\alpha G)$.*

PROOF. In view of 11.1.2, we need only show that there is a natural isomorphism of $K_0^G(A)$ with $K_0(L^1(G, A, \alpha))$. It suffices to prove the result for A unital, so we make this assumption. If (E, λ) is a projective (A, G, α)-module, we let $\Phi(E)$ be E with the $L^1(G, A, \alpha)$-module structure defined in 11.2.2. We will show that Φ is the desired isomorphism.

We first show that $\Phi(E)$ is a finitely generated projective module. Since Φ preserves direct sums, it suffices to show this for $E = V \otimes A$ (11.2.3), and we may in addition assume that the representation of G on V is irreducible. If χ is the normalized character of the complex conjugate $\tilde{V}$ of V, and p is a minimal projection in the matrix algebra $\chi L^1(G)\chi$, then $L^1(G)p \cong \tilde{V}$ as left $L^1(G)$-modules; so $V \cong (L^1(G)p)$. Define $\phi : (L^1(G)p) \otimes A \to pL^1(G, A, \alpha)$ by $\phi(f \otimes a) = f^* a$, where f^* is the adjoint of $f \in L^1(G) \subseteq L^1(G, A, \alpha)$ and a is regarded as a multiplier of $L^1(G, A, \alpha)$ in the evident way [Pedersen 1979, 7.6.3]. In other words, $\phi(f \otimes a)(h) = \overline{f(h^{-1})}\alpha_h(a)$. ϕ is linear since the two conjugations cancel out. We will show that ϕ is an $L^1(G)$-module isomorphism; thus $\Psi(V \otimes A)$ is projective.

To prove that ϕ is $L^1(G)$-linear, let $a \in A$, $f \in (L^1(G)p)$, $x \in L^1(G, A, \alpha)$, π the left regular representation of G. We have

$$\phi((f \otimes a)x)(g) = \phi\left[\int_G \alpha_{h^{-1}}(ax(h)) \otimes \pi_{h^{-1}}(f) \, dh\right](g)$$
$$= \int_G f(hg^{-1})\alpha_{gh^{-1}}(ax(h)) \, dh = (\phi(f \otimes a)x)(g).$$

To construct an inverse ψ for ϕ, let $\{\xi_1, \ldots, \xi_n\}$ be an orthonormal basis for $L^1(G)p$, $m_{ij}(g) = \langle \pi_g \xi_i, \xi_j \rangle$. We may assume $m_{11} = p$. Then

$$\psi(x) = \frac{1}{n} \sum_1^n m_{i1} \otimes x_i,$$

where

$$x_i = \int_G m_{i1}(h^{-1})\alpha_h^{-1}(x(h)) \, dh.$$

Thus Φ is a well defined homomorphism from $V^G(A)$ to $V(L^1(G, A, \alpha))$.

To show that Φ is a bijection, let E be an $L^1(G, A, \alpha)$-module of the form $p(\mathbb{C}^m \otimes L^1(G, A, \alpha))$; we may assume $p \leq p_i$ for some i (11.1.1). Make E and $\mathbb{C}^m \otimes p_n L^1(G, A, \alpha)$ into (A, G, α)-modules using the action of G and A as multipliers on $L^1(G, A, \alpha)$. If $p_i = \sum_{j=1}^n \chi_j$ is the decomposition of p_i into minimal projections, and V_j is the representation space corresponding to χ_j, then $\mathbb{C}^m \otimes p_n L^1(G, A, \alpha) \cong \bigoplus_{j=1}^n (\tilde{V}_j)^m \otimes A$ as (A, G, α)-modules, and is therefore finitely generated projective. E is a direct summand of this module. This process actually yields an inverse for Φ. $\qquad\square$

11.8. Module Structure on $K_0(A \times_\alpha G)$

Theorem 11.7.1 only establishes an isomorphism on the level of abelian groups. We now indicate how the $R(G)$-module structure on $K_0(A \times_\alpha G)$ corresponding to the natural one on $K_0^G(A)$ may be explicitly described.

Suppose V is a finite-dimensional representation space of G. We may assume V is a Hilbert space of dimension n and the representation π is unitary. Then $\mathbb{L}(V) \otimes A$ is a C^*-algebra with a canonical G-action, the diagonal action δ, and the embedding $a \to 1 \otimes a$ is equivariant; thus there is a homomorphism $\phi : A \times_\alpha G \to (\mathbb{L}(V) \otimes A) \times_\delta G$. But δ is exterior equivalent [Pedersen 1979, 8.11.3] to $\beta = 1 \otimes \alpha$, so $(\mathbb{L}(V) \otimes A) \times_\delta G$ is canonically isomorphic to $(\mathbb{M}_n \otimes A) \times_\beta G \cong M_n(A \times_\alpha G)$ via a homomorphism ψ. The standard embedding $\eta : A \times_\alpha G \to M_n(A \times_\alpha G)$ induces an isomorphism.

THEOREM 11.8.1. *The action of multiplication by $[V]$ on $K_0(A \times_\alpha G)$ given by $[V] \cdot x = (\eta_*^{-1} \circ \psi_* \circ \phi_*)(x)$ defines an $R(G)$-module structure on $K_0(A \times_\alpha G)$ which makes the isomorphism of* 11.7.1 *an $R(G)$-module isomorphism.*

The proof is straightforward but rather involved, and is omitted. See [Phillips 1987, 2.7].

EXAMPLE 11.8.2. Let H be a closed subgroup of G, and let $A = C(G/H)$ with obvious G-action by translation. Then $K_0^G(A) \cong R(H)$ as $R(G)$-modules, where $R(H)$ is viewed as an $R(G)$-module by restriction of representations from G to H.

If G is abelian, there is a much simpler description of the $R(G)$-module structure on $K_0(A \times_\alpha G)$. In this case, $\hat{G}$ is a (discrete) group, and $R(G)$ may be identified with the group ring of $\hat{G}$. There is a natural action of $\hat{G}$ on $A \times_\alpha G$, the dual action $\hat{\alpha}$, which defines an $R(G)$-module structure on $K_0(A \times_\alpha G)$.

PROPOSITION 11.8.3. *This $R(G)$-module structure coincides with the previously defined structure. So if $\gamma \in \hat{G}$, then the action of γ (regarded as an element of $R(G)$) on $K_0(A \times_\alpha G)$ is given by $\gamma \cdot x = (\hat{\alpha}_\gamma)_*(x)$.*

PROOF. Suppose A is unital. If (E, λ) is a projective (A, G, α)-module, then $\gamma \cdot E$ is E with G-action $(\gamma\lambda)_g(e) = \langle \gamma, g \rangle \lambda_g(e)$, so the corresponding $L^1(A, G, \alpha)$-

module $\Phi(\gamma \cdot E)$ is E with module structure

$$e\phi = \int_G \langle \gamma, g \rangle^{-1} \lambda_g^{-1}(e) \alpha_g^{-1}(\phi(g)) dg$$

But $\langle \gamma, g \rangle^{-1} \alpha_g^{-1}(\phi(g)) = \alpha_g^{-1}((\hat{\alpha}_\gamma^{-1}(\phi))(g))$, so this module structure coincides with the module structure on $(\hat{\alpha}_\gamma)_* \Phi(E)$.

If A is nonunital, then apply the above argument to A^+ to see that the module structures agree on $K_0^G(A) \subseteq K_0^G(A^+)$. $\qquad\square$

11.9. Properties of Equivariant K-Theory

Because of the natural isomorphism of $K_0^G(A)$ with $K_0(A \times_\alpha G)$, we have analogs of the most important properties of ordinary K-theory:

PROPOSITION 11.9.1 (STABILITY). *Let (A, G, α) be a covariant system. Then the morphism $\eta : A \to A \otimes \mathbb{K}$ sending a to $a \otimes p$ induces an isomorphism $\eta_* : K_0^G(A) \to K_0^G(A \otimes \mathbb{K})$ of $R(G)$-modules (where the action of G on $A \otimes \mathbb{K}$ is $\alpha \otimes 1$).*

Actually this result remains true even if the action of G on $\mathbb{K}$ is nontrivial, as long as p is chosen to be invariant (so that η is equivariant).

PROPOSITION 11.9.2 (CONTINUITY). *Let (A_i, G, α_i) be a directed system of covariant systems, and $(A, G, \alpha) = \varinjlim (A_i, G, \alpha_i)$ in the obvious sense. Then $K_0^G(A) \cong \varinjlim K_0^G(A_i)$ as $R(G)$-modules.*

DEFINITION 11.9.3. If (A, G, α) is a covariant system, set $K_1^G(A) = K_0^G(SA)$ (where G acts trivially on $C_0(\mathbb{R})$).

It is possible to give an alternative definition of $K_1^G(A)$ in terms of invertible elements (11.11.2).

THEOREM 11.9.4 (BOTT PERIODICITY). *There is a natural $R(G)$-module isomorphism $K_0^G(A) \cong K_1^G(SA)$ for any covariant system (A, G, α).*

There is an important generalization of this theorem, due to Atiyah [1968] (cf. 20.3.2):

THEOREM 11.9.5. *Let V be a finite-dimensional complex vector space with a representation of G, viewed as a locally compact G-space. Then $K_G^0(V) \cong R(G)$ and $K_G^1(V) = 0$. If A is a G-algebra, and $C_0(V) \otimes A$ is given the diagonal action, then $K_0^G(C_0(V) \otimes A) \cong K_0^G(A)$.*

THEOREM 11.9.6 (STANDARD EXACT SEQUENCE). *Let (A, G, α) be a covariant system, and let J be an invariant ideal of A. Then there are $R(G)$-module maps ∂_i making the following six-term sequence of $R(G)$-modules exact:*

$$
\begin{array}{ccccc}
K_0^G(J) & \longrightarrow & K_0^G(A) & \longrightarrow & K_0^G(A/J) \\
{\scriptstyle \partial_0}\uparrow & & & & \downarrow{\scriptstyle \partial_1} \\
K_1^G(A/J) & \longleftarrow & K_1^G(A) & \longleftarrow & K_1^G(J)
\end{array}
$$

This sequence is natural with respect to covariant maps of short exact sequences.

THEOREM 11.9.7. *If α and β are actions of G on A which are exterior equivalent, then there is a natural isomorphism $K_0^G(A, \alpha) \cong K_0^G(A, \beta)$.*

Direct proofs of these results can be given, but (except for 11.9.5, which is best done by Atiyah's method or using KK-theory) it is easiest to use the correspondence with the crossed products. The only thing which needs to be checked is that the maps in the exact sequence and the Bott map are $R(G)$-module maps, which is straightforward to verify.

11.10. Equivariant K-Theory for Noncompact Groups

Topologists have been almost exclusively concerned with compact groups in studying equivariant K-theory; the noncompact case has been a rather overlooked area. To the extent to which the subject has been considered, it has been a very difficult problem to give natural and useful definitions for equivariant K-theory for noncompact groups. There is one obvious starting point from 11.7: the equivariant K-theory of a covariant system could be *defined* to be the K-theory of the crossed product. These groups are sometimes called the "analytic" equivariant K-groups. The problem then becomes a matter of defining "topological" K-groups in some manner analogous to the compact case, and then showing that the topological groups coincide with the analytic ones. (For many applications the analytic groups are quite unnatural to work with and are often difficult to compute; a good topological definition, possibly involving classifying spaces, is usually much more useful.)

A reasonable definition of equivariant K-theory for *proper* actions on locally compact spaces can be given in terms of vector bundles (actually in terms of continuous fields of Hilbert spaces); see [Phillips 1989]. For the more general situation, see [Baum and Connes 1982] (cf. 24.4). It is difficult, for example, to develop a theory which will even satisfactorily treat the case of $\mathbb{R}$ acting trivially on a point.

An additional problem in the general case is the proper definition of the representation ring of a noncompact group.

KK-Theory can be used to shed much light on the situation of equivariant K-theory, including the definition of the representation ring. We will discuss equivariant KK-theory in Section 20.

11.11. EXERCISES AND PROBLEMS

11.11.1. Give the most general equivariant versions of all of the results of Chapter II. Some have analogs valid for arbitrary subgroups of the automorphism group, while others require averaging and so will be valid only for compact groups.

11.11.2. Give a definition of $K_1^G(A)$ in terms of invertible elements, and give direct proofs of the results of 11.9 (cf. [Phillips 1987]).

11.11.3. Let V be a finite-dimensional G-vector space, so that $B = \mathbb{L}(V)$ is a G-algebra. Let $B_n = B \otimes \cdots \otimes B$ (n factors). Show that $K_0^G(B_n) \cong R(G)$, $K_1^G(B_n) = 0$ for any n, and compute the $R(G)$-module map from $R(G)$ to itself induced by the map $B_n \cong B_n \otimes 1 \to B_{n+1}$. Let $A = \varinjlim B_n$; then A is a UHF algebra. Compute $K_0^G(A)$ using 11.9.2.

More generally, if G is a compact group and α is a product-type action on the UHF algebra A, then $K_0^G(A)$ can be given the structure of an ordered $R(G)$-module; this ordered module can be calculated in many cases and used to classify certain kinds of actions of compact groups on AF algebras.

These problems have been studied in [Handelman and Rossmann 1984] and [Wassermann 1989]; special cases were considered in [Fack and Maréchal 1979; 1981; Renault 1980].

K-Theory for Operator Algebras
MSRI Publications
Volume **5**, second edition, 1998

CHAPTER VI

MORE PRELIMINARIES

We have decided to collect all the preliminary results needed for *Ext*-theory and Kasparov theory into a single chapter, even though not all of the results will be needed immediately. We have done this since the three sections of this chapter are closely related and it is more efficient to do everything at once.

In Chapter VII, only parts of Sections 12 and 13 will be needed. Section 14 is not required (except for 15.13) until *KK*-theory (Sections 17ff.), which also requires all of Sections 12 and 13.

The reader who so desires may skip over this chapter, returning on an *ad hoc* basis as needed.

12. Multiplier Algebras

12.1. Introduction

Recall [Pedersen 1979, 3.12] that the *multiplier algebra* $M(A)$ of A is the maximal C^*-algebra containing A as an essential ideal. The *strict topology* on $M(A)$ is the topology generated by the seminorms $\|\|x\|\|_a = \|ax\| + \|xa\|$ for $a \in A$.

The *outer multiplier algebra* $Q(A)$ of A is the quotient $M(A)/A$. We will write Q for the Calkin algebra $Q(\mathbb{K}) = \mathbb{B}/\mathbb{K}$.

EXAMPLES 12.1.1. (a) If A is unital, then $M(A) = A$. $M(A)$ is always unital, so if A is nonunital $M(A) \neq A$. (In fact in this case $M(A)$ is generally much larger than A: for example, $M(A)$ is never separable if A is nonunital [Pedersen 1979, 3.12.12].)

(b) If $A = C_0(X)$, then $M(A) = C(\beta X)$, where βX is the Stone–Čech compactification of X. The strict topology on bounded subsets of $M(A)$ is the topology of uniform convergence on compact subsets of X.

(c) Generalizing (b), if $A = C_0(X) \otimes B$, then $M(A)$ is the set of strictly continuous functions from βX to $M(B)$ [Akemann et al. 1973].

(d) $M(\mathbb{K}) = \mathbb{B}$; the strict topology is the σ-strong-* topology.

PROPOSITION 12.1.2. *If x_i is a bounded sequence of self-adjoint elements in $M(A)$ and S is a total subset of A, then x_i converges strictly in $M(A)$ if and only if $x_i s$ is a norm-Cauchy sequence in A for all $s \in S$.*

The simple proof is left to the reader.

DEFINITION 12.1.3. The *stable multiplier algebra* $M^s(A)$ is the multiplier algebra of $A \otimes \mathbb{K}$. The *stable outer multiplier algebra* $Q^s(A)$ is the quotient $M(A \otimes \mathbb{K})/(A \otimes \mathbb{K})$.

Since $M(A \otimes B)$ contains a canonical copy of $M(A) \otimes M(B)$, we always have a copy of $1 \otimes \mathbb{B}$ inside $M^s(A)$ for any A. We also have $M_n(M^s(A)) \cong M^s(A)$ for all n. So $M^s(A)$ and $Q^s(A)$ are properly infinite (6.11).

12.2. K-Theory and Stable Multiplier Algebras

PROPOSITION 12.2.1. *If A is any C^*-algebra, $K_0(M^s(A)) = K_1(M^s(A)) = 0$.*

PROOF. Let v_i be a sequence of isometries in $1 \otimes \mathbb{B} \subseteq M^s(A)$ with orthogonal ranges. If p is any projection in $M^s(A)$, set $q = \sum v_i p v_i^*$. Set

$$w = \begin{bmatrix} 0 & 0 \\ v_1 & \sum v_{i+1} v_i^* \end{bmatrix} \begin{bmatrix} p & 0 \\ 0 & q \end{bmatrix}.$$

(The sums $\sum v_i p v_i^*$ and $\sum v_{i+1} v_i^*$ converge in the strict topology to multipliers of $A \otimes \mathbb{K}$.) Then $w^*w = \text{diag}(p, q)$ and $ww^* = \text{diag}(0, q)$; hence $[p] + [q] = [q]$ in $K_0(M^s(A))$. The same argument works in matrix algebras, so $K_0(M^s(A)) = 0$. A similar argument works for unitaries, showing that $K_1(M^s(A)) = 0$. $\square$

With some additional work, one can do better: the unitary group of $M^s(A)$ is always (norm-)contractible [Cuntz and Higson 1987]; cf. [Mingo 1987].

It is easy to see that $U_1(M^s(A))$ is path-connected in the *strict* topology:

PROPOSITION 12.2.2. *Let A be a C^*-algebra, and u a unitary in $M^s(A)$. Then there is a strictly continuous path (u_t) of unitaries in $M^s(A)$ with $u_0 = 1$ and $u_1 = u$.*

PROOF. Let (v_t), for $0 < t \leq 1$, be a σ-strong-$*$ continuous path of isometries in $\mathbb{B}$ with $p_t = v_t v_t^* \to 0$ strongly as $t \to 0$. For example, let $\mathbb{H} = L^2([0,1])$,

$$[v_t(f)](s) = \begin{cases} \frac{1}{\sqrt{t}} f(s/t) & \text{for } s \leq t, \\ 0 & \text{for } s > t. \end{cases}$$

(Here p_t is the projection onto $L^2([0,t])$.) Set $w_t = v_t \otimes 1$, $q_t = p_t \otimes 1$ in $\mathbb{B} \otimes M(A) \subseteq M(\mathbb{K} \otimes A)$, and let $u_t = w_t u w_t^* + (1 - q_t)$. $\square$

We now consider the K-theory exact sequence corresponding to the extension $0 \to A \otimes \mathbb{K} \to M^s(A) \to Q^s(A) \to 0$. Because of 12.2.1, the connecting maps from $K_i(Q^s(A))$ to $K_{1-i}(A)$ are isomorphisms, so we have:

COROLLARY 12.2.3. *If A is any C^*-algebra, then $K_i(A) \cong K_{1-i}(Q^s(A))$ for $i = 0, 1$.*

Actually, in computing $K_0(Q^s(A))$ it is not necessary to stabilize: by 6.11.7, the equivalence classes of infinite projections in $Q^s(A)$ form a complete set of representatives for *all* elements of $K_0(Q^s(A))$. If A is unital or commutative (and perhaps in general) the same is true for K_1 since $K_1(Q^s(A)) \cong U_1(Q^s(A))/U_1(Q^s(A))_0$ by [Mingo 1987] in this case.

12.2.4. The conclusion of Corollary 12.2.3 for $i = 0$ is sometimes taken as the definition of $K_0(A)$. The whole theory with the exception of the ordering on K_0 can be elegantly developed in this manner. In fact, considering K-theory as a special case of KK-theory leads to essentially this approach.

The groups $K_i(Q^s(A))$ can be alternately described in a way which motivates KK-theory. $K_1(Q^s(A)) \cong K_0(A)$ may be thought of as the group of equivalence classes of "Fredholm operators" in $M^s(A)$ (ones whose image in $Q^s(A)$ is invertible); $K_0(Q^s(A)) \cong K_1(A)$ is the group of equivalence classes of elements of $M^s(A)$ which are projections mod $A \otimes \mathbb{K}$. In each case the group operation is "orthogonal direct sum" (using the embedding $M^s(A) \oplus M^s(A) \subseteq M_2(M^s(A)) \cong M^s(A)$), and the equivalence relation is the one generated by homotopy and orthogonal addition of "degenerate elements" (actual invertible elements or projections). 3.4.6 is needed to prove the equivalence.

12.3. σ-Unital C^*-Algebras

We recall the definition of a strictly positive element of a C^*-algebra. A positive element $h \in A$ is *strictly positive* if $\phi(h) > 0$ for every state ϕ of A. A C^*-algebra has a strictly positive element if and only if it has a countable approximate identity [Pedersen 1979, 3.12.5], so in particular every separable C^*-algebra contains a strictly positive element. A C^*-algebra containing a strictly positive element will be called a *σ-unital C^*-algebra*.

PROPOSITION 12.3.1. *Let A be a C^*-algebra, and $h \in A_+$. Then h is strictly positive if and only if hA is dense in A.*

PROOF. If hA is not dense in A, then by [Dixmier 1969, 2.9.4] there is a state ϕ on A vanishing on hA. If u_λ is an approximate identity for A, then $\phi(hu_\lambda) = 0$, so $\phi(h) = 0$. Conversely, if $\phi(h) = 0$, then ϕ vanishes on hA by [Pedersen 1979, 3.1.3], so hA is not dense. $\qquad\square$

12.4. Kasparov's Technical Theorem

In this paragraph we prove a result about separation of orthogonal subalgebras of outer multiplier algebras, which will be needed for the development of Kasparov theory.

We first need a slight extension of [Pedersen 1979, 3.12.14]. Recall that if A is a C^*-subalgebra and Δ a linear subspace of a C^*-algebra B, then Δ *derives* A if $[a, d] = ad - da \in A$ for all $a \in A$, $d \in \Delta$. If Δ derives A, then an approximate identity u_λ for A is *quasicentral* for Δ if $\lim_\lambda [u_\lambda, d] = 0$ for all $d \in \Delta$.

PROPOSITION 12.4.1. *If Δ derives A and v_i is an approximate identity for A, then there is an approximate identity for A, contained in the convex hull of v_i, which is quasicentral for Δ.*

PROOF. The proof is identical to the proof of [Pedersen 1979, 3.12.14] (with B replaced by Δ) with the following modification in the proof that $M_{i\lambda} \neq \varnothing$. Fix $x_1, \ldots, x_n$ in Δ and let $C = \bigoplus_{k=1}^n B$. (B is the C^*-algebra containing A and Δ.) Working in B'' we know that $v_j \nearrow p$, where p is the identity of A'', regarded as a subalgebra of B''. We claim that p commutes with Δ. Δ clearly derives A'', so if $d \in \Delta$, $[p, d] = p[p, d] = [p, d]p$, and $[p, d] = [p^2, d] = p[p, d] + [p, d]p = 2[p, d]$, so $[p, d] = 0$. Now proceed as in [Pedersen 1979, 3.12.14]. □

We now prove a result which has become known as Kasparov's Technical Theorem.

THEOREM 12.4.2. *Let J be a σ-unital C^*-algebra. Let A_1 and A_2 be σ-unital C^*-subalgebras of $M(J)$, and Δ a separable subspace of $M(J)$. Suppose $A_1 \cdot A_2 \subseteq J$, and that Δ derives A_1. Then there are $M, N \in M(J)$ such that $0 \leq M \leq 1$, $N = 1 - M$, $M \cdot A_1 \subseteq J$, $N \cdot A_2 \subseteq J$, and $[M, \Delta] \subseteq J$.*

If one passes to the outer multiplier algebra, the result may be rephrased as follows: if B_1 and B_2 are orthogonal σ-unital C^*-subalgebras of the outer multiplier algebra of J, and Δ (separable) derives B_1, then there is a positive element M of $Q(J)$, of norm 1, commuting with Δ, which is a unit for B_2 and which is orthogonal to B_1.

If the subspace Δ is eliminated, and A_1 and A_2 are separable, the result follows almost trivially from [Pedersen 1979, 3.12.10], since the quotient map of $C^*(A_1, A_2)$ onto $C^*(B_1, B_2) \cong B_1 \oplus B_2$ extends to a map of $M(C^*(A_1, A_2)) \subseteq M(A)$ onto $M(C^*(B_1, B_2)) \cong M(B_1) \oplus M(B_2)$. (This observation is due to Cuntz.)

The result can fail if A_1 and A_2 are not σ-unital, even if B_1 and B_2 are commutative and $\Delta = 0$ [Choi and Christensen 1983].

So the difficulty all comes from the Δ. Kasparov's proof is extremely complicated; the elegant proof given here is due to N. Higson.

PROOF OF 12.4.2. We first reduce to the case where J and A_i are separable. Let $B_{1,1}$ be a separable C^*-subalgebra of A_1 containing a strictly positive element h_1 of A_1, and let $B_{1,n+1} = C^*(B_{1,n}, [B_{1,n}, \Delta])$. Then $B_1 = \overline{\bigcup B_{1,n}}$ is a separable C^*-subalgebra of A_1, and Δ derives B_1. Let B_2 be a separable C^*-subalgebra of A_2 containing a strictly positive element h_2 of A_2. Let $B_{0,1}$ be the C^*-subalgebra of J generated by a strictly positive element of J and $B_1 \cdot B_2$, and let $B_{0,n+1} = C^*(B_{0,n}, B_1 \cdot B_{0,n}, B_2 \cdot B_{0,n}, \Delta \cdot B_{0,n})$. Then if $B = \overline{\bigcup B_{0,n}}$, by [Pedersen 1979, 3.12.12], $M(B)$ can be identified with a subalgebra of $M(J)$ containing B_1, B_2, and Δ. If M is constructed for B, B_1, B_2, and Δ, then for $x \in A_1$ we have $Mx = \lim_n Mh_1^{1/n} x \in \overline{[B \cdot A_1]} \subseteq J$, and similarly $(1 - M)A_2 \subseteq J$.

So we may assume J, A_i are separable. Let X_i, Y, Z be compact sets of vectors of norm ≤ 1 (e.g. sequences converging to 0) which are total in A_i, Δ, and J respectively. By 12.4.1 there is an approximate identity (u_n) for A_1 such that

(a) for all n, $\|u_n x - x\| < 2^{-n}$ for all $x \in X$, and
(b) for all n, $\|[u_n, y]\| < 2^{-n}$ for all $y \in Y$.

There is also an approximate identity (v_n) for J such that

(c) for all n, $\|v_n w - w\| < 2^{-2n}$ for all w in the compact set $\{u_1, \ldots, u_n\} \cdot (Z \cup X_2)$ of J, and
(d) for all n, $\|[v_n, w]\|$ is small enough that $\|[b_n, w]\| < 2^{-n}$ for all $w \in X_1 \cup X_2 \cup Y \cup Z$, where $b_n = (v_n - v_{n-1})^{1/2}$.

It follows from property (c) that $\|b_n u_n w b_n\| < \sqrt{5} \cdot 2^{-n}$. If $z \in Z$, then $\|b_n u_n b_n z\| \leq \|b_n u_n z b_n\| + \|b_n u_n\| \cdot \|[b_n, z]\| < (\sqrt{5} + 1)2^{-n}$. So the series $\sum b_n u_n b_n z$ converges for all z. $b_n u_n b_n \leq b_n^2$, so the partial sums of the series $\sum b_n u_n b_n$ are bounded by 1. By 12.1.2, the series converges strictly to an element $N \in M(J)$ with $0 \leq N \leq 1$. Set $M = 1 - N$. To complete the proof, it suffices to show that $N \cdot X_2 \subseteq J$, $(1 - N) \cdot X_1 \subseteq J$, and $[N, Y] \subseteq J$. As above, if $x \in X_2$ we have $\|b_n u_n b_n x\| < (\sqrt{5} + 1)2^{-n}$, so the series $Nx = \sum b_n u_n b_n x$ is norm-convergent, and the terms are in J; so $N \cdot A_2 \subseteq J$. Similarly, if $x \in X_1$ we have

$$\|(b_n^2 - b_n u_n b_n)x\| \leq \|b_n(u_n x - x)b_n\| + \|b_n[b_n, x]\| + \|b_n u_n[b_n, x]\| < 3 \cdot 2^{-n},$$

so $\sum(b_n^2 - b_n u_n b_n)x$ converges; $\sum b_n^2 x$ converges to x, so the previous sum is $(1 - N)x$. All of the terms are in J, so $(1 - N)A_1 \subseteq J$. Finally, if $y \in Y$, using (b) and (d) we have

$$\|[b_n u_n b_n, y]\| \leq \|b_n u_n[b_n, y]\| + \|b_n[u_n, y]b_n\| + \|[b_n, y]u_n b_n\| < 3 \cdot 2^{-n},$$

so $[N, y]$ is a sum of a series in J. This completes the proof of Kasparov's Technical Theorem. $\qquad \square$

12.5. EXERCISES AND PROBLEMS

12.5.1. Prove directly that $K_0(A) \cong K_1(Q^s(A))$ without using the K-theory exact sequence or Bott periodicity, by constructing for each $[p] - [q] \in K_0(A)$ a partial isometry $v \in M^s(A)$ with $v^* v = 1 - p$, $vv^* = 1 - q$. Do the unital case first. Some of the arguments in [Kasparov 1980b, Theorem 6.3] will be helpful, as well as ideas from Chapter II.

12.5.2. If B is a C^*-algebra, the projections in B can be identified with the *-homomorphisms from $\mathbb{C}$ into B. If $C_0(\mathbb{R})$ is identified with the continuous functions on the unit circle which vanish at 1, and B is unital, there is a natural identification of the unitaries of B with the *-homomorphisms from $C_0(\mathbb{R})$ into B. If $B = Q^s(A)$, then for any C the set of unitary equivalence classes of

*-homomorphisms from C to B has an associative binary operation induced by diagonal sum using $M_2(B) \cong B$. The trivial homomorphisms (ones which lift to homomorphisms from C to $M^s(A)$) form a subsemigroup. (See Section 15.) Show that if $C = \mathbb{C}$ the quotient semigroup is a group isomorphic to $K_0(B)$, and that if $C = C_0(\mathbb{R})$ the quotient semigroup is a group isomorphic to $K_1(B)$.

13. Hilbert Modules

Although not absolutely necessary, it is most convenient to phrase the definitions and basic results of Kasparov theory in terms of Hilbert modules. The notion of Hilbert module goes back to Paschke [Paschke 1973], following related work of Kaplansky and Rieffel. We will use notation similar to that of [Kasparov 1980a]; we will work only with right modules.

13.1. Basic Definitions

DEFINITION 13.1.1. Let B be a C^*-algebra. A *pre-Hilbert module* over B is a right B-module E equipped with a B-valued "inner product", a function $\langle \cdot, \cdot \rangle : E \times E \to B$, with these properties:

(1) $\langle \cdot, \cdot \rangle$ is sesquilinear. (We make the convention that inner products are conjugate-linear in the first variable.)
(2) $\langle x, yb \rangle = \langle x, y \rangle b$ for all $x, y \in E, b \in B$.
(3) $\langle y, x \rangle = \langle x, y \rangle^*$ for all $x, y \in E$.
(4) $\langle x, x \rangle \geq 0$; if $\langle x, x \rangle = 0$, then $x = 0$.

For $x \in E$, put $\|x\| = \|\langle x, x \rangle\|^{1/2}$. This is a norm on E. If E is complete, E is called a *Hilbert module* over B. The closure of the span of $\{\langle x, y \rangle : x, y \in E\}$ is called the *support* of E, denoted $\langle E, E \rangle$. E is *full* if $\langle E, E \rangle = B$.

The completion of a pre-Hilbert module is a Hilbert module in the obvious way.

EXAMPLES 13.1.2. (a) B is itself a full Hilbert B-module with $\langle a, b \rangle = a^*b$. More generally, any (closed) right ideal of B is a Hilbert B-module.

(b) If E_i is a family of (pre-)Hilbert B-modules, the direct sum is a pre-Hilbert module with inner product $\langle \bigoplus x_i, \bigoplus y_i \rangle = \sum_i \langle x_i, y_i \rangle$.

(c) As a special case of (b), let $\mathbb{H}_B$ be the completion of the direct sum of a countable number of copies of B, i.e. $\mathbb{H}_B$ consists of all sequences (b_n) such that $\sum b_n^* b_n$ converges, with inner product $\langle (a_n), (b_n) \rangle = \sum_n a_n^* b_n$. $\mathbb{H}_B$ is called the *Hilbert space over B*. $\mathbb{H}_B$ is sometimes denoted B^∞; we can of course also form B^n. More generally, if E is any Hilbert module, we can form E^n and E^∞.

(d) A right B-rigged space in the sense of [Rieffel 1974] is a full pre-Hilbert B-module. A Morita equivalence bimodule between A and B [Rieffel 1974] is a pre-Hilbert B-module. Note that $\mathbb{H}_B$ is the standard equivalence bimodule between $B \otimes \mathbb{K}$ and B.

PROPOSITION 13.1.3 (CAUCHY–SCHWARTZ INEQUALITY). *For any $x, y \in E$, we have $\|\langle x, y \rangle\| \le \|x\| \|y\|$.*

PROOF. We may assume $y \ne 0$. Set $b = -\langle y, x \rangle / \|\langle y, y \rangle\|$, and use the fact that $\langle x + yb, \ x + yb \rangle \ge 0$. $\qquad \square$

13.2. Bounded Operators on Hilbert Modules

DEFINITION 13.2.1. Let E be a Hilbert B-module. $\mathbb{B}(E)$ is the set of all module homomorphisms $T : E \to E$ for which there is an adjoint module homomorphism $T^* : E \to E$ with $\langle Tx, y \rangle = \langle x, T^*y \rangle$ for all $x, y \in E$. $\mathbb{B}_B = \mathbb{B}(\mathbb{H}_B)$.

Actually, the assumption that T and T^* are module maps is unnecessary, since maps with adjoints are automatically module homomorphisms: we have

$$\langle T(xb), y \rangle = \langle xb, T^*y \rangle = b^* \langle x, T^*y \rangle = b^* \langle Tx, y \rangle = \langle (Tx)b, y \rangle$$

for all y, so $T(xb) = (Tx)b$. Even linearity is automatic for a map with an adjoint.

PROPOSITION 13.2.2. *Each operator in $\mathbb{B}(E)$ is bounded, and $\mathbb{B}(E)$ is a C^*-algebra with respect to the operator norm.*

PROOF. The existence of an adjoint implies that an operator in $\mathbb{B}(E)$ has closed graph, hence is bounded. The completeness and the C^*-axioms are shown exactly as in $\mathbb{B}(\mathbb{H})$, using 13.1.3. $\qquad \square$

$\mathbb{B}(E)$ does not contain all bounded module-endomorphisms of E in general; see [Paschke 1973, 2.5].

Note that a submodule of a Hilbert module, even of $\mathbb{H}_B$, need not be complemented in general, i.e. there is not generally a projection in $\mathbb{B}(E)$ onto the submodule. For example, a right ideal in B is rarely complemented.

There are some special "rank 1" operators on any Hilbert module: if $x, y \in E$, let $\theta_{x,y}$ be the operator defined by $\theta_{x,y}(z) = x\langle y, z \rangle$. We have $\theta_{x,y} \in \mathbb{B}(E)$; in fact, $\theta_{x,y}^* = \theta_{y,x}$. If $T \in \mathbb{B}(E)$, then $T\theta_{x,y} = \theta_{Tx,y}$; $\theta_{x,y}T = \theta_{x,T^*y}$, so the linear span of $\{\theta_{x,y}\}$ is an ideal in $\mathbb{B}(E)$. These are the "finite-rank" operators on E.

DEFINITION 13.2.3. $\mathbb{K}(E)$ is the closure of this linear span. It is a (closed) ideal in $\mathbb{B}(E)$. $\mathbb{K}_B = \mathbb{K}(\mathbb{H}_B)$.

EXAMPLES 13.2.4. (a) $\mathbb{K}(B) \cong B$ for any B; if B is unital, then also $\mathbb{B}(B) \cong B$.
(b) If E is a Hilbert module, then $\mathbb{B}(E^n) \cong \mathbb{M}_n \otimes \mathbb{B}(E)$; $\mathbb{K}(E^n) \cong \mathbb{M}_n \otimes \mathbb{K}(E)$.
    In particular, $\mathbb{K}(B^n) \cong M_n(B)$, and if B is unital $\mathbb{B}(B) \cong M_n(B)$.
(c) $\mathbb{K}_B \cong B \otimes \mathbb{K}$.
(d) If E is a (complete) $A - B$-equivalence bimodule, thought of as a Hilbert
    B-module as in 13.1.2(d), then $\mathbb{K}(E) \cong A$.

Proofs of these facts are straightforward and are left to the reader.

One can also define $\mathbb{B}(E_1, E_2)$ and $\mathbb{K}(E_1, E_2)$ for a pair of Hilbert B-modules E_1, E_2, in the same manner.

13.3. Regular Operators

We will sometimes have occasion to consider unbounded operators on Hilbert modules.

DEFINITION 13.3.1. A *regular operator* on a Hilbert B-module E is a densely defined operator T on E with densely defined adjoint T^*, such that $1 + T^*T$ has dense range in E.

As before, a regular operator must be B-linear. The domain of a regular operator is a B-module. Just as in $\mathbb{B}(\mathbb{H})$, if T is a regular operator, then $(1 + T^*T)^{-1}$ extends to a bounded operator (an element of $\mathbb{B}(E)$). Also, T^* is a regular operator, so we may define elements h and l in $\mathbb{B}(E)$ as the bounded extensions of $(1 + T^*T)^{-1/2}$ and $(1 + TT^*)^{-1/2}$ respectively. A routine calculation then shows the following.

PROPOSITION 13.3.2. *The operator Th extends to an operator F in $\mathbb{B}(E)$. F^* is the extension of T^*l, and $T = Fh^{-1}$, $T^* = F^*l^{-1}$.*

13.4. Hilbert Modules and Multiplier Algebras

The next theorem gives the fundamental relationship between Hilbert modules and multiplier algebras:

THEOREM 13.4.1. *Let E be a Hilbert B-module. Then the correspondence $T \in \mathbb{B}(E) \to (T_1, T_2) \in M(\mathbb{K}(E))$, where $T_1(\theta_{x,y}) = \theta_{Tx,y}$, $T_2(\theta_{x,y}) = \theta_{x,T^*y}$, defines an isomorphism of $\mathbb{B}(E)$ onto $M(\mathbb{K}(E))$.*

PROOF. We have $\|T_i\| \leq \|T\|$, so T_i defines a bounded operator on $\mathbb{K}(E)$; and it is easy to check that (T_1, T_2) is a double centralizer of $\mathbb{K}(E)$. If $T_i = 0$, then for any x $T_1(\theta_{x,Tx}) = \theta_{Tx,Tx} = 0$, so $Tx = 0$. So the map is injective. To show surjectivity, if $(T_1, T_2) \in M(\mathbb{K}(E))$, define

$$T(x) = \lim_{\varepsilon \to 0} T_1(\theta_{x,x})(x)[\langle x, x \rangle + \varepsilon]^{-1},$$

$$T^*(x) = \lim_{\varepsilon \to 0} [T_2(\theta_{x,x})]^*(x)[\langle x, x \rangle + \varepsilon]^{-1}.$$

(The limits exist because $T_1(S)^*T_1(S) \leq \|T_1\|^2 S^*S$ and $T_2(S)T_2(S)^* \leq \|T_2\|^2 SS^*$ for all $S \in \mathbb{K}(E)$.) Since for all $x \in E$ we have $x = \lim_{\varepsilon \to 0} \theta_{x,x}(x)[\langle x, x \rangle + \varepsilon]^{-1}$, it follows that

$$\langle x, T^*(y) \rangle = \lim_{\varepsilon \to 0} \langle x, [T_2(\theta_{y,y})]^*(y)[\langle y, y \rangle + \varepsilon]^{-1} \rangle$$
$$= \lim_{\varepsilon \to 0} \langle T_2(\theta_{y,y})\theta_{x,x}(x)[\langle x, x \rangle + \varepsilon]^{-1}, \ y[\langle y, y \rangle + \varepsilon]^{-1} \rangle$$
$$= \lim_{\varepsilon \to 0} \langle \theta_{y,y}T_1(\theta_{x,x})(x)[\langle x, x \rangle + \varepsilon]^{-1}, \ y[\langle y, y \rangle + \varepsilon]^{-1} \rangle$$
$$= \lim_{\varepsilon \to 0} \langle T_1(\theta_{x,x})(x)[\langle x, x \rangle + \varepsilon]^{-1}, \ \theta_{y,y}(y)[\langle y, y \rangle + \varepsilon]^{-1} \rangle = \langle T(x), y \rangle.$$

Finally, fix x and y; for $z \in E$ set $w = T_1(\theta_{x,y})(z) - \theta_{Tx,y}(z)$. Then for any u and v we have $\theta_{u,v}(w) = 0$, so $w = 0$. This is true for all z, so $T_1(\theta_{x,y}) = \theta_{Tx,y}$. Similarly $T_2(\theta_{x,y}) = \theta_{T^*x,y}$. $\qquad\square$

COROLLARY 13.4.2. $\mathbb{B}(B) = M(B)$, $\mathbb{B}(B^n) \cong M_n(M(B))$, and $\mathbb{B}(\mathbb{H}_B) \cong M^s(B)$.

13.5. Tensor Products of Hilbert Modules

If E_1 and E_2 are Hilbert modules over B_1 and B_2, respectively, and $\phi : B_1 \to \mathbb{B}(E_2)$ is a *-homomorphism, then we can form the tensor product $E_1 \otimes_\phi E_2$ as follows. Regard E_2 as a left B_1-module via ϕ, and form the algebraic tensor product $E_1 \odot_{B_1} E_2$; this is a right B_2-module. Define a B_2-valued pre-inner product on this algebraic tensor product by $\langle x_1 \otimes x_2, y_1 \otimes y_2 \rangle = \langle x_2, \phi(\langle x_1, y_1 \rangle_1) y_2 \rangle_2$, where $\langle \cdot, \cdot \rangle_i$ is the B_i-valued inner product on E_i.

DEFINITION 13.5.1. The completion of the algebraic tensor product with respect to this inner product (with vectors of length 0 divided out) is called the *tensor product* of E_1 and E_2, denoted $E_1 \otimes_\phi E_2$.

This tensor product is sometimes denoted $E_1 \otimes_{B_1} E_2$, but we prefer the more precise notation explicitly specifying the map ϕ.

There is a natural homomorphism from $\mathbb{B}(E_1)$ to $\mathbb{B}(E_1 \otimes_\phi E_2)$, which is injective if ϕ is; we will write $F \otimes 1$ for the image of F. However there is no homomorphism from $\mathbb{B}(E_2)$ to $\mathbb{B}(E_1 \otimes E_2)$ in general, i.e. there is no reasonable definition of $1 \otimes F$ in general. (If $F \in \mathbb{B}(E_2)$ commutes with $\phi(B_1)$, then there is a well-defined operator $1 \otimes F$.)

A related construction is the "outer (or external) tensor product": if E_1 and E_2 are Hilbert modules over B_1 and B_2, the algebraic tensor product $E_1 \odot E_2$ is a module over $B_1 \odot B_2$ in the obvious way, and its completion with respect to the inner product $\langle x_1 \otimes x_2, y_1 \otimes y_2 \rangle = \langle x_1, y_1 \rangle \otimes \langle x_2, y_2 \rangle$ makes $E_1 \otimes E_2$ into a Hilbert $(B_1 \otimes B_2)$-module. There is an embedding of $\mathbb{B}(E_1) \otimes \mathbb{B}(E_2)$ into $\mathbb{B}(E_1 \otimes E_2)$ in the obvious way; the restriction to $\mathbb{K}(E_1) \otimes \mathbb{K}(E_2)$ is an isomorphism onto $\mathbb{K}(E_1 \otimes E_2)$.

EXAMPLES 13.5.2. (a) If $\phi : B_1 \to B_2$ is a *-homomorphism, then $B_1 \otimes_\phi B_2$ is isomorphic to the closed right ideal $\overline{\phi(B_1) B_2}$ of B_2 generated by $\phi(B_1)$. So if ϕ is a unital homomorphism of unital C^*-algebras, or more generally if ϕ is *essential* in the sense that $\phi(B_1)$ contains an approximate identity for B_2, then $B_1 \otimes_\phi B_2$ may be identified with B_2.

(b) If $\phi : \mathbb{C} \to M(B)$ maps $\mathbb{C}$ to the scalars, then $\mathbb{H} \otimes_\phi B \cong \mathbb{H}_B$. We will sometimes write $\mathbb{H} \otimes_{\mathbb{C}} B$ for this tensor product.

(c) If $\phi : B_1 \to B_2$ is essential, we may identify $\mathbb{H}_{B_1} \otimes_\phi B_2$ with $\mathbb{H} \otimes_{\mathbb{C}} B_2 \cong \mathbb{H}_{B_2}$ in an evident way.

13.6. The Stabilization or Absorption Theorem

Before proving the absorption theorem for Hilbert modules, we need a fact about strictly positive elements in $\mathbb{K}(E)$:

PROPOSITION 13.6.1. *Let E be a Hilbert B-module and $T \in \mathbb{K}(E)_+$. Then T is strictly positive if and only if T has dense range.*

PROOF. If T is strictly positive then $\overline{T\mathbb{K}(E)} = \mathbb{K}(E)$. Since $\overline{\mathbb{K}(E)E} = E$, we have $\overline{TE} = \overline{T\mathbb{K}(E)E} = \overline{\mathbb{K}(E)E} = E$. If T has dense range, for any $x, y \in E$ choose a sequence z_n with $Tz_n \to x$. Then $\theta_{x,y} = \lim_n T\theta_{z_n,y} \in \overline{T\mathbb{K}(E)}$. So $T\mathbb{K}(E)$ is dense and T is strictly positive. $\qquad\square$

The next theorem, called the Stabilization or Absorption Theorem, is the most important technical result of this section. The theorem in this form is due to Kasparov [Kasparov 1980a]; but it is very similar to a theorem of Brown, Green, and Rieffel on Morita equivalence and stable isomorphism (13.7.1). The proof given here is due to Mingo and Phillips [Mingo and Phillips 1984].

THEOREM 13.6.2. *If E is a countably generated Hilbert B-module, then $E \oplus \mathbb{H}_B \cong \mathbb{H}_B$.*

PROOF. It suffices to prove that $\tilde{E} \oplus \mathbb{H}_B \cong \mathbb{H}_B$, where $\tilde{E}$ is E regarded as a Hilbert $\tilde{B}$-module, since $\overline{EB} = E$ and $\overline{\mathbb{H}_B B} = \mathbb{H}_B$. Thus we may assume B is unital.

Let $\{\eta_j\}$ be a bounded sequence of generators for E, with each generator repeated infinitely often. Let $\{\xi_j\}$ be the "standard" orthonormal basis for $\mathbb{H}_B$, i.e. ξ_j has a 1 in the j-th place and zeros elsewhere.

Define $T : \mathbb{H}_B \to E \oplus \mathbb{H}_B$ by $T(\xi_j) = 2^{-j}\eta_j \oplus 4^{-j}\xi_j$. Then $T = \sum 2^{-j}\theta_{(\eta_j + 2^{-j}\xi_j),\xi_j} \in \mathbb{K}(\mathbb{H}_B, E \oplus \mathbb{H}_B)$.

T is clearly one-to-one. Since each η_j is repeated infinitely often, $\eta_j \oplus 2^{-k}\xi_k \in \overline{T\mathbb{H}_B}$ for infinitely many k; thus $\eta_j \oplus 0$ and hence also $0 \oplus \xi_j$ are in $\overline{T\mathbb{H}_B}$, i.e. T has dense range. We also have $T^*T = S^*S + R^*R$, where $S(\xi_j) = 0 \oplus 4^{-j}\xi_j$, $R(\xi_j) = 2^{-j}\eta_j \oplus 0$. So $T^*T \geq S^*S = \mathrm{diag}(4^{-2}, 4^{-4}, 4^{-6}, \ldots)$; S^*S has dense range and is therefore strictly positive, so T^*T is strictly positive and thus has dense range. So T has a polar decomposition $T = U(T^*T)^{1/2}$, where U is a unitary in $\mathbb{B}(\mathbb{H}_B, E \oplus \mathbb{H}_B)$. [$U$ is defined by $U((T^*T)^{1/2}\xi) = T\xi$; $U^*(T\xi) = (T^*T)^{1/2}\xi$.] U implements an isomorphism between $\mathbb{H}_B$ and $E \oplus \mathbb{H}_B$. $\qquad\square$

COROLLARY 13.6.3. *If E is a Hilbert B-module, then E is countably generated if and only if $\mathbb{K}(E)$ has a strictly positive element.*

PROOF. As in the proof of 13.6.2 we may assume B is unital. Then by 13.6.2 there is a projection $P \in \mathbb{B}(\mathbb{H}_B)$ with $E \cong P\mathbb{H}_B$. Let (x_n) be the standard orthonormal basis for $\mathbb{H}_B$. Then $T = \sum 2^{-n}\theta_{x_n, x_n}$ is strictly positive by 13.6.1. $\mathbb{K}(E) \cong P\mathbb{K}(\mathbb{H}_B)P$, and PTP is strictly positive in $P\mathbb{K}(\mathbb{H}_B)P$. Conversely, if $\mathbb{K}(E)$ has a strictly positive element T, then T can be written as a convergent series $T = \sum \theta_{x_n, y_n}$. Since $\mathbb{K}(E)E$ is dense in E, (x_n) is a set of generators for E. $\qquad\square$

There is also an equivariant version of the Absorption Theorem (20.1.4).

13.7. EXERCISES AND PROBLEMS

13.7.1. (a) Use the stabilization theorem to prove that if B is σ-unital and E is a countably generated full Hilbert B-module, then $E^\infty \cong \mathbb{H}_B$ [Mingo and Phillips 1984, 1.9].

(b) Show that if A and B are C^*-algebras and there is a full Hilbert B-module E with $\mathbb{K}(E) \cong A$, there is a full Hilbert A-module E' with $\mathbb{K}(E') \cong B$. [Regard $\mathbb{K}(B, E)$ as a right $\mathbb{K}(E)$-module, and define $\langle x, y \rangle = x^*y$.] If there is such a module, A and B are said to be *(strongly) Morita equivalent* [Rieffel 1974]. If A is a full hereditary C^*-subalgebra of B, then the submodule $\overline{AB}$ of B gives a strong Morita equivalence between A and B.

(c) Show that if A and B are σ-unital C^*-algebras which are strongly Morita equivalent, then A and B are stably isomorphic, i.e. $A \otimes \mathbb{K} \cong B \otimes \mathbb{K}$ [Brown et al. 1977b] (cf. [Brown 1977]). [If $\mathbb{K}(E) \cong A$, then $\mathbb{K}(E^\infty) \cong A \otimes \mathbb{K}$.]

13.7.2. Prove the *Generalized Stinespring Theorem* [Kasparov 1980a, Theorem 3] on dilations of completely positive maps:

THEOREM. *Let A and B be C^*-algebras, with A separable and B σ-unital. Let $\phi : A \to M^s(B)$ be a completely positive contraction. Then there is a $*$-homomorphism $\rho : A \to M_2(M^s(B))$ such that*

$$\mathrm{diag}(1, 0)\rho(a)\,\mathrm{diag}(1, 0) = \mathrm{diag}(\phi(a), 0) \quad \text{for all } a \in A.$$

(a) Extend ϕ to a unital completely positive map from A^+ to $M^s(B)$ by [Choi and Effros 1976, 3.9]. Thus we may assume A is unital and $\phi(1) = 1$.

(b) Define a pre-inner product on the right B-module $A \odot \mathbb{H}_B$ by

$$\langle a_1 \otimes x_1, \, a_2 \otimes x_2 \rangle = \langle x_1, \, \phi(a_1^*a_2)x_2 \rangle_{\mathbb{H}_B}$$

Show that the completion of this module (with vectors of length zero divided out) is a countably generated Hilbert B-module E. There is a $*$-homomorphism $\omega : A \to \mathbb{B}(E)$ induced by the left action of A on $A \odot \mathbb{H}_B$.

(c) Define maps $S : \mathbb{H}_B \to A \odot \mathbb{H}_B$ and $T : A \odot \mathbb{H}_B \to \mathbb{H}_B$ by $S(x) = 1 \otimes x$, $T(a \otimes y) = \phi(a)y$. Show that for

$$\langle S(x), a \otimes y \rangle = \langle x, T(a \otimes y) \rangle \quad \text{for } x, y \in \mathbb{H}_B \text{ and } a \in A.$$

Since S is bounded, conclude that S and T extend to maps $W \in \mathbb{B}(\mathbb{H}_B, E)$ and $W^* \in \mathbb{B}(E, \mathbb{H}_B)$ which are adjoints of each other. $W^*\omega(a)W = \phi(a)$ for $a \in A$.

(d) $W^*W = 1$, so WW^* is a projection; $(WW^*)E \cong \mathbb{H}_B$. Under the chain of isomorphisms

$$E \oplus \mathbb{H}_B \cong [(WW^*)E \oplus (1 - WW^*)E] \oplus \mathbb{H}_B$$

$$\cong \mathbb{H}_B \oplus [(1 - WW^*)E \oplus \mathbb{H}_B] \cong \mathbb{H}_B \oplus \mathbb{H}_B$$

(the last of which uses the Stabilization Theorem), $W \oplus 0$ becomes $\mathrm{diag}(1, 0) \in M_2(M^s(B)) \cong \mathbb{B}(\mathbb{H}_B \oplus \mathbb{H}_B)$, and $\omega \oplus 0$ becomes a homomorphism $\rho : A \to$

$M_2(M^s(B))$. (If a unital ρ is desired, let ρ correspond to $\omega \oplus \sigma$, where $\sigma : A \to$ $\mathbb{B}(\mathbb{H}_B)$ is any unital homomorphism.)

14. Graded C^*-algebras

Kasparov's development of operator K-theory uses graded C^*-algebras. One of the things which makes [Kasparov 1980b] difficult to read is the great generality; he presents a theory which works equally well for complex, real, and "real" C^*-algebras, allowing a grading and a compact group action. The reader who is only interested in the special case of ordinary (trivially graded) complex C^*-algebras with no group action sometimes feels that the (often slight) technicalities necessary in the general case are an obstacle to understanding the case of interest.

When I first began to study [Kasparov 1980b], I decided to try to ignore the real and "real" cases, the group action, and the grading. I gradually came to appreciate the usefulness of considering graded C^*-algebras; some parts of the theory become much simpler and more natural when grading is included. For example, Bott Periodicity becomes a special case of the Takai Duality Theorem for $\mathbb{Z}_2$. Graded C^*-algebras also arise naturally in mathematical physics ("supersymmetry"). And graded C^*-algebras are really not such mysterious objects after all. Besides, if one is only interested in ordinary (ungraded) C^*-algebras, one only needs to consider two very simple special types of graded C^*-algebras.

In this section, we develop the general theory of graded C^*-algebras needed for KK-theory, and the important special cases for the KK-theory of ordinary C^*-algebras.

14.1. Basic Definitions

DEFINITION 14.1.1. Let A be a C^*-algebra. A $(\mathbb{Z}_2$-$)grading$ on A is a decomposition of A into a direct sum of two self-adjoint closed linear subspaces $A^{(0)}$ and $A^{(1)}$, such that if $x \in A^{(m)}$, $y \in A^{(n)}$, then $xy \in A^{(m+n)}$ (addition mod 2). An element of $A^{(n)}$ is said to be *homogeneous of degree* n. The degree of a homogeneous element a is denoted ∂a. If there is a self-adjoint unitary $g \in M(A)$ with $A^{(n)} = \{a \in A : g^*ag = (-1)^n a\}$, then the grading is called *even* and g is called a *grading operator* for the grading. If $A^{(1)} = 0$, the grading is *trivial*. A trivial grading is even with grading operator 1. A homomorphism $\phi : A \to B$ of graded C^*-algebras is a *graded homomorphism* if $\phi(A^{(n)}) \subseteq B^{(n)}$ for $n = 0$, 1. A C^*-subalgebra B of a graded C^*-algebra A is a *graded* C^*-*subalgebra* if $B = (B \cap A^{(0)}) + (B \cap A^{(1)})$, i.e. if $a^{(0)} + a^{(1)} \in B$ implies $a^{(0)}, a^{(1)} \in B$. In this case, the grading on A restricts to a grading on B. A state on A is *homogeneous* if it vanishes on $A^{(1)}$. Any state ϕ defines a homogeneous state ϕ_h by $\phi_h(a^{(0)} + a^{(1)}) = \phi(a^{(0)})$. If A is a graded C^*-algebra, the *graded commutator* of homogeneous elements a, b is $[a, b] = ab - (-1)^{\partial a \cdot \partial b} ba$. The graded commutator of general elements is then defined by linearity.

If A is graded, then $A^{(0)}$ is a C^*-subalgebra of A ($A^{(1)}$ is of course not a sub-algebra). Any C^*-algebra may be regarded as a trivially graded C^*-algebra. A grading on A extends uniquely to a grading on $\tilde{A}$.

A grading on A is nothing but a $\mathbb{Z}_2$-action on A : $A^{(0)}$ and $A^{(1)}$ are the eigenspaces for 1 and -1 respectively. That is, given an automorphism α of period 2, $A^{(0)} = \{a : \alpha(a) = a\}$, $A^{(1)} = \{a : \alpha(a) = -a\}$. If $a \in A$, the decomposition is $a = a^{(0)} + a^{(1)}$, where $a^{(0)} = (a + \alpha(a))/2$, $a^{(1)} = (a - \alpha(a))/2$. Conversely, given a grading, the corresponding $\mathbb{Z}_2$-action is given by $\alpha(a^{(0)} + a^{(1)}) = a^{(0)} - a^{(1)}$. A grading is even if and only if the corresponding $\mathbb{Z}_2$-action is inner.

A C^*-subalgebra of A is a graded subalgebra if and only if it is (globally) invariant under the grading automorphism.

EXAMPLES 14.1.2. (a) If A is any (ungraded) C^*-algebra, there is a grading on $M_2(A)$ with $M_2(A)^{(0)}$ the diagonal matrices and $M_2(A)^{(1)}$ the matrices with zero diagonal. This is an even grading with grading operator $\mathrm{diag}(1, -1)$, called the *standard even grading* on $M_2(A)$. By identifying $A \otimes \mathbb{K}$ with $M_2(A \otimes \mathbb{K})$, we obtain the *standard even grading* of $A \otimes \mathbb{K}$ (this grading is actually only determined up to conjugation by an inner automorphism homotopic to the identity).

(b) There is a *standard odd grading* on $A \oplus A$ for any (ungraded) A: $(A \oplus A)^{(0)} = \{(a, a) : a \in A\}$ and $(A \oplus A)^{(1)} = \{(a, -a) : a \in A\}$. If $A = \mathbb{C}$, we denote $\mathbb{C}^2$ with the standard odd grading by $\mathbb{C}_1$. ($\mathbb{C}_1$ is a complex Clifford algebra [Kasparov 1980b, § 2].) Note that $\mathbb{C}_1$ is isomorphic to the group C^*-algebra of $\mathbb{Z}_2$; the grading is given by the dual action of $\hat{\mathbb{Z}}_2 \cong \mathbb{Z}_2$.

(c) A grading on a C^*-algebra A induces a canonical grading on $M(A)$ in the obvious way. Thus, for any ungraded A, there is a standard even grading on $M^s(A)$ (determined up to inner automorphism homotopic to the identity).

The reader interested just in Kasparov theory for ordinary (ungraded) C^*-algebras need only be concerned with these examples.

Care must be exercised in applying identities with graded commutators. For example, it is not true in general that $[x, y] = -[y, x]$: if x has degree 1 then $[x, x] = 2x^2$. The next proposition gives some standard identities valid for graded commutators. The proof is a straightforward calculation checking case-by-case.

PROPOSITION 14.1.3. *Let A be a graded C^*-algebra, and x, y, z homogeneous elements of A. Then:*

(a) $[x, y] + (-1)^{\partial x \cdot \partial y}[y, x] = 0$.

(b) $[x, yz] = [x, y]z + (-1)^{\partial x \cdot \partial y} y[x, z]$.

(c) $(-1)^{\partial x \cdot \partial z}[[x, y], z] + (-1)^{\partial x \cdot \partial y}[[y, z], x] + (-1)^{\partial y \cdot \partial z}[[z, x], y] = 0$.

If x is self-adjoint and has degree 1, then $[x, x] \geq 0$. If y is also self-adjoint and degree 1, then $[x, y]$ is self-adjoint; the condition that $[x, y] \geq 0$ means that y "lines up with" x in an appropriate sense.

14.2. Graded Hilbert Modules

If B is a graded C^*-algebra, a graded Hilbert B-module is a Hilbert B-module E with a decomposition as a direct sum of subspaces $E^{(0)}$ and $E^{(1)}$ with $E^{(m)}B^{(n)} \subseteq E^{(m+n)}$ and $\langle E^{(m)}, E^{(n)} \rangle \subseteq B^{(m+n)}$. The elements of $B^{(0)}$ leave invariant the subspaces $E^{(0)}$ and $E^{(1)}$, and the elements of $B^{(1)}$ interchange them. There is a natural grading on $\mathbb{H}_B$, with $\mathbb{H}_B^{(n)}$ the set of sequences with all terms in $B^{(n)}$. A grading on E induces a grading on $\mathbb{B}(E)$ and $\mathbb{K}(E)$. More generally, one obtains a grading on the set of regular operators on E.

If $E = B$, the grading on $\mathbb{B}(B) \cong M(B)$ obtained in this way agrees with the grading defined in 14.1.2(c).

If E is a graded Hilbert B-module, we denote by E^{op} the graded Hilbert B-module obtained from E by interchanging $E^{(0)}$ and $E^{(1)}$. The grading on $\mathbb{B}(E)$ induced by E^{op} is the same as the grading induced by E.

We write $\hat{\mathbb{H}}_B$ for $\mathbb{H}_B \oplus \mathbb{H}_B^{op}$, where $\mathbb{H}_B$ has its natural grading. $\hat{\mathbb{H}}_B$ is isomorphic to $\mathbb{H}_B$ as a Hilbert B-module, but not in general as a graded Hilbert B-module. $\hat{\mathbb{H}}_B$ is in some sense the "universal" graded Hilbert B-module (14.6.1).

If $\mathbb{H}_B$ is trivially graded, then we say $\hat{\mathbb{H}}_B$ has standard even grading. The standard even grading on $\hat{\mathbb{H}}_B$ induces a standard even grading on $\mathbb{B}(\hat{\mathbb{H}}_B) \cong M^s(B)$. More generally, if E is trivially graded, then $E \oplus E^{op}$ induces a standard even grading on $M_2(\mathbb{B}(E))$.

14.3. Graded Homomorphisms

If A and B are graded C^*-algebras and $M_2(B)$ has the standard even grading, a graded homomorphism from A to $M_2(B)$ has the form

$$\phi = \begin{bmatrix} \phi_{11} & \phi_{12} \\ \phi_{21} & \phi_{22} \end{bmatrix},$$

where ϕ_{11}, ϕ_{22} vanish on $A^{(1)}$, ϕ_{12}, ϕ_{21} vanish on $A^{(0)}$. So if A is trivially graded, $\phi = \mathrm{diag}(\phi_{11}, \phi_{22})$.

If $B \oplus B$ has the standard odd grading, then a graded homomorphism from A to $B \oplus B$ is of the form $\phi = \phi_1 \oplus \phi_2$, where $\phi_2(a^{(0)} + a^{(1)}) = \phi_1(a^{(0)} - a^{(1)})$. If A is evenly graded with grading operator g, then conjugating the second coordinate by $\phi_2(g)$ yields that ϕ is conjugate to $\phi_1 \oplus \phi_1$, so ϕ may be identified with an (ordinary) homomorphism into B.

14.4. Graded Tensor Products

We must also introduce the notion of the graded (or skew-symmetric) tensor product of two graded C^*-algebras. The general construction takes a bit of work, but in the special cases of 14.1.2 the graded tensor product has a simple form.

Let A and B be graded C^*-algebras, and $A \odot B$ the algebraic tensor product. We define a new product and involution on $A \odot B$ by

$$(a_1 \,\hat{\otimes}\, b_1)(a_2 \,\hat{\otimes}\, b_2) = (-1)^{\partial b_1 \cdot \partial a_2}(a_1 a_2 \,\hat{\otimes}\, b_1 b_2),$$
$$(a \,\hat{\otimes}\, b)^* = (-1)^{\partial a \cdot \partial b}(a^* \hat{\otimes} b^*)$$

for homogeneous elementary tensors. The algebraic tensor product with this multiplication and involution is a $*$-algebra, denoted $A \hat{\odot} B$.

DEFINITION 14.4.1. $A \hat{\otimes}_{\max} B$ is the universal enveloping C^*-algebra of $A \hat{\odot} B$.

The representations of $A \hat{\otimes}_{\max} B$ are in natural one-one correspondence with the pairs (π, ρ) of representations of A and B on the same Hilbert space, with the property that for homogeneous a and b $\pi(a)\rho(b) = (-1)^{\partial a \cdot \partial b}\rho(b)\pi(a)$.

To show that there are sufficiently many such representations, note that if ϕ and ψ are homogeneous states on A and B respectively, then $\phi \hat{\otimes} \psi$ (defined in the usual way) is a state on $A \hat{\odot} B$, and the GNS representation from $\phi \hat{\otimes} \psi$ gives a C^*-seminorm on $A \hat{\odot} B$. The supremum of all such seminorms is a norm (proved just as for ordinary tensor products [Sakai 1971, 1.22.2]); the completion is called the *(minimal) graded tensor product of A and B*, denoted $A \hat{\otimes}_{\min} B$ or usually just $A \hat{\otimes} B$. It is a quotient of $A \hat{\otimes}_{\max} B$ in general.

If A or B is nuclear, then using ideas from [Effros and Lance 1977] it is not difficult to show that the quotient map from $A \hat{\otimes}_{\max} B$ to $A \hat{\otimes}_{\min} B$ is an isomorphism. This is the case we will be almost exclusively concerned with.

The grading on $A \hat{\otimes} B$ is the obvious one: homogeneous elementary tensors are homogeneous elements, with $\partial(a \hat{\otimes} b) = \partial a + \partial b$. The homogeneous elements of degree n are then limits of linear combinations of homogeneous elementary tensors of degree n. (Alternately, if α and β are the grading automorphisms on A and B, then $\alpha \hat{\otimes} \beta$ is the grading automorphism on $A \hat{\otimes} B$.)

It is straightforward to prove that $A \hat{\otimes}(B \hat{\otimes} C) \cong (A \hat{\otimes} B) \hat{\otimes} C$. The isomorphism $A \hat{\otimes} B \cong B \hat{\otimes} A$ is less obvious: $a \hat{\otimes} b \to (-1)^{\partial a \cdot \partial b} b \hat{\otimes} a$ does the trick. Analogous statements are true for $\hat{\otimes}_{\max}$.

14.4.2. We must also define the graded tensor product of Hilbert modules. If E_1 and E_2 are graded Hilbert modules over A and B respectively, and ϕ is a graded $*$-homomorphism from A to B, we define $E_1 \hat{\otimes}_{\phi} E_2$ to be the ordinary tensor product (13.5) with grading $\partial(x \hat{\otimes} y) = \partial x + \partial y$. The natural homomorphism from $\mathbb{B}(E_1)$ to $\mathbb{B}(E_1 \hat{\otimes}_{\phi} E_2)$ is a graded homomorphism.

There is no natural embedding of $\mathbb{B}(E_2)$ into $\mathbb{B}(E_1 \hat{\otimes}_{\phi} E_2)$. Even if $A = \mathbb{C}$ and ϕ is unital, the embedding of $\mathbb{B}(E_2)$ sending F to $1 \otimes F$ does not have good properties with respect to the grading. We will never consider such an embedding.

EXAMPLES 14.4.3. (a) If $\phi : A \to B$ is an essential graded homomorphism, then $A \hat{\otimes}_{\phi} B \cong B$, $\mathbb{H}_A \hat{\otimes}_{\phi} B \cong \mathbb{H}_B$, and $\hat{\mathbb{H}}_A \hat{\otimes}_{\phi} B \cong \hat{\mathbb{H}}_B$ as graded Hilbert B-modules.

(b) If E_1 is a Hilbert A-module and E_2 is a Hilbert B-module, and if $\phi : A \to \mathbb{B}(E_2)$ is a graded homomorphism, then $E_1^{op} \hat{\otimes}_{\phi} E_2 \cong E_1 \hat{\otimes}_{\phi} E_2^{op} \cong (E_1 \hat{\otimes}_{\phi} E_2)^{op}$ and $E_1^{op} \hat{\otimes}_{\phi} E_2^{op} \cong E_1 \hat{\otimes}_{\phi} E_2$ as graded Hilbert B-modules.

14.4.4. Although the recipe for the inner tensor product is exactly the same in the graded case as for ungraded algebras, the formula for the graded outer tensor product must be modified similarly to the way the graded tensor product of algebras is defined. If E_1 and E_2 are graded Hilbert modules over the graded C^*-algebras B_1 and B_2, respectively, we make the algebraic tensor product $E_1 \hat{\odot} E_2$ into a right $(B_1 \hat{\odot} B_2)$-module by $(x_1 \hat{\otimes} x_2)(b_1 \hat{\otimes} b_2) = (-1)^{\partial x_2 \cdot \partial b_1}(x_1 b_1 \hat{\otimes} x_2 b_2)$. The inner product is given by the formula

$$\langle x_1 \hat{\otimes} x_2, y_1 \hat{\otimes} y_2 \rangle = (-1)^{\partial x_2(\partial x_1 + \partial y_1)} \langle x_1, y_1 \rangle \hat{\otimes} \langle x_2, y_2 \rangle.$$

The grading is given by $\partial(x_1 \hat{\otimes} x_2) = \partial x_1 + \partial x_2$ as before. The embedding of $\mathbb{B}(E_1) \hat{\otimes} \mathbb{B}(E_2)$ into $\mathbb{B}(E_1 \hat{\otimes} E_2)$ is given by

$$(F_1 \hat{\otimes} F_2)(x_1 \hat{\otimes} x_2) = (-1)^{\partial F_2 \cdot \partial x_1} F_1(x_1) \hat{\otimes} F_2(x_2).$$

This embedding gives an isomorphism between $\mathbb{K}(E_1) \hat{\otimes} \mathbb{K}(E_2)$ and $\mathbb{K}(E_1 \hat{\otimes} E_2)$.

EXAMPLE 14.4.5. $\hat{\mathbb{H}}_B$ may be regarded as the internal tensor product $(\mathbb{H} \oplus \mathbb{H}^{op}) \hat{\otimes}_{\mathbb{C}} B$ or the external tensor product $\hat{\mathbb{H}} \hat{\otimes} B$. If the tensor product is internal, the natural isomorphism is the obvious one. However, if the tensor product is external, the natural isomorphism sends $(\xi \hat{\otimes} b, \eta \hat{\otimes} b)$ to $(\xi, -\eta) \hat{\otimes} b$. The embedding of $\mathbb{B}(B)$ into $\mathbb{B}(\hat{\mathbb{H}}_B)$, regarded as an external tensor product in this way, sends $b \in \mathbb{B}(B)^{(n)}$ to $\mathrm{diag}(1, (-1)^n) \hat{\otimes} b \in \mathbb{B}((\mathbb{H} \oplus \mathbb{H}^{op}) \hat{\otimes}_{\mathbb{C}} B)$. This is the only embedding which relates well to the natural embedding of $\mathbb{B}(\hat{\mathbb{H}})$.

The formulas for these graded tensor products appear daunting at first sight, but we will actually rarely need to make use of them. The next paragraph will clarify the structure of the graded tensor products we will need.

14.5. Structure of Graded Tensor Products

The next few results explain the structure of the graded tensor product in the special cases given in 14.1.2. For KK-theory of ordinary C^*-algebras, these are the only cases which must be considered.

PROPOSITION 14.5.1. If A is evenly graded, then $A \hat{\otimes} B \cong A \otimes B$ (and similarly $A \hat{\otimes}_{\max} B \cong A \otimes_{\max} B$). If A and B are both evenly graded with grading operators g and h respectively, then under this isomorphism $A \otimes B$ is evenly graded with grading operator $g \otimes h$.

PROOF. The isomorphism sends $a \hat{\otimes} b$ to $ag^{\partial b} \otimes b$. It is straightforward to check that this is an isomorphism of $A \hat{\odot} B$ onto $A \odot B$, and that it preserves tensor products of homogeneous states. It is also easy to check in the second case that conjugation of elementary tensors by $g \otimes h$ has the right effect. □

COROLLARY 14.5.2. If A is evenly graded and $\mathbb{M}_2$ has the standard even grading, then $A \hat{\otimes} \mathbb{M}_2 \cong M_2(A)$ with standard even grading. So if A is evenly graded and $\mathbb{K}$ has a standard even grading, then $A \hat{\otimes} \mathbb{K} \cong A \otimes \mathbb{K}$ with a standard even grading.

PROOF. Let g be a grading operator for A. Then $A \hat{\otimes} M_2 \cong M_2(A)$ with grading operator $\text{diag}(g, -g)$. The unitary $\frac{1}{2}\left[\begin{smallmatrix} 1+g & 1-g \\ 1-g & 1+g \end{smallmatrix}\right]$ conjugates $\text{diag}(g, -g)$ to $\text{diag}(1, -1)$, the grading operator for a standard even grading on $M_2(A)$. $\square$

Kasparov theory really only uses graded C^*-algebras of the form $A \hat{\otimes} \mathbb{K}$, where $\mathbb{K}$ has the standard even grading. 14.5.2 shows that for the purposes of KK-theory evenly graded C^*-algebras behave exactly like trivially graded ones.

COROLLARY 14.5.3. *If A is evenly graded, then $A \hat{\otimes} \mathbb{C}_1 \cong A \oplus A$ with the standard odd grading (14.1.2(b)).*

PROOF. Under the isomorphism $A \hat{\otimes} \mathbb{C}_1 \cong A \oplus A$, the grading becomes

$$(A \oplus A)^{(0)} = \{(a^{(0)}+a^{(1)}, \quad a^{(0)}-a^{(1)}) \mid a^{(n)} \in A^{(n)}\},$$
$$(A \oplus A)^{(1)} = \{(a^{(0)}+a^{(1)}, \quad -a^{(0)}+a^{(1)}) \mid a^{(n)} \in A^{(n)}\}.$$

If α is the grading automorphism of A, then the automorphism $1 \oplus \alpha$ of $A \oplus A$ converts this grading into the standard odd grading. $\square$

PROPOSITION 14.5.4. *Let A be a graded C^*-algebra, with corresponding $\mathbb{Z}_2$-action α. Then $A \hat{\otimes} \mathbb{C}_1 \cong A \times_\alpha \mathbb{Z}_2$. The grading on $A \hat{\otimes} \mathbb{C}_1$ corresponds to $\alpha \cdot \hat{\alpha}$, where $\hat{\alpha}$ is the dual action on the crossed product.*

PROOF. The representations of $A \hat{\otimes} \mathbb{C}_1$ are in one-one correspondence with the twisted covariant pairs (π, ρ) as in 14.4. $\mathbb{C}_1$ is generated by the self-adjoint unitary $\varepsilon = (1, -1)$, which is the nontrivial element of $\mathbb{Z}_2$ in $C^*(\mathbb{Z}_2)$. For any pair (π, ρ) we must have $\rho(\varepsilon)\pi(a) = \pi(\alpha(a))\rho(\varepsilon)$, exactly the covariance condition for a representation of $A \times_\alpha \mathbb{Z}_2$. $\square$

Since α is inner on $A \times_\alpha \mathbb{Z}_2 = A \otimes \mathbb{C}_1$, the crossed products by $\hat{\alpha}$ and by $\alpha \cdot \hat{\alpha}$ are isomorphic. So we obtain

COROLLARY 14.5.5. $\mathbb{C}_1 \hat{\otimes} \mathbb{C}_1 \cong M_2$ *with its standard even grading (14.1.2(a)).*

This is really the Takai Duality Theorem for $\mathbb{Z}_2$. The general form of Takai Duality for $\mathbb{Z}_2$ comes from taking the graded tensor product of both sides of the equation in 14.5.5 with a general graded C^*-algebra A and using associativity of graded tensor products.

14.5.6. In view of 14.5.5, we may define $\mathbb{C}_n$ as follows: if $n = 2m$ is even, set $\mathbb{C}_n = M_{2^m}$ with standard even grading, and if $n = 2m+1$, then $\mathbb{C}_n = M_{2^m} \oplus M_{2^m}$ with standard odd grading. Then $\mathbb{C}_p \hat{\otimes} \mathbb{C}_q \cong \mathbb{C}_{p+q}$ for all p, q. ($\mathbb{C}_n$ is the complex Clifford algebra associated with an n-dimensional complex vector space.)

14.6. Miscellaneous Theorems

Finally, there are graded analogs of the Stabilization Theorem (13.6.2) and of Kasparov's Technical Theorem (12.4.2):

THEOREM 14.6.1 (STABILIZATION THEOREM). *Let B be a graded C^*-algebra and E a countably generated graded Hilbert B-module. Give $\hat{\mathbb{H}}_B$ its natural grading (14.2). Then $E \oplus \hat{\mathbb{H}}_B \cong \hat{\mathbb{H}}_B$ as graded Hilbert B-modules.*

The proof is a simple modification of the proof of 13.6.2: let $\eta_j = \eta_j^{(0)} + \eta_j^{(1)}$, and let $\xi_j^{(0)}$ be the "standard" orthonormal basis for $\mathbb{H}_B$, $\xi_j^{(1)}$ the corresponding "standard" basis for $\mathbb{H}_B^{op}$. Set $T(\xi_j^{(n)}) = 2^{-j}\eta_j^{(n)} \oplus 4^{-j}\xi_j^{(n)}$. Note that T is of degree 0.

THEOREM 14.6.2 (KASPAROV'S TECHNICAL THEOREM). *Let J be a σ-unital graded C^*-algebra. Let A_1 and A_2 be σ-unital graded C^*-subalgebras of $M(J)$ (with the induced grading from J), and Δ a separable graded linear subspace of $M(J)$. Suppose $A_1 \cdot A_2 \subseteq J$ and that Δ derives A_1 (in the graded sense, i.e. using the graded commutator). Then there are $M, N \in M(J)^{(0)}$ such that $0 \leq M \leq 1$, $N = 1 - M$, $M \cdot A_1 \subseteq J$, $N \cdot A_2 \subseteq J$, and $[M, \Delta] \subseteq J$.*

For the proof, note that a σ-unital graded C^*-algebra always contains a homogeneous strictly positive element of degree 0. For if h is strictly positive and α is the grading automorphism, then $h + \alpha(h)$ is strictly positive of degree 0. In the proof one should always use strictly positive elements and approximate identities of degree 0. Note that a graded commutator is just an ordinary commutator when one of the components is homogeneous of degree 0. With these observations, the proof is identical to the proof of 12.4.2.

14.7. EXERCISES AND PROBLEMS

14.7.1. Let B be a graded C^*-algebra. Let $\hat{\mathbb{H}}_B$ be as in (14.2). Then $\mathbb{K}(\hat{\mathbb{H}}_B) \cong B \hat{\otimes} \mathbb{K}$ as graded C^*-algebras, where $\mathbb{K}$ has a standard even grading.

14.7.2. Formulate and prove graded analogs of the Brown–Green–Rieffel Theorem (13.7.1) and the Generalized Stinespring Theorem (13.7.2).

K-Theory for Operator Algebras
MSRI Publications
Volume **5**, second edition, 1998

CHAPTER VII

THEORY OF EXTENSIONS

In this chapter, we will develop the Brown–Douglas–Fillmore (BDF) theory of extensions, and the generalization due to Kasparov.

Extension theory is important in many contexts, since it describes how more complicated C^*-algebras can be constructed out of simpler "building blocks". Some of the most important applications of extension theory are:

(1) Structure of type I C^*-algebras, group C^*-algebras, and crossed products.
(2) Classification of essentially normal operators.
(3) Index theory for elliptic pseudodifferential operators.
(4) Using associated homological invariants to distinguish between C^*-algebras (often simple C^*-algebras).

These applications will be discussed in more detail later.

We will not follow the historical development of the theory very closely; in fact, much of what we do will be in reverse historical order. We have chosen to do things this way since much of the general theory discussed in Section 15 must be done in essentially the same way anyway for BDF theory, and it is no more difficult to do things in full generality.

We will not go into all the ramifications of BDF theory; the interested reader may consult [Douglas 1980] for a more complete treatment. [Rosenberg 1982a] is also recommended as a good overall source, including Kasparov's theory. Other references include [Brown 1976; Baum and Douglas 1982b; Valette 1982].

15. Basic Theory of Extensions

In this section, we will develop the basics of extension theory. Our treatment will (approximately) follow Kasparov, although many of the basic ideas come directly from the work of Busby and Brown–Douglas–Fillmore.

15.1. Basic Definitions

DEFINITION 15.1.1. Let A and B be C^*-algebras. An *extension of A by B* is a short exact sequence

$$0 \longrightarrow B \xrightarrow{j} E \xrightarrow{q} A \longrightarrow 0$$

of C^*-algebras.

There is some nonuniformity of terminology concerning extensions: sometimes such a sequence is called an extension of B by A. We have adopted what seems to have become the dominant terminology, especially since it matches up nicely with the notation of Kasparov theory.

The goal of extension theory is, given A and B, to classify all extensions of A by B up to a suitable notion of equivalence.

EXAMPLE 15.1.2. We examine a very simple example to get an idea of what is involved. Let $A = \mathbb{C}$, $B = C_0((0,1))$. There are four possible choices of E : $C_0((0,1)) \oplus \mathbb{C}$, $C_0((0,1])$, $C_0([0,1))$, and $C(S^1)$. Each has an obvious associated exact sequence. It is not clear at this point whether we should regard the extensions corresponding to $C_0((0,1])$ and $C_0([0,1))$ as being the "same" or "different".

15.2. The Busby Invariant

The key to analyzing extensions is the so-called Busby invariant. Busby [1968] was the first to study extensions of C^*-algebras; his work did not attract the attention it deserved until the development of BDF theory several years later. The Busby invariant is based on an earlier, purely algebraic construction of Hochschild.

Given an extension $0 \to B \to E \to A \to 0$, B sits as an ideal of E. Hence there is a *-homomorphism σ from E into $M(B)$ [Pedersen 1979, 3.12.8]. σ is injective if and only if B is essential in E. If we compose σ with the quotient map $\pi : M(B) \to Q(B)$, we obtain a *-homomorphism τ from $E/B \cong A$ to $Q(B)$.

DEFINITION 15.2.1. τ is the Busby invariant of the extension $0 \to B \to E \to A \to 0$.

τ is injective if and only if B is essential in A.

EXAMPLE 15.2.2. In the situation of 15.1.2, $M(B) \cong C(\beta\mathbb{R})$, and $Q(B) \cong C(\beta\mathbb{R} \setminus \mathbb{R})$. $\beta\mathbb{R} \setminus \mathbb{R}$ has two components. The Busby invariant corresponding to the four extensions is the map from $\mathbb{C}$ to $Q(B)$ sending 1 to 0, the characteristic function of the component at $+\infty$, the characteristic function of the component at $-\infty$, and 1 respectively.

We will see in the next paragraph that an extension can be recovered from its Busby invariant. We will often identify an extension with its Busby invariant, so we will frequently think of an extension as a *-homomorphism into an outer multiplier algebra instead of as an exact sequence.

15.3. Pullbacks

We now discuss a general construction which will be of use in this chapter and also in several places in Kasparov theory.

Suppose A_1, A_2, B are C^*-algebras, and ϕ_i is a *-homomorphism from A_i to B. We seek a C^*-algebra P and *-homomorphisms ψ_i from P to A_i making the following diagram commutative:

$$
\begin{array}{ccc}
P & \xrightarrow{\ \psi_2\ } & A_2 \\
\Big\downarrow{\psi_1} & & \Big\downarrow{\phi_2} \\
A_1 & \xrightarrow{\ \phi_1\ } & B
\end{array}
$$

and which is universal in the sense that if C is any C^*-algebra and $\omega_i : C \to A_i$ satisfies $\phi_1 \circ \omega_1 = \phi_2 \circ \omega_2$, then there is a *-homomorphism $\theta : C \to P$ such that $\omega_i = \psi_i \circ \theta$.

Any such P is obviously unique up to isomorphism commuting with the ψ_i.

One way of constructing P is as $\{(a_1, a_2) \mid \phi_1(a_1) = \phi_2(a_2)\} \subseteq A_1 \oplus A_2$.

DEFINITION 15.3.1. P is called the *pullback* of (A_1, A_2) along (ϕ_1, ϕ_2).

EXAMPLES 15.3.2. (a) Let $0 \to B \to E \to A \to 0$ be a short exact sequence of C^*-algebras. Form the Busby invariant $\tau : A \to Q(B)$. Then E is naturally isomorphic to the pullback of $(A, M(B))$ along (τ, π).

(b) Let $\phi : A \to B$ be a *-homomorphism. Let $\pi_0 : C_0([0,1), B) \to B$ be evaluation at 0. Then the pullback of $(A, C_0([0,1), B))$ along (ϕ, π_0) is called the *mapping cone* of ϕ, denoted C_ϕ.

15.3.2(a) shows how an extension can be recovered from its Busby invariant.

Mapping cones will be important in deriving exact sequences in KK-theory; general pullbacks will be considered again in the Mayer–Vietoris sequence in Section 21.

15.4. Equivalence

If we want to have a reasonable classification of the extensions of A by B, we need a suitable notion of equivalence. There are several obvious candidates. Throughout, we fix A and B, and consider two extensions

$$0 \longrightarrow B \xrightarrow{j_1} E_1 \xrightarrow{q_1} A \longrightarrow 0$$

and

$$0 \longrightarrow B \xrightarrow{j_2} E_2 \xrightarrow{q_2} A \longrightarrow 0$$

with associated Busby invariants τ_1 and τ_2.

(1) Strong isomorphism (called "strong equivalence" in [Busby 1968] and [Rosenberg 1982a]): there is a *-isomorphism γ making the following diagram commute:

$$
\begin{array}{ccccccccc}
0 & \longrightarrow & B & \xrightarrow{\ j_1\ } & E_1 & \xrightarrow{\ q_1\ } & A & \longrightarrow & 0 \\
 & & \Big\| & & \Big\downarrow{\gamma} & & \Big\| & & \\
0 & \longrightarrow & B & \xrightarrow{\ j_2\ } & E_2 & \xrightarrow{\ q_2\ } & A & \longrightarrow & 0
\end{array}
$$

(2) Weak isomorphism (called "weak equivalence" in [Busby 1968] and [Rosenberg 1982a]): there are *-isomorphisms α, β, γ making the following diagram commute:

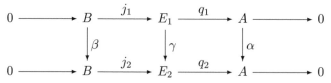

(3) Strong (unitary) equivalence: there is a unitary $u \in M(B)$ such that $\tau_2(a) = \pi(u)\tau_1(a)\pi(u)^*$ for all $a \in A$.

(4) Weak (unitary) equivalence: there is a unitary $v \in Q(B)$ such that $\tau_2(a) = v\tau_1(a)v^*$ for all $a \in A$.

(5) Homotopy equivalence: the homomorphisms $\tau_i : A \to Q(B)$ are homotopic.

(6) Stable equivalence: will be discussed in 15.6.3.

One could also consider a very weak equivalence relation regarding the two extensions as equivalent if E_1 and E_2 are isomorphic (or just homotopy equivalent) without regard to how the isomorphism respects A and B. Such an equivalence, while of interest in some situations, is rather unnatural from the point of view of studying extensions.

It follows from 15.3.1(a) and the uniqueness of pullbacks that two extensions are strongly isomorphic if and only if their Busby invariants coincide. Thus the Busby invariant exactly determines the strong isomorphism class of an extension.

It is trivial to check that strong isomorphism $\Longrightarrow$ strong equivalence $\Longrightarrow$ weak equivalence; if the unitary group of $M(B)$ is connected (in particular, if B is a σ-unital stable algebra (12.2.1)), then strong equivalence $\Longrightarrow$ homotopy equivalence.

EXAMPLES 15.4.1. (a) Again returning to the situation of 15.1.2, all four extensions are distinct under all the equivalence relations except weak isomorphism. The extensions corresponding to $C_0((0,1])$ and $C_0([0,1))$ are weakly isomorphic.

(b) We examine extensions of $\mathbb{M}_n$ by $\mathbb{K}$. Any homomorphism τ from $\mathbb{M}_n$ to Q can be lifted to $\mathbb{B}$ as follows. Let $\{\bar{e}_{ij}\}$ be the matrix units in $\tau(\mathbb{M}_n)$. First lift the $\{\bar{e}_{ii}\}$ to orthogonal projections p_{ii} in $\mathbb{B}$. Let $w \in \mathbb{B}$ with $\pi(w) = \bar{e}_{1i}$, and let e_{1i} be the partial isometry in the polar decomposition of $p_{11}wp_{ii}$. Then e_{1i} is a partial isometry in $\mathbb{B}$ with $\pi(e_{1i}) = \bar{e}_{1i}$, $e_{1i}^*e_{1i} \le p_{ii}$, $e_{1i}e_{1i}^* \le p_{11}$. If we set $e_{ij} = e_{1i}^*e_{1j}$, then $\{e_{ij}\}$ form a lifted set of matrix units.

Note that even if $\sum \bar{e}_{ii} = 1_Q$, we cannot in general arrange that $p = \sum e_{ii} = 1_{\mathbb{B}}$, i.e. τ will not generally have a unital lifting. p is a projection of finite codimension k; by adding on a homomorphism from $\mathbb{M}_n$ to $(1-p)\mathbb{B}(1-p) \cong \mathbb{M}_k$, we may arrange that $0 \le k < n$. This number k is well defined by the extension and is called the *defect* of τ.

Now let τ_1 and τ_2 be extensions; we want to determine when they are strongly or weakly equivalent. Let $\{e_{ij}^m\}$ be a lifted set of matrix units for τ_m, and let

u_1 be a partial isometry with $u_1^* u_1 = e_{11}^1$ and $u_1 u_1^* = e_{11}^2$. Set $u_i = e_{i1}^2 u_1 e_{1i}^1$, $u = \sum u_i$. Then u is a partial isometry, and $\pi(u)$ conjugates τ_1 to τ_2.

If τ_1 and τ_2 are both nonunital, then u can be enlarged to a unitary, so τ_1 and τ_2 are strongly equivalent. If τ_1 and τ_2 are both unital, then u is a Fredholm operator whose index is (defect τ_1)−(defect τ_2). $\pi(u)$ is unitary, so τ_1 and τ_2 are weakly equivalent, but if the defects are different they cannot be strongly equivalent. If one extension is unital and the other nonunital, then they obviously cannot be weakly equivalent.

So we have shown that there are exactly 3 weak equivalence classes and $n + 2$ strong equivalence classes of extensions of $\mathbb{M}_n$ by $\mathbb{K}$:

(1) The nonessential extension $\mathbb{K} \oplus \mathbb{M}_n (\tau = 0)$.
(2) The essential nonunital extensions (τ injective and nonunital).
(3) The essential unital extensions of defect k ($0 \le k < n$).

Strong and weak isomorphism coincide with strong equivalence in this case.

It turns out that strong and weak isomorphism are not very tractable equivalence relations on extensions in general; the other relations are much more amenable to analysis by methods of noncommutative topology.

Instead of considering all extensions, one could restrict to essential extensions (ones with injective Busby invariant) and/or, if A is unital, one could restrict to unital extensions (ones in which τ is unital). In special cases, one might restrict attention to extensions which were of some other special form (cf. 15.12). The same equivalence relations make sense for such restricted classes.

15.5. Trivial Extensions

We say that an extension is *trivial* if the Busby invariant $\tau : A \to Q(B)$ lifts to a *-homomorphism from A to $M(B)$. This will be the case if and only if the associated exact sequence *splits*, i.e. if there is a cross-section *-homomorphism $s : A \to E$ such that $q \circ s = 1_A$.

EXAMPLES 15.5.1. (a) When $A = \mathbb{C}$, $B = C_0((0,1))$ (15.1.2), the extensions corresponding to $C_0((0,1)) \oplus \mathbb{C}$ and $C(S^1)$ are trivial, while the other two are nontrivial.
(b) All extensions of $\mathbb{M}_n$ by $\mathbb{K}$ are trivial.

In general, there are many trivial extensions. For example, one can always form the "most trivial extension" $A \oplus B$; this is the extension with Busby invariant 0. If B is unital, then this is the only extension, since $Q(B) = 0$. However, if B is nonunital, $M(B)$ is large, and one would expect many other extensions. If B is stable, then for any separable A there is an injective *-homomorphism from A into $M(B)$ whose image does not intersect B; hence there are essential trivial extensions in this case. If A is separable and unital, and B is stable, one can construct both unital and nonunital essential trivial extensions in this way.

It is remarkable that, if $B = \mathbb{K}$ and A is separable, up to strong equivalence the variations described above are the only possible ones: by a theorem of Voiculescu (15.12.3), two essential trivial extensions are strongly equivalent if and only if they are either both unital or both nonunital.

The example of $\mathbb{M}_n$ by $\mathbb{K}$ shows that we must be careful in stating what is meant by a unital trivial extension. We will say that a trivial extension τ is *strongly unital* if τ lifts to a unital *-homomorphism from A to $M(B)$. Note that if there is a unital *-homomorphism from A to $\mathbb{C}$ (in particular, if A is commutative), then any unital trivial extension is strongly unital.

The term "trivial" is used for these extensions since we will need to regard them as trivial elements for the additive structure described below, in order to obtain a reasonable algebraic structure (a group in good cases).

15.6. Additive Structure

In this paragraph, we assume B is stable. Then for any A there is an additive structure on the set of strong (or weak) equivalence classes of extensions of A by B. Fix an isomorphism of $\mathbb{K}$ with $M_2(\mathbb{K})$; this isomorphism induces an isomorphism $B \cong M_2(B)$ and hence isomorphisms $M(B) \cong M_2(M(B))$, $Q(B) \cong M_2(Q(B))$. These isomorphisms are called *standard isomorphisms* and are uniquely determined up to unitary equivalence.

DEFINITION 15.6.1. If τ_1, τ_2 are extensions of A by B, then the *sum* $\tau_1 \oplus \tau_2$ is the extension whose Busby invariant is $\tau_1 \oplus \tau_2 : A \to Q(B) \oplus Q(B) \subseteq M_2(Q(B)) \cong Q(B)$, where the last isomorphism is a standard isomorphism.

We have cheated slightly here: the sum of two extensions is well defined only up to strong equivalence. Thus we should really define the sum on strong equivalence classes, giving a binary operation on the set of strong equivalence classes.

PROPOSITION 15.6.2. *Addition is a well defined binary operation on the set* **Ext**(A, B) *of strong equivalence classes of extensions of A by B, and is associative and commutative. So* **Ext**(A, B) *is a commutative semigroup. The following form subsemigroups of* **Ext**(A, B):

(a) *the classes of trivial extensions;*
(b) *the classes of essential extensions;*
(c) *if A is unital, the unital extensions;*
(d) *if A is unital, the strongly unital trivial extensions.*

(*Some of these subsemigroups may be empty in general.*) *Addition also respects weak equivalence and homotopy equivalence; so there are semigroups* **Ext**$_w(A, B)$, **Ext**$_h(A, B)$ *which are quotients of* **Ext**(A, B) (*sometimes called* **Ext**$_s(A, B)$).

The simple proof is left to the reader.

DEFINITION 15.6.3. We define $Ext(A, B) = Ext_s(A, B)$ as the quotient of **Ext**(A, B) by the subsemigroup of trivial extensions. Similarly, $Ext_w(A, B)$ and

$Ext_h(A, B)$ are defined as the quotients of $\mathbf{Ext}_w(A, B)$ and $\mathbf{Ext}_h(A, B)$ by the subsemigroup of trivial extensions. For $* = s, w, h$, we define $Ext_*^e(A, B)$ as the quotient of the subsemigroup of essential extensions by the subsemigroup of essential trivial extensions. If A is unital, we define $Ext_*^u(A, B)$, as the quotient of the subsemigroup of unital extensions by that of strongly unital trivial extensions, and $Ext_*^{eu}(A, B)$ as the quotient of the subsemigroup of essential unital extensions by that of essential strongly unital trivial extensions. (Some of these quotient semigroups may not be defined if the denominator semigroup is empty.)

Each of these semigroups has an identity (the class of trivial extensions). $Ext_w(A, B)$ is a quotient of $Ext_s(A, B)$ and $Ext_h(A, B)$ is a quotient of $Ext_w(A, B)$. For $* = e, u, eu$, $Ext_w^*(A, B)$ is a quotient of $Ext_s^*(A, B)$ and $Ext_h^*(A, B)$ is a quotient of $Ext_w^*(A, B)$. For $* = s, w, h$, there is a homomorphism from $Ext_*^u(A, B)$ to $Ext_*(A, B)$ and one from $Ext_*^{eu}(A, B)$ to $Ext_*^e(A, B)$.

The equivalence relation on extensions that they represent the same element of $Ext_s(A, B)$ [resp. Ext_w, Ext_h] is called *stable strong* [resp. *stable weak, stable homotopy*] *equivalence*.

We will show presently that most of these semigroups coincide under mild hypotheses. The only difference in most cases involves unital extensions.

PROPOSITION 15.6.4. *We always have* $Ext(A, B) = Ext_w(A, B)$.

PROOF. If τ_1 and τ_2 are weakly equivalent via a unitary $v \in Q(B)$, then $\tau_1 \oplus 0$ and $\tau_2 \oplus 0$ are weakly equivalent via $\mathrm{diag}(v, v^{-1})$. $\mathrm{diag}(v, v^{-1})$ lifts to a unitary in $M_2(M(B))$ by 3.4.1.   □

Note that $Ext_s^u(A, B)$ and $Ext_w^u(A, B)$ do *not* coincide in general: see 15.6.6(a).

PROPOSITION 15.6.5. *If there exists an essential trivial extension of A by B, then* $Ext_*^e(A, B) = Ext_*(A, B)$ *for* $* = s, w, h$. *If A is unital, and there is an essential unital trivial extension of A by B, then the same is true for* Ext_*^u *and* Ext_*^{eu}. *In particular, if A is separable, every stable equivalence class is represented by an essential extension (unital if A is unital).*

EXAMPLES 15.6.6. (a) Consider extensions of $\mathbb{M}_n$ by $\mathbb{K}$. From 15.5.1(b) we have that all extensions are trivial, so $Ext(\mathbb{M}_n, \mathbb{K}) = 0$. However, we have that $Ext_s^u(\mathbb{M}_n, \mathbb{K}) \cong \mathbb{Z}_n$ (at least as a set, and it is easy to check that the group operation is the same). $Ext_w^u(\mathbb{M}_n, \mathbb{K})$ is trivial.

(b) Consider extensions of $C(S^1)$ by $\mathbb{K}$. By 15.6.5 we need only consider essential extensions. It is also only necessary to consider unital extensions: if $\tau : C(S^1) \to Q$ is an extension, set $p = \tau(1)$, and let P be a projection in $\mathbb{B}$ with $\pi(P) = p$. P is an infinite-rank projection (unless $\tau = 0$), so if τ' is the restriction of τ to $P\mathbb{H}$, τ is strongly equivalent to $\tau' \oplus 0$.

A unital extension of $C(S^1)$ by $\mathbb{K}$ is just a choice of a unitary in Q. It is clear that $Ext_h(C(S^1), \mathbb{K}) = Ext_h^u(C(S^1), \mathbb{K}) \cong \mathbb{Z}$ via Fredholm index. With a bit of work, we can show the same for $Ext(C(S^1), \mathbb{K})$ and $Ext_s^u(C(S^1), \mathbb{K})$: if

v is a unitary in Q, then v can be written as $\pi(V)$, where V is an isometry or coisometry in $\mathbb{B}$. By the von Neumann–Wold decomposition [Halmos 1967, Problem 118] V is unitarily equivalent to $U^n \oplus W$ (if V is an isometry) or $U^{*n} \oplus W$ (if V is a coisometry), where U is the unilateral shift and W is unitary. Thus any extension is stably strongly equivalent to one coming from powers of the unilateral shift; these are obviously classified by Fredholm index.

(c) Let $A = Q$, $B = \mathbb{K}$. There is one nonessential extension and a myriad of essential extensions in this case. The only trivial extension is the nonessential extension, so the semigroups $Ext_*^e(Q, \mathbb{K})$, $Ext_*^u(Q, \mathbb{K})$, $Ext_*^{eu}(Q, \mathbb{K})$ are not even defined. Calculating $Ext(Q, \mathbb{K})$ and $Ext_h(Q, \mathbb{K})$ amounts to classifying unital endomorphisms of Q up to unitary equivalence or homotopy, a problem which appears hopeless at the moment (it is unknown, for example, whether or not Q has outer automorphisms). The semigroup $Ext(Q, \mathbb{K})$ is obviously very bad, however; it fails to have any invertible elements except the identity, and it may very well fail to have cancellation.

This example shows the pathology which can occur if we allow A to be non-separable. To obtain a reasonable theory, we will for the most part only consider extensions where A is separable.

If B is not stable, we can define $Ext(A, B)$, etc., to be $Ext(A, B \otimes \mathbb{K})$. It is easy to check that if B is already stable, this definition of $Ext(A, B)$ agrees with the previous one. If A is separable, the other groups also agree.

It is possible to define an analogous group structure on the set of extensions of A by B even if B is not stable, as long as $B \cong M_2(B)$; this idea, with $B = O_2$, plays a crucial role in Kirchberg's classification of purely infinite C^*-algebras [Kirchberg 1998].

15.7. Inverses

Not much is known in general about the semigroup $Ext(A, B)$. It can be very pathological as shown in 15.6.6(c). Even if A is separable, it is not known in general whether the semigroup has cancellation.

We can, however, give a nice description of the invertible classes. In good cases (e.g. if A is separable and nuclear) we will show that every class is invertible, i.e. $Ext(A, B)$ is an abelian group.

The key observation is as follows. Suppose τ is an invertible extension, i.e. there is an extension τ^{-1} such that $\tau \oplus \tau^{-1}$ is trivial. Then $\tau \oplus \tau^{-1} : A \to M_2(Q^s(B))$ lifts to a *-homomorphism

$$\phi = \begin{bmatrix} \phi_{11} & \phi_{12} \\ \phi_{21} & \phi_{22} \end{bmatrix} : A \to M_2(M^s(B))$$

ϕ_{11} (and also ϕ_{22}), being the compression of a *-homomorphism, must be a completely positive contraction from A to $M^s(B)$, and $\pi \circ \phi_{11} = \tau$. So if τ is invertible, then τ has a completely positive lifting to $M^s(B)$.

The converse is true by the Generalized Stinespring Theorem (13.7.1) if A is separable: if τ has a completely positive contractive lifting ϕ_{11} to $M^s(B)$, then ϕ_{11} can be dilated to a homomorphism

$$\phi = (\phi_{ij}) : A \to M_2(M^s(B)),$$

which may be chosen unital if A is unital. $\pi \circ \phi_{11} = \tau$ is a homomorphism, so (it is easily checked) $\pi \circ \phi_{22}$ is also a homomorphism from A to $Q^s(B)$. $\pi \circ \phi_{22}$ is an inverse for τ. Thus we have proved

THEOREM 15.7.1. *If A is separable, an extension $\tau : A \to Q^s(B)$ defines an invertible element of $Ext(A, B)$ if and only if τ lifts to a completely positive contraction from A to $M^s(B)$. If A is unital and τ is unital, then τ defines an invertible element of $Ext_s^u(A, B)$ if and only if τ lifts to a (not necessarily unital) completely positive contraction from A to $M^s(B)$. (Such an extension is called semisplit.)*

Actually, the requirement that the completely positive lifting be a contraction is not necessary: if there exists a completely positive lifting, there exists a completely positive contractive lifting [Cuntz and Skandalis 1986, 2.5].

There is a separable C^*-algebra A such that $Ext(A, \mathbb{C})$ is not a group [Anderson 1978].

15.7.2. Let A be separable. If we consider the group $Ext(A, B)^{-1}$ of invertible elements of $Ext(A, B)$, we may express the elements as pairs (ϕ, P), where ϕ is a *-homomorphism from A to $M^s(B)$ and P is a projection in $M^s(B)$ which commutes with $\phi(A)$ mod $B \otimes \mathbb{K}$. A pair is trivial if and only if P actually commutes with $\phi(A)$. Since ϕ and P are only determined up to "compact perturbation" (modulo $B \otimes \mathbb{K}$), we must regard two pairs which agree mod $B \otimes \mathbb{K}$ as identical. Strong equivalence corresponds to unitary equivalence of pairs, and sum to direct sum of pairs. Thus the group $Ext(A, B)$ is isomorphic to the quotient of the semigroup of equivalence classes of such pairs, under the equivalence relation generated by unitary equivalence and "compact perturbation", with direct sum, modulo the subsemigroup of classes of exact (trivial) pairs.

15.8. Nuclear C^*-Algebras

We digress to give a very brief survey of nuclear C^*-algebras. For more details, and proofs of the results, see [Lance 1982] and the references therein.

Nuclear C^*-algebras are a very important class of C^*-algebras, large enough to include most C^*-algebras which arise "naturally"; yet nuclear C^*-algebras have nice structural properties which are of great technical use in applications. The following two theorems, which are an amalgamation of results of Choi, Connes, Effros, Haagerup, and Lance, summarize the most important properties of the class of nuclear C^*-algebras.

THEOREM 15.8.1. *Let A be a C^*-algebra. The following conditions are equivalent*:

(1) *For every C^*-algebra B, the algebraic tensor product $A \odot B$ has a unique C^*-cross norm.*

(2) *The identity map from A to A can be approximated pointwise in norm by completely positive finite-rank contractions.*

(3) *A^{**} is an injective ("hyperfinite") von Neumann algebra.*

(4) *A is C^*-amenable: every derivation from A into a dual normal Banach A-module is inner.*

A C^*-algebra satisfying these conditions is called nuclear.

The term "nuclear" comes from condition (1) in analogy with the use of the same term in topological vector space theory. Condition (2) is an analog of (although not the same as) the metric approximation property for Banach spaces. Condition (3) says that $\pi(A)''$ is injective for any representation π of A; since the injective von Neumann algebras are the "next step up" from the type I von Neumann algebras in terms of complexity, the class of nuclear C^*-algebras is one step up from the type I C^*-algebras. Condition (4) is an analogy with amenable groups, and says that nuclear C^*-algebras are "cohomologically trivial".

THEOREM 15.8.2. *The class of nuclear C^*-algebras contains all type I C^*-algebras (in particular, all commutative or finite-dimensional C^*-algebras and $\mathbb{K}$), and is closed under stable isomorphism, quotients, extensions, inductive limits, tensor products, and crossed products by amenable groups. So all inductive limits of type I C^*-algebras (in particular, all AF algebras) are nuclear. If G is a locally compact group which is amenable or connected, then $C^*(G)$ is nuclear. (Conversely, if G is discrete and $C_r^*(G)$ is nuclear, then G is amenable.)*

It is not known whether the class of nuclear C^*-algebras is the smallest class of C^*-algebras closed under the operations described in the theorem. See 22.3.4-22.3.5 for more comments on this matter.

It is not true that a C^*-subalgebra of a nuclear C^*-algebra is always nuclear [Choi 1979]. In fact, every non-type-I C^*-algebra contains a nonnuclear C^*-subalgebra [Blackadar 1985a].

For our immediate purposes, the following lifting theorem is the main feature of nuclear C^*-algebras. This theorem is due to Choi and Effros [1976], with an important special case due to Arveson [1977].

THEOREM 15.8.3. *Let A be a nuclear C^*-algebra, D any C^*-algebra, J a (closed two-sided) ideal of D, $\pi : D \to D/J$ the quotient map. Let $\psi : A \to D/J$ be a completely positive contraction (e.g. a *-homomorphism). Then there is a completely positive contraction $\phi : A \to D$ with $\psi = \pi \circ \phi$.*

COROLLARY 15.8.4. *If A is a separable nuclear C^*-algebra and B is any C^*-algebra, then $Ext(A, B)$ is a group. If A is unital, then $Ext_s^u(A, B)$ is also a group.*

Effros and Haagerup [1985] have characterized those C^*-algebras A for which $Ext(A, \mathbb{C})$ is a group; the class is strictly larger than the class of nuclear C^*-algebras, but does not (yet) have a good intrinsic definition.

Theorem 15.8.3 remains true if we assume D is nuclear and A is arbitrary. However, this is irrelevant for our purposes since $M^s(B)$ is never nuclear, even if B is. (Even $\mathbb{B}$ is nonnuclear.)

15.9. Functoriality

It is obvious that Ext is contravariantly functorial in the first variable, i.e. if $f : A_1 \to A_2$ is a *-homomorphism, then there is a semigroup homomorphism $f^* : \mathbf{Ext}(A_2, B) \to \mathbf{Ext}(A_1, B)$, defined by $f^*[\tau] = [\tau \circ f]$. This homomorphism drops to a homomorphism, also denoted f^*, from $Ext_*(A_2, B)$ to $Ext_*(A_1, B)$ (where $* = s, h$); if f is injective, the same is true for Ext^e_*; if the A_i are unital and f is unital, the same holds for Ext^u_*. Thus, for fixed B, $Ext(\,\cdot\,, B)$, etc., is a contravariant functor from C^*-algebras to abelian semigroups, and is a contravariant functor from separable nuclear C^*-algebras to abelian groups.

Functoriality in B is a bit harder. If $g : B_1 \to B_2$ is a *-homomorphism, and B_1 is σ-unital, then $\mathbb{H}_{B_1} \otimes_g B_2$ is a countably generated Hilbert B_2-module, and is therefore a direct summand of $\mathbb{H}_{B_2}$. The map g thus induces a *-homomorphism from $M^s(B_1)$ to $M^s(B_2)$ which sends $B_1 \otimes \mathbb{K}$ into $B_2 \otimes \mathbb{K}$, and hence a *-homomorphism $\tilde{g}$ from $Q^s(B_1)$ to $Q^s(B_2)$. If $[\tau] \in \mathbf{Ext}(A, B_1)$, set $g_*[\tau] = \tilde{g} \circ \tau$. It is not difficult to see that the class of $\tilde{g} \circ \tau$ in $Ext(A, B_2)$ depends only on the class of τ in $Ext(A, B_1)$, so that g_* is well defined. g_* is obviously a homomorphism. Thus $Ext(A, \cdot)$ is a covariant functor from σ-unital C^*-algebras to abelian semigroups.

15.10. Homotopy Invariance

It has been a difficult problem to prove that $Ext(A, B)$ is homotopy invariant. One may ask about homotopy invariance in each variable of Ext, meaning that $f_0, f_1 : A_1 \to A_2$ [resp. $g_0, g_1 : B_1 \to B_2$] homotopic implies $f_0^* = f_1^*$, [resp. $g_{0*} = g_{1*}$].

Homotopy invariance has been proved for successively more general cases; the present state of knowledge, while still incomplete, is quite satisfactory.

Brown, Douglas, and Fillmore proved that $Ext(A, \mathbb{C})$ is homotopy invariant in A for A separable, commutative, and unital. O'Donovan [1977] and Salinas [1977] extended this result to the case where A is separable and quasidiagonal. Finally, Kasparov proved that $Ext(A, B)^{-1}$ is always homotopy invariant in both variables if A is separable and B is σ-unital. So $Ext(A, B)$ is at least homotopy invariant in both variables if A is separable nuclear and B is σ-unital.

We will not give any proofs of homotopy invariance here. Kasparov's proof of the most general result known uses KK-theory, and will be deferred until 18.5.4.

Homotopy invariance in the second variable can be phrased as follows:

COROLLARY 15.10.1. *Let A be separable and B σ-unital. Then the quotient map from $Ext(A, B)^{-1}$ to $Ext_h(A, B)^{-1}$ is an isomorphism. So if A is nuclear, the map from $Ext(A, B)$ to $Ext_h(A, B)$ is an isomorphism.*

15.11. Bott Periodicity, Exact Sequences

The story with Bott periodicity and exact sequences is much the same historically as homotopy invariance: the results were proved by BDF in the case of $Ext(C(X), \mathbb{C})$, for X a compact metric space, and extended by Pimsner, Popa, and Voiculescu [Pimsner et al. 1979; 1980] to $Ext(A, \mathbb{C})$ for A separable nuclear. The most general results known are due to Kasparov:

THEOREM 15.11.1 (BOTT PERIODICITY). *If A is separable and B is σ-unital, then $Ext(S^2 A, B)^{-1}$ and $Ext(A, S^2 B)^{-1}$ are naturally isomorphic to $Ext(A, B)^{-1}$.*

PROOF. See 19.2.2.   □

THEOREM 15.11.2. *If A is separable nuclear, J is an ideal in A, and B is σ-unital, then there is a natural six-term cyclic exact sequence*

$$
\begin{array}{ccccc}
Ext(A/J, B) & \xrightarrow{\ q^* \ } & Ext(A, B) & \xrightarrow{\ j^* \ } & Ext(J, B) \\
{\scriptstyle \partial}\big\uparrow & & & & \big\downarrow{\scriptstyle \partial} \\
Ext(SJ, B) & \xleftarrow{\ j^* \ } & Ext(SA, B) & \xleftarrow{\ q^* \ } & Ext(S(A/J), B)
\end{array}
$$

where $j : J \to A$ is the inclusion and $q : A \to A/J$ the quotient map. There is a similar exact sequence in B for fixed A.

PROOF. See 19.5.4.   □

Because of these results, we sometimes write $Ext_1(A, B) = Ext(A, B)$ and

$$Ext_0(A, B) = Ext(SA, B) \cong Ext(A, SB).$$

The reason for this choice of indexing will become clear when we consider K-homology (16.3) and KK-theory.

15.12. Absorbing Extensions

In good cases, it can be shown that the stabilization obtained by dividing out $\mathbf{Ext}(A, B)$ by the subsemigroup of trivial extensions is unnecessary or at least can be accomplished in a much cleaner manner.

DEFINITION 15.12.1. *Let A and B be C^*-algebras. An extension τ of A by B is absorbing if it is strongly equivalent to $\tau \oplus \sigma$, for any trivial extension σ. If A is unital, τ is unital-absorbing if it is strongly equivalent to $\tau \oplus \sigma$, for any strongly unital trivial extension σ.*

Note that if σ is an absorbing trivial extension, then $\tau \oplus \sigma$ is an absorbing extension for any τ. Also note that even if A is unital, an absorbing extension of A by B is necessarily nonunital.

PROPOSITION 15.12.2. *The classes of absorbing extensions form a (possibly empty) subsemigroup of* **Ext**(A, B); *the restriction of the quotient map*

$$\textbf{Ext}(A, B) \to Ext(A, B)$$

to the absorbing classes is injective. If there exists an absorbing trivial extension, then the map from the absorbing classes to $Ext(A, B)$ *is also surjective. If A is unital and there exists a unital-absorbing trivial extension, then the quotient map from the classes of unital-absorbing extensions to* $Ext_s^u(A, B)$ *is also an isomorphism.*

We have the following theorem of Voiculescu [1976]. See [Arveson 1977] for a good exposition and proof. The case $A = C(X)$ has a more elementary proof, which was done earlier by BDF; see [Douglas 1980, Theorems 4 and 5].

THEOREM 15.12.3. *If A is separable, every essential nonunital extension of A by $\mathbb{K}$ is absorbing. If A is separable unital, every essential unital extension of A by $\mathbb{K}$ is unital-absorbing.*

So if A is separable, the map from the subsemigroup of **Ext**$(A, \mathbb{C})$ *consisting of the essential nonunital extensions, to* $Ext(A, \mathbb{C})$, *is an isomorphism; if A is unital, the map from the subsemigroup of* **Ext**$(A, \mathbb{C})$ *consisting of the essential unital extensions, to* $Ext_s^u(A, \mathbb{C})$, *is an isomorphism.*

Kasparov generalized part of Voiculescu's theorem [Kasparov 1975, Theorem 6; 1980b, 1.16], obtaining a somewhat weaker result for more general B:

THEOREM 15.12.4. *Let A be a separable [resp. separable unital] C^*-algebra and B a σ-unital C^*-algebra. Suppose either A or B is nuclear. Let τ be an essential nonunital [resp. essential strongly unital] trivial extension of A by $\mathbb{K}$. Regard τ as an extension of A by $B \otimes \mathbb{K}$ through the identification $1 \otimes \mathbb{B} \subseteq M(B \otimes \mathbb{K})$. Then τ is absorbing [resp. unital-absorbing].*

It cannot be true in general that every essential nonunital extension of A by $B \otimes \mathbb{K}$ is absorbing; even an essential trivial extension is not necessarily absorbing. If D is a quotient of B, then there is an induced quotient map from $M^s(B)$ onto $M^s(D)$ [Pedersen 1979, 3.12.10], inducing a quotient map $\rho : Q^s(B) \to Q^s(D)$. If A is separable, a necessary condition that τ be absorbing is that $\rho_*(\tau)$ be an essential extension of A by $D \otimes \mathbb{K}$ for each such D. Such an extension is called *residually essential*.

EXAMPLE 15.12.5. Let $A = \mathbb{C}$, $B = C_0([0, 1))$. Define an extension of A by B by $\tau(1) = 1 \otimes p \in C([0, 1], \mathbb{K}) \subseteq M^s(B)$, where p is a one-dimensional projection in $\mathbb{K}$. Then τ is an essential nonunital trivial extension which is not residually essential. (Consider the quotient map of B to $D = \mathbb{C}$ sending f to $f(0)$.)

It is not known whether every residually essential nonunital extension is absorbing. Rosenberg and Schochet [1981] showed that if A and B are commutative, $B = C_0(Y)$ with Y^+ finite-dimensional, then every residually essential extension is absorbing. It would be interesting to have an intrinsic characterization of absorbing extensions.

Actually, both Voiculescu's and Kasparov's theorems contain additional results about approximating one extension by unitary conjugates of another. We have picked out the parts of the theorems which are most relevant for our purposes.

15.13. Extensions of Graded C^*-Algebras

If A and B are graded C^*-algebras, we may also define $Ext(A, B)$ in the same manner as before, only requiring all homomorphisms to be graded. Thus one wants to classify short exact sequences of graded C^*-algebras, the Busby invariant is a graded homomorphism from A to $Q(B)$ with its induced grading, etc. We will consider extensions of graded algebras in 17.6.5. Unlike the ungraded (trivially graded) case, however, there is not a good correspondence between graded Ext and KK, and it is unclear how far the theory of graded Ext can be taken in analogy with the ungraded theory.

15.14. Extensions and K-Theory

We know from 15.8.4 that $Ext(\mathbb{C}, B)$ and $Ext(C_0(\mathbb{R}), B)$ are groups for any B. Actually these groups are nothing but the K-groups of B.

The correspondence is easily seen using 12.2.3. An extension of $\mathbb{C}$ by $B \otimes \mathbb{K}$ is exactly a choice of a projection in $Q^s(B)$. A trivial extension is a projection in $Q^s(B)$ which lifts to a projection in $M^s(B)$; these are exactly the projections in $Q^s(B)$ whose class in $K_0(Q^s(B))$ is 0. In both $Ext(\mathbb{C}, B)$ and $K_0(Q^s(B))$ the group operation is direct sum, and the equivalence relation is unitary equivalence and addition of degenerate elements. Thus $Ext(\mathbb{C}, B)$ is naturally isomorphic to $K_0(Q^s(B))$, which by 12.2.3 is naturally isomorphic to $K_1(B)$ via the connecting map in the long exact sequence.

It is harder to see that $Ext(C_0(\mathbb{R}), B) \cong K_1(Q^s(B)) \cong K_0(B)$. It is clear that $Ext_h(C_0(\mathbb{R}), B) \cong K_1(Q^s(B))$, since extensions of $C_0(\mathbb{R})$ by B correspond naturally to unitaries in $Q^s(B)$. There does not seem to be any good way to show that $Ext(C_0(\mathbb{R}), B) \cong K_0(B)$ except by invoking 15.10.1.

EXAMPLE 15.14.1. We examine $Ext(\mathbb{C}, C_0(\mathbb{R}))$. We know from above that $Ext(\mathbb{C}, C_0(\mathbb{R})) \cong K^{-1}(\mathbb{R}) \cong \mathbb{Z}$. The exact sequence

$$0 \to C_0(\mathbb{R}) \to C_0(\mathbb{R} \cup \{+\infty\}) \to \mathbb{C} \to 0$$

defines a generator. If we wish to describe this extension in standard form via its Busby invariant τ, let f be a continuous function on $\mathbb{R}$ with $\lim_{t \to +\infty} f(t) = 1$ and $\lim_{t \to -\infty} f(t) = 0$, let p be a one-dimensional projection in $\mathbb{K}$. Regard $f \otimes p$ as an element of $M(C_0(\mathbb{R})) \otimes \mathbb{K} \subseteq M(C_0(\mathbb{R}) \otimes \mathbb{K}) = M^s(C_0(\mathbb{R}))$, and define $\tau(z) = \pi(zf \otimes p)$, where $\pi : M^s(C_0(\mathbb{R})) \to Q^s(C_0(\mathbb{R}))$ is the quotient map.

The inverse of this extension is the extension defined by

$$0 \to C_0(\mathbb{R}) \to C_0(\{-\infty\} \cup \mathbb{R}) \to \mathbb{C} \to 0.$$

As a related matter, we examine the homomorphism from $Ext_w^u(A, B)$ to $Ext(A, B)$ (in the case where A is unital). This map cannot be surjective in general, since $Ext_w^u(\mathbb{C}, B)$ (and also $Ext_s^u(\mathbb{C}, B)$) is always trivial while $Ext(\mathbb{C}, B) \cong K_1(B)$ may not be. $K_1(B)$ is the only obstruction, however:

PROPOSITION 15.14.2. *Let A be a separable unital C^*-algebra, and B a C^*-algebra with $K_1(B) = 0$. Then the map $Ext_w^u(A, B) \to Ext(A, B)$ is an isomorphism.*

PROOF. The proof is a jazzed-up version of the argument of 15.6.6(b). The simple argument is left to the reader.   □

16. Brown–Douglas–Fillmore Theory and Other Applications

16.1. We have seen that if A is separable and nuclear and B is σ-unital, then the multitude of semigroups of the previous section are reduced to at most three: $Ext(A, B)$ and (if A is unital) $Ext_s^u(A, B)$ and $Ext_w^u(A, B)$. If A is commutative then even these last two coincide. Furthermore, these semigroups are abelian groups.

Brown, Douglas, and Fillmore [Brown et al. 1973; 1977a] made the first careful study of these groups in the case $B = \mathbb{C}$; they were almost exclusively interested in the case $A = C(X)$ for a compact metrizable space X.

DEFINITION 16.1.1. If A is a C^*-algebra, $Ext(A) = Ext(A, \mathbb{C})$; if A is unital, $Ext_s(A) = Ext_s^u(A, \mathbb{C})$. (Sometimes $Ext(A)$ is called $Ext_w(A)$.) If X is a locally compact Hausdorff space, then $Ext(X) = Ext(C_0(X))$; if X is compact, then $Ext_s(X) = Ext_s(C(X))$.

Note that by 15.14.2 $Ext(A) \cong Ext_w^u(A)$ if A is separable and unital; if A is separable, unital, and commutative, then all the groups coincide.

As described in Section 15, BDF proved that $Ext(X)$ is a homotopy-invariant *covariant* functor on the category of compact metrizable spaces, and proved Bott Periodicity and the six-term cyclic exact sequence, along with the special case of Voiculescu's theorem on absorbing extensions. The main theoretical consequence of these general results is that $Ext(X) \cong K_1(X)$, the first K-homology group of X. Other important consequences include the classification of essentially normal operators and the structure of some naturally occurring C^*-algebras.

16.2. Essentially Normal Operators

We begin with the simplest case of extensions. An operator $T \in \mathbb{B}(\mathbb{H})$ is *essentially normal* if $T^*T - TT^* \in \mathbb{K}$. In other words, if $t = \pi(T)$ is the image in the Calkin algebra, then T is essentially normal if t is normal. The basic question studied by BDF was: under what conditions could T be written as $N+K$, where N is normal and K compact? More generally, given two essentially normal operators T_1 and T_2, under what conditions is T_1 unitarily equivalent to a compact perturbation of T_2?

For the second question, one obvious necessary condition is that T_1 and T_2 have the same essential spectrum (the spectrum of the image in the Calkin algebra). So the question may be rephrased: given a compact subset X of $\mathbb{C}$, denote by $EN(X)$ the set of essentially normal operators with essential spectrum X. Given $T_1, T_2 \in EN(X)$, under what conditions is T_1 unitarily equivalent to a compact perturbation of T_2?

The problem is translated into an extension problem by noting that if $T \in EN(X)$, then $C^*(t, 1) \cong C(X)$; so if we set $\mathbb{A}(T) = C^*(T, \mathbb{K}, 1)$, $\mathbb{A}(T)$ corresponds naturally to an extension of $C(X)$ by $\mathbb{K}$. The question of whether T_1 is unitarily equivalent to a compact perturbation of T_2 is exactly the question of whether the corresponding extensions are strongly equivalent, i.e. whether they represent the same element of $Ext(X)$.

The main theorem in the classification of essentially normal operators is the following:

THEOREM 16.2.1. *If* $X \subseteq \mathbb{C}$, $Ext(X) \cong [\mathbb{C} \setminus X, \mathbb{Z}]$, *the group of homotopy classes of continuous functions of compact support from* $\mathbb{C} \setminus X$ *to* $\mathbb{Z}$. *Thus* $Ext(X) \cong \prod \mathbb{Z}$, *with one factor for each bounded component of* $\mathbb{C} \setminus X$. *The isomorphism sends the class of* $T \in EN(X)$ *to* $\prod \text{Index}(T - \lambda_n 1)$, *where* λ_n *is in the n-th bounded component of* $\mathbb{C} \setminus X$. (*This* Index *is constant on connected components of* $\mathbb{C} \setminus X$ *and vanishes on the unbounded component.*)

This is actually a special case of the Universal Coefficient Theorem (16.3.3, 23.1.1): if X is a compact subset of $\mathbb{C}$, then $K^1(X) \cong \pi^1(X) = [X, S^1] \cong \oplus \mathbb{Z}$ with one summand for each bounded component of $\mathbb{C} \setminus X$; and $K^0(X)$ is torsion-free.

COROLLARY 16.2.2. *An essentially normal operator* T *can be written* $T = N + K$, *with* N *normal and* K *compact, if and only if* $\text{Index}(T - \lambda 1) = 0$ *for all* λ *not in* $\sigma_e(T)$. *If* $T_1, T_2 \in EN(X)$, *then* T_1 *is unitarily equivalent to a compact perturbation of* T_2 *if and only if* $\text{Index}(T_1 - \lambda 1) = \text{Index}(T_2 - \lambda 1)$ *for all* $\lambda \notin X$.

COROLLARY 16.2.3. *If* X *is a compact subset of* $\mathbb{C}$ *with connected complement, then any essentially normal operator with essential spectrum* X *can be written as* (*normal*) + (*compact*), *and any two essentially normal operators with essential spectrum* X *are unitarily equivalent up to compact perturbation. In particular, any* $T \in EN(X)$ *is a compact perturbation of a diagonalizable normal operator* (*one with an orthonormal basis of eigenvectors*).

In the case $X \subseteq \mathbb{R}$ (the case of essentially self-adjoint operators), 16.2.3 applies and is known as the Weyl-von Neumann Theorem. Berg later showed that any two normal operators with the same essential spectrum are unitarily equivalent up to compact perturbation; Voiculescu's Theorem may be regarded as a generalization of this fact.

EXAMPLE 16.2.4. Let X be the "Hawaiian earring" formed as the union of all the circles of radius $1/n$ centered at $(1/n, 0)$. Then $Ext(X)$ is the full direct product of a countable number of copies of Z; thus $Ext(X)$ is uncountable. To obtain an explicit essentially normal operator corresponding to the sequence $(s_n) \in \prod_{\mathbb{N}} Z$, take $\bigoplus(\frac{1}{n} + \frac{1}{n}U^{d_n})$, where $d_n = s_n - s_{n-1}$, U is the unilateral shift, $U^k = U^{*|k|}$ for $k < 0$, and $U^0 = V$, the bilateral shift.

16.3. *Ext* as *K*-Homology

Complex K-theory is an extraordinary cohomology theory on compact Hausdorff spaces, and it has been of great interest to find concrete realizations of the corresponding homology theory, called K-homology. If X is a finite complex, $K_*(X)$ can be defined by Spanier–Whitehead duality by embedding X as a subset of a sphere of large odd dimension and taking the K-theory of the complement. This definition is very clumsy to use in practice.

Atiyah [1968] proposed an operator-theoretic definition by generalizing the notion of an elliptic operator.

DEFINITION 16.3.1. If X is a compact space, $\mathrm{Ell}(X)$ is the set of triples (σ_0, σ_1, T), where σ_i is a representation of $C(X)$ on a Hilbert space $\mathbb{H}_i$ and T is a Fredholm operator from $\mathbb{H}_0$ to $\mathbb{H}_1$ with $T\sigma_0(f) - \sigma_1(f)T$ compact for all $f \in C(X)$.

The standard example is where T is an elliptic pseudodifferential operator of degree 0 between two smooth vector bundles E_0 and E_1 over X. If E_i is given a smooth Hermitian structure, T defines a Fredholm operator between the L^2-spaces which almost intertwines the action of $C(X)$ by multiplication.

There is a binary operation on $\mathrm{Ell}(X)$ given by orthogonal direct sum.

Atiyah defined a map from $\mathrm{Ell}(X)$ to $K_0(X)$ using a slant product, and showed that in the case of an actual pseudodifferential operator, the image in $K_0(X)$ (which can be called the "analytic index" of the operator) coincides with the "topological index" defined by means of the Chern character. He was able to conclude that the map from $\mathrm{Ell}(X)$ to $K_0(X)$ is surjective if X is a finite complex.

Atiyah was not able to give a description of the proper equivalence relation on $\mathrm{Ell}(X)$ to make the quotient equal $K_0(X)$. A byproduct of the BDF work was to give a characterization. Actually, there are several ways of describing the equivalence relation; one way is that it is the equivalence relation on triples generated by unitary equivalence, orthogonal direct sum with a degenerate triple (where T is invertible and actually intertwines σ_0 and σ_1), and homotopy of T keeping the σ_i fixed. We will study this situation in great detail when we develop KK-theory.

Once we have K_0, we can define the higher K-homology groups by suspension.

16.3.2. If A is any C^*-algebra, there is a pairing of $Ext(A)$ and $K_1(A)$ which generalizes the pairing of 16.2.1 and which is of fundamental importance. An extension τ of A by $\mathbb{K}$ extends uniquely to a unital *-homomorphism from A^+ to Q, and hence a unital homomorphism $\tilde{\tau}$ from $M_n(A^+)$ to $M_n(Q) \cong Q$. If u is a unitary in $M_n(A^+)$, define $\langle \tau, u \rangle = \mathrm{Index}(\tilde{\tau}(u)) \in \mathbb{Z}$. It is easily seen that $\langle \tau, u \rangle$ depends only on the class of τ in $Ext(A)$ and the class of u in $K_1(A)$. Thus each $\tau \in Ext(A)$ defines a homomorphism $\gamma_\infty(\tau)$ from $K_1(A)$ to $\mathbb{Z}$. So we get a homomorphism

$$\gamma_\infty : Ext(A) \to \mathrm{Hom}(K_1(A), \mathbb{Z})$$

If $\tau \in \ker(\gamma_\infty)$, then in the K-theory exact sequence

$$0 = K_1(\mathbb{K}) \to K_1(E) \to K_1(A) \xrightarrow{\partial} K_0(\mathbb{K}) \to K_0(E) \to K_0(A) \to K_1(\mathbb{K}) = 0$$

corresponding to the extension $0 \to \mathbb{K} \to E \to A \to 0$ of τ, the map from $K_1(E)$ to $K_1(A)$ is surjective so $\partial = 0$, i.e. we get a short exact sequence

$$0 \to \mathbb{Z} = K_0(\mathbb{K}) \to K_0(E) \to K_0(A) \to 0$$

which defines an element of $\mathrm{Ext}^1_{\mathbb{Z}}(K_0(A), \mathbb{Z})$ (where $\mathrm{Ext}^1_{\mathbb{Z}}$ is the derived functor of the Hom-functor in homological algebra).

Brown [1984] proved that for $A = C(X)$, this procedure defines an isomorphism between $\ker(\gamma_\infty)$ and $\mathrm{Ext}^1_{\mathbb{Z}}(K^0(X), \mathbb{Z})$, thus obtaining the *Universal Coefficient Theorem*:

THEOREM 16.3.3. *If X is a compact metric space, the above procedure yields an exact sequence*

$$0 \to \mathrm{Ext}^1_{\mathbb{Z}}(K^0(X), \mathbb{Z}) \to Ext(X) \to \mathrm{Hom}(K^1(X), \mathbb{Z}) \to 0$$

This sequence is natural in X, and splits unnaturally.

Rosenberg and Schochet [1986; 1987] have obtained far-reaching generalizations of this theorem, which will be the subject of Section 23.

Using this theorem, one can obtain an abstract isomorphism between $Ext(X)$ and $K_1(X)$. A more explicit correspondence with Atiyah's theory is described in [Douglas 1980, Chapter 5]; we will not get into this now since it will fall out of our work on KK-theory. To get an idea of how the correspondence works, and as motivation for KK-theory, note the similarity of Atiyah's definition of $K_0(X)$ with the description of $Ext(X)$ in 15.7.2 (note that in Atiyah's definition it suffices to consider triples with $\mathbb{H}_0 = \mathbb{H}_1$), and compare the definitions with the definitions of $K^0(X)$ and $K^1(X)$ given in 12.2.4.

16.3.4. Because of the isomorphism of $Ext(X)$ with $K_1(X)$, the groups $Ext_i(A)^{-1}$ are sometimes called the *K-homology* groups of the C^*-algebra A, denoted $K^i(A)$ (the index conventions require an upper index for a contravariant functor). The term "K-homology" does not really seem to be appropriate, since Ext_* is actually a *cohomology* theory on the category of separable nuclear C^*-algebras; we will not use the term "K-homology" in the sequel, although the notation $K^i(A)$ ($i = 0, 1$) will be useful.

16.4. EXERCISES AND PROBLEMS

16.4.1. Let $t_1, \ldots, t_n$ be commuting normal elements in Q. We can ask various lifting questions such as:

(1) Can we lift $t_1, \ldots, t_n$ to commuting $T_1, \ldots, T_n \in \mathbb{B}$?
(2) Can we lift $t_1, \ldots, t_n$ to $T_1, \ldots, T_n \in \mathbb{B}$ such that $C^*(T_i)$ commutes with $C^*(T_j)$ for $i \neq j$? (Equivalently, do there exist $T_1, \ldots, T_n$ such that T_i *-commutes with T_j for $i \neq j$, i.e. T_i commutes with T_j and T_j^*?)
(3) Can we lift $t_1, \ldots, t_n$ to commuting normal operators $T_1, \ldots, T_n \in \mathbb{B}$?

These questions are successively stronger (question (3) is stronger than (2) by the Fuglede theorem).

The questions can be rephrased: if $T_1, \ldots, T_n$ are essentially commuting essentially normal operators, when can $T_1, \ldots, T_n$ be perturbed by compacts to yield commuting (*-commuting, commuting normal) operators?

(a) If $X \subseteq \mathbb{C}^n$ is the joint essential spectrum of $(t_1, \ldots, t_n)$, then the canonical isomorphism $C(X) \cong C^*(t_1, \ldots, t_n, 1)$ defines an element of $Ext(X)$. The answer to (3) is yes if and only if this extension is trivial.

Conversely, if $X \subseteq \mathbb{C}^n$ and $\tau \in Ext(X)$, then τ defines a natural set of commuting normal elements $(\tau(f_1), \ldots, \tau(f_n))$ in Q, where f_i is the i-th coordinate function in $C(X)$.

(b) Let $X \subseteq \mathbb{C}^2$ be homeomorphic to $\mathbb{RP}^2$, and let τ be a nontrivial element of $Ext(X) \cong \mathbb{Z}_2$. Then τ defines a commuting pair t_1, t_2 of normal elements of Q such that $t_1 \oplus t_1$ and $t_2 \oplus t_2$ can be lifted to commuting normal operators, but t_1 and t_2 cannot be. This type of torsion phenomenon is a typical example of the difficulties which arise only in the case of more than one operator.

(c) If $t_1 = u$, the image of the unilateral shift in Q, and $t_2 = 1_Q$, then the answer to (2) is obviously yes, although the answer to (3) is no. If $t_1 = t_2 = u$, the answer to (1) is yes and the answer to (2) is no. Questions (1) and (2) (for $n = 2$) are studied in [Davidson 1982]; the extensions for which the answer to (2) is yes form a subsemigroup $L(X)$ of $Ext(X)$ which is a group in all known cases, and which can be characterized in many instances. $L(X)$ in general depends on how X is embedded in $\mathbb{C}^2$.

16.4.2. An [essentially] n-normal operator is an operator unitarily equivalent to one of the form $[T_{ij}] \in \mathbb{B}(\mathbb{H} \otimes \mathbb{C}^n)$, where the T_{ij}, for $1 \leq i, j \leq n$, are [essentially] commuting [essentially] normal operators. Essentially n-normal operators correspond to matrices $[t_{ij}] \in M_n(Q)$, where the t_{ij} are commuting normal elements in Q.

The fundamental question, not yet solved in full generality, is to classify essentially n-normal operators up to either weak or strong equivalence (just as with

$\mathbb{M}_n$, the two notions differ in general), and in particular to determine which essentially n-normal operators are compact perturbations of n-normal operators.

(a) The reducing matricial spectrum of an n-normal operator $T = [T_{ij}]$ is the set

$$R_n(T) = \{a \in \mathbb{M}_n \mid \text{there is a *-homomorphism } \pi : C^*(T) \to \mathbb{M}_n \text{ with } \pi(T) = a\}.$$

The reducing essential matricial spectrum of an essentially n-normal operator $T = [T_{ij}]$ is the set

$$R_n^e(T) = \{a \in \mathbb{M}_n \mid \text{there is a *-homomorphism } \pi : C^*(t) \to \mathbb{M}_n \text{ with } \pi(t) = a\}.$$

(b) If T is essentially n-normal and $X = R_n^e(T)$, then T defines an extension of $M_n(C(X)) \cong C(X, \mathbb{M}_n)$. Conversely, any extension of $M_n(C(X))$ defines an essentially n-normal operator T with $R_n^e(T) = X$. The equivalence classes of essentially n-normal operators with reducing essential matricial spectrum X form a group closely related to $Ext(X)$; for example, there are exact sequences.

Essentially n-normal operators have been primarily studied by Salinas [1979; 1982a; 1982b]. The case $n = 2$ (essentially binormal operators) is somewhat simpler, and is treated in [McGovern et al. 1981].

16.4.3. Let G be the real "$(ax+b)$-group", the group

$$\left\{ \begin{bmatrix} a & b \\ 0 & 1 \end{bmatrix} : a, b \in \mathbb{R}, \, a \neq 0 \right\}.$$

(a) Show that $C^*(G)^+$ is an extension of $C(X)$ by $\mathbb{K}$, where X is a "figure 8".

(b) Calculate the index of suitable operators in order to show that the element of $Ext(X) \cong \mathbb{Z}^2$, when identified as in 16.2.1, is $(1, 1)$.

(c) Conclude that the extension does not split (or even stably split).

This example is due to Diep [1974], and was one of the first applications of BDF theory. Green [1977] and Rosenberg [1976] subsequently used BDF theory to get information about the structure of the C^*-algebras of other solvable Lie groups. Groups such as the Heisenberg group are much more difficult to analyze, and are not yet completely understood; Kasparov [1975] has used more high-powered Ext-theory to obtain some results. (Voiculescu [1981] also obtained related results.) In fact, the Heisenberg group C^*-algebra has been one of the main motivating forces for the development of Ext-theory beyond the BDF work.

16.4.4. Let X and Y be locally compact and second countable. Show that

$$Ext(C_0(X), C_0(Y)) \cong RK^0(F(X^+) \wedge Y^+),$$

where $F(X^+)$ is the functional Spanier–Whitehead dual spectrum of X^+ and RK^0 is representable K-theory [Rosenberg and Schochet 1981].

16.4.5. Show by an argument similar to 15.4.2(b) that $Ext(O_n) \cong \mathbb{Z}_{n-1}$, and that $Ext_s(M_k(O_n)) \cong \mathbb{Z} \oplus \mathbb{Z}_g$, where g is the greatest common divisor of k and $n - 1$. Show also that $Ext_0(O_n) = 0$. More generally, $Ext(O_A) \cong \mathbb{Z}^n / (1 - A)\mathbb{Z}^n$, $Ext_0(O_A) \cong \ker(1 - A)$.

The calculation of $Ext(O_n)$ was done independently by Brown, Paschke–Salinas, and Pimsner–Popa, and the results of 10.11.8(c) were proved [Paschke and Salinas 1979], before the K-theory of O_n was known.

16.4.6. (a) Let $\tau : A \to Q^s(B)$ be an extension. Define $\alpha_\tau : K_0(\tau(A)') \to Ext(A, B)$, where $\tau(A)'$ is the commutant of $\tau(A)$ in $Q^s(B)$, by the formula $\alpha_\tau([p])(a) = p(1 \otimes \tau(a))$ for p a projection in $M_n(\tau(A)') \cong (1_n \otimes \tau(A))' \subseteq \mathbb{M}_n \otimes Q^s(B) \cong Q^s(B)$. Show that α_τ is a well-defined homomorphism. Define a similar homomorphism from $K_1(\tau(A)') \cong K_0(S(\tau(A)'))$ to $Ext_0(A, B) \cong Ext(SA, B)$.

(b) If τ is absorbing, show that α_τ is an isomorphism. In particular, if A is separable and $B = \mathbb{C}$, and τ is faithful and nonunital, then α_τ is an isomorphism. This may be regarded as a "noncommutative Spanier–Whitehead duality."

(c) Use (b) to prove the following "poor man's Pimsner–Voiculescu exact sequence" for Ext (cf. 10.2.1):

THEOREM. *If A is separable nuclear and B is σ-unital, and $\alpha \in \mathrm{Aut}(A)$, then there is a connecting map ∂ making the following sequence exact:*

$$Ext_0(A \times_\alpha \mathbb{Z}, B) \xrightarrow{i^*} Ext_0(A, B) \xrightarrow{1 - \alpha^*} Ext_0(A, B)$$

$$\downarrow \partial$$

$$Ext(A, B) \xleftarrow{1 - \alpha^*} Ext(A, B) \xleftarrow{i^*} Ext(A \times_\alpha \mathbb{Z}, B)$$

These results are due to Valette [1983], based on earlier work of Paschke [1981].

(The other vertical map can be filled in to give a true Pimsner–Voiculescu exact sequence for Ext: see 19.6.1. Pimsner and Voiculescu obtained this exact sequence for $B = \mathbb{K}$ in their original paper [1980a].)

16.4.7. (a) Let A be an AF algebra. Show directly that

$$Ext(A) \cong \mathrm{Ext}_{\mathbb{Z}}^1(K_0(A), \mathbb{Z})$$

(cf. 16.3.3). So $Ext(A)$ is trivial if and only if $K_0(A)$ is a free abelian group.

(b) If A is a unital AF algebra, show that $Ext_s(A) \cong \mathrm{Ext}_{\mathbb{Z}}^1(K_0(A)/\langle[1_A]\rangle, \mathbb{Z})$. So $Ext_s(A)$ is trivial if and only if $K_0(A)$ is free and A is not a matrix algebra over some other C^*-algebra. For example, if A is the CAR algebra, then $Ext_s(A)$ is isomorphic to the additive group of 2-adic integers.

The groups $Ext_s(A)$ were calculated in [Pimsner and Popa 1978; Pimsner 1979]. Handelman [1982] calculated $Ext(A, B)$ and $Ext_s(A, B)$ for more general pairs of AF algebras (cf. 23.15.3).

16.4.8. The strong operator topology induces a natural topology on $Ext(X)$ making it into a (not necessarily Hausdorff) topological group. The closure of the identity is the maximal divisible subgroup. If $X = \varprojlim X_n$ with X_n a finite complex, then the closure of the identity is the kernel of the natural map from $Ext(X)$ to $\prod Ext(X_n)$. The kernel of γ_∞ (16.3.2) is the maximal compact subgroup; it is the torsion subgroup if X is a finite complex. The maximal Hausdorff quotient of $Ext(X)$ gives the Čech K-homology of X.

Characterize which X have $Ext(X)$ compact or discrete. Generalize the results of this problem to $Ext(A)$ and $Ext(A, B)$. Show that if A is the CAR algebra, then $Ext(A)$ is topologically isomorphic to the (compact) additive group of 2-adic integers.

K-Theory for Operator Algebras
MSRI Publications
Volume **5**, second edition, 1998

CHAPTER VIII

KASPAROV'S *KK*-THEORY

17. Basic Theory

One of the principal features of Kasparov theory is the great generality of the definitions. In addition to the fact that the theory works for real and "real" C^*-algebras as well as for complex ones, and allows a locally compact group action, the definitions are made in such a way that the objects are extremely general.

This generality is both an advantage and disadvantage. The advantage is that the objects of KK-theory can be recognized in a variety of quite different situations, leading to broad applicability. Some of the contexts in which KK-theory arises are given below, and others in section 24. On the other hand, the generality of the objects creates a certain lack of concreteness which appears to make the theory difficult to understand.

It is difficult to motivate the most general and useful form of the definition from the point of view taken so far in these notes, although some motivation is given in 12.2.4, 15.7.2, and 16.3. The best motivation comes from the theory of pseudodifferential operators (17.1.2(e)). So rather than trying to give any more motivation at first, we will simply plunge in and see later that all the K-groups and Ext-groups considered so far are special cases of KK-groups.

We will begin with the general definitions, and then present some simplifications which can be made in the theory which lead to some concrete descriptions of the objects from various points of view. The two principal points of view we will take are the Fredholm Module Picture, which is essentially Kasparov's original viewpoint, and the Quasihomomorphism Picture due to Cuntz.

In order to simplify matters, we will not consider real or "real" C^*-algebras at all (cf. 19.9.7), and group actions will only be treated in section 20. We will work with graded C^*-algebras for reasons explained in section 14.

The theory works well only with C^*-algebras with countable approximate identities (although the basic definitions can be made in general), so we will restrict to this case whenever convenient. In fact, for some of the theory we will have to limit ourselves to separable C^*-algebras.

17.1. Kasparov Modules

DEFINITION 17.1.1. Let A and B be graded C^*-algebras. $\mathbb{E}(A, B)$ is the set of all triples (E, ϕ, F), where E is a countably generated graded Hilbert module over B, ϕ is a graded *-homomorphism from A to $\mathbb{B}(E)$, and F is an operator in $\mathbb{B}(E)$ of degree 1, such that $[F, \phi(a)]$, $(F^2 - 1)\phi(a)$, and $(F - F^*)\phi(a)$ are all in $\mathbb{K}(E)$ for all $a \in A$. The elements of $\mathbb{E}(A, B)$ are called *Kasparov modules* for (A, B). [$\mathbb{E}(A, B)$ has nothing to do with E-theory (Section 25).] $\mathbb{D}(A, B)$ is the set of triples in $\mathbb{E}(A, B)$ for which $[F, \phi(a)]$, $(F^2 - 1)\phi(a)$, and $(F - F^*)\phi(a)$ are 0 for all a. The elements of $\mathbb{D}(A, B)$ are called *degenerate Kasparov modules*.

Sometimes the ϕ is suppressed and (E, F) is simply regarded as an (A, B)-bimodule. We will usually avoid this notation, although it can be useful in simplifying formulas.

There is a binary operation on $\mathbb{E}(A, B)$ given by direct sum; $\mathbb{D}(A, B)$ is closed under this operation.

EXAMPLES 17.1.2. (a) Let $\phi : A \to B$ be a graded *-homomorphism. Then $(B, \phi, 0)$ is a Kasparov (A, B)-module. More generally, if $\phi : A \to B \hat{\otimes} \mathbb{K}$ is a graded *-homomorphism, then $(\hat{\mathbb{H}}_B, \phi, 0)$ is a Kasparov (A, B)-module, where $\mathbb{K}(\hat{\mathbb{H}}_B)$ is identified with $B \hat{\otimes} \mathbb{K}$ in the standard way. [Actually, $(\hat{\mathbb{H}}_B, \phi, F)$ is a Kasparov (A, B)-module for any $F \in \mathbb{B}_B$ of degree 1.] We could also associate to ϕ the Kasparov (A, B)-module $\left(B \oplus B^{op}, \phi \oplus 0, \begin{bmatrix} 0 & 1 \\ 1 & 0 \end{bmatrix}\right)$. So there are two canonical ways of associating elements of $\mathbb{E}(A, B)$ to each graded *-homomorphism from A to B (which will give the same KK-element when the equivalence relation is divided out). In fact, the elements of $\mathbb{E}(A, B)$ may be regarded as "generalized homomorphisms" from A to B (17.6).

(b) Given a split exact sequence

$$0 \longrightarrow B \xrightarrow{\;\;j\;\;} D \underset{q}{\overset{s}{\rightleftarrows}} A \longrightarrow 0$$

of graded C^*-algebras, we associate a Kasparov (D, B)-module

$$\left(B \oplus B^{op},\, \omega \oplus \omega \circ s \circ q,\, \begin{bmatrix} 0 & 1 \\ 1 & 0 \end{bmatrix}\right),$$

where ω is the canonical homomorphism from D to $M(B) \cong \mathbb{B}(B)$ (15.2). This Kasparov module (or its equivalence class) is called the *splitting morphism* of the exact sequence. In the special case $D \cong B \oplus A$, the splitting morphism agrees with the second type of $\mathbb{E}(D, B)$-element associated to the projection of D onto B as in part (a).

More generally, if

$$0 \longrightarrow B \hat{\otimes} \mathbb{K} \xrightarrow{\;\;j\;\;} D \underset{q}{\overset{s}{\rightleftarrows}} A \longrightarrow 0$$

is a split exact sequence, we can associate the Kasparov (D, B)-module

$$\left(\mathbb{H}_B \oplus \mathbb{H}_B^{op},\, \omega \oplus \omega \circ s \circ q,\, \begin{bmatrix} 0 & 1 \\ 1 & 0 \end{bmatrix}\right),$$

which will be called the splitting morphism.

The splitting morphism generally depends on the cross section chosen, even when the equivalence relation is divided out.

(c) If (E_i, ϕ_i, F_i) is a Kasparov (A_i, B)-module, for $i = 1, 2$, then $(E_1 \oplus E_2,\ \phi_1 \oplus \phi_2,\ F_1 \oplus F_2)$ is a Kasparov $(A_1 \oplus A_2, B)$-module. Similarly, if (E_i, ϕ_i, F_i) is a Kasparov (A, B_i)-module, make E_i into a Hilbert $B_1 \oplus B_2$-module by letting the other algebra act trivially. Then $(E_1 \oplus E_2,\ \phi_1 \oplus \phi_2,\ F_1 \oplus F_2)$ is a Kasparov $(A,\ B_1 \oplus B_2)$-module.

(d) If $0 \to B \to D \to A \to 0$ is an invertible extension of A by B, i.e. the Busby invariant $\tau : A \to Q(B)$ dilates to a *-homomorphism

$$\phi = \begin{bmatrix} \phi_{11} & \phi_{12} \\ \phi_{21} & \phi_{22} \end{bmatrix} : A \to M_2(M(B))$$

(15.7), we associate to this extension the module $((B \oplus B) \hat{\otimes} \mathbb{C}_1,\ \phi \hat{\otimes} 1,\ \left[\begin{smallmatrix} 1 & 0 \\ 0 & -1 \end{smallmatrix}\right] \hat{\otimes} \varepsilon)$ in $\mathbb{E}(A,\ B \hat{\otimes} \mathbb{C}_1)$.

(e) Let M be a smooth closed manifold (compact without boundary). Fix a Riemannian metric on M. Let $V^{(0)}$ and $V^{(1)}$ are vector bundles over M, and $P : C^\infty(V^{(0)}) \to C^\infty(V^{(1)})$ be an elliptic pseudodifferential operator of degree 0. $V^{(0)}$ and $V^{(1)}$ can be given Hermitian structures so that P extends to an essentially unitary Fredholm operator from $L^2(V^{(0)})$ to $L^2(V^{(1)})$. Let $\mathbb{H} = L^2(V^{(0)}) \oplus L^2(V^{(1)})$, with $\mathbb{H}^{(i)} = L^2(V^{(i)})$. Let $\phi : C(M) \to \mathbb{B}(\mathbb{H})$ denote the action as multiplication operators. Then

$$\left(\mathbb{H},\ \phi,\ \begin{bmatrix} 0 & Q \\ P & 0 \end{bmatrix} \right) \in \mathbb{E}(C(M), \mathbb{C})$$

where Q is a parametrix for P.

This example is the original motivating example for KK-theory, and is crucial for many of the applications described in Section 24.

(f) As a related example, let M be a complete Riemannian manifold. Give the cotangent bundle its natural almost complex structure, and let $D = \bar{\partial} + \bar{\partial}^*$ be the Dolbeault operator on smooth forms with compact support. Let $\mathbb{H}$ be the Hilbert space of L^2-forms of bidegree $(0, *)$ on T^*M, graded by decomposing into forms of even and odd degree. Then D is an essentially self-adjoint operator on $\mathbb{H}$ of degree 1, and $F = D(1+D^2)^{1/2}$ makes $(\mathbb{H}, \phi, F)$ into a Kasparov $(C_0(T^*M), \mathbb{C})$-module (where ϕ is again action as multiplication operators). The equivalence class of this element is called the *Dolbeault element* of M, denoted $[\bar{\partial}_M]$.

(g) Another related example which will be very important in applications is the Gysin or "shriek" map. The construction is a bit technical; we will outline it here for readers familiar with the terminology. Let V and W be smooth manifolds, not necessarily compact, and let $f : V \to W$ be a smooth map, not necessarily proper. Let d be a Riemannian metric on W such that whenever $d(p, q) < 1$, there is a unique tangent vector $X(p, q) \in T_p W$ with $\|X(p, q)\| < 1$

and $exp_p X(p,q) = q$, i.e. there is a unique geodesic between p and q of length less than one. Set $V_p = \{x \in V \mid d(f(x), p) < 1\}$.

If f is K-oriented, i.e. there is a spin^c-structure on $TV \oplus f^*(TW)$, let $S^\pm$ be the corresponding spinors. We can then form the corresponding Hilbert space $\mathbb{H}_p = L^2(V_p, S^+) \oplus L^2(V_p, S^-)$. Give $\mathbb{H}_p$ the grading with $\mathbb{H}_p^{(0)} = L^2(V_p, S^+) \oplus 0$ and $\mathbb{H}_p^{(1)} = 0 \oplus L^2(V_p, S^-)$. The $\mathbb{H}_p$ can be made into a continuous field of Hilbert spaces over W; the set of continuous sections vanishing at infinity forms a Hilbert $C_0(W)$-module E in an obvious way.

There is a "Dirac operator" D_p on V_p, an elliptic pseudodifferential operator of degree 0 with principal symbol

$$\sigma^P(x, \xi) = \mu\left(\left\langle M(f(x), p)^{1/2}\xi, \ (1 - M(f(x), p))^{1/2} \frac{X(f(x), p)}{\|X(f(x), p)\|}\right\rangle\right),$$

where $\xi \in T_x(V)$ is a unit vector, M is a smooth function on $W \times W$ equal to 1 on a neighborhood of the diagonal and with support contained in $\{(p, q) \mid d(p, q) < 1\}$, and μ denotes Clifford multiplication. D_p defines an operator from $L^2(V_p, S^+)$ to $L^2(V_p, S^-)$.

The family (D_p) defines an operator T from $E^{(0)}$ to $E^{(1)}$. If ϕ is the action of $C_0(V)$ on E by multiplication, then $\left(E, \phi, \left[\begin{smallmatrix} 0 & T^* \\ T & 0 \end{smallmatrix}\right]\right)$ defines a Kasparov $(C_0(V), C_0(W))$-module. The equivalence class of this module is denoted $f!$ and read "f shriek".

If f is a proper map, then f defines a homomorphism from $C_0(W)$ to $C_0(V)$ and hence a Kasparov $(C_0(W), C_0(V))$-module. In this case, $f!$ is a sort of "inverse" to f. If f is not proper, there is no obvious way to construct a Kasparov $(C_0(W), C_0(V))$-module.

Shriek maps define a "wrong-way functoriality" (i.e. a *covariant* functoriality) from spaces with K-oriented maps to C^*-algebras, in contrast to the ordinary (contravariant) functoriality from spaces with proper maps to C^*-algebras.

The shriek map construction can be done more generally: if V_1, V_2 are smooth manifolds with foliations F_1, F_2 respectively, and f is a K-oriented smooth foliation map from (V_1, F_1) to (V_2, F_2), then f defines a shriek map $f! \in \mathbb{E}(C^*(V_1/F_1), C^*(V_2/F_2))$.

It is beyond the scope of these notes to give a complete treatment of shriek maps; the interested reader should see [Connes 1982] for a thorough discussion, including the foliation case. (See [Moore and Schochet 1988] and [Hilsum and Skandalis 1987] for further details.)

We will see many more interesting examples later.

There is another way of phrasing the definition of a Kasparov module which is sometimes useful. If D is a C^*-subalgebra of $\mathbb{B}(E)$, set

$$^c D = \{T \in \mathbb{B}(E) \mid TD \subseteq \mathbb{K}(E)\},$$

$$D^c = \{T \in \mathbb{B}(E) \mid DT \subseteq \mathbb{K}(E)\}.$$

Then ^cD is a closed left ideal of $\mathbb{B}(E)$, $D^c = (^cD)^*$ is a closed right ideal, and $^cD^c = {}^cD \cap D^c$ is a hereditary C^*-subalgebra of $\mathbb{B}(E)$ containing $\mathbb{K}(E)$.

PROPOSITION 17.1.3. *Let E be a graded Hilbert B-module, $\phi : A \to \mathbb{B}(E)$ a graded $*$-homomorphism, and $F \in \mathbb{B}(E)$ of degree 1. Then the following conditions are equivalent, where $D = \phi(A)$:*

(1) $F^2 - 1$, $F - F^*$, $[F, \phi(A)]$ *are all contained in ^cD.*
(2) $F^2 - 1$, $F - F^*$, $[F, \phi(A)]$ *are all contained in D^c.*
(3) $F^2 - 1$, $F - F^*$, $[F, \phi(A)]$ *are all contained in $^cD^c$.*
(4) $(E, \phi, F) \in \mathbb{E}(A, B)$.

PROOF. (4) $\Longrightarrow$ (1), (3) $\Longrightarrow$ (1), (3) $\Longrightarrow$ (2) are trivial. To prove that (1) $\Longrightarrow$ (3), note that

$$\phi(a)(F^2 - 1) = [(F^{*2} - 1)\phi(a^*)]^*$$
$$= [(F^2 - 1)\phi(a^*) - F(F - F^*)\phi(a^*) - F^*(F - F^*)\phi(a^*)]^*;$$

the other parts are similar.

For (1) $\Longrightarrow$ (4), we need to show that if $[F, \phi(a)]\phi(b) \in \mathbb{K}(E)$ for all a, $b \in A$, then $[F, \phi(A)] \subseteq \mathbb{K}(E)$. If a and b are homogeneous, then

$$[F, \phi(ab)] = \pm\phi(a)[F, \phi(b)] \pm [F, \phi(a)]\phi(b)$$
$$= \pm([F^*, \phi(b^*)]\phi(a^*))^* \pm [F, \phi(a)]\phi(b)$$
$$= \pm([F, \phi(b^*)]\phi(a^*))^* \pm ([F - F^*, \phi(b^*)]\phi(a^*))^* \pm [F, \phi(a)]\phi(b) \in \mathbb{K}(E).$$

(The signs are chosen according to 14.1.3(b).) Linear combinations of products of homogeneous elements are dense in A (actually they fill up all of A since every positive element of A is a product of two elements of A). $\qquad\square$

17.2. Equivalence Relations

DEFINITION 17.2.1. Two triples (E_0, ϕ_0, F_0) and (E_1, ϕ_1, F_1) are *unitarily equivalent* if there is a unitary in $\mathbb{B}(E_0, E_1)$, of degree 0, intertwining the ϕ_i and F_i. Unitary equivalence is denoted $\approx_u$.

DEFINITION 17.2.2. A *homotopy* connecting (E_0, ϕ_0, F_0) and (E_1, ϕ_1, F_1) is an element (E, ϕ, F) of $\mathbb{E}(A, IB)$ for which $(E \hat{\otimes}_{f_i} B, f_i \circ \phi, f_{i*}(F)) \approx_u (E_i, \phi_i, F_i)$, where f_i, for $i = 0, 1$, is the evaluation homomorphism from IB to B. Homotopy respects direct sums. Homotopy equivalence is denoted $\sim_h$. If $E_0 = E_1$, a *standard homotopy* is a homotopy of the form $E = C([0, 1], E_0)$ (which is a Hilbert IB-module in the obvious way), $\phi = (\phi_t)$, $F = (F_t)$, where $t \to F_t$ and $t \to \phi_t(a)$ are strong-$*$-operator continuous for each a. This is not the most general homotopy (although any homotopy can be converted into one in standard form using the stabilization theorem). It follows immediately from 12.2.2 that if $(\hat{\mathbb{H}}_B, \phi_0, F_0) \approx_u (\hat{\mathbb{H}}_B, \phi_1, F_1)$, then $(\hat{\mathbb{H}}_B, \phi_0, F_0)$ and $(\hat{\mathbb{H}}_B, \phi_1, F_1)$ are connected by a standard homotopy. A standard homotopy where in addition ϕ_t is constant and F_t is norm-continuous is called an *operator homotopy*.

The next proposition illustrates the very general nature of the equivalence relation $\sim_h$.

PROPOSITION 17.2.3. *If* $(E, \phi, F) \in \mathbb{D}(A, B)$, *then* (E, ϕ, F) *is homotopic to the* 0-*module.*

PROOF. $C_0([0, 1), E)$ is a Hilbert (IB)-module in the obvious way (in the language of section 13-14, $C_0([0, 1))$ is a trivially graded Hilbert $C([0, 1])$-module, and $C_0([0, 1), E)$ is the external tensor product $C_0([0, 1)) \hat{\otimes} E$), and

$$(C_0([0, 1), E), 1 \hat{\otimes} \phi, 1 \hat{\otimes} F) \in \mathbb{D}(A, IB) \subseteq \mathbb{E}(A, IB)$$

is the desired homotopy.   □

There are three natural equivalence relations on $\mathbb{E}(A, B)$ in addition to $\sim_h$:

DEFINITION 17.2.4. $\sim_{oh}$ is the equivalence relation on $\mathbb{E}(A, B)$ generated by operator homotopy and addition of degenerate elements, i.e. $(E_0, \phi_0, F_0) \sim_{oh} (E_1, \phi_1, F_1)$ if there are $(E_i', \phi_i', F_i') \in \mathbb{D}(A, B)$ such that

$$(E_0, \phi_0, F_0) \oplus (E_0', \phi_0', F_0') \quad \text{and} \quad (E_1, \phi_1, F_1) \oplus (E_1', \phi_1', F_1')$$

are operator-homotopic (up to unitary equivalence). $\sim_{cp}$ is the equivalence relation on $\mathbb{E}(A, B)$ generated by unitary equivalence, "compact perturbation" of F, and addition of degenerate elements. (E, ϕ, F') is a "compact perturbation" of (E, ϕ, F) if $(F - F')\phi_1(a) \in \mathbb{K}(E)$ for all $a \in A$. (The quotation marks reflect the fact that for F' to be a "compact perturbation" of F it is not necessary that $F - F'$ itself be in $\mathbb{K}(E)$.) $\sim_c$ is a "stabilized" version of $\sim_{cp}$: $(E_0, \phi_0, F_0) \sim_c (E_1, \phi_1, F_1)$ if and only if there are unitarily equivalent (E_0', ψ_0, G_0) and (E_1', ψ_1, G_1) with $(E_0, \phi_0, F_0) \oplus (E_0', \psi_0, G_0) \sim_{cp} (E_1, \phi_1, F_1) \oplus (E_1', \psi_1, G_1)$.

$\sim_c$ is called "homology" in [Kasparov 1980b, § 7] and "cobordism" in [Cuntz and Skandalis 1986] (17.10).

We will show later that $\sim_h$, $\sim_{oh}$, and $\sim_c$ all coincide when A is separable and B is σ-unital.

The next two propositions are good practice exercises in working with the equivalence relations.

PROPOSITION 17.2.5. *Let* (E, ϕ, F) *and* (E, ϕ, F') *belong to* $\mathbb{E}(A, B)$, *with* F' *a* "compact perturbation" *of* F. *Then* (E, ϕ, F) *and* (E, ϕ, F') *are operator homotopic.*

PROOF. The straight line segment from F to F' is an operator homotopy.   □

COROLLARY 17.2.6. $\sim_h$, $\sim_{oh}$, *and* $\sim_{cp}$ *are successively stronger equivalence relations.*

PROOF. Use 17.2.3 and 17.2.5.   □

For future reference, we need a generalization of 17.2.5. If (E, ϕ, F) is a Kasparov (A, B)-module, then the graded commutator $[F, F]$ is approximately equal to 2 times the identity (more precisely, $\phi(a)[F, F]\phi(a)^* = 2\phi(aa^*)$ mod $\mathbb{K}(E)$ for all $a \in A$). Suppose also $(E, \phi, F') \in \mathbb{E}(A, B)$. If $\phi(a)[F, F']\phi(a)^* \geq 0$ mod $\mathbb{K}(E)$ for all $a \in A$, we may regard F and F' as being "close"; this condition will be satisfied, for example, if F' is a "compact perturbation" of F or if $\|F - F'\| < 1$ (or, more generally, if $\|(F - F')$ mod $\mathbb{K}(E)\| < 1$).

The next proposition, due to Connes and Skandalis, is related in spirit to 4.6.6. The proof is an easy introduction to some of the techniques used in working with KK-theory.

PROPOSITION 17.2.7. *Let (E, ϕ, F) and (E, ϕ, F') be Kasparov (A, B)-modules, with $\phi(a)[F, F']\phi(a)^* \geq 0$ mod $\mathbb{K}(E)$ for all $a \in A$. Then (E, ϕ, F) and (E, ϕ, F') are operator homotopic.*

PROOF. Set

$$C = \{T \in \mathbb{B}(E) \mid [T, \phi(a)] \in \mathbb{K}(E) \text{ for all } a \in A\},$$
$$J = \{T \in C \mid T\phi(a) \in \mathbb{K}(E) \text{ for all } a \in A\}.$$

Then $[F, F'] \in C$, and $[F, F'] \geq 0$ mod J. Set $[F, F'] = P + K$, where $P \in C$, $P \geq 0$, and $K \in J$, P and K of degree 0. P commutes with F and F' mod J since $F^2 = F'^2 = 1$ mod J. For $0 \leq t \leq \pi/2$, set

$$F_t = (1 + \cos t \sin t\, P)^{-1/2}(\cos t\, F + \sin t\, F').$$

We have $F_t \in C$, $F_t - F_t^* \in J$, $F_t^2 - 1 \in J$. Thus $(E, \phi, F_t) \in \mathbb{E}(A, B)$ for all t, giving the desired operator homotopy. $\square$

17.3. The KK-Groups

All of the equivalence relations respect direct sums; thus the set of equivalence classes has an induced binary operation in each case. The operations (also called direct sum or just sum) are associative and commutative, since each equivalence relation is weaker than unitary equivalence.

DEFINITION 17.3.1. $KK(A, B) = KK_h(A, B)$ is the set of equivalence classes of $\mathbb{E}(A, B)$ under $\sim_h$. $KK_{oh}(A, B)$ (resp. $KK_{cp}(A, B)$, $KK_c(A, B)$) is the set of equivalence classes of $\mathbb{E}(A, B)$ under $\sim_{oh}$ (resp. $\sim_{cp}$, $\sim_c$). $KK^1(A, B) = KK(A, B \hat{\otimes} \mathbb{C}_1)$ (14.1.2(b)), and similarly for $KK_{oh}^1(A, B)$, $KK_c^1(A, B)$, and $KK_{cp}^1(A, B)$. More generally, we set $KK^n(A, B) = KK(A, B \hat{\otimes} \mathbb{C}_n)$ (14.5.6), and similarly for $KK_{oh}^n(A, B)$, $KK_c^n(A, B)$. Note that $KK^0(A, B) = KK(A, B)$. In each case, the set is an abelian semigroup under direct sum.

$KK_{oh}(A, B)$ is denoted $\widetilde{KK}(A, B)$ in [Skandalis 1984]; $KK_c(A, B)$ is denoted $\overline{KK}(A, B)$ in [Cuntz and Skandalis 1986].

There are obvious surjective homomorphisms $KK_{cp}(A, B) \to KK_{oh}(A, B) \to KK(A, B)$.

REMARKS 17.3.2.   (a) There is a technical problem with the definition of $KK(A,B)$ if B is not σ-unital: B may not have enough countably generated Hilbert modules. Kasparov allows the Hilbert B-module $\hat{\mathbb{H}}_B$ to be used for $\mathbb{E}(A,B)$ whether or not it is countably generated; thus for B not σ-unital our definition of $KK(A,B)$ (which is taken from [Skandalis 1984]) need not agree with his.

(b) Kasparov's definition of $KK(A,B)$ requires dividing by $\mathbb{D}(A,B)$ as well as taking homotopy equivalence classes. But it is unnecessary to divide by $\mathbb{D}(A,B)$ because of 17.2.3.

PROPOSITION 17.3.3. $KK(A,B)$ and $KK_{oh}(A,B)$ are abelian groups; and $KK_c(A,B)$ and $KK_{cp}(A,B)$ are abelian semigroups with identity.

PROOF. It is clear that any two degenerate elements are equivalent under $\sim_{cp}$, and that the class of degenerate elements forms an identity. It remains to show the existence of inverses. If $(E,\phi,F) \in \mathbb{E}(A,B)$, let $\tilde{\phi} : A \to \mathbb{B}(E^{op})$ be defined by $\tilde{\phi}(a^{(0)} + a^{(1)}) = \phi(a^{(0)} - a^{(1)})$. Then

$$(E,\phi,F) \oplus (E^{op}, \tilde{\phi}, -F) \sim_{oh} \left(E \oplus E^{op}, \begin{bmatrix} \phi & 0 \\ 0 & \tilde{\phi} \end{bmatrix}, \begin{bmatrix} 0 & 1 \\ 1 & 0 \end{bmatrix} \right)$$

via the operator homotopy

$$\left(C([0,1], E \oplus E^{op}), \begin{bmatrix} \phi & 0 \\ 0 & \tilde{\phi} \end{bmatrix}, \begin{bmatrix} F\cos t & \sin t \\ \sin t & -F\cos t \end{bmatrix} \right). \qquad \square$$

$KK_c(A,B)$ is also a group if A is separable, a fact which is difficult to prove directly but which follows immediately from the fact that $\sim_{oh}$ and $\sim_c$ coincide. We defer the proof until 17.10.

It turns out that $KK_{cp}(A,B)$ does not have cancellation in general. There is a natural surjective homomorphism from $KK_c(A,B)$ to the cancellation semigroup of $KK_{cp}(A,B)$; since $KK_{oh}(A,B)$ is a group, there is an induced surjective homomorphism from $KK_c(A,B)$ to $KK_{oh}(A,B)$ (i.e. $\sim_c$ is a stronger equivalence relation than $\sim_{oh}$).

The reader should try to avoid being confused by this multitude of equivalence relations and groups, since they all coincide (at least in the cases of interest). It is useful in relating the Kasparov groups to K-groups and Ext-groups, and also for some applications, to have the equivalence relation expressed in different forms. The agreement of the different relations (especially the agreement of $\sim_h$ and $\sim_{oh}$) is quite a deep result. The philosophical point of view to take is that the definition of the Kasparov groups is not very sensitive to the equivalence relation used: $\sim_c$ is the strongest "reasonable" relation (for which there is hope of obtaining a group), and $\sim_h$ the weakest "reasonable" one.

EXAMPLE 17.3.4. We analyze these groups in detail for $A = B = \mathbb{C}$. An element of $\mathbb{E}(\mathbb{C}, \mathbb{C})$ is a module of the form $\alpha = (\mathbb{H}, \phi, F)$, where $\mathbb{H}$ is a graded Hilbert space (of finite or countably infinite dimension). We may write $\mathbb{H}$ as $\mathbb{H}_0 \oplus \mathbb{H}_1^{op}$, where $\mathbb{H}_0$ and $\mathbb{H}_1$ are trivially graded. Then $\phi(1)$ is a projection of degree 0, i.e. $\phi(1) = \mathrm{diag}(P, Q)$ for projections P and Q; and F is of the form $\left[\begin{smallmatrix} 0 & S \\ T & 0 \end{smallmatrix}\right]$ since F is of degree 1. The conditions that $(\mathbb{H}, \phi, F)$ be a Kasparov module are that $(ST - 1)P$, $(TS - 1)Q$, $(S - T^*)P$, $(Q - S^*)Q$, $PS - SQ$, $TP - QT$ are all compact; the module is degenerate if all are 0.

α is a "compact perturbation" of the module $(\mathbb{H}, \phi, \phi(1)F\phi(1))$; i.e.

$$\alpha \sim_{cp} \left(\mathbb{H}_0 \oplus \mathbb{H}_1^{op}, \begin{bmatrix} P & 0 \\ 0 & Q \end{bmatrix}, \begin{bmatrix} 0 & PSQ \\ QTP & 0 \end{bmatrix} \right)$$

$$\approx_u \left(P\mathbb{H}_0 \oplus Q\mathbb{H}_1^{op}, \begin{bmatrix} P & 0 \\ 0 & Q \end{bmatrix}, \begin{bmatrix} 0 & PSQ \\ QTP & 0 \end{bmatrix} \right) \oplus ((1-P)\mathbb{H}_0 \oplus (1-Q)\mathbb{H}_1^{op}, 0, 0)$$

$$\sim_{cp} \left(P\mathbb{H}_0 \oplus Q\mathbb{H}_1^{op}, \begin{bmatrix} P & 0 \\ 0 & Q \end{bmatrix}, \begin{bmatrix} 0 & PSQ \\ QTP & 0 \end{bmatrix} \right),$$

since $((1 - P)\mathbb{H}_0 \oplus (1 - Q)\mathbb{H}_1^{op}, 0, 0) \in \mathbb{D}(\mathbb{C}, \mathbb{C})$. Thus the equivalence class of α in $KK_{cp}(\mathbb{C}, \mathbb{C})$ can be represented by a module of the form

$$\beta = \left(\tilde{\mathbb{H}}_0 \oplus \tilde{\mathbb{H}}_1^{op}, \ 1, \ \begin{bmatrix} 0 & \tilde{S} \\ \tilde{T} & 0 \end{bmatrix} \right)$$

(i.e. with ϕ unital). For such a module, $\tilde{T}$ is essentially a unitary operator from $\tilde{\mathbb{H}}_0$ to $\tilde{\mathbb{H}}_1$, and $\tilde{S}$ is essentially $\tilde{T}^*$. By performing another "compact perturbation", we may assume $\tilde{T}$ is either an isometry or coisometry, and that $\tilde{S} = \tilde{T}^*$. If $\tilde{T}$ is unitary, then β is degenerate. If $\tilde{T}$ is a proper coisometry (i.e. $\tilde{T}\tilde{T}^* = 1$, $\tilde{T}^*\tilde{T} \neq 1$), set $\tilde{P} = 1 - \tilde{T}^*\tilde{T}$; then $\tilde{P}$ is a projection in $\mathbb{B}(\tilde{\mathbb{H}}_0)$ of finite rank n, and β is unitarily equivalent to

$$(\tilde{P}\tilde{\mathbb{H}}_0, 1, 0) \oplus \left((1 - \tilde{P})\tilde{\mathbb{H}}_0 \oplus \tilde{\mathbb{H}}_1^{op}, \ 1, \ \begin{bmatrix} 0 & \tilde{T}^* \\ \tilde{T} & 0 \end{bmatrix} \right).$$

The second module is degenerate, and the first module is isomorphic to n times the module obtained from the identity map $\mathbb{C} \to \mathbb{C}$ as in 17.1.2(a) (or, equivalently, the first module is isomorphic to the module coming from the unital map $\mathbb{C} \to \mathbb{M}_n$). Similarly, if $\tilde{T}$ is a proper isometry, set $\tilde{Q} = 1 - \tilde{T}\tilde{T}^* \in \mathbb{B}(\tilde{\mathbb{H}}_1)$. Then β is unitarily equivalent to

$$(\tilde{Q}\tilde{\mathbb{H}}_1^{op}, 1, 0) \oplus \left(\tilde{\mathbb{H}}_0 \oplus (1 - \tilde{Q})\tilde{\mathbb{H}}_1^{op}, \ 1, \ \begin{bmatrix} 0 & \tilde{T}^* \\ \tilde{T} & 0 \end{bmatrix} \right),$$

which is $\sim_{cp}$-equivalent to the negative of a homomorphism. Thus the map $\mathbb{Z} \to KK_{cp}(\mathbb{C}, \mathbb{C})$ sending 1 to the class of $1_{\mathbb{C}}$ is surjective, i.e. $KK_{cp}(A, B)$ is a cyclic group generated by $[1_{\mathbb{C}}]$.

There is an inverse map from $KK_{oh}(\mathbb{C}, \mathbb{C})$ to $\mathbb{Z}$ defined by sending the original module $(\mathbb{H}_0 \oplus \mathbb{H}_1^{op}, \left[\begin{smallmatrix} P & 0 \\ 0 & Q \end{smallmatrix}\right], \left[\begin{smallmatrix} 0 & S \\ T & 0 \end{smallmatrix}\right])$ to the Fredholm index of QTP (this makes sense as $(\dim P\mathbb{H}_0 - \dim Q\mathbb{H}_1)$ if $P\mathbb{H}_0$ and $Q\mathbb{H}_1$ are finite-dimensional). This map is well defined because an operator homotopy must preserve this index. The map is obviously surjective. Thus we get a sequence of surjective maps

$$\mathbb{Z} \to KK_{cp}(\mathbb{C}, \mathbb{C}) \to KK_c(\mathbb{C}, \mathbb{C}) \to KK_{oh}(\mathbb{C}, \mathbb{C}) \to \mathbb{Z}$$

whose composition is the identity map, so all the maps are isomorphisms.

It is more difficult to analyze $KK(\mathbb{C}, \mathbb{C})$ because of the very general nature of $\sim_h$; but we will see later that $KK(\mathbb{C}, \mathbb{C})$ is also $\mathbb{Z}$.

17.4. Standard Simplifications

We now discuss some simplifications in the definition.

PROPOSITION 17.4.1. *If B is σ-unital, then in the definition of $KK_c(A, B)$ (hence also for $KK_{oh}(A, B)$ and $KK(A, B)$) it suffices to consider only those triples (E, ϕ, F) where $E = \hat{\mathbb{H}}_B$.*

PROOF. The triple $(\hat{\mathbb{H}}_B, 0, 0)$ is in $\mathbb{D}(A, B)$, so (E, ϕ, F) has the same image in $KK_c(A, B)$ as $(E \oplus \hat{\mathbb{H}}_B, \phi \oplus 0, F \oplus 0)$. We have $E \oplus \hat{\mathbb{H}}_B \cong \hat{\mathbb{H}}_B$ by the stabilization theorem (14.6.1). □

PROPOSITION 17.4.2. *If $(E, \phi, F) \in \mathbb{E}(A, B)$, then there is a "compact perturbation" (E, ϕ, G) of (E, ϕ, F) with $G = G^*$. So in the definition of $KK_c(A, B)$ (hence also for $KK_{oh}(A, B)$ and $KK(A, B)$), it suffices to consider only those triples (E, ϕ, F) where $F = F^*$, and "compact perturbations", homotopies, and operator homotopies may be taken within this class.*

PROOF. If $(E, \phi, F) \in \mathbb{E}(A, B)$, then so are (E, ϕ, F^*) and $(E, \phi, (F + F^*)/2)$, which are "compact perturbations" of (E, ϕ, F). The same procedure may be applied to a homotopy [resp. operator homotopy] (E_t, ϕ_t, F_t) from (E_0, ϕ_0, F_0) to (E_1, ϕ_1, F_1) to yield a homotopy [resp. operator homotopy] $(E_t, \phi_t, (F_t + F_t^*)/2)$ from $(E_0, \phi_0, (F_0 + F_0^*)/2)$ to $(E_1, \phi_1, (F_1 + F_1^*)/2)$. □

PROPOSITION 17.4.3. *If $(E, \phi, F) \in \mathbb{E}(A, B)$, then there is a "compact perturbation" (E, ϕ, G) of (E, ϕ, F) with $G = G^*$, $\|G\| \le 1$. If A is unital, we may in addition obtain a G with $G^2 - 1 \in \mathbb{K}(E)$. So in the definition of $KK_c(A, B)$ (hence also for $KK_{oh}(A, B)$ and $KK(A, B)$), it suffices to consider only those triples where $F = F^*$, $\|F\| \le 1$. If A is unital, we may in addition assume $F^2 - 1 \in \mathbb{K}(E)$. "Compact perturbations", homotopies, and operator homotopies may be taken to lie within this class.*

PROOF. Let $(E, \phi, F) \in \mathbb{E}(A, B)$. We may assume $F = F^*$ by 17.4.2. Let g be the continuous function on $\mathbb{R}$ with $g(x) = -1$ for $x \le -1$, $g(x) = x$ for $-1 \le x \le 1$, and $g(x) = 1$ for $x \ge 1$. Set $G = g(F)$. Then $(E, \phi, G) \in \mathbb{E}(A, B)$, and is a "compact perturbation" of (E, ϕ, F). If A is unital, let $P = \phi(1)$, and replace G by $PGP + (1 - P)$. The same procedures may be applied to a Kasparov module for (A, IB) implementing a homotopy or operator homotopy to yield a homotopy or operator homotopy between the perturbed modules. □

Actually, we can require $F^2 - 1 \in \mathbb{K}(E)$ even if A is nonunital (17.6).

The simplifications of the three preceding propositions may be done independently and simultaneously. We will usually make these simplifications, sometimes without comment. They were not built into the definition since in some applications of Kasparov theory (e.g. 17.1.2(e)) elements of $\mathbb{E}(A, B)$ arise naturally which are not of the special form.

There are two other simplifications which can be made in addition to 17.4.1-17.4.3. These additional simplifications *cannot* be made simultaneously. A choice of one or the other leads to the two standard pictures of $KK(A, B)$.

17.5. Fredholm Picture of $KK(A, B)$

Suppose A is unital. Then in the definition of $KK_c(A, B)$ (hence also of $KK_{oh}(A, B)$ and $KK(A, B)$) it suffices to consider only those triples (E, ϕ, F) with ϕ unital [apply 17.4.3 and replace E by PE and F by PFP.] If B is σ-unital and there is a unital (graded) homomorphism from A to $\mathbb{B}(\hat{\mathbb{H}}_B)$ (in particular, if A has a nonzero representation on a separable Hilbert space), we may in addition restrict to the case $E = \hat{\mathbb{H}}_B$ (with ϕ unital). Again, "compact perturbations", homotopies, or operator homotopies may always be chosen to lie within this class.

If A is only σ-unital, in the definition of $KK(A, B)$ we may still restrict to triples (E, ϕ, F) where ϕ is essential. However, the proof is much more complicated, and will be deferred until 18.3.6.

17.5.1. We examine in more detail the case where A is unital and trivially graded and B is σ-unital and trivially graded. We may identify $\mathbb{B}(\hat{\mathbb{H}}_B)$ with $M_2(M^s(B))$, with the diagonal-off diagonal grading as in 14.1.2(a). With this identification, $\phi = \mathrm{diag}(\phi_0, \phi_1)$, where ϕ_i is a unital homomorphism from A to $M^s(B)$, and

$$ F = \begin{bmatrix} 0 & T^* \\ T & 0 \end{bmatrix}, $$

for some $T \in M^s(B)$ with $\|T\| \leq 1$. The intertwining conditions for F and ϕ become $T^*T - 1$, $TT^* - 1$, $T\phi_1(a) - \phi_0(a)T \in B \otimes \mathbb{K}$ for all $a \in A$. The equivalence relation $\sim_h$ is homotopy of triples (ϕ_0, ϕ_1, T) (with strong-operator continuity); $\sim_{oh}$ [resp. $\sim_c$] is generated by unitary equivalence, norm-homotopy [resp. compact perturbation] of T, and adding on degenerate triples (where T is a unitary exactly intertwining ϕ_0 and ϕ_1).

A Kasparov module in the Fredholm picture is often denoted by a triple $(E^{(0)} \oplus E^{(1)}, \phi^{(0)} \oplus \phi^{(1)}, T)$, where $T \in \mathbb{B}(E^{(0)}, E^{(1)})$ is an operator with the right algebraic and intertwining properties as above. Modules expressed in this way give the *Fredholm picture* of $KK(A, B)$.

17.5.2. Under the same assumptions on A and B, we get a similar description of $KK^1(A, B)$. Now $\mathbb{B}(\hat{\mathbb{H}}_B \hat{\otimes} \mathbb{C}_1) \cong M^s(B) \hat{\otimes} \mathbb{C}_1 \cong M^s(B) \oplus M^s(B)$ with standard odd grading. ϕ is of the form $\psi \oplus \psi$, and $F = T \oplus (-T)$ for some unital $\psi : A \to$

$M^s(B)$ and $T \in M^s(B)$ with $T = T^*$, $\|T\| \leq 1$. The commutation relations become $T^2 - 1$ and $T\psi(a) - \psi(a)T \in B \otimes \mathbb{K}$ for all a. The degenerate pairs (ϕ, T) have T a self-adjoint unitary commuting with $\psi(A)$.

So the Fredholm picture of modules for $KK^1(A, B)$, for A, B trivially graded, is as triples (E, ψ, T), where E is a (trivially graded) Hilbert B-module, ψ is a unital *-homomorphism from A to $\mathbb{B}(E)$, and T is a self-adjoint, essentially unitary operator on E essentially commuting with $\psi(A)$. (Sometimes it is convenient to only require T to be essentially self-adjoint in this picture.)

17.5.3. If A is not unital, we can still use the Fredholm picture for Kasparov (A, B)-modules; in the descriptions of the previous two paragraphs, we must replace the commutation conditions by $(T^*T - 1)\phi_1(a)$ and $(TT^* - 1)\phi_2(a) \in B \otimes \mathbb{K}$ for $KK(A, B)$, and similarly for KK^1. We can actually arrange for $T^*T - 1$, $TT^* - 1$ and (for KK^1) $T - T^*$ to be compact (or even 0) by 17.6; but Kasparov modules often arise naturally expressed in the Fredholm picture form which do not have such strong properties.

17.5.4. A particularly simple and important case is when $A = \mathbb{C}$. Since there is only one unital homomorphism from $\mathbb{C}$ to $M^s(B)$, the ϕ can be eliminated entirely in this case. Thus the descriptions of $KK(\mathbb{C}, B)$ and $KK^1(\mathbb{C}, B)$ become

$$KK(\mathbb{C}, B) \cong \{[T] : T \in M^s(B), \ T^*T - 1, \ TT^* - 1 \in B \otimes \mathbb{K}\},$$

$$KK^1(\mathbb{C}, B) \cong \{[T] : T \in M^s(B), \ T = T^*, \ T^2 - 1 \in B \otimes \mathbb{K}\}.$$

The elements of $\mathbb{E}(\mathbb{C}, B)$ are thus (up to equivalence) just the preimages of unitaries in $Q^s(B)$. The equivalence relation in $KK_{oh}(\mathbb{C}, B)$ is homotopy and orthogonal composition with degenerate elements (unitaries in $M^s(B)$). Since any homotopy in $Q^s(B)$ can be lifted to a homotopy in $M^s(B)$ (3.4.6), we have proved:

PROPOSITION 17.5.5. *If B is a trivially graded σ-unital C^*-algebra, then*

$$KK_{oh}(\mathbb{C}, B) \cong K_1(Q^s(B)) \cong K_0(B).$$

Similarly, the elements of $KK^1_{oh}(\mathbb{C}, B)$ can be identified with self-adjoint elements in $M^s(B)$ with unitary image in $Q^s(B)$. These in turn may be identified with projections in $Q^s(B)$, so that

PROPOSITION 17.5.6. *If B is a trivially graded σ-unital C^*-algebra, then*

$$KK^1_{oh}(\mathbb{C}, B) \cong K_0(Q^s(B)) \cong K_1(B).$$

Note that the identification of $K_0(Q^s(B))$ with $K_1(B)$ requires Bott periodicity of K-theory, as well as the triviality of $K_*(M^s(B))$, so the identification of $KK^1_{oh}(\mathbb{C}, B)$ with $K_1(B)$ requires K-theory Bott periodicity. The identification of $KK_{oh}(\mathbb{C}, B)$ with $K_0(B)$ does not require K-theory Bott periodicity, only triviality of $K_*(M^s(B))$ and exactness of the connecting map ∂ of 8.3. The identification of $KK^i_{oh}(\mathbb{C}, B)$ with $K_{1-i}(Q^s(B))$ is elementary in both cases.

There is another related identification. Any *-homomorphism ψ from $C_0((0,1))$ into a unital C^*-algebra defines a unitary $\psi(f) + 1$, where $f(t) = e^{2\pi i t} - 1$, and conversely any unitary u defines a homomorphism by sending f to $u - 1$. Two homomorphisms are homotopic if and only if the corresponding unitaries are in the same connected component. Thus elements of $KK^1(C_0(\mathbb{R}), B)(B$ trivially graded) are represented by triples $(\mathbb{H}_B, U, T)$, where U is a unitary in $\mathbb{B}_B \cong M^s(B)$ and T is a self-adjoint, essentially unitary operator in $M^s(B)$ essentially commuting with U. If we set $P = (T+1)/2$, then P is essentially a projection. Set $\omega(U, P) = \pi(PUP + 1 - P)$, where π is the quotient map from $M^s(B)$ to $Q^s(B)$. Then $\omega(U, P)$ is a unitary in $Q^s(B)$, and $\omega(U, P)$ and $\omega(V, Q)$ are in the same component if and only if (U, P) and (V, Q) are operator homotopic. (U, P) is degenerate if P is an actual projection exactly commuting with U. If u is a unitary in $Q^s(B)$, let U be a lift of $\mathrm{diag}(u, u^{-1})$ in $M_2(M^s(B))$; then $(U, \mathrm{diag}(1,0))$ has $\mathrm{diag}(u, 1)$ as image; so we have proved

PROPOSITION 17.5.7. *If B is a trivially graded σ-unital C^*-algebra, then*

$$KK^1_{oh}(C_0(\mathbb{R}), B) \cong K_1(Q^s(B)) \cong K_0(B).$$

The idea behind the Fredholm Picture is to push as much of the nontrivial information of a triple as possible into the operator F. This can be done completely, however, only in the case $A = \mathbb{C}$. The Fredholm Picture represents KK-elements as "generalized elliptic operators" (cf. 17.1.2(e)).

17.6. Quasihomomorphism or Cuntz Picture of $KK(A, B)$

In this picture, we put all the nontrivial information of a triple into the homomorphism ϕ; KK-elements will be represented as "generalized homomorphisms." We will assume throughout that B is σ-unital.

Let $(E, \phi, F) \in \mathbb{E}(A, B)$, with $F = F^*$ and $\|F\| \le 1$, as in 17.4.3. Then $(E^{op}, 0, -F) \in \mathbb{D}(A, B)$, and $(E \oplus E^{op}, \phi \oplus 0, F \oplus (-F))$ is a "compact perturbation" of $(E \oplus E, \phi \oplus 0, G)$, where

$$G = \begin{bmatrix} F & (1-F^2)^{1/2} \\ (1-F^2)^{1/2} & -F \end{bmatrix}.$$

G is a self-adjoint unitary. Thus we need only consider triples where $F = F^* = F^{-1}$. Again, homotopies or operator homotopies may be taken to stay within this class.

17.6.1. We may further simplify the module (E, ϕ, F), where $F = F^* = F^{-1}$, by adding on the degenerate element $(E^{op}, 0, -F)$ to obtain $(E \oplus E^{op}, \phi \oplus 0, F \oplus (-F))$. This is unitarily equivalent to $(E \oplus E^{op}, (\mathrm{ad}\, U) \circ (\phi \oplus 0), \begin{bmatrix} 0 & 1 \\ 1 & 0 \end{bmatrix})$, where $U = \frac{1}{\sqrt{2}} \begin{bmatrix} 1 & -F \\ F & 1 \end{bmatrix}$. (Note that U is a unitary of degree 0.) So every KK_c-equivalence class can be represented by a module of the form $(E \oplus E^{op}, \psi, \begin{bmatrix} 0 & 1 \\ 1 & 0 \end{bmatrix})$ for a suitable graded Hilbert B-module E. Thus we may eliminate the F entirely.

By adding on the degenerate module $(\mathbb{H}_B \oplus \mathbb{H}_B^{op}, 0, \left[\begin{smallmatrix} 0 & 1 \\ 1 & 0 \end{smallmatrix}\right])$, and using the Stabilization Theorem (14.6.1), we may represent any KK_c-element by a module of the form $(\mathbb{H}_B \oplus \mathbb{H}_B^{op}, \phi, \left[\begin{smallmatrix} 0 & 1 \\ 1 & 0 \end{smallmatrix}\right])$, where $\mathbb{H}_B$ has its natural grading.

17.6.2. If B is trivially graded, we have

$$\phi = \begin{bmatrix} \phi_{00} & \phi_{01} \\ \phi_{10} & \phi_{11} \end{bmatrix},$$

where $\phi_{00}(a) - \phi_{11}(a)$ and $\phi_{10}(a) + \phi_{01}(a)$ are in $B \hat{\otimes} \mathbb{K}$ for all $a \in A$.

If A is also trivially graded, then $\phi_{10} = \phi_{01} = 0$, and $\phi^{(0)} = \phi_{00}$ and $\phi^{(1)} = \phi_{11}$ are homomorphisms from A to $M^s(B)$. So

$$KK(A, B) = \{[\phi^{(0)}, \phi^{(1)}] \mid \phi^{(i)} : A \to M^s(B), \phi^{(0)}(a) - \phi^{(1)}(a) \in B \otimes \mathbb{K} \text{ for all } a\}.$$

Thus, if A and B are trivially graded, the elements of $KK(A, B)$, etc., may be taken to be equivalence classes of *quasihomomorphisms* from A to $B \otimes \mathbb{K}$: pairs $(\phi^{(0)}, \phi^{(1)})$ of homomorphisms into $M^s(B)$ which agree mod $B \otimes \mathbb{K}$. Degenerate elements are quasihomomorphisms with $\phi^{(0)} = \phi^{(1)}$.

More generally, a *prequasihomomorphism* from A to B is a triple $(\phi^{(0)}, \phi^{(1)}, \mu)$, where $\phi^{(i)}$ are homomorphisms from A to a C^*-algebra D containing an ideal J, such that $\phi^{(0)}(a) - \phi^{(1)}(a) \in J$ for all $a \in A$, and μ is a homomorphism from J to B. The prequasihomomorphism is a *quasihomomorphism* if J is essential in D and μ is an embedding (sometimes it is also assumed that D is generated by $\phi^{(0)}(A)$ and $\phi^{(1)}(A)$ and that J is generated by $(\phi^{(0)} - \phi^{(1)})(A)$). It is easy to see that quasihomomorphisms in this sense exactly correspond to homomorphisms from qA (10.11.13) to B.

Any prequasihomomorphism from A to B defines an element of $\mathbb{E}(A, B)$ by setting $E = J \otimes_\mu B$ with trivial grading; there is then a canonical homomorphism $\psi : D \to M(J) = \mathbb{B}(J) \to \mathbb{B}(E)$. Then $(E \oplus E^{op}, \psi \circ \phi^{(0)} \oplus \psi \circ \phi^{(1)}, \left[\begin{smallmatrix} 0 & 1 \\ 1 & 0 \end{smallmatrix}\right]) \in \mathbb{E}(A, B)$.

17.6.3. The equivalence relations (except $\sim_{oh}$) are easy to describe in the setting where A and B are trivially graded. Unitary equivalence corresponds to conjugating ϕ_0 and ϕ_1 by the same unitary, and compact perturbation corresponds to conjugating ϕ_1 by a unitary which is a compact perturbation of the identity. Homotopy can be cleanly described as follows. Suppose $[\phi_0^{(0)}, \phi_0^{(1)}] \sim_h [\phi_1^{(0)}, \phi_1^{(1)}]$ via an element of $\mathbb{E}(A, IB)$. The procedure of 17.6.1 can be applied to this homotopy module to convert it into a module of the form $(\mathbb{H}_{IB} \oplus \mathbb{H}_{IB}^{op}, \psi^{(0)} \oplus \psi^{(1)}, \left[\begin{smallmatrix} 0 & 1 \\ 1 & 0 \end{smallmatrix}\right])$, i.e. a quasihomomorphism from A to IB. This quasihomomorphism gives a homotopy from $[\phi_0^{(0)}, \phi_0^{(1)}] \oplus$ (degenerate) to $[\phi_1^{(0)}, \phi_1^{(1)}] \oplus$ (degenerate). Thus $\sim_h$ is the equivalence relation generated by addition of degenerates and homotopies given by paths $(\phi_t^{(0)})$, $(\phi_t^{(1)})$, strictly continuous in t for each a, such that $(\phi_t^{(0)}, \phi_t^{(1)})$ is a quasihomomorphism for each t. Thus $KK(A, B)$ can be identified with $[qA, B \otimes \mathbb{K}]$ (9.4.4, 10.11.13).

17.6.4. Now let us examine the Cuntz picture of KK^1, again assuming A and B are trivially graded. As in the Fredholm picture, $\mathbb{B}(\hat{\mathbb{H}}_B \hat{\otimes} \mathbb{C}_1) \cong M^s(B) \oplus M^s(B)$ with the standard odd grading. ϕ is of the form $\psi \oplus \psi$ and $F = T \oplus (-T)$, where now T is a self-adjoint unitary almost commuting with ψ. If $P = 2T - 1$, then P is a projection almost commuting with ψ. Thus we may write

$$KK^1(A, B) = \{ [(\psi, P)] \mid \psi : A \to M^s(B), \ P = P^* = P^2 \in M^s(B),$$
$$P\psi(a) - \psi(a)P \in B \otimes \mathbb{K} \text{ for all } a \in A \}.$$

We will sometimes use this picture also allowing P to be a projection mod $B \hat{\otimes} \mathbb{K}$ instead of an honest projection in $M^s(B)$.

Using the stabilization theorem, we may assume that $\psi : A \to M_2(M^s(B))$, $P = \mathrm{diag}(1, 0)$. Expressed differently, every element of $KK_c^1(A, B)$ can be represented by a module of the form $\big((\mathbb{H}_B \oplus \mathbb{H}_B) \hat{\otimes} \mathbb{C}_1, \ \psi \hat{\otimes} 1, \ \left[\begin{smallmatrix} 1 & 0 \\ 0 & -1 \end{smallmatrix}\right] \hat{\otimes} \varepsilon\big)$ (cf. 17.1.2(d)).

The equivalence relations are describable as before. Unitary equivalence has the obvious meaning. "Compact perturbation" means replacing P by P', where $(P - P')\psi(a) \in B \otimes \mathbb{K}$ for all a. A homotopy is an element of $\mathbb{E}(A, IB)$, which can be described as a pair (ψ, P) in the same manner. An operator homotopy corresponds to a path (P_t); by the unitary version of 4.3.3 we can find a path (U_t) of unitaries with $U_t P_t U_t^* = P_0$, and we can use the path U_t to transfer the operator homotopy to a homotopy (ψ_t, P_0), where $\psi_t = U_t^* \psi U_t$. Thus $\sim_{oh}$ is the equivalence relation generated by this type of homotopy, along with unitary equivalence and addition of degenerate elements. A degenerate element is a pair (ψ, P) where P commutes with $\psi(A)$.

Each pair (ψ, P) defines an extension τ by $\tau(a) = \pi(P\psi(a)P)$, where $\pi : M^s(A) \to Q^s(A)$ is the quotient map. The relation $\sim_c$ is exactly the same as the defining relation for $Ext(A, B)$, so we obtain an injective homomorphism from $KK_c^1(A, B)$ to $Ext(A, B)$. The range consists of invertible extensions, since the pair $(\psi, 1 - P)$ defines an inverse for (ψ, P) in $Ext(A, B)$. The argument of 15.7.2 shows that every invertible extension occurs in this manner, and describes the inverse map $Ext(A, B)^{-1} \to KK_c^1(A, B)$. So we have proved:

PROPOSITION 17.6.5. *Let A and B be trivially graded, with B σ-unital. Then $KK_c^1(A, B) \cong Ext(A, B)^{-1}$. If A is separable and nuclear, then $KK_c^1(A, B) \cong Ext(A, B)$.*

17.6.6. If A and B (with B σ-unital) are not trivially graded, there is still a homomorphism from $Ext(A, B)^{-1}$, the group of invertible grading-preserving extensions of A by $B \hat{\otimes} \mathbb{K}$ (15.13), to $KK^1(A, B)$, defined in the same way as in 15.7.2. However, this map is not an isomorphism in general. For example, $Ext(\mathbb{C}, \mathbb{C}_1)$ is trivial, but $KK^1(\mathbb{C}, \mathbb{C}_1) \cong \mathbb{Z}$.

17.7. Additivity

It follows immediately from the constructions of 17.1.2 that $KK(A_1 \oplus A_2, B) \cong KK(A_1, B) \oplus KK(A_2, B)$ and $KK(A, B_1 \oplus B_2) \cong KK(A, B_1) \oplus KK(A, B_2)$ under the obvious maps. The same is true for KK_c and KK_{oh}.

Additivity in the first variable also holds for countable direct sums (19.7); the proof requires the equality of $\sim_h$ and $\sim_{oh}$. KK is not countably additive in the second variable in general; but it is for certain special A.

17.8. Functoriality

If $f : A_1 \to A_2$ is a graded homomorphism, then for any B there is an induced map $\mathbb{E}(A_2, B) \to \mathbb{E}(A_1, B)$ given by $(E, \phi, F) \to (E, \phi \circ f, F)$. This map respects direct sums and all three equivalence relations, and so defines homomorphisms $f^* : KK(A_2, B) \to KK(A_1, B)$, etc. Thus, for fixed B, $KK(\cdot, B)$, $KK_{oh}(\cdot, B)$, and $KK_c(\cdot, B)$ are contravariant functors from C^*-algebras to abelian groups.

If $g : B_1 \to B_2$ is a graded homomorphism, then for any A there is a map $\mathbb{E}(A, B_1) \to \mathbb{E}(A, B_2)$ given by $(E, \phi, F) \to (E \hat{\otimes}_g B_2, \phi \hat{\otimes} 1, F \hat{\otimes} 1)$. This map also respects direct sums and all three equivalence relations, and so defines homomorphisms $g_* : KK(A, B_1) \to KK(A, B_2)$, etc. Thus, for fixed A, $KK(A, \cdot)$, $KK_{oh}(A, \cdot)$, and $KK_c(A, \cdot)$ are covariant functors from C^*-algebras to abelian groups.

Combining the previous paragraphs, KK, KK_{oh}, and KK_c are *bifunctors* from pairs of C^*-algebras to abelian groups, contravariant in the first variable and covariant in the second.

DEFINITION 17.8.1. Let $f : A \to B \hat{\otimes} \mathbb{K}$ be a graded *-homomorphism. Denote the element of $KK(A, B)$ (or KK_c, or KK_{oh}) corresponding to f as in 17.1.2(a) by $\boldsymbol{f}$.

EXAMPLES 17.8.2. (a) If $f : A \to B$ is a graded homomorphism, then $\boldsymbol{f} \in KK(A, B)$ is represented by $(B, f, 0)$. Let E be the graded Hilbert (IB)-submodule $\{h : [0, 1] \to B \mid h(1) \in \overline{f(A)B}\}$ of $C([0, 1], B)$; then $(E, 1 \otimes f, 0)$ gives a homotopy between $(B, f, 0)$ and $(\overline{f(A)B}, f, 0)$, so $(\overline{f(A)B}, f, 0)$ also represents $\boldsymbol{f}$ in $KK(A, B)$. It is not clear at this point that this module also represents $\boldsymbol{f}$ in $KK_{oh}(A, B)$ unless f is *essential* in the sense that $f(A)$ generates B as a closed right ideal.

(b) If $f : A \to D$ and $g : D \to B$ are graded homomorphisms, then $f^*(\boldsymbol{g})$ and $\boldsymbol{g} \circ \boldsymbol{f}$ are both represented by $(B, g \circ f, 0) \in \mathbb{E}(A, B)$, so $f^*(\boldsymbol{g}) = \boldsymbol{g} \circ \boldsymbol{f}$ in $KK_c(A, B)$. $g_*(\boldsymbol{f})$ is represented by $(\overline{g(D)B}, g \circ f, 0) \in \mathbb{E}(A, B)$, so $g_*(\boldsymbol{f}) = f^*(\boldsymbol{g}) = \boldsymbol{g} \circ \boldsymbol{f}$ in $KK(A, B)$. If g is essential, then also $g_*(\boldsymbol{f}) = f^*(\boldsymbol{g}) = \boldsymbol{g} \circ \boldsymbol{f}$ in $KK_c(A, B)$.

(c) Let $i : A \hat{\otimes} \mathbb{K} \to A \hat{\otimes} \mathbb{K}$ be the identity map, and $\boldsymbol{i}$ the corresponding element of $KK_c(A \hat{\otimes} \mathbb{K}, A)$. Let $h : A \to A \hat{\otimes} \mathbb{K}$ send a to $a \hat{\otimes} p$ for a one-dimensional projection p of degree 0. Then $h^*(\boldsymbol{i}) = \boldsymbol{1}_A$ and $h_*(\boldsymbol{i}) = \boldsymbol{1}_{A \hat{\otimes} \mathbb{K}}$. (Standard representatives give the modules of 17.2.1(a) plus degenerate modules.)

(d) Let

$$0 \longrightarrow B \xrightarrow[j]{} D \mathrel{\substack{\xrightarrow{s} \\ \longleftarrow \\ q}} A \longrightarrow 0$$

be a split exact sequence of graded C^*-algebras, and let π_s be the element of $KK(D, B)$ defined by the Kasparov module of 17.1.2(b). Then $j^*(\pi_s) = \mathbf{1}_B$. It is not true in general that $j_*(\pi_s) = \mathbf{1}_D$.

(e) In the situation of (d), $j \oplus s$ gives a graded homomorphism from $B \oplus A$ to $D \oplus D \subseteq M_2(D)$. We will show later (19.9.1) that the corresponding element $\boldsymbol{j} \oplus \boldsymbol{s}$ of $KK(B \oplus A, D)$ is invertible. (The inverse is basically $\pi_s \oplus \boldsymbol{q}$.)

If we regard a homomorphism $f : A \to B$ as the element $\boldsymbol{f} \in KK(A, B)$, then the functoriality of KK may be regarded as a statement that an element of $KK(A, D)$ may be composed with an element of $KK(D, B)$ to yield an element of $KK(A, B)$, provided that one of the elements comes from an actual homomorphism. The intersection product extends this law of composition to arbitrary KK-elements, preserving associativity.

It is an interesting fact that any quasihomomorphism can be canonically factored into an ordinary homomorphism and a splitting morphism; so in the trivially graded case any KK-element can be expressed in terms of an ordinary homomorphism and a splitting morphism.

PROPOSITION 17.8.3. *Let $[\phi^{(0)}, \phi^{(1)}]$ be a quasihomomorphism from A to B. Then there is a canonically associated split exact sequence*

$$0 \longrightarrow B \otimes \mathbb{K} \xrightarrow[j]{} D \mathrel{\substack{\xrightarrow{s} \\ \longleftarrow \\ q}} A \longrightarrow 0$$

and a *-homomorphism $f : A \to D$ such that $[\phi^{(0)}, \phi^{(1)}] = f^*(\pi_s)$. D is separable if A and B are separable.

PROOF. The $\phi^{(i)}$ are homomorphisms from A to $M^s(B)$ which agree mod $B \hat{\otimes} \mathbb{K}$. Let $D = \{(a, \phi^{(0)}(a) + b) \mid a \in A, \, b \in B \hat{\otimes} \mathbb{K}\} \subseteq A \oplus M^s(B)$. (The same algebra is obtained by considering pairs $(a, \phi^{(1)}(a) + b)$.) D contains an ideal $\{(0, b) \mid b \in B \hat{\otimes} \mathbb{K}\}$ isomorphic to $B \hat{\otimes} \mathbb{K}$; the quotient is isomorphic to A via projection q onto the first coordinate. $s : A \to D$ defined by $s(a) = (a, \phi^{(1)}(a))$ is a cross section. Let $f : A \to D$ be defined by $f(a) = (a, \phi^{(0)}(a))$; then it is easily checked that $f^*(\pi_s) = [\phi^{(0)}, \phi^{(1)}]$. $\square$

COROLLARY 17.8.4. *Let A and B be trivially graded C^*-algebras, and $\boldsymbol{x} \in KK(A, B)$. Then there is a split extension D of B and a *-homomorphism $f : A \to D$ such that $\boldsymbol{x} = f^*(\pi_s)$, where $\pi_s \in KK(D, B)$ is the splitting morphism. If A and B are separable we may assume D is separable.*

17.8.5. There is another related functorial construction. If A, B, D are C^*-algebras, there is a map from $\mathbb{E}(A, B)$ to $\mathbb{E}(A \hat{\otimes} D, B \hat{\otimes} D)$ given by $(E, \phi, F) \to$

$(E \hat{\otimes} D, \phi \hat{\otimes} 1, F \hat{\otimes} 1)$. Again, this map respects direct sums and the equivalence relation and so induces homomorphisms $\tau_D : KK(A, B) \to KK(A \hat{\otimes} D, B \hat{\otimes} D)$, etc. The homomorphism τ_D is natural in each variable.

Combining this construction with the previous ones, if $h : D_1 \to D_2$, we can define $\hat{\otimes} h = (1 \hat{\otimes} h)_* \circ \tau_{D_1} : KK(A, B) \to KK(A \hat{\otimes} D_1, B \hat{\otimes} D_2)$.

One must be slightly careful with functoriality here. If $h : D_1 \to D_2$ is essential, then for any Hilbert B-module E we may identify $(E \hat{\otimes} D_1) \hat{\otimes}_{1 \hat{\otimes} h} (B \hat{\otimes} D_2)$ with $E \hat{\otimes} D_2$, and then we have the obvious equality $(1 \hat{\otimes} h)_* \circ \tau_{D_1} = (1 \hat{\otimes} h)^* \circ \tau_{D_2}$. So this functoriality holds in KK_{oh} for *essential* maps. The difficulty is similar to the one in 17.8.2(b).

PROPOSITION 17.8.6. *For any homomorphism $h : D_1 \to D_2$, we have $(1 \hat{\otimes} h)_* \circ \tau_{D_1} = (1 \hat{\otimes} h)^* \circ \tau_{D_2}$ in $KK(A \hat{\otimes} D_1, B \hat{\otimes} D_2)$.*

PROOF. Let $(E, \phi, F) \in \mathbb{E}(A, B)$, and set $J = D_1 \hat{\otimes}_h D_2$; then J is the closed right ideal of D_2 generated by $h(D_1)$. Then $(1 \hat{\otimes} h)_* \circ \tau_{D_1}(E, \phi, F) = (E \hat{\otimes}_{\mathbb{C}} J, \phi \hat{\otimes} h, F \hat{\otimes} 1)$ and $(1 \hat{\otimes} h)^* \circ \tau_{D_2}(E, \phi, F) = (E \hat{\otimes}_{\mathbb{C}} D_2, \phi \hat{\otimes} h, F \hat{\otimes} 1)$; we must show that these modules are homotopic. Let $\tilde{E} = \{f : [0, 1] \to D_2 \mid f(1) \in J\}$; then $\tilde{E}$ is a Hilbert D_2-module in the obvious way. Let $\tilde{h}$ be the action of D_1 on $\tilde{E}$ given by multiplication by constant functions (via h). Then $(E \hat{\otimes}_{\mathbb{C}} \tilde{E}, \phi \hat{\otimes} \tilde{h}, F \hat{\otimes} 1)$ gives the desired homotopy. $\square$

PROPOSITION 17.8.7. *For any A and B, $\tau_{\mathbb{K}} : KK(A, B) \to KK(A \hat{\otimes} \mathbb{K}, B \hat{\otimes} \mathbb{K})$ is an isomorphism, and similarly for KK_{oh} and KK_c. ($\mathbb{K}$ has standard even grading.)*

PROOF. The inverse map sends the module $(E, \phi, F) \in \mathbb{E}(A \hat{\otimes} \mathbb{K}, B \hat{\otimes} \mathbb{K})$ to $(E \hat{\otimes}_i \mathbb{H}_B, \phi \circ h, F \hat{\otimes} 1)$, where $i : B \hat{\otimes} \mathbb{K} \to \mathbb{B}(\mathbb{H}_B) \cong M(B \hat{\otimes} \mathbb{K})$ and $h : A \to A \hat{\otimes} \mathbb{K}$ are as in 17.8.2(b). It is routine to verify that this map has the required properties. $\square$

COROLLARY 17.8.8. *For any A and B, and for any m and n, there are natural isomorphisms*

$$KK(A, B) \cong KK(A \hat{\otimes} \mathbb{M}_n, B \hat{\otimes} \mathbb{M}_m) \cong KK(A \hat{\otimes} \mathbb{K}, B) \cong KK(A, B \hat{\otimes} \mathbb{K})$$
$$\cong KK(A \hat{\otimes} \mathbb{K}, B \hat{\otimes} \mathbb{K}).$$

The same holds for KK_{oh} and KK_c. $\mathbb{M}_n$ and $\mathbb{K}$ can have any (even) grading, including the trivial grading. So we have natural isomorphisms $KK^n(A, B) \cong KK^{n+2}(A, B)$ for all n and for all A, B.

COROLLARY 17.8.9 (FORMAL BOTT PERIODICITY). *For any A and B the map $\tau_{\mathbb{C}_1} : KK(A, B) \to KK(A \hat{\otimes} \mathbb{C}_1, B \hat{\otimes} \mathbb{C}_1)$ is an isomorphism, and similarly for KK_{oh} and KK_c. So for any A and B there are natural isomorphisms*

$$KK^1(A, B) \cong KK(A \hat{\otimes} \mathbb{C}_1, B)$$

and

$$KK(A, B) \cong KK^1(A, B \hat{\otimes} \mathbb{C}_1) \cong KK^1(A \hat{\otimes} \mathbb{C}_1, B) \cong KK(A \hat{\otimes} \mathbb{C}_1, B \hat{\otimes} \mathbb{C}_1),$$

and similarly for KK_{oh}, KK_c.

PROOF. Use 17.7.2 and 14.5.5 to conclude that $\tau_{\mathbb{C}_1} \circ \tau_{\mathbb{C}_1} = \tau_{\mathbb{M}_2}$ is an isomorphism. Then $\tau_{\mathbb{C}_1}$ has $\tau_{\mathbb{M}_2}^{-1} \circ \tau_{\mathbb{C}_1}$ as inverse. □

We will use Formal Bott Periodicity to prove usual Bott Periodicity in Section 19. We need the intersection product first.

17.9. Homotopy Invariance

We can use the functoriality of 17.8 to prove that KK is homotopy-invariant in each variable.

PROPOSITION 17.9.1. (a) *Let* $g_0, g_1 : D \to B$ *be homotopic homomorphisms. Then, for any* A, $g_{0*} = g_{1*} : KK(A, D) \to KK(A, B)$.
(b) *Let* $f_0, f_1 : A \to D$ *be homotopic homomorphisms. Then, for any* B, $f_0^* = f_1^* : KK(D, B) \to KK(A, B)$.

PROOF. (a) Let $g : D \to IB$ be a homotopy between g_0 and g_1. Then, if $(E, \phi, F) \in \mathbb{E}(A, D)$, the map $g_*(E, \phi, F) \in \mathbb{E}(A, IB)$ gives a homotopy between $g_{0*}(E, \phi, F)$ and $g_{1*}(E, \phi, F)$.

(b) Let $f : A \to ID$ be a homotopy between f_0 and f_1. If $(E, \phi, F) \in \mathbb{E}(D, B)$, then $f^*(\tau_{C([0,1])}(E, \phi, F)) \in \mathbb{E}(A, IB)$ gives a homotopy between $f_0^*(E, \phi, F)$ and $f_1^*(E, \phi, F)$. □

Homotopy invariance of KK_{oh} and KK_c is much more delicate (it is in effect the statement that $\sim_{oh}$ and $\sim_c$ coincide with $\sim_h$), and cannot be shown without the intersection product (see 18.5.3).

17.10. Cobordism and Isomorphism of KK_{oh} and KK_c

We now prove that the natural map from $KK_c(A, B)$ to $KK_{oh}(A, B)$ is an isomorphism (i.e. that $\sim_c$ and $\sim_{oh}$ coincide) when A is separable. The proof does not require use of the intersection product, although it does require one (easy) application of Kasparov's Technical Theorem (in 17.10.5).

The proof entails another interesting notion, cobordism. The notion of cobordism in Kasparov theory is due to Cuntz and Skandalis [1986].

DEFINITION 17.10.1. Let $(E, \phi, F) \in \mathbb{E}(A, B)$, and p a projection in $\mathbb{B}(E)$, of degree 0, commuting with $\phi(A)$. Then $(E, \phi, F)_p$ is the Kasparov module (pE, ψ, pFp), where $\psi(a) = p\phi(a) = p\phi(a)p$.

DEFINITION 17.10.2. Two elements (E_0, ϕ_0, F_0) and (E_1, ϕ_1, F_1) of $\mathbb{E}(A, B)$ are *cobordant* if there is an $(E, \phi, F) \in \mathbb{E}(A, B)$ and a partial isometry $v \in \mathbb{B}(E)$ of degree 0, commuting with $\phi(A)$, such that

(i) $[v, F]\phi(a) \in \mathbb{K}(E)$ for all a,

(ii) $(E, \phi, F)_{1-vv^*}$ is unitarily equivalent to (E_0, ϕ_0, F_0), and

(iii) $(E, \phi, F)_{1-v^*v}$ is unitarily equivalent to (E_1, ϕ_1, F_1).

The module (E, ϕ, F, v) may be regarded as a "Kasparov module with boundary $(E, \phi, F)_{1-vv^*} - (E, \phi, F)_{1-v^*v}$." The term "cobordism" is in analogy with a similar use in topology.

PROPOSITION 17.10.3. *Cobordism is an equivalence relation compatible with direct sums.*

PROOF. Everything is obvious except transitivity. Suppose that (E', ϕ', F', v') and (E'', ϕ'', F'', v'') are Kasparov modules with boundary, and that

$$u \in \mathbb{B}\big((1 - v'^*v')E', \ (1 - v''v''^*)E''\big)$$

gives a unitary equivalence between $(E', \phi', F')_{1-v'^*v'}$ and $(E'', \phi'', F'')_{1-v''v''^*}$. Let $v = (v' \oplus v'') + u \in \mathbb{B}(E' \oplus E'')$. Then $(E' \oplus E'', \ \phi' \oplus \phi'', \ F' \oplus F'', \ v' \oplus v'')$ gives a cobordism between $(E', \phi', F')_{1-v'v'^*}$ and $(E'', \phi'', F'')_{1-v''^*v''}$. $\qquad\square$

PROPOSITION 17.10.4. *Cobordism is exactly the equivalence relation $\sim_c$.*

PROOF. Unitarily equivalent modules are cobordant (with $v = 0$). If (E, ϕ, F) is degenerate, let v be an isometry of degree -1 on a separable Hilbert space $\mathbb{H}$; then $(\mathbb{H} \hat{\otimes} E, 1 \otimes \phi, 1 \otimes F, v \otimes 1)$ gives a cobordism between (E, ϕ, F) and the 0-module. If $(E, \phi, F) \in \mathbb{E}(A, B)$ and F' is a "compact perturbation" of F, then $\big(E \oplus E, \ \phi \oplus \phi, \ F \oplus F', \ \big[\begin{smallmatrix} 0 & 0 \\ 1 & 0 \end{smallmatrix}\big]\big)$ defines a cobordism between (E, ϕ, F) and (E, ϕ, F'). So $\sim_c$ implies cobordism. Conversely, let (E, ϕ, F, v) define a cobordism between (E_0, ϕ_0, F_0) and (E_1, ϕ_1, F_1). Then (E, ϕ, F) is a "compact perturbation" of $(E, \phi, F)_{vv^*} \oplus (E, \phi, F)_{1-vv^*}$ and also of $(E, \phi, F)_{v^*v} \oplus (E, \phi, F)_{1-v^*v}$, so $(E_1, \phi_1, F_1) \oplus (E, \phi, F)_{v^*v}$ is a "compact perturbation" of $(E_0, \phi_0, F_0) \oplus (E, \phi, F)_{vv^*}$. But $(E, \phi, F)_{v^*v}$ and $(E, \phi, F)_{vv^*}$ are unitarily equivalent, so $(E_0, \phi_0, F_0) \sim_c (E_1, \phi_1, F_1)$. $\qquad\square$

Before proving the main result, we need two lemmas on extending operator homotopies (mod $\mathbb{K}$) from ideals. The proof in spirit is similar to the proof of 17.2.7.

LEMMA 17.10.5. *Let A be a graded separable C^*-algebra, I a closed two-sided ideal of B. Let E be a countably generated Hilbert B-module for some B, and $\phi : A \to \mathbb{B}(E)$ a $*$-homomorphism. Set*

$$C = \{x \in \mathbb{B}(E) \mid [x, \phi(a)] \in \mathbb{K}(E) \text{ for all } a \in A\},$$

$$K = \{x \in \mathbb{B}(E) \mid x\phi(a) \in \mathbb{K}(E) \text{ for all } a \in A\},$$

$$D = \{x \in \mathbb{B}(E) \mid [x, \phi(a)] \in \mathbb{K}(E) \text{ for all } x \in I\},$$

$$L = \{x \in \mathbb{B}(E) \mid x\phi(a) \in \mathbb{K}(E) \text{ for all } x \in I\}.$$

These are all graded C^*-subalgebras of $\mathbb{B}(E)$; K is an ideal of C, L an ideal of D. Then $D = C + L$.

PROOF. Let $x \in D$ be homogeneous. Set $J = \mathbb{K}(E)$, $A_1 = \mathbb{C} \cdot 1 + \mathbb{K}(E)$, A_2 the C^*-subalgebra of $\mathbb{B}(E)$ generated by $\mathbb{K}(E)$ and $\{[x, \phi(a)] \mid a \in A\}$, Δ the subspace generated by x and $\phi(A)$. Apply Kasparov's Technical Theorem (14.6.2) to get M and N; we have $Mx \in L$, $Nx \in C$. □

LEMMA 17.10.6. *Let A and B be graded C^*-algebras, with A separable, and let $(E, \phi, F) \in \mathbb{E}(A, B)$. Let I be an ideal in A. Suppose there is an operator homotopy (E, ϕ, F_t) in $\mathbb{E}(I, B)$ with $F_0 = F$. Then there is an operator homotopy (E, ϕ, G_t) in $\mathbb{E}(A, B)$ with $G_0 = F$ and $(G_t - F_t)\phi(a) \in \mathbb{K}(E)$ for all t and $a \in I$.*

PROOF. Let C, K, D, L be as in 17.10.5. Then $F \in C$, $F_t \in D$ for all t. Let $q : C/K \to D/L$ be the map induced by the inclusion of C into D; by 17.10.5 q is surjective. If f_t is the image of F_t in D/L and g_0 the image of F in C/K, then the f_t and g_0 are self-adjoint unitaries of degree 1. There is a continuous path $\langle g_t \rangle$ of preimages in C/K, consisting of self-adjoint unitaries of degree 1 (break up $[0, 1]$ into small subintervals on which f_t is nearly constant; successively lift each interval to an interval of invertible self-adjoint elements of degree 1, and take the unitaries in the polar decompositions). Then let (G_t) be a continuous lift of (g_t) in C with $G_0 = F$ and each G_t of degree 1 (3.4.6). □

THEOREM 17.10.7. *If A is separable and B is σ-unital, then $\sim_c$ and $\sim_{oh}$ coincide on $\mathbb{E}(A, B)$.*

PROOF. Let u be the unilateral shift on a (trivially graded) Hilbert space $\mathbb{H}$, and T the C^*-algebra generated by u (9.4.2). Then T contains $\mathbb{K}(\mathbb{H})$; the projection $p = 1 - uu^*$ is a one-dimensional projection. Let (E, ϕ, F) be a Kasparov (A, B)-bimodule which is operator homotopic to a degenerate element (E, ϕ, F'). The restriction of $(E \hat{\otimes}_{\mathbb{C}} \mathbb{H}, \phi \hat{\otimes} 1_T, F' \hat{\otimes} 1) \in \mathbb{E}(A \hat{\otimes} T, B)$ to a Kasparov $(A \hat{\otimes} \mathbb{K}(\mathbb{H}), B)$-bimodule is operator homotopic to $(E \hat{\otimes}_{\mathbb{C}} \mathbb{H}, \phi \hat{\otimes} 1_{\mathbb{K}(\mathbb{H})}, F \hat{\otimes} 1)$. By 17.10.6 there is a $G \in \mathbb{B}(E \hat{\otimes}_{\mathbb{C}} \mathbb{H})$ such that $(E \hat{\otimes}_{\mathbb{C}} \mathbb{H}, \phi \hat{\otimes} 1_T, G) \in \mathbb{E}(A \hat{\otimes} T, B)$ and

$$(G - F \hat{\otimes} 1)(\phi \hat{\otimes} 1)(x) \in \mathbb{K}(E \hat{\otimes}_{\mathbb{C}} \mathbb{H})$$

for all $x \in A \hat{\otimes} \mathbb{K}(\mathbb{H})$.

If ψ is the map of $A \cong A \hat{\otimes} 1 \subseteq A \hat{\otimes} T$ into $\mathbb{B}(E \hat{\otimes}_{\mathbb{C}} \mathbb{H})$ defined by $\phi \hat{\otimes} 1_T$, then $(E \hat{\otimes}_{\mathbb{C}} \mathbb{H}, \psi, G, 1 \hat{\otimes} u)$ defines a cobordism between

$$\left(E \hat{\otimes}_{\mathbb{C}} p\mathbb{H}, (1 \hat{\otimes} p)\psi(1 \hat{\otimes} p), (1 \hat{\otimes} p)G(1 \hat{\otimes} p) \right)$$

and the 0-module. But, by 17.10.4, (E, ϕ, F) is cobordant to

$$\left(E \hat{\otimes}_{\mathbb{C}} p\mathbb{H}, (1 \hat{\otimes} p)\psi(1 \hat{\otimes} p), (1 \hat{\otimes} p)G(1 \hat{\otimes} p) \right).$$ □

17.11. Unbounded Kasparov Modules

It is sometimes convenient to define Kasparov modules using an unbounded operator in place of F.

Baaj and Julg [1983] showed how to define $KK(A, B)$ in terms of unbounded Kasparov modules. We call this approach the *Baaj–Julg picture of $KK(A, B)$*.

DEFINITION 17.11.1. Let A and B be graded C^*-algebras. An *unbounded Kasparov module* for (A, B) is a triple (E, ϕ, D) where E is a Hilbert B-module, $\phi : A \to \mathbb{B}(E)$ is a graded *-homomorphism, and D is a self-adjoint regular operator on E, homogeneous of degree 1, such that

(i) $(1+D^2)^{-1}\phi(a)$ extends to an element of $\mathbb{K}(E)$ for all $a \in A$, and
(ii) the set of $a \in A$ such that $[D, \phi(a)]$ is densely defined and extends to an element of $\mathbb{B}(E)$, is dense in A.

It is possible to relax somewhat the condition that D be actually self-adjoint; in applications such modules arise naturally which are not exactly of the above form.

Denote the set of all unbounded Kasparov modules for (A, B) by $\Psi_1(A, B)$. $\Psi_1(A, B)$ is a semigroup under orthogonal direct sum. (The notation suggests that an unbounded Kasparov module is a sort of pseudodifferential operator of degree 1 from A to B.)

EXAMPLE 17.11.2. Let M be a compact Riemannian manifold and P an elliptic pseudodifferential operator of degree 1 between smooth vector bundles $V^{(0)}$ and $V^{(1)}$ over M. Then, as in 17.2.1(d),

$$\left(L^2(V^{(0)}) \oplus L^2(V^{(1)}), \quad \phi, \quad \begin{bmatrix} 0 & P^* \\ P & 0 \end{bmatrix} \right)$$

gives an unbounded Kasparov $(C(M), \mathbb{C})$-module.

The advantage of considering unbounded Kasparov modules in this situation is that it is much easier to take tensor products of elliptic pseudodifferential operators of degree 1 than it is for operators of degree 0 (cf. 18.9).

Define a map $\beta : \Psi_1(A, B) \to \mathbb{E}(A, B)$ as follows. If (E, ϕ, D) is an unbounded Kasparov module, then $D(1+D^2)^{-1}$ extends to an operator $F \in \mathbb{B}(E)$.

PROPOSITION 17.11.3. $(E, \phi, F) \in \mathbb{E}(A, B)$.

PROOF. F is self-adjoint and of degree 1. Next, $1 - F^2$ is the extension of $(1+D^2)^{-1}$, so $(F^2-1)\phi(a) \in \mathbb{K}(E)$ for all a. It is a little harder to show that $[F, \phi(a)] \in \mathbb{K}(E)$. If a is such that $[D, \phi(a)]$ extends to an element of $\mathbb{B}(E)$ (such a are dense in A), then we have

$$[F, \phi(a)] = [D, \phi(a)](1+D^2)^{-1/2} + D[(1+D^2)^{-1/2}, \phi(a)]$$

By 17.1.3 it suffices to show that $[F, \phi(a)] \in {}^c\phi(A)$, i.e. $[F, \phi(a)]\phi(b) \in \mathbb{K}(E)$ for all $b \in A$. $(1+D^2)^{-1} \in {}^c\phi(A)^c$, so $(1+D^2)^{-1/2} \in {}^c\phi(A)^c$; thus $[D, \phi(a)](1+D^2)^{1/2}\phi(b) \in \mathbb{K}(E)$ for any b. Also, since

$$(1+D^2)^{-1/2} = \frac{1}{\pi} \int_0^\infty \lambda^{-1/2}(1+D^2+\lambda)^{-1} \, d\lambda$$

with a uniformly converging integral, and since $[D, \phi(a)]$ is bounded, the integral

$$\frac{1}{\pi} \int_0^\infty \lambda^{-1/2} D[(1+D^2+\lambda)^{-1}, \phi(a)] \phi(b) \, d\lambda$$

converges uniformly to $D[(1+D^2)^{-1/2}, \phi(a)]\phi(b)$. Also,

$$D[(1+D^2+\lambda)^{-1}, \phi(a)] \phi(b) = D(1+D^2+\lambda)^{-1/2}(1+D^2+\lambda)^{-1/2}\phi(ab)$$
$$-[D, \phi(a)](1+D^2+\lambda)^{-1}\phi(b) \pm \phi(a)(1+D^2+\lambda)^{-1/2}D(1+D^2+\lambda)^{-1/2}\phi(b).$$

These terms are all in $\mathbb{K}(E)$ since $D(1+D^2+\lambda)^{-1/2} \in \mathbb{B}(E)$ and

$$(1+D^2+\lambda)^{-1/2} \in {}^c\phi(A)^c. \qquad \square$$

THEOREM 17.11.4. *If A is separable, then the map from $\Psi_1(A, B)$ to $KK_c(A, B)$ is surjective.*

PROOF. Let (E, ϕ, F) be a Kasparov module, with $F = F^*$ and $F^2 = 1$. Let (a_j) be a sequence in A which is total. Let h be a strictly positive element of $\mathbb{K}(E)$ of degree 0, which commutes with F (e.g. $h_0 + Fh_0F$ for h_0 strictly positive of degree 0).

We need a lemma reminiscent of the Kasparov Technical Theorem (14.6.2).

LEMMA 17.11.5. *There is an $l \in C^*(h)$, strictly positive of degree 0, such that, for all j,*

(i) *$[F, \phi(a_j)]l^{-1}$ extends to an element of $\mathbb{B}(E)$, and*
(ii) *$[l^{-1}, \phi(a_j)]$ is defined on the domain of l^{-1} and extends to an element of $\mathbb{B}(E)$.*

PROOF. By [Pedersen 1979, 3.12.14] (cf. 12.4.1), there exists an approximate identity u_n for $\mathbb{K}(E)$, contained in $C^*(h)$, quasicentral for $\phi(A)$, with the property that $u_{n+1}u_n = u_n$ for all n. By passing to a subsequence we may assume that $\|[F, \phi(a_j)]d_n\| < 2^{-2n}$ and $\|[d_n, \phi(a_j)]\| < 2^{-2n}$ for $1 \le j \le n$, where $d_n = u_{n+1} - u_n$. Set $X = \widehat{C^*(h)} \cong \sigma(h)\setminus\{0\}$, and let X_n be the support of u_n. Then $\langle X_n \rangle$ is an increasing sequence of compact subsets of X with $X = \bigcup X_n$. The sum $\sum 2^n d_n$ converges pointwise on X to an unbounded function r; $r \ge 2^n$ on $X \setminus X_n$. Then $l = r^{-1}$ defines an element of $C^*(h)$. For any j the sequences $\sum_n 2^n[F, \phi(a_j)]d_n$ and $\sum_n 2^n[d_n, \phi(a_j)]$ are norm-convergent. $\qquad \square$

PROOF OF 17.11.4 (CONT.). Let l be as in 17.11.5, and set $D = Fl^{-1}$. Then $D = D^*$ and $(1+D^2)^{-1}$ extends to $l^2(1+l^2)^{-1} \in \mathbb{K}(E)$. For all j, we have that $[D, \phi(a_j)] = [F, \phi(a_j)]l^{-1} + F[l^{-1}, \phi(a_j)]$ extends to an element of $\mathbb{B}(E)$. Thus $(E, \phi, D) \in \Psi_1(A, B)$. $\beta(E, \phi, D) = (E, \phi, G)$, where $G = D(1+D^2)^{-1/2} = F(1+l^2)^{-1/2}$. G is a "compact perturbation" of F. $\qquad \square$

Thus every KK-element can be represented by an unbounded Kasparov module. We leave to the reader the task of appropriately formulating the equivalence relations on $\Psi_1(A, B)$ corresponding to the standard relations on $\mathbb{E}(A, B)$; see [Baaj and Julg 1983; Hilsum 1985].

18. The Intersection Product

This section contains the heart of Kasparov theory, the intersection product (often called the Kasparov product). This product generalizes composition and tensor product of *-homomorphisms, and also the cup and cap products of topological K-theory. The intersection product is the device used to prove all the deeper properties of the KK-groups.

18.1. Description of the Product

The most general form of the product is a pairing

$$\hat{\otimes}_D : KK(A_1, B_1 \hat{\otimes} D) \times KK(D \hat{\otimes} A_2, B_2) \to KK(A_1 \hat{\otimes} A_2, B_1 \hat{\otimes} B_2).$$

However, we will restrict attention first to the special case where $B_1 = A_2 = \mathbb{C}$, from which the general case can be derived. That is, we will define a map

$$\hat{\otimes}_D : KK(A, D) \times KK(D, B) \to KK(A, B).$$

If $x \in KK(A, D)$ and $y \in KK(D, B)$, we will write $x \hat{\otimes}_D y$, $x \cap y$, xy, or $y \circ x$ for the product, which lies in $KK(A, B)$. The various notations reflect the different interpretations of the product from the standard points of view.

18.2. Outline of the Construction

We will define the product as follows. Given $x \in KK(A, D)$, $y \in KK(D, B)$, choose representatives $(E_1, \phi_1, F_1) \in \mathbb{E}(A, D)$ and $(E_2, \phi_2, F_2) \in \mathbb{E}(D, B)$; then define $(E, \phi, F) \in \mathbb{E}(A, B)$ by $E = E_1 \hat{\otimes}_{\phi_2} E_2$, $\phi = \phi_1 \hat{\otimes}_{\phi_2} 1$, $F = F_1 \# F_2$, where $F_1 \# F_2$ is a suitable combination of F_1 and F_2.

The entire technical difficulty is in finding a suitable definition of $F_1 \# F_2$. In special cases it is clear what to do. For example, if x is induced by a homomorphism $f : A \to D$, we may choose $E_1 = D$, $\phi_1 = f$, $F_1 = 0$; then $E \cong E_2$, and we may take $F = F_2$. More generally, if x comes from a homomorphism $f : A \to D \hat{\otimes} \mathbb{K}$, we may take $E_1 = \mathbb{H}_D$, $\phi_1 = f$, $F_1 = 0$. Then $E \cong \mathbb{H} \hat{\otimes}_{\mathbb{C}} E_2$, and we may take $F = 1 \hat{\otimes} F_2$. On the other hand, if y is induced by a homomorphism $g : D \to B$, we may take $E_2 = B$, $\phi_2 = g$, $F_2 = 0$. Then we may take $F = F_1 \hat{\otimes} 1$. The situation when y comes from $g : D \to B \hat{\otimes} \mathbb{K}$ is similar. In each case we end up with the composition defined in 17.8.

In general, F will be a combination of $F_1 \hat{\otimes} 1$ and $1 \hat{\otimes} F_2$ (when these are suitably defined). The coefficients will come from a judiciously chosen "partition of unity". It is a quite delicate matter to choose these coefficients in general so that (E, ϕ, F) is actually a Kasparov module for (A, B) with the right properties. Roughly speaking, we will want to take F to be of the form

$$\left[\frac{1 \hat{\otimes} F_2^2}{F_1^2 \hat{\otimes} 1 + 1 \hat{\otimes} F_2^2} \right]^{1/2} (F_1 \hat{\otimes} 1) + \left[\frac{F_1^2 \hat{\otimes} 1}{F_1^2 \hat{\otimes} 1 + 1 \hat{\otimes} F_2^2} \right]^{1/2} (1 \hat{\otimes} F_2)$$

We need to make sense out of this expression.

EXAMPLES 18.2.1. To get an idea of what is involved in the general construction, let us look at two simple examples.

(a) Suppose $A = D = B = \mathbb{C}$,

$$(E_1, \phi_1, F_1) = \left(\mathbb{H}_1 \oplus \mathbb{H}_1^{op}, 1, \begin{bmatrix} 0 & V^* \\ V & 0 \end{bmatrix} \right),$$

$$(E_2, \phi_2, F_2) = \left(\mathbb{H}_2 \oplus \mathbb{H}_2^{op}, 1, \begin{bmatrix} 0 & W^* \\ W & 0 \end{bmatrix} \right)$$

as in 17.3.4, with V and W Fredholm partial isometries. Then $E = E_1 \,\hat{\otimes}_{\phi_2}\, E_2 \cong (\mathbb{H}_1 \oplus \mathbb{H}_1^{op}) \,\hat{\otimes}\, (\mathbb{H}_2 \oplus \mathbb{H}_2^{op})$ as an external tensor product, so we may define $F_1 \,\hat{\otimes}\, 1$ and $1 \,\hat{\otimes}\, F_2$ as in 14.4.4. Under the isomorphism of this external tensor product with the internal tensor product $(\mathbb{H}_1 \oplus \mathbb{H}_1^{op}) \,\hat{\otimes}_{\mathbb{C}}\, (\mathbb{H}_2 \oplus \mathbb{H}_2^{op}) \cong [(\mathbb{H}_1 \oplus \mathbb{H}_1^{op}) \,\hat{\otimes}\, \mathbb{H}_2] \oplus [(\mathbb{H}_1 \oplus \mathbb{H}_1^{op}) \,\hat{\otimes}\, \mathbb{H}_2^{op}]$, these operators become

$$F_1 \hat{\otimes} 1 = \left[\begin{array}{cc|cc} 0 & V^* \hat{\otimes} 1 & & \\ V \hat{\otimes} 1 & 0 & & 0 \\ \hline & & 0 & V^* \hat{\otimes} 1 \\ & 0 & V \hat{\otimes} 1 & 0 \end{array} \right],$$

$$1 \hat{\otimes} F_2 = \left[\begin{array}{cc|cc} & & 1 \hat{\otimes} W^* & 0 \\ & 0 & 0 & -1 \hat{\otimes} W^* \\ \hline 1 \hat{\otimes} W & 0 & & \\ 0 & -1 \hat{\otimes} W & & 0 \end{array} \right].$$

We collect together the homogeneous spaces by interchanging the second and fourth summands to obtain

$$(\mathbb{H}_1 \otimes \mathbb{H}_2 \oplus \mathbb{H}_1^{op} \otimes \mathbb{H}_2^{op}) \oplus (\mathbb{H}_1 \otimes \mathbb{H}_2^{op} \oplus \mathbb{H}_1^{op} \otimes \mathbb{H}_2);$$

the operators become

$$F_1 \hat{\otimes} 1 = \left[\begin{array}{cc|cc} & & 0 & V^* \otimes 1 \\ & 0 & V \otimes 1 & 0 \\ \hline 0 & V^* \otimes 1 & & \\ V \otimes 1 & 0 & & 0 \end{array} \right],$$

$$1 \hat{\otimes} F_2 = \left[\begin{array}{cc|cc} & & 1 \otimes W^* & 0 \\ & 0 & 0 & -1 \otimes W \\ \hline 1 \otimes W & 0 & & \\ 0 & -1 \otimes W^* & & 0 \end{array} \right].$$

If we form $F = F_1 \,\hat{\otimes}\, 1 + 1 \,\hat{\otimes}\, F_2$, then F is a self-adjoint Fredholm operator, but $F^2 - 1 \notin \mathbb{K}$, in fact $F^2 = \mathrm{diag}(V^*V \otimes 1 + 1 \otimes W^*W, \; VV^* \otimes 1 + 1 \otimes WW^*, \; V^*V \otimes 1 + 1 \otimes WW^*, \; VV^* \otimes 1 + 1 \otimes W^*W)$.

We can make sense out of the operators

$$M = \frac{F_1^2 \,\hat{\otimes}\, 1}{F_1^2 \,\hat{\otimes}\, 1 + 1 \,\hat{\otimes}\, F_2^2} \quad \text{and} \quad N = \frac{1 \,\hat{\otimes}\, F_2^2}{F_1^2 \,\hat{\otimes}\, 1 + 1 \,\hat{\otimes}\, F_2^2}$$

by setting $P_1 = V^*V$, $Q_1 = VV^*$, $P_2 = W^*W$, $Q_2 = WW^*$; then M should be an operator which is

$$1 \text{ on } (P_1\mathbb{H}_1 \oplus Q_1\mathbb{H}_1^{op}) \otimes ((1 - P_2)\mathbb{H}_2 \oplus (1 - Q_2)\mathbb{H}_2^{op}),$$
$$0 \text{ on } ((1 - P_1)\mathbb{H}_1 \oplus (1 - Q_1)\mathbb{H}_1^{op}) \otimes (P_2\mathbb{H}_2 \oplus Q_2\mathbb{H}_2^{op}),$$
$$\tfrac{1}{2} \text{ on } (P_1\mathbb{H}_1 \oplus Q_1\mathbb{H}_1^{op}) \otimes (P_2\mathbb{H}_2 \oplus Q_2\mathbb{H}_2^{op}).$$

(It doesn't matter how M is defined on the finite-dimensional subspace

$$((1 - P_1)\mathbb{H}_1 \oplus (1 - Q_1)\mathbb{H}_1^{op}) \otimes ((1 - P_2)\mathbb{H}_2 \oplus (1 - Q_2)\mathbb{H}_2^{op})$$

except that we want $0 \leq M \leq 1$; for symmetry take M to be $\frac{1}{2}$ on this subspace.)
Then let $N = 1 - M$; it follows that $F' = M^{1/2}(F_1 \,\hat{\otimes}\, 1) + N^{1/2}(1 \,\hat{\otimes}\, F_2)$ will have all the right properties to make $(E, 1, F)$ an element of $\mathbb{E}(\mathbb{C}, \mathbb{C})$, and this module is a natural candidate for the product.

This example is due to R. Douglas.

(b) We consider essentially the same example from a different point of view. This time, take the modules to be

$$\left(\mathbb{H}_i \oplus \mathbb{H}_i^{op}, \begin{bmatrix} P_i & 0 \\ 0 & Q_i \end{bmatrix}, \begin{bmatrix} 0 & 1 \\ 1 & 0 \end{bmatrix} \right)$$

for $i = 1, 2$, where P_i, Q_i are projections with $P_i - Q_i$ compact. Now $E \cong (\mathbb{H}_1 \oplus \mathbb{H}_1^{op}) \,\hat{\otimes}\, (P_2\mathbb{H}_2 \oplus Q_2\mathbb{H}_2^{op})$ as an external tensor product.

There is a problem in even defining $1 \,\hat{\otimes}\, F_2$ in this case. To give a satisfactory definition, write E as

$$\left(1 \,\hat{\otimes}\, \begin{bmatrix} P_2 & 0 \\ 0 & Q_2 \end{bmatrix} \right) [(\mathbb{H}_1 \oplus \mathbb{H}_1^{op}) \,\hat{\otimes}\, (\mathbb{H}_2 \oplus \mathbb{H}_2^{op})].$$

Then $1 \,\hat{\otimes}\, F_2$ can be defined as

$$\left(1 \,\hat{\otimes}\, \begin{bmatrix} P_2 & 0 \\ 0 & Q_2 \end{bmatrix} \right) \left(1 \,\hat{\otimes}\, \begin{bmatrix} 0 & 1 \\ 1 & 0 \end{bmatrix} \right) \left(1 \,\hat{\otimes}\, \begin{bmatrix} P_2 & 0 \\ 0 & Q_2 \end{bmatrix} \right) = 1 \,\hat{\otimes}\, \begin{bmatrix} 0 & P_2Q_2 \\ Q_2P_2 & 0 \end{bmatrix}.$$

$F_1 \,\hat{\otimes}\, 1$ makes sense as $\begin{bmatrix} 0 & 1 \\ 1 & 0 \end{bmatrix} \,\hat{\otimes}\, 1$.

Now suppose for simplicity that $P_2 = Q_2 = 1$ (the second module is then degenerate). If we expand the tensor product and collect together homogeneous subspaces as in (a), the operators become

$$F_1 \hat{\otimes} 1 = \left[\begin{array}{cc|cc} & & 0 & 1\otimes1 \\ \multicolumn{2}{c|}{0} & 1\otimes1 & 0 \\ \hline 0 & 1\otimes1 & & \\ 1\otimes1 & 0 & \multicolumn{2}{c}{0} \end{array} \right],$$

$$1\hat{\otimes}F_2 = \left[\begin{array}{cc|cc} & & 1\otimes1 & 0 \\ \multicolumn{2}{c|}{0} & 0 & -1\otimes1 \\ \hline 1\otimes1 & 0 & & \\ 0 & -1\otimes1 & \multicolumn{2}{c}{0} \end{array} \right].$$

Now we have $(F_1\hat{\otimes}1+1\hat{\otimes}F_2)^2 = F_1^2\hat{\otimes}1+1\hat{\otimes}F_2^2 = 2$, so if we blindly follow the previous formulas for M and N we get $M = N = \frac{1}{2}$, so $F = \frac{1}{\sqrt{2}}(F_1\hat{\otimes}1+1\hat{\otimes}F_2)$. However, unless $P_1 = Q_1$, $\left(E, \left[\begin{smallmatrix} P_1 & 0 \\ 0 & Q_1 \end{smallmatrix}\right], F\right)$ is not a Kasparov $(\mathbb{C}, \mathbb{C})$-module since

$$[F, \phi(1)] = \frac{1}{\sqrt{2}} \left[\begin{array}{cc} 0 & (Q_1-P_1)\otimes1 \\ (P_1-Q_1)\otimes1 & 0 \end{array} \right] \notin \mathbb{K}(E).$$

So we must take the formula

$$M = \frac{F_1^2 \hat{\otimes} 1}{F_1^2 \hat{\otimes} 1 + 1 \hat{\otimes} F_2^2}$$

with a grain of salt; a more delicate choice of M is necessary in this case.

Of course, in these particular situations there are simple ways of avoiding the difficulties by modifying the modules; however, these two examples (particularly the second one) illustrate the problems involved in making a general definition of the product.

18.3. Connections

The first problem is to make sense out of the operators $F_1 \hat{\otimes} 1$ and $1 \hat{\otimes} F_2$. There is no problem giving $F_1 \hat{\otimes} 1$ the obvious meaning (14.4.2), but there is no obvious meaning for $1 \hat{\otimes} F_2$ (even if F_2 commutes with $\phi(D)$). The notion of connection, due to Connes and Skandalis [1984], is designed to handle this situation.

The construction of the product using connections is a great improvement over Kasparov's original method, since the connection method makes it feasible to explicitly calculate products of KK-elements which arise naturally in applications.

Suppose E_1 is a countably generated Hilbert D-module, E_2 a countably generated Hilbert B-module, $\psi : D \to \mathbb{B}(E_2)$ is a graded *-homomorphism, and $F_2 \in \mathbb{B}(E_2)$ has the property that $[F_2, \psi(D)] \subseteq \mathbb{K}(E_2)$. (Recall that all commutators are graded commutators.) We seek an operator $F \in \mathbb{B}(E)$ $(E = E_1 \hat{\otimes}_\psi E_2)$ which "plays the role of $1 \hat{\otimes} F_2$ up to compacts." More specifically, we make the following definitions.

For any $x \in E_1$ there is an operator $T_x \in \mathbb{B}(E_2, E)$ defined by $T_x(y) = x \,\hat{\otimes}\, y$. T_x^* is defined by $T_x^*(z \,\hat{\otimes}\, y) = \psi(\langle x, z \rangle)y$.

DEFINITION 18.3.1. An operator $F \in \mathbb{B}(E)$ is called an F_2-*connection* for E_1 (or an F_2-*connection on* E) if, for any $x \in E_1$,

$$T_x \circ F_2 - (-1)^{\partial x \cdot \partial F_2} F \circ T_x \subseteq \mathbb{K}(E_2, E),$$

$$F_2 \circ T_x^* - (-1)^{\partial x \cdot \partial F_2} T_x^* \circ F \subseteq \mathbb{K}(E, E_2).$$

The commutation conditions of 18.3.1 may be more cleanly described as follows. For $x \in E_1$, define

$$\tilde{T}_x = \begin{bmatrix} 0 & T_x^* \\ T_x & 0 \end{bmatrix} \in \mathbb{B}(E_2 \oplus E).$$

Similarly, if $F \in \mathbb{B}(E)$, set $\tilde{F} = F_2 \oplus F \in \mathbb{B}(E_2 \oplus E)$. Then F is an F_2-connection if and only if $[\tilde{T}_x, \tilde{F}] \in \mathbb{K}(E_2 \oplus E)$ for all $x \in E_1$.

EXAMPLES 18.3.2. (a) If $[F_2, \psi(D)] = 0$, then $1 \,\hat{\otimes}\, F_2$ makes sense in $\mathbb{B}(E)$, and is an F_2-connection on E.

(b) If D is unital and $E_1 = \hat{\mathbb{H}}_D$, then there is a standard connection for any F_2: in this case, there is an isomorphism $\hat{\mathbb{H}}_D \,\hat{\otimes}_\psi\, E_2 \cong \hat{\mathbb{H}} \,\hat{\otimes}\, E_2$ (regarded as an external tensor product), and the operator F corresponding to $1 \,\hat{\otimes}\, F_2$ under this isomorphism is an F_2-connection for $\hat{\mathbb{H}}_D$. This connection is called the *Grassmann connection*. If $\hat{\mathbb{H}}_D \,\hat{\otimes}_\phi\, E_2$ is regarded as the internal tensor product $\hat{\mathbb{H}} \,\hat{\otimes}_{\mathbb{C}}\, E_2$, then the Grassmann connection becomes $\operatorname{diag}(1, -1) \otimes F_2$; in other words, $F(x \,\hat{\otimes}\, y) = (-1)^{\partial F \cdot \partial x}(x \,\hat{\otimes}\, F_2 y)$.

(c) Suppose $F_2 = 0$. A 0-connection on E is an operator $F \in \mathbb{B}(E)$ with the property that FT and TF are in $\mathbb{K}(E)$ for all $T \in \mathbb{K}(E_1) \,\hat{\otimes}\, 1$. To prove this, use the fact that $\mathbb{K}(E_1) \,\hat{\otimes}\, 1$ is generated by the rank one operators $\theta_{x,y} \,\hat{\otimes}\, 1 = T_x \circ T_y^*$.

PROPOSITION 18.3.3. *For any* E_1, E_2, ψ, F_2 *as above, there exists an* F_2-*connection for* E_1. *In fact, if* $t \to F_2^t$ *is a norm-continuous path of operators on* E_2, *all commuting with* $\psi(D)$ (*in the graded sense*) *mod* $\mathbb{K}(E_2)$, *then there is a norm-continuous path* F^t, *where* F^t *is an* F_2^t-*connection for* E_1. *If the* F_2^t *are homogeneous of degree* n, *then the* F^t *may be chosen homogeneous of degree* n; *if* F_2^t *is self-adjoint then* F^t *may be chosen self-adjoint.*

PROOF. This is an easy consequence of the stabilization theorem and 18.3.2(b). We can write $E_1 = P\hat{\mathbb{H}}_D$ as Hilbert $\tilde{D}$-modules, where P is a projection of degree 0 in $\mathbb{B}(\hat{\mathbb{H}}_D)$. Then $E_1 \,\hat{\otimes}_\psi\, E_2 \cong E_1 \,\hat{\otimes}_\psi\, E_2 \cong P\hat{\mathbb{H}}_D \,\hat{\otimes}_\psi\, E_2 \cong (P \,\hat{\otimes}\, 1)(\hat{\mathbb{H}}_D \,\hat{\otimes}_\psi\, E_2)$. Let G^t be the Grassmann connection of F_2^t on $\hat{\mathbb{H}}_D \,\hat{\otimes}_\psi\, E_2$, and set

$$F^t = (P \,\hat{\otimes}\, 1)G^t(P \,\hat{\otimes}\, 1). \qquad \square$$

The connections defined in the proof of 18.3.3 are also called Grassmann connections.

PROPOSITION 18.3.4. *Let* E_1, E_2, ψ, F_2, F_2' *be as above.*

(a) *If F is an F_2-connection, then F^* is an F_2^*-connection, $F^{(0)}$ is an $F_2^{(0)}$-connection, and $F^{(1)}$ is an $F_2^{(1)}$-connection.*

(b) *If F is an F_2-connection and F' is an F_2'-connection, then $F + F'$ is an $(F_2 + F_2')$-connection and FF' is an (F_2F_2')-connection.*

(c) *The set of all F_2-connections is an affine space parallel to the space of all 0-connections.*

(d) *If F is an F_2-connection and F and F_2 are normal operators, and f is a continuous function, then $f(F)$ is an $f(F_2)$-connection.*

(e) *If F is an F_2-connection and $T \in \mathbb{K}(E_1) \hat{\otimes} 1$, then $[T, F] \in \mathbb{K}(E)$.*

(f) *Let E_3 be a Hilbert C-module and $\omega : B \to \mathbb{B}(E_3)$, and $F_3 \in \mathbb{B}(E_3)$ with $[F_3, \omega(B)] \subseteq \mathbb{K}(E_3)$. If F_{23} is an F_3-connection on $E_2 \hat{\otimes}_\omega E_3$, and F is an F_{23}-connection on $E = E_1 \hat{\otimes}_\psi (E_2 \hat{\otimes}_\omega E_3)$, then F is an F_3-connection on $E \cong (E_1 \hat{\otimes}_\psi E_2) \hat{\otimes}_\omega E_3$.*

PROOF. (a), (b), (c), (d), and (f) are obvious, and (e) follows from an argument similar to 18.3.2(c). ☐

The next proposition, which will be useful in proving associativity of the intersection product, is a good practice computation using connections and graded commutators.

PROPOSITION 18.3.5. *Let $(E_1, \phi_1, F_1) \in \mathbb{E}(A, D_1)$, $(E_2, \phi_2, F_2) \in \mathbb{E}(D_1, D_2)$, $(E_3, \phi_3, F_3) \in \mathbb{E}(D_2, B)$. Set $E_{12} = E_1 \hat{\otimes}_{\phi_2} E_2$, $E_{23} = E_2 \hat{\otimes}_{\phi_3} E_3$, $E = E_1 \hat{\otimes}_{\phi_2} E_{23} \cong E_{12} \hat{\otimes}_{\phi_3} E_3$. Let G_{12} be an F_2-connection on E_{12} and G_{23} an F_3-connection on E_{23}. If G is a G_{23}-connection on E, then $[G_{12} \hat{\otimes} 1, G]$ is an $[F_2 \hat{\otimes} 1, G_{23}]$-connection on E (where $G_{12} \hat{\otimes} 1 \in \mathbb{B}(E)$ is defined using the isomorphism $E \cong E_{12} \hat{\otimes}_{\phi_3} E_3$).*

PROOF. We may assume that the elements are homogeneous by 18.3.4(a). Set $H = [G_{12} \hat{\otimes} 1, G]$, $H_{23} = [F_2 \hat{\otimes} 1, G_{23}]$. If $x \in E_1$, let $\tilde{T}_x \in \mathbb{B}(E_{23} \oplus E)$ be as in 18.3.1, and let $\tilde{G} = G_{23} \oplus G \in \mathbb{B}(E_{23} \oplus E)$, $\tilde{G}_{12} = F_2 \oplus G_{12} \in \mathbb{B}(E_2 \oplus E_{12})$, $\tilde{G}_{12} \hat{\otimes} 1 = (F_2 \hat{\otimes} 1) \oplus (G_{12} \hat{\otimes} 1) \in \mathbb{B}(E_{23} \oplus E)$, $\tilde{H} = H_{23} \oplus H \in \mathbb{B}(E_{23} \oplus E)$. Then $\tilde{H} = [\tilde{G}_{12} \hat{\otimes} 1, \tilde{G}]$.

We must show $[\tilde{H}, \tilde{T}_x] \in \mathbb{K}(E_{23} \oplus E)$ for any $x \in E_1$. We have

$$[[\tilde{G}_{12} \hat{\otimes} 1, \tilde{G}], \tilde{T}_x] = \pm[[\tilde{T}_x, \tilde{G}_{12} \hat{\otimes} 1], \tilde{G}] \pm [[\tilde{G}, \tilde{T}_x], \tilde{G}_{12} \hat{\otimes} 1],$$

where the signs are determined according to 14.1.3(c). The second term is in $\mathbb{K}(E_{23} \oplus E)$ since $[\tilde{G}, \tilde{T}_x] \in \mathbb{K}(E_{23} \oplus E)$ (because G is a connection). For the same reason, $[\tilde{T}_x, \tilde{G}_{12} \otimes 1] \in \mathbb{K}(E_2 \oplus E_{12}) \hat{\otimes} 1 \subseteq \mathbb{B}(E_{23} \oplus E)$, so the first term is in $\mathbb{K}(E_{23} \oplus E)$ by 18.3.4(e) and (f). ☐

As an application of the notion of connection, we have the following result, which will be useful in relating functoriality of KK with the intersection product. The proof is a generalization of the construction done in 17.8.2(a).

PROPOSITION 18.3.6. *Let A and B be graded C^*-algebras, with A σ-unital. If $(E, \phi, F) \in \mathbb{E}(A, B)$, then there is an $(E', \phi', F') \in \mathbb{E}(A, B)$ with ϕ' essential and $(E, \phi, F) \sim_h (E', \phi', F')$.*

PROOF. Let E_0 be the submodule $\overline{\phi(A)E}$ of E. We will show that $(E_0, \phi, H) \sim_h (E, \phi, F)$ for suitable $H \in \mathbb{B}(E_0)$. If E_0 is complemented in E, the proof is easy: if $P \in \mathbb{B}(E)$ is the projection onto E_0, then (E, ϕ, F) is a "compact perturbation" of $(E_0, \phi, PFP) \oplus ((1 - P)E, 0, 0)$. Since E_0 is not necessarily complemented, we must use an alternate argument. Let $\bar{E} = C([0, 1], E)$ as a graded Hilbert (IB)-module, and let $\bar{E}_0$ be the submodule $\{f : [0, 1] \to E \mid f(1) \in E_0\}$. We will find a homotopy of the form $(\bar{E}_0, \psi, G) \in \mathbb{E}(A, IB)$.

Let $J = \{f : [0, 1] \to \tilde{A} \mid f(1) \in A\}$; J is an ideal in $I\tilde{A}$. Let $\omega : A \to J$ embed as constant functions. If J is regarded as a graded Hilbert $(I\tilde{A})$-module, then $\bar{E}_0 \cong J \hat{\otimes}_{1\hat{\otimes}\phi} \bar{E}$, where $\tilde{\phi} : \tilde{A} \to \mathbb{B}(E)$ is the unital extension of ϕ. Let $\bar{F} = 1 \otimes F \in \mathbb{B}(\bar{E})$, and let G be a Grassmann $\bar{F}$-connection on $\bar{E}_0$. Then $(\bar{E}_0, \omega \hat{\otimes} 1, G) \in \mathbb{E}(A, IB)$. The restriction to 0 gives a Kasparov (A, B)-module of the form $(\tilde{A} \hat{\otimes}_\phi E, j, G_0)$, where $j : A \to \tilde{A}$ is the inclusion; when $\tilde{A} \hat{\otimes}_\phi E$ is identified with E this triple becomes (E, ϕ, F_0), where F_0 is a compact perturbation of F. Similarly, the restriction to 1 gives a Kasparov (A, B)-module of the form $(A \hat{\otimes}_\phi E, i \hat{\otimes} 1, G_1)$; under the isomorphism $A \hat{\otimes}_\phi E \cong E_0$ this triple becomes (E, ϕ, H) for some H. $\square$

There does not seem to be any way to prove the analogous statement for $\sim_{oh}$ or $\sim_c$ except to use the fact that these equivalence relations coincide with $\sim_h$.

18.4. Construction of the Product

We are now in a position to construct the product of two Kasparov modules. We will show that if (E_1, ϕ_1, F_1) is a Kasparov (A, D)-module and (E_2, ϕ_2, F_2) is a Kasparov (D, B)-module, then there is an F_2-connection F for E_1 making (E, ϕ, F) a Kasparov (A, B)-module and for which $[F_1 \hat{\otimes} 1, F]$ is "small". Such a connection is unique up to operator homotopy and is compatible with the equivalence relation $\sim_h$, so we obtain a pairing $KK(A, D) \times KK(D, B) \to KK(A, B)$.

The proofs require the use of Kasparov's Technical Theorem (14.6.2) in several places.

DEFINITION 18.4.1. Let F be an F_2-connection on E. (E, ϕ, F) is called a *Kasparov product* for (E_1, ϕ_1, F_1) and (E_2, ϕ_2, F_2) if

(a) (E, ϕ, F) is a Kasparov (A, B)-module
(b) for all $a \in A$, $\phi(a)[F_1 \hat{\otimes} 1, F]\phi(a)^* \geq 0 \bmod \mathbb{K}(E)$.

The set of all F such that (E, ϕ, F) is a Kasparov product is denoted $F_1 \#_D F_2$.

Note that the commutator in (b) is a graded commutator and the terms are of degree 1; modulo $\mathbb{K}(E)$ the terms are self-adjoint unitaries (on the support of $\phi(A)$), so the commutator is a self-adjoint operator of norm at most 2 mod

$\mathbb{K}(E)$, i.e. the essential spectrum is contained in $[-2, 2]$. Condition (b) requires that the spectrum be contained in $[0, 2]$ on the image of the support of $\phi(A)$; it could be relaxed to merely require that the essential spectrum be contained in $[\lambda, 2]$ on the support of $\phi(A)$, for some $\lambda > -2$. This is the condition that $[F_1 \hat{\otimes} 1, F]$ be "small". (We want the commutator to stay away from -2 instead of $+2$, so that F "lines up with" $F_1 \hat{\otimes} 1$ instead of $-(F_1 \hat{\otimes} 1)$.) See the comments preceding 17.2.7.

EXAMPLES 18.4.2. (a) Consider the modules $(D, f, 0) \in \mathbb{E}(A, D)$ and

$$(E_2, \phi_2, F_2) \in \mathbb{E}(D, B),$$

where $(D, f, 0)$ corresponds to a homomorphism $f : A \to D$ as in 17.1.2(a). Assume that ϕ_2 is essential (by 18.3.6 we can always arrange this within any KK-equivalence class). Then $D \hat{\otimes}_{\phi_2} E_2 \cong E_2$, and $F_2 \in 0 \#_D F_2$. So $(E_2, \phi_2 \circ f, F_2)$ is a Kasparov product for $(D, f, 0)$ and (E_2, ϕ_2, F_2). This Kasparov product represents $f^*(\boldsymbol{y})$, where $\boldsymbol{y}$ is the class of (E_2, ϕ_2, F_2).

(b) Consider $(E_1, \phi_1, F_1) \in \mathbb{E}(A, D)$ and $(B, g, 0) \in \mathbb{E}(D, B)$, where $(B, g, 0)$ corresponds to a homomorphism $g : D \to B$. Set $E = E_1 \hat{\otimes}_g B$. Then $F_1 \hat{\otimes} 1$ is a 0-connection, and $(E, \phi_1 \hat{\otimes} 1, F_1 \hat{\otimes} 1)$ is a Kasparov product for (E_1, ϕ_1, F_1) and $(B, g, 0)$. This Kasparov product represents $g_*(\boldsymbol{x})$, where $\boldsymbol{x}$ is the class of (E_1, ϕ_1, F_1).

(c) Let $\alpha = \left(\hat{\mathbb{H}}_D, \phi_1, \left[\begin{smallmatrix} 0 & 1 \\ 1 & 0 \end{smallmatrix} \right] \right) \in \mathbb{E}(A, D)$ and $\beta = \left(\hat{\mathbb{H}}_B, \phi_2, \left[\begin{smallmatrix} 0 & 1 \\ 1 & 0 \end{smallmatrix} \right] \right) \in \mathbb{E}(D, B)$ be in the standard form of the Cuntz picture (17.6.1), and assume furthermore that ϕ_2 is essential. Then $\hat{\mathbb{H}}_D \hat{\otimes}_{\phi_2} \hat{\mathbb{H}}_B \cong \hat{\mathbb{H}} \hat{\otimes}_{\mathbb{C}} \hat{\mathbb{H}}_B$, and the Grassmann connection $G \cong \left[\begin{smallmatrix} 1 & 0 \\ 0 & -1 \end{smallmatrix} \right] \hat{\otimes} \left[\begin{smallmatrix} 0 & 1 \\ 1 & 0 \end{smallmatrix} \right]$ commutes (in the graded sense) with $F_1 \hat{\otimes} 1 \cong \left[\begin{smallmatrix} 0 & 1 \\ 1 & 0 \end{smallmatrix} \right] \hat{\otimes} 1$. Thus, if $(\hat{\mathbb{H}}_D \hat{\otimes}_{\phi_2} \hat{\mathbb{H}}_B, \phi_1 \hat{\otimes} 1, G)$ is a Kasparov (A, B)-module, it is automatically a Kasparov product for α and β.

   In particular, if A, D, B are trivially graded, $(\phi_1^{(0)}, \phi_1^{(1)})$ is a quasihomomorphism from A to D, and $(\phi_2^{(0)}, \phi_2^{(1)})$ is a quasihomomorphism from D to B, and $\phi_2^{(0)}$ and $\phi_2^{(1)}$ extend to homomorphisms from $C = \phi_1^{(0)}(A) + D \subseteq M^s(D)$ to $M^s(B)$ which agree mod B (so that $(\phi_2^{(0)}, \phi_2^{(1)})$ defines a quasihomomorphism from C to B), then the quasihomomorphism $(\phi_2^{(0)} \circ \phi_1^{(0)} \oplus \phi_2^{(1)} \circ \phi_1^{(1)}, \phi_2^{(1)} \circ \phi_1^{(0)} \oplus \phi_2^{(0)} \circ \phi_1^{(1)})$ from A to B is a Kasparov product for $(\phi_1^{(0)}, \phi_1^{(1)})$ and $(\phi_2^{(0)}, \phi_2^{(1)})$.

   We will see later that we can always arrange for such extendibility if the algebras are trivially graded (18.13.1). However, if either A or B is trivially graded and the other has a standard odd grading, in general we cannot expect extendibility, because if $(\hat{\mathbb{H}}_D \hat{\otimes}_{\phi_2} \hat{\mathbb{H}}_B, \phi_1 \hat{\otimes} 1, G) \in \mathbb{E}(A, B)$ for the Grassmann connection, it not only represents a Kasparov product for $\left(\hat{\mathbb{H}}_D, \phi_1, \left[\begin{smallmatrix} 0 & 1 \\ 1 & 0 \end{smallmatrix} \right] \right)$ and $\left(\hat{\mathbb{H}}_B, \phi_2, \left[\begin{smallmatrix} 0 & 1 \\ 1 & 0 \end{smallmatrix} \right] \right)$, but also represents a product for $\left(\hat{\mathbb{H}}_D, \phi_1, \left[\begin{smallmatrix} 0 & -1 \\ -1 & 0 \end{smallmatrix} \right] \right)$ and $\left(\mathbb{H}_B, \phi_2, \left[\begin{smallmatrix} 0 & 1 \\ 1 & 0 \end{smallmatrix} \right] \right)$; in this case, $\left(\hat{\mathbb{H}}_D, \phi_1, \left[\begin{smallmatrix} 0 & -1 \\ -1 & 0 \end{smallmatrix} \right] \right)$ represents the negative of $\left(\hat{\mathbb{H}}_D, \phi_1, \left[\begin{smallmatrix} 0 & 1 \\ 1 & 0 \end{smallmatrix} \right] \right)$ in $KK(A, D)$, so extendibility can only occur if the product is the 0-element of $KK(A, B)$.

(d) Let A, D and B be graded C^*-algebras; let $\alpha = (E, \phi_1, F) \in \mathbb{E}(A, D)$ with $F = F^*$ and $\|F\| \leq 1$ (17.4.3), and let $\beta = (E_2, \phi_2, F_2) \in \mathbb{E}(D, B)$. Let G be any F_2-connection of degree 1 on $E = E_1 \hat{\otimes}_{\phi_2} E_2$. Set

$$F = F_1 \hat{\otimes} 1 + ((1 - F_1^2)^{1/2} \hat{\otimes} 1)G.$$

Then $F^2 - 1$ and $F - F^*$ are in $\mathbb{K}(E)$ and $[F, F_1 \hat{\otimes} 1] \geq 0 \bmod \mathbb{K}(E)$, because G (graded) commutes mod $\mathbb{K}(E)$ with $F_1 \hat{\otimes} 1$ by 18.3.4(e), and $G^2 - 1$ and $G - G^*$ are 0-connections. Suppose $[F, \phi(A)] \subseteq \mathbb{K}(E)$, where $\phi = \phi_1 \hat{\otimes} 1$ (this will not generally hold, but it will, for example, if $[F_1, \phi_1(A)] = 0$). Then $(E, \phi, F) \in \mathbb{E}(A, B)$.

(E, ϕ, F) will not generally be a Kasparov product for α and β since F is not generally an F_2-connection. However, we will show later (18.10.1) that (E, ϕ, F) is operator homotopic to a Kasparov product for α and β if A is separable and D σ-unital.

THEOREM 18.4.3. *If A is separable and D is σ-unital, then there is a Kasparov product for (E_1, ϕ_1, F_1) and (E_2, ϕ_2, F_2), which is unique up to operator homotopy. If F_1 and F_2 are self-adjoint, then there is a self-adjoint $F \in F_1 \#_D F_2$.*

PROOF. Let G be any F_2-connection of degree 1. Let $J = \mathbb{K}(E)$; $A_1 = \mathbb{K}(E_1) \hat{\otimes} 1 + \mathbb{K}(E)$; A_2 the C^*-subalgebra of $\mathbb{B}(E)$ generated by $G^2 - 1$, $[G, \phi(A)]$, $G - G^*$, and $[G, F_1 \hat{\otimes} 1]$; and Δ the subspace spanned by $F_1 \hat{\otimes} 1$, G, and $\phi(A)$. It is obvious that $F_1 \hat{\otimes} 1$ and $\phi(A)$ derive A_1, and so does G by 18.3.4(e); thus Δ derives A_1. $G^2 - 1$ and $G - G^*$ are 0-connections, so they multiply A_1 into J. $[G, \phi(a)]T = [G, \phi(a)T] - \phi(a)[G, T]$ for $T \in \mathbb{K}(E_1) \hat{\otimes} 1$, $a \in A$, so $[G, \phi(a)]$ multiplies A_1 into J by 18.3.4(e). Similarly, $[G, F_1 \hat{\otimes} 1]$ multiplies A_1 into J, so $A_1 \cdot A_2 \subseteq J$. Thus the hypotheses of the Technical Theorem are satisfied, so we obtain M and N of degree 0 with $0 \leq M, N \leq 1$, $M + N = 1$, and MA_1, NA_2, and $[N, \Delta]$ all contained in $\mathbb{K}(E)$.

Put $F = M^{1/2}(F_1 \hat{\otimes} 1) + N^{1/2}G$. Since M is a 0-connection, $M^{1/2}$ is also a 0-connection by 18.3.4(d). We have $[M^{1/2}, F_1 \hat{\otimes} 1] \in \mathbb{K}(E)$, so $M^{1/2}(F_1 \hat{\otimes} 1)$ is a 0-connection. $N = 1 - M$ is a 1-connection, so $N^{1/2}$ is also a 1-connection by 18.3.4(d); thus by 18.3.4(b) F is an F_2-connection.

We show that $(E, \phi, F) \in \mathbb{E}(A, B)$. Since M and N commute with $F_1 \hat{\otimes} 1$, G, $\phi(A)$, and each other mod $\mathbb{K}(E)$, for $a \in A$ we have

$$(F^2 - 1)\phi(a) = (M(F_1^2 \hat{\otimes} 1) + M^{1/2}N^{1/2}[G, F_1 \hat{\otimes} 1] + NG^2 - 1)\phi(a) \bmod \mathbb{K}(E)$$

$$= (M(F_1^2 - 1) + NG^2 - M - N)\phi(a) \bmod \mathbb{K}(E)$$

(since $N[G, F_1 \hat{\otimes} 1] \in \mathbb{K}(E)$); this in turn equals

$$M((F_1^2 - 1) \hat{\otimes} 1)\phi(a) + N(G^2 - 1)\phi(a) \bmod \mathbb{K}(E)$$

since $((F_1^2 - 1) \hat{\otimes} 1)\phi(a) \in A_1$, $G^2 - 1 \in A_2$. Similarly, $(F - F^*)\phi(a)$ and $[F, \phi(a)]$ are in $\mathbb{K}(E)$.

Also, $[F_1 \hat{\otimes} 1, F] = M^{1/2}[F_1 \hat{\otimes} 1, F_1 \hat{\otimes} 1] = 2M^{1/2}(F_1^2 \hat{\otimes} 1) \mod \mathbb{K}(E)$, and $(F_1^2 \hat{\otimes} 1)\phi(a) = \phi(a)$ for $a \in A$, so $\phi(a)[F_1 \hat{\otimes} 1, F]\phi(a^*) = 2\phi(a)M^{1/2}\phi(a^*) \geq 0$ $\mod \mathbb{K}(E)$. Thus $F \in F_1 \#_D F_2$.

If F_1 and F_2 are self-adjoint, and a self-adjoint F is desired, set

$$F = M^{1/4}(F_1 \hat{\otimes} 1)M^{1/4} + N^{1/4}GN^{1/4}.$$

This F is a "compact perturbation" of the previous F.

To prove uniqueness, let $F, F' \in F_1 \#_D F_2$. Set $J = \mathbb{K}(E)$; $A_1 = \mathbb{K}(E_1) \hat{\otimes} 1 + \mathbb{K}(E)$; A_2 the C^*-algebra generated by $[F_1 \hat{\otimes} 1, F]$, $[F_1 \hat{\otimes} 1, F']$, and $F - F'$; and Δ the subspace spanned by $\phi(A)$, $F_1 \hat{\otimes} 1$, F, F'. Apply the Technical Theorem to obtain M and N, and set $F'' = M^{1/2}(F_1 \hat{\otimes} 1) + N^{1/2}F$. By an argument essentially identical to that in the existence proof, we have that $F'' \in F_1 \#_D F_2$. Also, for all $a \in A$ we have $\phi(a)[F, F'']\phi(a)^* \geq 0$ and $\phi(a)[F', F'']\phi(a)^* \geq 0$ $\mod \mathbb{K}(E)$. So (E, ϕ, F) and (E, ϕ, F') are both operator homotopic to (E, ϕ, F'') by 17.2.7, completing the proof of 18.4.3.   □

THEOREM 18.4.4. *If A is separable and D is σ-unital, the Kasparov product defines a bilinear function $\hat{\otimes}_D : KK(A, D) \times KK(D, B) \to KK(A, B)$.*

PROOF. The Kasparov product of two Kasparov modules is uniquely determined up to homotopy; it is also clear that it respects direct sums in either variable. We only need to show it is well defined on $KK(A, D) \times KK(D, B)$. Suppose $(E_2, \phi_2, F_2) \in \mathbb{E}(D, IB)$ gives a homotopy between the Kasparov (D, B)-modules (E_2^0, ϕ_2^0, F_2^0) and (E_2^1, ϕ_2^1, F_2^1); then a Kasparov product of $(E_1, \phi_1, F_1) \in \mathbb{E}(A, D)$ and (E_2, ϕ_2, F_2) gives a homotopy between a Kasparov product of (E_1, ϕ_1, F_1) and (E_2^0, ϕ_2^0, F_2^0) and a Kasparov product of (E_1, ϕ_1, F_1) and (E_2^1, ϕ_2^1, F_2^1). Similarly, if $(E_1, \phi_1, F_1) \in \mathbb{E}(A, ID)$ gives a homotopy between Kasparov (A, D)-modules (E_1^0, ϕ_1^0, F_1^0) and (E_1^1, ϕ_1^1, F_1^1), and $(E_2, \phi_2, F_2) \in \mathbb{E}(D, B)$, then a Kasparov product of (E_1, ϕ_1, F_1) and $\tau_{C([0,1])}(E_2, \phi_2, F_2)$ gives a homotopy between a Kasparov product for (E_1^0, ϕ_1^0, F_1^0) and (E_2, ϕ_2, F_2) and a Kasparov product for (E_1^1, ϕ_1^1, F_1^1) and (E_2, ϕ_2, F_2).   □

18.5. Isomorphism of KK_h and KK_{oh}

Using the product and one additional application of the Technical Theorem, we can prove that $\sim_h$ and $\sim_{oh}$ coincide when A and B are small.

We first prove the result in a very special case by explicitly constructing an operator homotopy.

LEMMA 18.5.1. *Let $f_t : C([0,1]) \to \mathbb{C}$ be evaluation at t. Then $\boldsymbol{f}_0 = \boldsymbol{f}_1$ in $KK_{oh}(C([0,1]), \mathbb{C})$.*

PROOF. We may represent $\boldsymbol{f}_t$ by any module of the form

$$\left(\mathbb{H} \oplus \mathbb{H}^{op}, \; f_t \cdot 1, \; \begin{bmatrix} 0 & T^* \\ T & 0 \end{bmatrix} \right),$$

where $\mathbb{H}$ is a trivially graded Hilbert space and T is an essentially unitary Fredholm operator on $\mathbb{H}$ of index 1. The idea will be as follows. Let $\mathbb{H} = L^2[0,3]$ and $\hat{\mathbb{H}} = \mathbb{H} \oplus \mathbb{H}^{op}$, and define $\phi : C([0,1]) \to \mathbb{B}(\hat{\mathbb{H}})$ by letting $\phi(g)$ be multiplication by $\tilde{g}$, where $\tilde{g}(t) = g(0)$ for $0 \le t \le 1$, $\tilde{g}(t) = g(t-1)$ for $1 \le t \le 2$, $\tilde{g}(t) = g(1)$ for $2 \le t \le 3$. Let P be the projection onto $L^2[0,1/2] \oplus L^2[0,1/2]$ and Q the projection onto $L^2[5/2,3] \oplus L^2[5/2,3]$. We seek operators U and V of the above form $\begin{bmatrix} 0 & T^* \\ T & 0 \end{bmatrix}$ commuting with P and Q respectively, such that $(1-P)U$ and $(1-Q)V$ are self-adjoint unitaries on $(1-P)\hat{\mathbb{H}}$ and $(1-Q)\hat{\mathbb{H}}$ respectively, commuting with the action of $C([0,1])$, and such that U and V are connected by a path of essentially self-adjoint isometries essentially commuting with $\phi(C([0,1]))$. Then $(P\hat{\mathbb{H}}, P\phi, PU)$ represents $\boldsymbol{f}_0$, $(Q\hat{\mathbb{H}}, Q\phi, QV)$ represents $\boldsymbol{f}_1$, $((1-P)\hat{\mathbb{H}}, (1-P)\phi, (1-P)U)$ and $((1-Q)\hat{\mathbb{H}}, (1-Q)\phi, (1-Q)V)$ are degenerate, and $(P\hat{\mathbb{H}}, P\phi, PU) \oplus ((1-P)\hat{\mathbb{H}}, (1-P)\phi, (1-P)U) = (\hat{\mathbb{H}}, \phi, U)$ is operator homotopic to $(Q\hat{\mathbb{H}}, Q\phi, QV) \oplus ((1-Q)\hat{\mathbb{H}}, (1-Q)\phi, (1-Q)V) = (\hat{\mathbb{H}}, \phi, V)$.

We define U and V as follows. It is convenient to work on $L^2[0,2\pi]$ instead of $L^2[0,3]$. Define $D \in \mathbb{B}(L^2[0,2\pi])$ using the basis $1, \ldots, \cos nx, \ldots, \sin nx$, $\ldots$, by $D(1) = 0$, $D(\cos nx) = -\sin nx$, $D(\sin nx) = \cos nx$. Then $D = -D^*$, $D^2 + 1 \in \mathbb{K}$, and D commutes mod $\mathbb{K}$ with multiplication operators. (To see this, note that $[D, M_{\sin nx}]$ and $[D, M_{\cos nx}]$ are finite-rank for all n by standard trigonometric identities, where M_h is the multiplication operator corresponding to h.)

Let $\mathbb{S} = \{h \in C([0,2\pi]) \mid -1 \le h(t) \le 1, h(0) = 1, h(2\pi) = -1\}$. For $h \in \mathbb{S}$, set $T_h = M_h - (1 - M_h^2)^{1/2}D$. T_h is essentially unitary and essentially commutes with multiplication operators. If h_0 and h_1 are in $\mathbb{S}$, then T_{h_0} and T_{h_1} are connected by a norm-continuous path of operators of the same form. We calculate the Fredholm index of T_h. It suffices to consider the case $h = \cos x/2$. Then $T_h(1) = \cos x/2$, $T_h(\cos nx) = \cos((n - \frac{1}{2})x)$, $T_h(\sin nx) = \sin((n - \frac{1}{2})x)$. So $\ker T_h$ is spanned by $1 - \cos x$. On the other hand, T_h is surjective, i.e. the closed span of $\{\cos((n - \frac{1}{2})x), \sin((n - \frac{1}{2})x)\}$ equals $L^2[0,2\pi]$. To see this, for $f \in L^2[0,2\pi]$, define $\tilde{f} \in L^2[0,2\pi]$ by

$$\tilde{f}(x) = \begin{cases} f(2x) & \text{for } 0 \le x \le \pi, \\ -f(2x - 2\pi) & \text{for } \pi < x \le 2\pi. \end{cases}$$

The Fourier series for $\tilde{f}$ involves only $\cos nx$ and $\sin nx$ for n odd. (I am indebted to J. Rosenberg for this calculation.) Thus the index of T_h is 1.

Define

$$U_h = \begin{bmatrix} 0 & T_h^* \\ T_h & 0 \end{bmatrix} \in \mathbb{B}(L^2[0,2\pi] \oplus L^2[0,2\pi]).$$

Then U_h essentially commutes with multiplication operators. Let h_0 be a function in $\mathbb{S}$ with $h_0(t) = -1$ for $t \ge \pi/3$, h_1 a function in $\mathbb{S}$ with $h_1(t) = 1$ for $t \le 5\pi/3$, and set $U = U_{h_0}$, $V = U_{h_1}$. $\square$

Let B be any graded C^*-algebra, and $g_0, g_1 : IB \to B$ the evaluation homomorphisms, regarded as elements $(B, g_i, 0)$ of $\mathbb{E}(IB, B)$. By tensoring the Kasparov $(C([0,1]), \mathbb{C})$-module ω of 18.5.1 with $\mathbf{1}_B$, we obtain a Kasparov (IB, B)-module ω_B giving an operator homotopy between $g_0 \oplus$ (degenerate) and $g_1 \oplus$ (degenerate).

Our goal is to show that if $\alpha_0 = (E_0, \phi_0, F_0)$ and $\alpha_1 = (E_1, \phi_1, F_1)$ are homotopic elements of $\mathbb{E}(A, B)$, then we can add on degenerates to make the α_i operator homotopic. A natural approach is to take a module $\alpha = (E, \phi, F) \in \mathbb{E}(A, IB)$ giving a homotopy between α_0 and α_1, and form a Kasparov product β for α and ω_B. Then β gives a homotopy from $\alpha_0 \oplus \gamma_0$ to $\alpha_1 \oplus \gamma_1$ of the form $(\tilde{E}, \tilde{\phi}, F_t)$, where each of γ_0 and γ_1 is a Kasparov product of α and a degenerate module.

There are two difficulties with this approach: (1) we must show that we can choose β so that (F_t) is norm-continuous (for a general Kasparov product (F_t) will be only strictly continuous); (2) however we choose β, we need to show that γ_i is operator homotopic to a degenerate.

The second difficulty is easy to resolve: if $(E_1, \phi_1, F_1) \in \mathbb{E}(A, D)$ and $(E_2, \phi_2, F_2) \in \mathbb{D}(D, B)$, then if G is any Grassmann F_2-connection for E_1 we have that $G \in F_1 \#_D F_2$, and $(E_1 \hat{\otimes}_{\phi_2} E_2, \phi_1 \hat{\otimes} 1, G) \in \mathbb{D}(A, B)$. Any other Kasparov product is operator homotopic to this one by 18.4.3.

The following argument shows that we can find a Kasparov product with the F_t norm-continuous. This argument is due to Skandalis [1984].

LEMMA 18.5.2. *Let A, D, B be graded C^*-algebras with A separable and D σ-unital. Let $(E_1, \phi_1, F_1^n) \in \mathbb{E}(A, D)$ and $(E_2, \phi_2, F_2^n) \in \mathbb{E}(D, B)(n = 0, 1)$, with (E_i, ϕ_i, F_i^0) and (E_i, ϕ_i, F_i^1) operator homotopic for $i = 1, 2$. Then there is an operator homotopy between a Kasparov product for (E_1, ϕ_1, F_1^0) and (E_2, ϕ_2, F_2^0) and a Kasparov product for (E_1, ϕ_1, F_1^1) and (E_2, ϕ_2, F_2^1).*

PROOF. If (E_1, ϕ_1, F_1^t) and (E_2, ϕ_2, F_2^t) are operator homotopies, choose a norm-continuous path G^t, where G^t is an F_2^t-connection for E_1 as in 18.3.3. Let $J = \mathbb{K}(E)$; $A_1 = \mathbb{K}(E_1) \hat{\otimes} 1 + \mathbb{K}(E)$; A_2 the C^*-algebra generated by $[G^t, \phi(A)]$, $(G^t)^2 - 1$, $G^t - G^{t*}$, and $[F_1^t \hat{\otimes} 1, G^t]$ for all t; and Δ the subspace spanned by $\phi(A)$, $F_1^t \hat{\otimes} 1$, and G^t for all t. Apply the Technical Theorem to obtain M and N, and set $F^t = M^{1/2}(F_1^t \hat{\otimes} 1) + N^{1/2}G^t$. As in the proof of 18.4.2, check that $(E, \phi, F^t) \in \mathbb{E}(A, B)$ for all t. $\quad\square$

THEOREM 18.5.3. *If A is separable and B is σ-unital, then the relations $\sim_h$ and $\sim_{oh}$ coincide on $\mathbb{E}(A, B)$.*

PROOF. Combine 18.5.2 with the preceding comments, and recall that Kasparov products are unique up to operator homotopy. $\quad\square$

COROLLARY 18.5.4. *If A and B are trivially graded with A separable and B σ-unital, then $K_i(B) \cong KK^i(\mathbb{C}, B)(i = 0, 1)$ and $Ext(A, B)^{-1} \cong KK^1(A, B)$. So $Ext(\,\cdot\,,\,\cdot\,)^{-1}$ is homotopy invariant in each variable (15.10).*

An alternate approach, taken in [Skandalis 1984], is to define the product on KK_{oh} as well as on KK_h right from the beginning. 18.5.2 essentially says that the product is well defined on KK_{oh}. I feel that the approach we have taken here is slightly simpler from an expository point of view.

18.6. Associativity

The associativity of the intersection product originally appeared rather mysterious; Kasparov's original proof, while a technical masterpiece, was extremely difficult to understand and gave no insight into why the result is true. The simplified argument given here is due to Skandalis [1984]. Unfortunately, this proof also gives only limited insight.

If all algebras are trivially graded, there is an elegant alternate proof due to Cuntz [1983a] (18.13.1). This proof is more conceptual since it reduces associativity of the intersection product to associativity of composition of homomorphisms. However, Cuntz' proof cannot be generalized to the case where the algebras are not trivially graded.

THEOREM 18.6.1. *Let A, B, D_1, D_2 be graded C^*-algebras with A and D_1 separable and D_2 σ-unital. Let $\boldsymbol{x} \in KK(A, D_1)$, $\boldsymbol{y} \in KK(D_1, D_2)$, $\boldsymbol{z} \in KK(D_2, B)$. Then*

$$\boldsymbol{x} \,\hat{\otimes}_{D_1} (\boldsymbol{y} \,\hat{\otimes}_{D_2} \boldsymbol{z}) = (\boldsymbol{x} \,\hat{\otimes}_{D_1} \boldsymbol{y}) \,\hat{\otimes}_{D_2} \boldsymbol{z}.$$

PROOF. Represent $\boldsymbol{x}$, $\boldsymbol{y}$, $\boldsymbol{z}$ by (E_1, ϕ_1, F_1), (E_2, ϕ_2, F_2), (E_3, ϕ_3, F_3) respectively, with F_1, F_2, F_3 self-adjoint. Let $E_{12} = E_1 \,\hat{\otimes}_{\phi_2} E_2$, $E_{23} = E_2 \,\hat{\otimes}_{\phi_3} E_3$, $E = E_1 \,\hat{\otimes}_{\phi_2 \hat{\otimes} 1} E_{23} \cong E_{12} \,\hat{\otimes}_{\phi_3} E_3$, $\phi = \phi_1 \,\hat{\otimes}\, 1 \,\hat{\otimes}\, 1$. Let $F_{12} \in F_1 \,\#_{D_1} F_2$, $F_{23} \in F_2 \,\#_{D_2} F_3$, $F \in F_1 \,\#_{D_1} F_{23}$, with F_{12}, F_{23}, F self-adjoint. Then $(E_{12}, \phi_1 \hat{\otimes} 1, F_{12})$ represents $\boldsymbol{x} \hat{\otimes}_{D_1} \boldsymbol{y}$, $(E_{23}, \phi_2 \hat{\otimes} 1, F_{23})$ represents $\boldsymbol{y} \hat{\otimes}_{D_2} \boldsymbol{z}$, and (E, ϕ, F) represents $\boldsymbol{x} \,\hat{\otimes}_{D_1} (\boldsymbol{y} \,\hat{\otimes}_{D_2} \boldsymbol{z})$.

$[F_{12} \,\hat{\otimes}\, 1, F]$ is an $[F_2 \,\hat{\otimes}\, 1, F_{23}]$-connection on E by 18.3.5, so $[F_{12} \hat{\otimes} 1, F]_+$ is an $[F_2 \,\hat{\otimes}\, 1, F_{23}]_+$-connection for E_1 by 18.3.4(d). (By x_+ we denote the positive part of x. Note that $[F_{12} \hat{\otimes} 1, F]$ is self-adjoint.) Since $\phi_2(d)[F_2 \hat{\otimes} 1, F_{23}]\phi(d)^* \geq 0$ mod $\mathbb{K}(E_{23})$ for all $d \in D_1$, we have that $[F_{12} \,\hat{\otimes}\, 1, F]_-$ is a 0-connection for E_1. We also have that $[F_{12} \,\hat{\otimes}\, 1, F]$ (and hence $[F_{12} \,\hat{\otimes}\, 1, F]_-$) is a 0-connection for E_{12} by 18.3.4(e).

Set $J = \mathbb{K}(E)$, $A_1 = \mathbb{K}(E) + \mathbb{K}(E_{12} \,\hat{\otimes}\, 1) + \mathbb{K}(E_1 \,\hat{\otimes}\, 1 \,\hat{\otimes}\, 1)$, A_2 the C^*-algebra generated by $[F_{12} \,\hat{\otimes}\, 1, F]_-$ and $[F_1 \,\hat{\otimes}\, 1 \,\hat{\otimes}\, 1, F]$, and Δ the subspace spanned by $\phi(A)$, F, $F_1 \,\hat{\otimes}\, 1 \,\hat{\otimes}\, 1$, and $F_{12} \,\hat{\otimes}\, 1$. Apply the Technical Theorem to obtain M and N, and set $F' = M^{1/2}(F_1 \,\hat{\otimes}\, 1) + N^{1/2}F$. As in the proof of 18.4.3, $(E, \phi, F') \in \mathbb{E}(A, B)$, and $\phi(a)[F, F']\phi(a)^* \geq 0$ mod $\mathbb{K}(E)$ for all $a \in A$, so (E, ϕ, F') also represents $\boldsymbol{x} \,\hat{\otimes}_{D_1} (\boldsymbol{y} \,\hat{\otimes}_{D_2} \boldsymbol{z})$ by 17.2.7. On the other hand, $F - F'$ is a 0-connection on E_{12} since M multiplies $\mathbb{K}(E_{12}) \,\hat{\otimes}\, 1$ into $\mathbb{K}(E)$ and commutes mod $\mathbb{K}(E)$ with $F_1 \,\hat{\otimes}\, 1 \,\hat{\otimes}\, 1$. Since F is an F_3-connection for E_{12} (18.3.4(f)), F' is also an F_3-connection for E_{12}. Also, $\phi(a)[F_{12} \,\hat{\otimes}\, 1, F']\phi(a)^* \geq 0$ mod $\mathbb{K}(E)$ for $a \in A$ since $[F_{12} \,\hat{\otimes}\, 1, F'] = N^{1/2}[F_{12} \,\hat{\otimes}\, 1, F]$ mod $\mathbb{K}(E)$ and $N^{1/2}$ multiplies $[F_{12} \,\hat{\otimes}\, 1, F]_-$ into $\mathbb{K}(E)$. Thus $F' \in F_{12} \,\#_{D_2} F_3$, i.e. (E, ϕ, F') represents $(\boldsymbol{x} \,\hat{\otimes}_{D_1} \boldsymbol{y}) \,\hat{\otimes}_{D_2} \boldsymbol{z}$. $\qquad \square$

18.7. Functoriality

The next proposition summarizes the important functoriality properties of the product.

PROPOSITION 18.7.1. *Let A, A', B, B', D, D' be graded C^*-algebras, with A and A' separable and D and D' σ-unital, and let $f : A' \to A$, $g : B \to B'$, $h : D \to D'$ be graded *-homomorphisms. Let $\boldsymbol{x} \in KK(A, D)$, $\boldsymbol{y} \in KK(D, B)$, $\boldsymbol{z} \in KK(D', B)$. Then:*

(a) $f^*(\boldsymbol{x} \hat{\otimes}_D \boldsymbol{y}) = f^*(\boldsymbol{x}) \hat{\otimes}_D \boldsymbol{y}.$

(b) $h_*(\boldsymbol{x}) \hat{\otimes}_{D'} \boldsymbol{z} = \boldsymbol{x} \hat{\otimes}_D h^*(\boldsymbol{z}).$

(c) $g_*(\boldsymbol{x} \hat{\otimes}_D \boldsymbol{y}) = \boldsymbol{x} \hat{\otimes}_D g_*(\boldsymbol{y}).$

(d) $\boldsymbol{1}_A \hat{\otimes}_A \boldsymbol{x} = \boldsymbol{x} \hat{\otimes}_D \boldsymbol{1}_D = \boldsymbol{x}$.

PROOF. To prove (a), note that if (E, ϕ, F) is a Kasparov product for (E_1, ϕ_1, F_1) and (E_2, ϕ_2, F_2), then $(E, \phi \circ f, F)$ is a Kasparov product for $(E_1, \phi_1 \circ f, F_1)$ and (E_2, ϕ_2, F_2).

For (b), let $(E_1, \phi_1, F_1) \in \mathbb{E}(A, D)$ and $(E_2, \phi_2, F_2) \in \mathbb{E}(D', B)$ represent $\boldsymbol{x}$ and $\boldsymbol{z}$ respectively. Then $\boldsymbol{x} \hat{\otimes}_D h^*(\boldsymbol{z})$ is represented by $(E_1 \hat{\otimes}_{\phi_2 \circ h} E_2, \phi_1 \hat{\otimes} 1, F)$, where F is an F_2-connection for E_1. There is an obvious isomorphism $E_1 \hat{\otimes}_{\phi_2 \circ h} E_2 \cong (E_1 \hat{\otimes}_h D') \hat{\otimes}_{\phi_2} E_2$, which carries $\phi_1 \hat{\otimes} 1$ to $(\phi_1 \hat{\otimes} 1) \hat{\otimes} 1$ and sends F to an F_2-connection F' for $(E_1 \hat{\otimes}_h D')$; $((E_1 \hat{\otimes}_h D') \hat{\otimes}_{\phi_2} E_2, (\phi_1 \hat{\otimes} 1) \hat{\otimes} 1, F')$ is clearly a Kasparov product for $(E_1 \hat{\otimes}_h D', \phi_1 \hat{\otimes} 1, F_1 \hat{\otimes} 1)$ and (E_2, ϕ_2, F_2) and thus represents $h_*(\boldsymbol{x}) \hat{\otimes}_{D'} \boldsymbol{z}$.

(c) is similar to (a), and (d) is obvious. □

This functoriality may be regarded as a special case of the associativity of the product because of the following proposition, which is an immediate corollary of 18.3.6 and 18.4.2(a)-(b):

PROPOSITION 18.7.2. *Let A, D, B be graded C^*-algebras, with A separable and D σ-unital, and let $f : A \to D$ and $g : D \to B$ be graded homomorphisms. Then:*

(a) *For any $\boldsymbol{x} \in KK(A, D)$, $g_*(\boldsymbol{x}) = \boldsymbol{x} \hat{\otimes}_D \boldsymbol{g}$ in $KK(A, B)$.*
(b) *For any $\boldsymbol{y} \in KK(D, B)$, $f^*(\boldsymbol{y}) = \boldsymbol{f} \hat{\otimes}_D \boldsymbol{y}$ in $KK(A, B)$.*

18.8. Ring Structure on $KK(A, A)$

If A is a separable C^*-algebra, it follows immediately from 18.6.1 and 18.7.1(d) that $KK(A, A)$ is a unital ring under intersection product.

PROPOSITION 18.8.1. $KK(\mathbb{C}, \mathbb{C}) \cong \mathbb{Z}$ *as a ring.*

PROOF. We know from 17.3.4 that $KK(\mathbb{C}, \mathbb{C}) \cong \mathbb{Z}$ as a group, and that any KK-element arises as a difference of homomorphisms. The bilinearity of the product and the fact that it agrees with composition for homomorphisms shows that it must be the ordinary one on $\mathbb{Z}$. □

More generally, if A is an AF algebra, then $KK(A, A)$ is the endomorphism ring of $K_0(A)$ (23.15.2). $KK(A, A)$ can be calculated in a great many cases using the Universal Coefficient Theorem (23.11.1).

18.9. General Form of the Product

We now develop the most general form of the intersection product. Let A_1, A_2, B_1, B_2, and D be graded C^*-algebras with A_1 and A_2 separable and B_1 and D σ-unital, and let $\boldsymbol{x} \in KK(A_1, B_1 \,\hat{\otimes}\, D)$, $\boldsymbol{y} \in KK(D \,\hat{\otimes}\, A_2, B_2)$. We define $\boldsymbol{x} \,\hat{\otimes}_D\, \boldsymbol{y}$ to be the composite

$$(\boldsymbol{x} \,\hat{\otimes}\, 1_{A_2}) \,\hat{\otimes}_E\, (1_{B_1} \,\hat{\otimes}\, \boldsymbol{y}) = \tau_{A_2}(\boldsymbol{x}) \,\hat{\otimes}_E\, \tau_{B_1}(\boldsymbol{y}),$$

where $E = B_1 \,\hat{\otimes}\, D \,\hat{\otimes}\, A_2$ (17.8.5). This makes sense since $\tau_{A_2}(\boldsymbol{x}) \in KK(A_1 \,\hat{\otimes}\, A_2, B_1 \,\hat{\otimes}\, D \,\hat{\otimes}\, A_2)$ and $\tau_{B_1}(\boldsymbol{y}) \in KK(B_1 \,\hat{\otimes}\, D \,\hat{\otimes}\, A_2, B_1 \,\hat{\otimes}\, B_2)$.

The next proposition summarizes the functoriality properties of the general form of the product. In each case we will assume, without specific mention, that all C^*-algebras satisfy the appropriate size restrictions (separable or σ-unital) necessary to make the products defined.

PROPOSITION 18.9.1. (a) *The intersection product* $\hat{\otimes}_D : KK(A_1, B_1 \,\hat{\otimes}\, D) \times KK(D \,\hat{\otimes}\, A_2, B_2) \to KK(A_1 \,\hat{\otimes}\, A_2, B_1 \,\hat{\otimes}\, B_2)$ *is bilinear, contravariantly functorial in* A_1 *and* A_2 *and covariantly functorial in* B_1 *and* B_2.
(b) *If* $h : D_1 \to D_2$, $\boldsymbol{x} \in KK(A_1, B_1 \,\hat{\otimes}\, D_1)$, $\boldsymbol{y} \in KK(D_2 \,\hat{\otimes}\, A_2, B_2)$, *then* $(1 \,\hat{\otimes}\, h)_*(\boldsymbol{x}) \,\hat{\otimes}_{D_2}\, \boldsymbol{y} = \boldsymbol{x} \,\hat{\otimes}\, (h \,\hat{\otimes}\, 1)^*(\boldsymbol{y})$.
(c) *If* $\boldsymbol{x} \in KK(A_1, B_1 \,\hat{\otimes}\, D)$ *and* $\boldsymbol{y} \in KK(D \,\hat{\otimes}\, A_2, B_2)$, *then for any* D_1 *we have*

$$\tau_{D_1}(\boldsymbol{x} \,\hat{\otimes}_D\, \boldsymbol{y}) = \tau_{D_1}(\boldsymbol{x}) \,\hat{\otimes}_E\, \tau_{D_1}(\boldsymbol{y}),$$

where $E = D \,\hat{\otimes}\, D_1$.
(d) *If* $\boldsymbol{x} \in KK(A_1, B_1 \,\hat{\otimes}\, D_1 \,\hat{\otimes}\, D)$ *and* $\boldsymbol{y} \in KK(D \,\hat{\otimes}\, D_2 \,\hat{\otimes}\, A_2, B_2)$, *then*

$$\tau_{D_2}(\boldsymbol{x}) \,\hat{\otimes}_E\, \tau_{D_1}(\boldsymbol{y}) = \boldsymbol{x} \,\hat{\otimes}_D\, \boldsymbol{y},$$

where $E = D_1 \,\hat{\otimes}\, D \,\hat{\otimes}\, D_2$.

If $D = \mathbb{C}$, we have a way of forming the tensor product of two KK-elements which generalizes the tensor product of two homomorphisms. One may construct this tensor product somewhat more simply using unbounded Kasparov modules [Baaj and Julg 1983]: if (E_1, ϕ_1, D_1) and (E_2, ϕ_2, D_2) are unbounded Kasparov modules (17.9) for $\boldsymbol{x}$ and $\boldsymbol{y}$ respectively, then $(E_1 \,\hat{\otimes}\, E_2, \phi_1 \,\hat{\otimes}\, \phi_2, D_1 \,\hat{\otimes}\, 1 + 1 \,\hat{\otimes}\, D_2)$ (suitably interpreted) is an unbounded Kasparov module for $\boldsymbol{x} \,\hat{\otimes}\, \boldsymbol{y}$. Details are left to the reader.

It is instructive to see why the same approach does not work to define the tensor product of two ordinary Kasparov modules (the operator $F_1 \,\hat{\otimes}\, 1 + 1 \,\hat{\otimes}\, F_2$ does not have the right algebraic and intertwining properties in general). The difficulty is similar (and closely related) to the fact that a tensor product of elliptic pseudodifferential operators of degree 0 is not in general an operator of the same type. The construction of the intersection product in this case is exactly the same as the sharp product of elliptic operators of Atiyah and Singer.

18.10. Products on KK^1

As another special case of the general form of the product, we can define a pairing $\hat{\otimes}_D : KK^i(A, D) \times KK^j(D, B) \to KK^{i+j}(A, B)$ (addition mod 2). To define the map $KK^1(A, D) \times KK^1(D, B) \to KK(A, B)$, for example, let $\boldsymbol{x} \hat{\otimes}_D \boldsymbol{y} = \boldsymbol{x} \hat{\otimes}_E \tau_{\mathbb{C}_1} \boldsymbol{y}$, where $E = D \hat{\otimes} \mathbb{C}_1$; the others are similar.

There is a particularly nice interpretation of this product in a special case. We first note a preliminary result of independent interest (cf. 18.4.2(d)).

PROPOSITION 18.10.1. *Let A, D, B be graded C^*-algebras, with A separable and D σ-unital. Let $\alpha = (E_1, \phi_1, F_1) \in \mathbb{E}(A, D)$, $\beta = (E_2, \phi_2, F_2) \in \mathbb{E}(D, B)$, with $F_1 = F_1^*$ and $\|F_1\| \leq 1$. Let G be an F_2-connection on $\tilde{E} = E_1 \hat{\otimes}_{\phi_2} E_2$. Set $\phi = \phi_1 \hat{\otimes} 1$ and*

$$F = F_1 \hat{\otimes} 1 + ((1 - F_1^2)^{1/2} \hat{\otimes} 1)G.$$

If $[F, \phi(A)] \subseteq \mathbb{K}(E)$, then $\gamma = (E, \phi, F) \in \mathbb{E}(A, B)$, and γ is operator-homotopic to a Kasparov product for α and β, i.e. $[\gamma] = [\alpha] \hat{\otimes}_D [\beta]$ in $KK(A, B)$.

PROOF. If $[F, \phi(A)] \subseteq \mathbb{K}(E)$, then by 18.4.2(d) $(E, \phi, F) \in \mathbb{E}(A, B)$, and satisfies all properties of a Kasparov product except that F is not necessarily an F_2-connection. We can construct a Kasparov product (E, ϕ, F'), with $F' = M^{1/2}(F_1 \hat{\otimes} 1) + N^{1/2}G$, as in 18.4.3. (E, ϕ, F) is operator-homotopic to (E, ϕ, F') via

$$F_t = [tM + (1 - t)]^{1/2}(F_1 \hat{\otimes} 1) + [tN + (1 - t)((1 - F_1^2) \hat{\otimes} 1)]^{1/2}G.$$

It is routine to check that $(E, \phi, F_t) \in \mathbb{E}(A, B)$ for $0 \leq t \leq 1$. □

We now show that the intersection product on KK^1 generalizes the pairing between K-theory and K-homology. Let B be a trivially graded σ-unital C^*-algebra. If p is a projection in $Q(B)$, then p defines an element $\boldsymbol{x} \in KK^1(\mathbb{C}, B)$ as in 17.6.4. Similarly, if $\boldsymbol{y} \in KK^1(B, \mathbb{C})$, then $\boldsymbol{y}$ may be regarded as an invertible extension τ of B by $\mathbb{K}$ (i.e. $\tau : B \to Q$) by 17.6.4. If $x \in M(B)_+$ with $q(x) = p$, then $e^{2\pi i x}$ is a unitary in $\tilde{B}$, so $\tilde{\tau}(e^{2\pi i x})$ is a unitary in Q, where $\tilde{\tau}$ is the unital extension of τ to $\tilde{B}$.

THEOREM 18.10.2. *Under the standard identification of $KK(\mathbb{C}, \mathbb{C})$ with $\mathbb{Z}$, $\boldsymbol{x} \hat{\otimes}_B \boldsymbol{y}$ becomes Index $\tilde{\tau}(e^{2\pi i x})$.*

PROOF. Let $s \in M(B)$ with $s = s^*$, $\|s\| \leq 1$, $q(s) = 2p - 1$. Set $t = (1 - s^2)^{1/2} \in B$, and $v = t + is$. Then v is a unitary in $M(B)$, and $v^2 \in \tilde{B}$ is homotopic to $e^{2\pi i x}$ in $U_1(\tilde{B})$. The element $\boldsymbol{x}$ is represented by $(B \hat{\otimes} \mathbb{C}_1, 1 \hat{\otimes} 1, s \hat{\otimes} \varepsilon)$, and $\boldsymbol{y}$ is represented by a module of the form

$$((\mathbb{H} \oplus \mathbb{H}) \hat{\otimes} \mathbb{C}_1, \phi \hat{\otimes} 1, \begin{bmatrix} 1 & 0 \\ 0 & -1 \end{bmatrix} \hat{\otimes} \varepsilon),$$

as in 17.6.4. The map $\phi : B \to \mathbb{B}(\mathbb{H} \oplus \mathbb{H})$ is of the form

$$\phi(b) = \begin{bmatrix} \tau(b) & \theta(b) \\ \omega(b) & \sigma(b) \end{bmatrix}$$

and $\theta(b)$, $\omega(b)$ are compact for $b \in B$. We may assume τ and ϕ are essential to simplify notation. ϕ extends to a unital homomorphism from $M(B)$ to $\mathbb{B}(\mathbb{H} \oplus \mathbb{H})$, also denoted ϕ, but the off-diagonal terms of $\phi(m)$ are not necessarily compact for $m \in M(B)$.

To form the product, we tensor the above $\boldsymbol{y}$-module with $\mathbb{C}_1$ to obtain

$$\big((\mathbb{H} \oplus \mathbb{H}) \hat{\otimes} (\mathbb{C}_1 \hat{\otimes} \mathbb{C}_1), (\phi \hat{\otimes} 1) \hat{\otimes} 1, (\begin{bmatrix} 1 & 0 \\ 0 & -1 \end{bmatrix} \hat{\otimes} \varepsilon) \hat{\otimes} 1\big) \in \mathbb{E}(B \hat{\otimes} \mathbb{C}_1, M_2).$$

We further tensor this module with the standard module $(\mathbb{C} \oplus \mathbb{C}^{op}, id, 0) \in \mathbb{E}(M_2, \mathbb{C})$ to obtain an element of $\mathbb{E}(B \hat{\otimes} \mathbb{C}_1, \mathbb{C})$.

Under the identification of $\mathbb{C}_1 \hat{\otimes} \mathbb{C}_1$ with M_2, $\varepsilon \hat{\otimes} 1$ corresponds to $\begin{bmatrix} 0 & 1 \\ 1 & 0 \end{bmatrix}$ and $1 \hat{\otimes} \varepsilon$ becomes $\begin{bmatrix} 0 & -i \\ i & 0 \end{bmatrix}$, so we obtain the module

$$\big(E, \psi, \begin{bmatrix} 0 & r \\ r & 0 \end{bmatrix}\big) \in \mathbb{E}(B \hat{\otimes} \mathbb{C}_1, \mathbb{C}),$$

where $E = (\mathbb{H} \oplus \mathbb{H}) \oplus (\mathbb{H} \oplus \mathbb{H})^{op}$,

$$\psi(b \hat{\otimes} 1) = \begin{bmatrix} \phi(b) & 0 \\ 0 & \phi(b) \end{bmatrix}, \quad \psi(1 \hat{\otimes} \varepsilon) = \begin{bmatrix} 0 & -i1 \\ i1 & 0 \end{bmatrix}, \quad r = \begin{bmatrix} 1 & 0 \\ 0 & -1 \end{bmatrix}.$$

We must compute the product of $(B \hat{\otimes} \mathbb{C}_1, 1 \hat{\otimes} 1, s \hat{\otimes} \varepsilon)$ and $\big(E, \psi, \begin{bmatrix} 0 & r \\ r & 0 \end{bmatrix}\big)$. By 18.10.1, the product is given by $((B \hat{\otimes} \mathbb{C}_1) \hat{\otimes}_\psi E, 1, F)$, where

$$F = (s \hat{\otimes} \varepsilon) \hat{\otimes} 1 + ((1 - (s \hat{\otimes} \varepsilon)^2)^{1/2} \hat{\otimes} 1)(1 \hat{\otimes} \begin{bmatrix} 0 & r \\ r & 0 \end{bmatrix})$$

Since ϕ is essential, we may identify $(B \hat{\otimes} \mathbb{C}_1) \hat{\otimes}_\psi E$ with E, and F becomes

$$\begin{bmatrix} 0 & -i\phi(s) \\ i\phi(s) & 0 \end{bmatrix} + \begin{bmatrix} \phi(t) & 0 \\ 0 & \phi(t) \end{bmatrix} \begin{bmatrix} 0 & r \\ r & 0 \end{bmatrix}$$

$$= \begin{bmatrix} & 0 & \begin{matrix} -i\tau(s) & -i\theta(s) \\ -i\omega(s) & -i\sigma(s) \end{matrix} \\ \begin{matrix} i\tau(s) & i\theta(s) \\ i\omega(s) & i\sigma(s) \end{matrix} & 0 \end{bmatrix} + \begin{bmatrix} & 0 & \begin{matrix} \tau(t) & -\theta(t) \\ \omega(t) & -\sigma(t) \end{matrix} \\ \begin{matrix} \tau(t) & -\theta(t) \\ \omega(t) & -\sigma(t) \end{matrix} & 0 \end{bmatrix}$$

$$= \begin{bmatrix} & 0 & \begin{matrix} \tau(v^*) & -\theta(v) \\ \omega(v^*) & -\sigma(v) \end{matrix} \\ \begin{matrix} \tau(v) & -\theta(v^*) \\ \omega(v) & -\sigma(v^*) \end{matrix} & 0 \end{bmatrix}.$$

The element of $KK(\mathbb{C}, \mathbb{C}) \cong \mathbb{Z}$ corresponding to the product is

$$\text{Index} \begin{bmatrix} \tau(v) & -\theta(v^*) \\ \omega(v) & -\sigma(v^*) \end{bmatrix}.$$

To calculate this index, note first that $\phi(v^2) = [\phi(v)]^2$, i.e.

$$\begin{bmatrix} \tau(v^2) & \theta(v^2) \\ \omega(v^2) & \sigma(v^2) \end{bmatrix} = \begin{bmatrix} \tau(v)^2 + \theta(v)\omega(v) & \tau(v)\theta(v) + \theta(v)\sigma(v) \\ \omega(v)\tau(v) + \sigma(v)\omega(v) & \omega(v)\theta(v) + \sigma(v)^2 \end{bmatrix}.$$

Since $v^2 \in \tilde{B}$, the off-diagonal terms are compact. Also, $\phi(v)\phi(v^*) = 1$, so

$$\begin{bmatrix} 1 & 0 \\ 0 & 1 \end{bmatrix} = \begin{bmatrix} \tau(v)\tau(v^*) + \theta(v)\omega(v^*) & \tau(v)\theta(v^*) + \theta(v)\sigma(v^*) \\ \omega(v)\tau(v^*) + \sigma(v)\omega(v^*) & \omega(v)\theta(v^*) + \sigma(v)\sigma(v^*) \end{bmatrix}.$$

Finally, $\phi(v)$ is unitary, so

$$\mathrm{Index}\begin{bmatrix} \tau(v) & -\theta(v^*) \\ \omega(v) & -\sigma(v^*) \end{bmatrix} = \mathrm{Index}\begin{bmatrix} \tau(v) & \theta(v) \\ \omega(v) & \sigma(v) \end{bmatrix}\begin{bmatrix} \tau(v) & -\theta(v^*) \\ \omega(v) & -\sigma(v^*) \end{bmatrix}$$

$$= \mathrm{Index}\begin{bmatrix} \tau(v)^2 + \theta(v)\omega(v) & -\tau(v)\theta(v^*) - \theta(v)\sigma(v^*) \\ \omega(v)\tau(v) + \sigma(v)\omega(v) & -\omega(v)\theta(v^*) - \sigma(v)\sigma(v^*) \end{bmatrix}.$$

By the preceding discussion, this matrix equals

$$\begin{bmatrix} \tau(v^2) & 0 \\ k & -1 \end{bmatrix},$$

with k compact, so the index equals $\mathrm{Index}\,\tau(v^2) = \mathrm{Index}\,\tau(e^{2\pi i x})$. $\square$

COROLLARY 18.10.3. *Let B be σ-unital and trivially graded. Identify $KK^1(\mathbb{C}, B)$ with $K_1(B)$ and $KK^1(B, \mathbb{C})$ with $K^1(B) = \mathrm{Ext}(B)^{-1}$ in the standard way. Then the pairing of $K_1(B)$ with $K^1(B)$ using the intersection product agrees with the pairing defined in 16.3.2.*

Note that the identification of $K_1(B)$ with $KK^1(\mathbb{C}, B)$ in 18.10.3 requires K-theory Bott periodicity; but 18.10.2 does not (in fact, 18.10.2 requires nothing outside of the KK-theory of Sections 17 and 18).

18.11. Extendibility of KK-Elements

If $f : A_1 \to A_2$ is a graded *-homomorphism and $x \in KK(A_1, B)$, we say x *factors through* f if there is $y \in KK(A_2, B)$ with $x = f^*(y)$. If f is an inclusion map, we say x *extends to* A_2 if x factors through f.

The y is not in general unique. There is a partially defined functor $f_!$ (read f lower shriek!) from $KK(A_1, B)$ to $KK(A_2, B)$, defined on the set of elements which factor uniquely through f. In some cases $f_!$ is defined everywhere, or can be naturally extended to a larger set of elements; when this is possible there is a sort of "wrong-way functoriality" associated with f. Normally $f_!$ comes from left intersection product with an element $f! \in KK(A_2, A_1)$.

Shriek maps occur more generally; see 17.1.2(g) for a discussion of shriek maps in geometry.

One could also consider factorizations and shriek maps on the right.

PROPOSITION 18.11.1. *If*

$$0 \longrightarrow J \xrightarrow{\quad j \quad} A \overset{s}{\underset{}{\rightleftarrows}} A/J \longrightarrow 0$$

is a split exact sequence of separable graded C^*-algebras, and B is any graded C^*-algebra, then any element of $KK(J,B)$ is extendible to A.

PROOF. Suppose $x \in KK(J,B)$. Let $\pi_s \in KK(A,J)$ be the splitting morphism (17.1.2(b)), and set $y = \pi_s \hat{\otimes}_J x$. We have $j^*(y) = j^*(\pi_s) \hat{\otimes}_J x = 1_J \hat{\otimes}_J x = x$. $\square$

PROPOSITION 18.11.2. *Let A be separable, and let $\eta : A \to A \hat{\otimes} \mathbb{K}$ send a to $a \hat{\otimes} p$, where p is a one-dimensional projection of degree 0. Then every element of $KK(A,B)$ (for any B) factors through η, i.e. extends to $A \hat{\otimes} \mathbb{K}$.*

PROOF. Let i be the $KK(A \hat{\otimes} \mathbb{K}, A)$-element corresponding to the identity map i on $A \hat{\otimes} \mathbb{K}$ as in 17.1.2(a). Then $\eta^*(i) = 1_A$. Set $y = i \hat{\otimes}_A x$; then $\eta^*(y) = \eta^*(i) \hat{\otimes}_A x = 1_A \hat{\otimes}_A x = x$. $\square$

Note that to prove extendibility, it is *not* enough to simply show that the homomorphism ϕ of $(E, \phi, F) \in \mathbb{E}(J,B)$ extends to a graded *-homomorphism from A to $\mathbb{B}(E)$, since the extended ϕ may no longer have the right intertwining properties with F.

From one point of view, the result of 18.11.1 is a fundamental fact; the construction of the product and the proof of its properties could be proved fairly easily for trivially graded algebras from 17.8.4 assuming split extendibility. Even if a direct proof of extendibility were available, however, the construction of the product in the manner we have done it would be of use in explicitly constructing products in applications.

It should be noted that the proof of 17.2.7, due to Connes and Skandalis, really gives an extendibility result.

18.12. Recapitulation

We summarize the principal facts we have proved so far about the Kasparov groups.

18.12.1. $KK^n(n = 0,1)$ is a bifunctor from pairs (A,B) of graded C^*-algebras to abelian groups. This functor is well behaved only when A is separable and B is σ-unital; throughout the rest of this paragraph we assume without mention that all C^*-algebras satisfy these size restrictions.

18.12.2. The elements of $KK^n(A,B)$ are equivalence classes of (A,B)-bimodules. The bimodules considered may be expressed in various standard forms; for example, in the Fredholm module picture (17.5), the Cuntz picture (17.6), or the Baaj–Julg picture (17.11). The equivalence relation used is not at all critical; any equivalence relation weaker than $\sim_c$ and stronger than $\sim_h$ will do.

18.12.3. KK^n is homotopy invariant in each variable (17.9.1).

18.12.4. If B is trivially graded, then $KK^n(\mathbb{C}, B)$ is naturally isomorphic to $K_n(B)$ for $n = 0, 1$ (17.5.5, 17.5.6).

18.12.5. If A and B are trivially graded, then $KK^1(A, B)$ is naturally isomorphic to $Ext(A, B)^{-1}$. If A is nuclear, then $KK^1(A, B)$ is naturally isomorphic to $Ext(A, B)$.

18.12.6. KK^n is a stable invariant in each variable, i.e. there are natural isomorphisms

$$KK^n(A, B) \cong KK^n(A \,\hat{\otimes}\, \mathbb{K}, B) \cong KK^n(A, B \,\hat{\otimes}\, \mathbb{K}) \cong KK^n(A \,\hat{\otimes}\, \mathbb{K}, B \,\hat{\otimes}\, \mathbb{K}).$$

18.12.7. There is a product

$$\hat{\otimes}_D : KK^n(A_1, B_1 \,\hat{\otimes}\, D) \times KK^m(D \,\hat{\otimes}\, A_2, B_2) \to KK^{n+m}(A_1 \,\hat{\otimes}\, A_2, B_1 \,\hat{\otimes}\, B_2).$$

This product is associative and functorial in all possible senses. The product generalizes composition and tensor product of *-homomorphisms, cup and cap products, tensor product of elliptic pseudodifferential operators, and the pairing between K-theory and "K-homology".

18.13. EXERCISES AND PROBLEMS

18.13.1. Combine 18.4.2(c) and 18.11.1 to give a simplified proof of associativity of the intersection product in the trivially graded case, as follows.

(a) Let $(\phi_1^{(0)}, \phi_1^{(1)})$ be a quasihomomorphism from A to D_1. By adding on a degenerate quasihomomorphism, we may assume that $\phi_1^{(0)}(A) \cap (D_1 \otimes \mathbb{K}) = 0$. So $C_1 = \phi_1^{(0)}(A) + (D \otimes \mathbb{K}) = \phi_1^{(1)}(A) + (D \otimes \mathbb{K}) \subseteq M^s(D_1)$ is a split extension of $\phi_1^{(0)}(A)$ by $D \otimes \mathbb{K}$.

(b) If $(\phi_2^{(0)}, \phi_2^{(1)})$ is a quasihomomorphism from D_1 to D_2, then $(\phi_2^{(0)}, \phi_2^{(1)})$ can be extended to a quasihomomorphism, also denoted $(\phi_2^{(0)}, \phi_2^{(1)})$, from C_1 to D_2 by 18.11.1 and 18.11.2. Then by 18.4.2(c) the intersection product is represented by the quasihomomorphism $(\phi_2^{(0)} \circ \phi_1^{(0)} \oplus \phi_2^{(1)} \circ \phi_1^{(1)}, \phi_2^{(1)} \circ \phi_1^{(0)} \oplus \phi_2^{(0)} \circ \phi_1^{(1)})$.

(c) By adding on more degenerates, we may assume $\phi_2^{(0)}(C_1) \cap (D_2 \otimes \mathbb{K}) = 0$, so $C_2 = \phi_2^{(0)}(C_1) + (D_2 \otimes \mathbb{K})$ is a split extension of $(D_2 \otimes \mathbb{K})$. If $(\phi_3^{(0)}, \phi_3^{(1)})$ is a quasihomomorphism from D_2 to B, extend to a quasihomomorphism from C_2 to B, also denoted $(\phi_3^{(0)}, \phi_3^{(1)})$. Then both intersection products

$$([\phi_1^{(0)}, \phi_1^{(1)}] \,\hat{\otimes}_{D_1}\, [\phi_2^{(0)}, \phi_2^{(1)}]) \,\hat{\otimes}_{D_2}\, [\phi_3^{(0)}, \phi_3^{(1)}]$$

and

$$[\phi_1^{(0)}, \phi_1^{(1)}] \,\hat{\otimes}_{D_1}\, ([\phi_2^{(0)}, \phi_2^{(1)}] \,\hat{\otimes}_{D_2}\, [\phi_3^{(0)}, \phi_3^{(1)}])$$

are represented by the quasihomomorphism

$$\big(\phi_3^{(0)} \circ \phi_2^{(0)} \circ \phi_1^{(0)} \oplus \phi_3^{(0)} \circ \phi_2^{(1)} \circ \phi_1^{(1)} \oplus \phi_3^{(1)} \circ \phi_2^{(1)} \circ \phi_1^{(0)} \oplus \phi_3^{(1)} \circ \phi_2^{(0)} \circ \phi_1^{(1)},$$

$$\phi_3^{(1)} \circ \phi_2^{(0)} \circ \phi_1^{(0)} \oplus \phi_3^{(1)} \circ \phi_2^{(1)} \circ \phi_1^{(1)} \oplus \phi_3^{(0)} \circ \phi_2^{(1)} \circ \phi_1^{(0)} \oplus \phi_3^{(0)} \circ \phi_2^{(0)} \circ \phi_1^{(1)}\big).$$

18.13.2. Define a new equivalence relation on Kasparov modules as follows.

(a) Let $T = C^*(u)$ be the Toeplitz algebra (9.4.2), and T' the subalgebra of $T \oplus T$ generated by $w = u \oplus u^*$. Let $\lambda : T' \to \mathbb{C}$ be the homomorphism sending w to 1, and let $\hat{I} = \ker \lambda$. Define $j_0, j_1 : \mathbb{C} \to \hat{I}$ by $j_0(1) = (1-uu^*) \oplus 0 = 1-ww^*$ and $j_1(1) = 0 \oplus (1 - u^*u) = 1 - w^*w$. Give $\hat{I}$ the trivial grading.

(b) If u is represented in the standard way on $l^2(\mathbb{N})$, then T' is naturally isomorphic to the subalgebra of $\mathbb{B}(l^2(\mathbb{Z}))$ generated by the bilateral shift and $\mathbb{K}$. Hence there is an essential split extension

$$0 \to \mathbb{K} \to \hat{I} \to C_0(\mathbb{R}) \to 0$$

j_0 and j_1 map $\mathbb{C}$ onto orthogonal one-dimensional projections, so j_0 and j_1 are homotopic.

(c) Two Kasparov (A, B)-modules (E_0, ϕ_0, F_0) and (E_1, ϕ_1, F_1) are said to be *cohomotopic* if there is $(E, \phi, F) \in \mathbb{E}(A \hat{\otimes} \hat{I}, B)$ with $j_i^*(E, \phi, F) \approx_u (E_i, \phi_i, F_i)$ for $i = 0, 1$. Let $\sim_{ch}$ be the equivalence relation generated by cohomotopy and addition of unitarily equivalent elements.

(d) From (b) we have that cohomotopy implies homotopy. On the other hand, if $\beta = (E, \phi, F')$ is a "compact perturbation" of $\alpha = (E, \phi, F)$, then the module $(E \oplus E, \psi, F \oplus F')$ is a cohomotopy between α and β, where $\psi(a \otimes 1) = \operatorname{diag}(\phi(a), \phi(a))$ and $\psi(1 \otimes w) = \left[\begin{smallmatrix} 0 & 0 \\ 1 & 0 \end{smallmatrix}\right]$. Thus $\sim_{ch}$ is stronger than $\sim_c$ and weaker than $\sim_h$, so if A is separable and B σ-unital $\sim_{ch}$ coincides with $\sim_c, \sim_{oh}, \sim_h$.

Cohomotopy was introduced by Cuntz and Skandalis [1986] in an attempt to reverse the asymmetric roles of A and B in $KK(A, B)$. This asymmetry is illustrated, for example, by the much greater difficulty of the second Puppe sequence (19.4.3) compared to the first one. There is a "dual Puppe sequence" [Cuntz and Skandalis 1986, 4.2].

18.13.3. [Cuntz 1987] (a) If A is a C^*-algebra, let qA be as in 10.11.13, and inductively let $q^n A = q(q^{n-1} A)$. Prove:

THEOREM. *If A is separable, then there is a $*$-homomorphism $\phi : qA \to M_2(q^2 A)$ such that $\phi \circ \pi_0$ is homotopic to $j_{q^2 A}$ and $\pi_0^{(2)} \circ \phi$ is homotopic to j_{qA}. [Here $j_B : B \to M_2(B)$ sends x to $\operatorname{diag}(x, 0)$, $\pi_0 : q^2 A \to qA$ is as in 10.11.13, and $\pi_0^{(2)} = \pi_0 \otimes id_{M_2} : M_2(q^2 A) \to M_2(qA)$.] So up to stabilization by 2×2 matrices, $q^2 A$ is homotopy equivalent to A. In particular, $q^2 A \otimes \mathbb{K}$ is homotopy equivalent to $qA \otimes \mathbb{K}$.*

(b) Using this theorem, the isomorphism $KK(A, B) \cong [qA, B \otimes \mathbb{K}]$ (17.6.3), and the isomorphism $[A, B \otimes \mathbb{K}] \cong [A \otimes \mathbb{K}, B \otimes \mathbb{K}]$ valid for any A, B, the natural composition

$$[qA, B \otimes \mathbb{K}] \times [qB, C \otimes \mathbb{K}] \to [q^2 A, C \otimes \mathbb{K}]$$

given by $[\phi][\psi] = [(\psi \otimes id_{\mathbb{K}}) \circ q\phi]$ gives an associative product $KK(A, B) \times KK(B, C) \to KK(A, C)$ for separable trivially graded A, B, C. Show that this coincides with the Kasparov product.

This approach gives perhaps the most natural and straightforward way of defining the KK-groups, constructing the product, and proving its properties. It should be noted, however, that although this approach is overall technically much simpler than the approach we have presented in the text, the proof of the crucial Theorem 18.13.3 is not elementary and relies on an application of G. Pedersen's derivation lifting theorem [1979, 8.6.15] which is related to, and of comparable difficulty to, Kasparov's Technical Theorem.

19. Further Structure in KK-Theory

In this section, we use the intersection product to develop all the further properties of the KK-groups.

We will always assume, often without explicit mention, that all C^*-algebras considered satisfy the appropriate size restrictions (separable or σ-unital) necessary to define the relevant products. Readers not interested in the utmost generality may simply require all C^*-algebras to be separable.

To conserve space, we will let S and C denote the (trivially graded) C^*-algebras $C_0(\mathbb{R}) \cong C_0((0, 1))$ and $C_0([0, 1))$ respectively. [Do not confuse C with the complex numbers $\mathbb{C}$.]

19.1. KK-Equivalence

The important notion of KK-equivalence is basic both in this section and in Chapter IX.

DEFINITION 19.1.1. An element $x \in KK(A, B)$ is a KK-equivalence if there is $y \in KK(B, A)$ with $xy = 1_A$, $yx = 1_B$. A and B are KK-equivalent if there exists a KK-equivalence in $KK(A, B)$.

If $x \in KK(A, B)$ is a KK-equivalence, then for any D, $x \hat{\otimes}_B (\cdot) : KK(B, D) \to KK(A, D)$ and $(\cdot) \hat{\otimes}_A x : KK(D, A) \to KK(D, B)$ are isomorphisms; these isomorphisms are natural in D by associativity. So KK-equivalent C^*-algebras "behave identically" with regard to KK-theory. In particular, if A and B are trivially graded, right multiplication by x gives an isomorphism from $K_i(A) \cong KK^i(\mathbb{C}, A)$ to $K_i(B) \cong KK^i(\mathbb{C}, B)$, and left multiplication by y gives an isomorphism from $Ext_i(A)^{-1} \cong KK^i(A, \mathbb{C})$ to $Ext_i(B)^{-1} \cong KK^i(B, \mathbb{C})$.

If $x \in KK(A, B)$ is a KK-equivalence with inverse y, then $y \hat{\otimes}_A (\cdot) \hat{\otimes}_A x$ is a ring-isomorphism from $KK(A, A)$ onto $KK(B, B)$.

Note that the notion of KK-equivalence is only defined for separable C^*-algebras.

EXAMPLES 19.1.2. (a) For any A, A, $M_n(A)$, and $A \hat{\otimes} \mathbb{K}$ are all KK-equivalent (17.8.8).

(b) Homotopy-equivalent C^*-algebras are KK-equivalent in the obvious way.

(c) If A and B are KK-equivalent (via $\boldsymbol{x}$), then for any D $A \,\hat{\otimes}\, D$ and $B \,\hat{\otimes}\, D$ are KK-equivalent (via $\tau_D(\boldsymbol{x})$).

(d) If $0 \to A \to D \to B \to 0$ is a *split* exact sequence of C^*-algebras, then D is KK-equivalent to $A \oplus B$ by 19.9.1.

(e) If A and B are AF algebras, then A and B are KK-equivalent if and only if their dimension groups are isomorphic as abstract groups, ignoring the order structure (23.15.2).

(f) The argument in 9.4.2 shows that the Toeplitz algebra T is KK-equivalent to $\mathbb{C}$: $\boldsymbol{q} \in KK(T, \mathbb{C})$ is a KK-equivalence, with inverse $\boldsymbol{j} \in KK(\mathbb{C}, T)$. More generally, B and $B \otimes T$ are KK-equivalent for any B. Another generalization is given in 19.9.2.

(g) The argument in 10.11.11·shows that if $A *_D B$ is an amalgamated free product, and there are retractions of A and B onto D, then $A *_D B$ is KK-equivalent to the pullback $P = A \oplus_D B$. In fact, $\boldsymbol{k} \in KK(A *_D B, P)$ is a KK-equivalence with inverse $\boldsymbol{f} - \boldsymbol{g}$. In particular, $C^*(\mathbb{F}_2)$ is KK-equivalent to $C(X_2)$, where X_2 is a figure 8, and more generally $C^*(\mathbb{F}_n)$ is KK-equivalent to $C(X_n)$, where X_n consists of n circles joined at a point (the one-point compactification of the disjoint union of n copies of $\mathbb{R}$). The condition that there are retractions onto D can be relaxed [Germain 1997].

(h) For many locally compact groups G, including free groups, the quotient map $\lambda : C^*(G) \to C_r^*(G)$ is a KK-equivalence (20.9). So $C_r^*(\mathbb{F}_n)$ is also KK-equivalent to $C(X_n)$.

(i) If A is any (separable) C^*-algebra, and qA is as in 10.11.13, then the KK-element in $KK(qA, A)$ corresponding to $\pi_0 : qA \to A$ is a KK-equivalence.

Later in this section we will give more important examples of KK-equivalence.

We will show in Chapter IX that KK-equivalence is quite weak. In fact, for "nice" trivially graded A and B, A and B are KK-equivalent if and only if $K_i(A) \cong K_i(B)$ for $i = 0, 1$ (Universal Coefficient Theorem).

A C^*-algebra A is called *K-contractible* if $KK(A, A) = 0$. This implies that $KK(A, B) = KK(B, A) = 0$ for all B.

EXAMPLES 19.1.3. (a) Any contractible C^*-algebra is K-contractible. In particular, $CB = C_0([0, 1), B)$ is K-contractible for any B.

(b) If $0 \longrightarrow J \longrightarrow A \overset{q}{\longrightarrow} A/J \longrightarrow 0$ is an exact sequence of graded C^*-algebras for which the exact sequence of KK-theory in either variable holds (19.5.7), and if the quotient map is a KK-equivalence, then J is K-contractible. In particular, if $T_0 = C^*(u - 1)$ is as in 9.4.2, then $B \,\hat{\otimes}\, T_0$ is K-contractible for any B.

19.2. Bott Periodicity

We define $x \in KK(\mathbb{C}_1, S)$ and $y \in KK(S, \mathbb{C}_1)$ as follows. To define x, regard $KK(\mathbb{C}_1, S)$ as $KK^1(\mathbb{C}, S) \cong Ext(\mathbb{C}, S)$. Then x is represented by the extension

$$0 \to S \to C \to \mathbb{C} \to 0$$

We may alternately interpret x as the element of $K_1(S)$ corresponding to the unitary $u(t) = e^{2\pi i t}$ for $0 < t < 1$.

$y \in KK^1(S, \mathbb{C}) \cong Ext(S)$ is the extension

$$0 \to \mathbb{K} \to C^*(v-1) \to S \to 0$$

where v is a coisometry of Fredholm index 1, e.g. the adjoint of the unilateral shift. (The homomorphism $\tau : S \to Q$ sends f to $q(v)-1$, where $f(t) = e^{2\pi i t} - 1$.)

We know from earlier sections that $KK^1(\mathbb{C}, S)$ and $KK^1(S, \mathbb{C})$ are both isomorphic to $\mathbb{Z}$ and that x and y are generators; but we will not need to make use of this fact.

THEOREM 19.2.1. x *is a KK-equivalence, with inverse y.*

COROLLARY 19.2.2 (BOTT PERIODICITY). *For any A and B, we have*

$$KK^1(A, B) \cong KK(A, SB) \cong KK(SA, B)$$

and $KK(A, B) \cong KK^1(A, SB) \cong KK^1(SA, B) \cong KK(S^2A, B) \cong KK(A, S^2B) \cong KK(SA, SB)$. The isomorphisms are induced by intersection product with a fixed element, and are therefore natural in A and B.

PROOF. Follows immediately from 19.2.1 and 17.8.9. □

COROLLARY 19.2.3. *For any A, A and S^2A are KK-equivalent. In particular, $\mathbb{C}$ and $C_0(\mathbb{R}^2)$ are KK-equivalent.*

Of course, the Bott Periodicity Theorem of K-theory is a special case of 19.2.2. 19.2.2 also implies Bott Periodicity for Ext (at least for those C^*-algebras for which Ext is a group).

We will prove the theorem in stages. We first show that $xy = 1_{\mathbb{C}_1}$. Then we will give an argument using Bott Periodicity for K-theory to conclude that $yx = 1_S$. Then in the course of proving the Thom Isomorphism we will give another proof that $yx = 1_S$ not depending on K-theory Bott periodicity, thus giving a self-contained treatment.

LEMMA 19.2.4. *We have $xy = 1_{\mathbb{C}_1}$.*

PROOF. This follows immediately from 18.10.2 and 17.8.9. Note that the application of the argument of 18.10.2 does not depend on the identification of $KK^1(\mathbb{C}, S)$ with $K_1(S)$ using K-theory Bott Periodicity. □

It is not too difficult to calculate yx if the elements are expressed in terms of Dirac operators (19.9.3). But since Dirac operators and geometric arguments are foreign to operator algebraists such as the author and many potential readers, we will give an argument which completely eliminates the need to calculate yx.

It follows immediately from 19.2.2 that yx is a nonzero idempotent in the ring $KK(S, S)$. The proof of 19.2.1 would be complete if we could show that 1_S is the only nonzero idempotent in this ring.

If we assume we know $K_1(S) = \mathbb{Z}$ (which requires K-theory Bott periodicity), along with the K-theory exact sequence and 12.2.1, we can give an argument as follows. From 17.5.7 it follows that $KK^1(S, S^2) \cong K_1(Q^s(S^2)) \cong K_0(S^2) \cong K_1(S) \cong \mathbb{Z}$. The map $KK(S, S) \cong KK^1(S, S \,\hat{\otimes}\, \mathbb{C}_1) \to KK^1(S, S^2) \cong \mathbb{Z}$ obtained by right multiplication by x is injective by 19.2.4, so $KK(S, S)$ is a cyclic group and thus as a ring must have only one nonzero idempotent.

19.2.5. If $x_n = x \hat{\otimes} \cdots \hat{\otimes} x \in KK(\mathbb{C}_1 \hat{\otimes} \cdots \hat{\otimes} \mathbb{C}_1, S \hat{\otimes} \cdots \hat{\otimes} S) = KK(\mathbb{C}_n, C_0(\mathbb{R}^n)) \cong K_n(C_0(\mathbb{R}^n))$ and similarly $y_n = y \,\hat{\otimes}\, \cdots \,\hat{\otimes}\, y \in K^n(C_0(\mathbb{R}^n))$, then we have $x_n \hat{\otimes}_{C_0(\mathbb{R}^n)} y_n = 1_{\mathbb{C}_n}$ and $y_n \hat{\otimes}_{\mathbb{C}_n} x_n = 1_{C_0(\mathbb{R}^n)}$. If $n = 2$, we get an invertible element $x_2 \in K_0(C_0(\mathbb{R}^2))$ called the *Bott element*. It can be explicitly described as a quasihomomorphism $(\phi^{(0)}, \phi^{(1)})$ from $\mathbb{C}$ to $M_2(C_0(\mathbb{R}^2))$ by setting $\phi^{(0)}(1) = p$, $\phi^{(1)}(1) = q$, where p and q are defined in 9.2.10. The Bott map of 9.1 is given by intersection product with the Bott element.

19.2.6. If A and B are trivially graded, we may further identify $KK(A, B) \cong KK^1(A, SB)$ with $Ext(A, SB)^{-1}$. There is a nice description of the extension corresponding to an element z of $KK(A, B)$: represent z by a Kasparov module $\left(\mathbb{H}_B \oplus \mathbb{H}_B^{op}, \phi^{(0)} \oplus \phi^{(1)}, \left[\begin{smallmatrix} 0 & 1 \\ 1 & 0 \end{smallmatrix}\right]\right)$ as in the Cuntz picture, and then the extension corresponds to "putting A at each end" of $S(B \otimes \mathbb{K})$ via $\phi^{(0)}$ and $\phi^{(1)}$ respectively. More specifically, the map $\phi : A \to C([0, 1], M(B \otimes \mathbb{K})) \subseteq M(S(B \otimes \mathbb{K}))$ defined by $[\phi(a)](t) = t\phi^{(1)}(a) + (1 - t)\phi^{(0)}(a)$ drops to a *-homomorphism from A to $Q^s(SB)$ and hence defines an (invertible) extension of A by $SB \otimes \mathbb{K}$. The connecting maps from $K_i(A)$ to $K_{1-i}(SB) \cong K_i(B)$ are just the maps induced by right multiplication by z.

19.3. Thom Isomorphism

Just as in the case of K-theory, Bott Periodicity has a generalization to crossed products by $\mathbb{R}$. The result is called the "Thom Isomorphism", although it is actually the KK generalization of Connes' analog of the Thom Isomorphism of algebraic topology (10.2.2). This result is due to Fack and Skandalis [1981]. For a true generalization of the topological Thom isomorphism, see 19.9.4.

The Thom Isomorphism can be defined and its properties proved only using 19.2.4 and not the full force of Bott Periodicity; as a special case we obtain the full Bott Periodicity theorem. The proof of the Thom Isomorphism for KK is actually simpler and more elegant than the proof for K-theory, now that we have the powerful machinery of Section 18.

In this subsection, we work only with *trivially graded* C^*-algebras.

Recall that if α is a continuous action of $\mathbb{R}$ on A, then there are canonical homomorphisms ϕ and ψ from A and $C^*(\mathbb{R}) \cong S$, respectively, into $M(A \times_\alpha \mathbb{R})$. If f is any bounded complex-valued continuous function on $\mathbb{R}$, thought of as an element of $M(C^*(\mathbb{R}))$, then f defines a multiplier F_f of $A \times_\alpha \mathbb{R}$ in the evident way. (To prove this, note that linear combinations of elements of the form $\phi(a) \cdot \psi(g)$, where $a \in A$ and $g \in C^*(\mathbb{R})$, are dense in $A \times_\alpha \mathbb{R}$, and similarly linear combinations of elements of the form $\psi(g) \cdot \phi(a)$ are dense.)

DEFINITION 19.3.1. If f is a continuous complex-valued function on $\mathbb{R}$ for which $\lim_{t \to +\infty} f(t) = 1$ and $\lim_{t \to -\infty} f(t) = -1$, the corresponding element $F_f \in M(A \times_\alpha \mathbb{R})$ is called a *Thom operator* on $A \times_\alpha \mathbb{R}$.

Any two Thom operators differ by an element of $C^*(\mathbb{R})$, so if F_f and F_g are Thom operators, then $(F_f - F_g)\phi(a) \in A \times_\alpha \mathbb{R}$ for all $a \in A$.

PROPOSITION 19.3.2. *If F_f is a Thom operator, then $[F_f, \phi(A)] \subseteq A \times_\alpha \mathbb{R}$.*

PROOF. If the result is true for one Thom operator, it is true for all. Choose f to be the function $f(s) = \frac{1}{\pi} \int_{-s}^{s} \frac{\sin t}{t} dt$. Then f is the Fourier transform of the function g, where $g(t) = i/t$ for $|t| \leq 1$, $g(t) = 0$ for $|t| > 1$. So if $A \times_\alpha \mathbb{R}$ is represented as the completion of the twisted convolution algebra $L^1(\mathbb{R}, A)$, and $a \in A$ is C^∞ for the action, we have that $[T_f, \phi(a)] = h$, where

$$
h(t) = \begin{cases} i\,\dfrac{a - \alpha_t(a)}{t} & \text{for } |t| \leq 1, \\[2mm] 0 & \text{for } |t| > 1. \end{cases}
$$

We have $h \in L^1(\mathbb{R}, A) \subseteq A \times_\alpha \mathbb{R}$. Such a are dense in A. □

So $(A \times_\alpha \mathbb{R}, \phi, F_f)$ can be identified with a Kasparov $(A, (A \times_\alpha \mathbb{R}) \hat{\otimes} \mathbb{C}_1)$-module as in 17.6.4; such a Kasparov module is called a *Thom module* for α. Any two such modules are "compact perturbations" by the remark after 19.3.1.

DEFINITION 19.3.3. The class of $(A \times_\alpha \mathbb{R}, \phi, F_f)$ in $KK^1(A, A \times_\alpha \mathbb{R})$ is called the *Thom element* of (A, α), denoted t_α.

EXAMPLES 19.3.4. (a) Let $A = \mathbb{C}$, $\alpha = \iota$ the trivial action. Then $\mathbb{C} \times_\iota \mathbb{R} \cong C_0(\mathbb{R})$. If $C_0(\mathbb{R})$ is identified with $C_0(0, 1)$ by identifying $+\infty$ with 0 and $-\infty$ with 1, $t_\iota = x$ of 19.2.

(b) Let $A = C_0(\mathbb{R})$, $\hat{\iota}$ the action of $\mathbb{R}$ on $C_0(\mathbb{R})$ by translation. Then $A \times_{\hat\iota} \mathbb{R} \cong \mathbb{K}$ by Takai duality (or by the Heisenberg commutation relations). The element $t_{\hat\iota}$ of $KK^1(C_0(\mathbb{R}), \mathbb{K})$ corresponds to the element $y \in KK(S, \mathbb{C}_1)$ of 19.2.

PROPOSITION 19.3.5. (a) *If $g : (A, \alpha) \to (B, \beta)$ is an equivariant homomorphism, inducing the homomorphism $h : A \times_\alpha \mathbb{R} \to B \times_\beta \mathbb{R}$, then $h_*(t_\alpha) = g^*(t_\beta)$.*

(b) *If γ is the action $\alpha \otimes 1$ on $A \otimes B$, then*

$$t_\gamma = t_\alpha \,\hat{\otimes}\, 1_B = \tau_B(t_\alpha)$$

in $KK^1(A \otimes B, (A \otimes B) \times_\gamma \mathbb{R}) \cong KK^1(A \otimes B, (A \times_\alpha \mathbb{R}) \otimes B)$.

(c) *If $(A \times_\alpha \mathbb{R}) \times_{\hat{\alpha}} \mathbb{R}$ is identified with $A \otimes \mathbb{K}$ in the standard way using Takai duality, then $t_{\hat{\alpha}} = t_\alpha \,\hat{\otimes}\, 1_\mathbb{K}$.*

PROOF. Obvious.   □

THEOREM 19.3.6 (THOM ISOMORPHISM). *If A is a separable trivially graded C^*-algebra with a continuous action α of $\mathbb{R}$, then $A \times_\alpha \mathbb{R}$ is KK-equivalent to SA. The element t_α is an isomorphism.*

The key technical lemma in the proof is the following.

LEMMA 19.3.7. *Let α be an action of $\mathbb{R}$ on a separable C^*-algebra A, and let $\hat{\alpha}$ be the dual action on $A \times_\alpha \mathbb{R}$. Then the element $u_\alpha = t_\alpha \,\hat{\otimes}_{A \times_\alpha \mathbb{R}}\, t_{\hat{\alpha}}$ of $KK(A, (A \times_\alpha \mathbb{R}) \times_{\hat{\alpha}} \mathbb{R}) \cong KK(A, A \otimes \mathbb{K}) \cong KK(A, A)$ is independent of α.*

PROOF. Define an action β of $\mathbb{R}$ on IA by $[\beta_t(\phi)](s) = \alpha_{st}(\phi(s))$. If $f_s : IA \to A$ is evaluation at s, then f_s is an equivariant homomorphism from (IA, β) to (A, α^s), where $\alpha_t^s = \alpha_{st}$. $\alpha^1 = \alpha$, and $\alpha^0 = \iota$, the trivial action. Let $g_s : IA \times_\beta \mathbb{R} \to A \times_{\alpha^s} \mathbb{R}$ and $h_s : (IA \times_\beta \mathbb{R}) \times_{\hat{\beta}} \mathbb{R} \to (A \times_{\alpha^s} \mathbb{R}) \times_{\hat{\alpha}^s} \mathbb{R}$ be the induced homomorphisms from f_s. As a homomorphism from $IA \otimes \mathbb{K}$ to $A \otimes \mathbb{K}$ under the standard identifications of Takai duality, we have $h_s = f_s \otimes 1$. Thus h_0 and h_1 are homotopic as homomorphisms from $IA \otimes \mathbb{K}$ to $A \otimes \mathbb{K}$, so $v_s = h_{s*}(u_\beta) \in KK(IA, A \otimes \mathbb{K})$ is independent of s.

We have $g_{s*}(t_\beta) = f_s^*(t_{\alpha^s})$ and $h_{s*}(t_{\hat{\beta}}) = g_s^*(t_{\hat{\alpha}^s})$. So $v_s = t_\beta \,\hat{\otimes}_{IA \times_\beta \mathbb{R}}\, h_{s*}(t_{\hat{\beta}}) = t_\beta \,\hat{\otimes}_{IA \times_\beta \mathbb{R}}\, g_s^*(t_{\hat{\alpha}^s}) = g_{s*}(t_\beta) \,\hat{\otimes}_{A \times_{\alpha^s} \mathbb{R}}\, t_{\hat{\alpha}^s} = f_s^*(t_{\alpha^s}) \,\hat{\otimes}_{A \times_{\alpha^s} \mathbb{R}}\, t_{\hat{\alpha}^s} = f_s^*(v_s)$ by functoriality. If $d : A \to IA$ embeds A as constant functions, we have $f_s \circ d = 1_A$, so $u_{\alpha^s} = d^*(f_s^*(t_{\alpha^s})) \,\hat{\otimes}_{A \times_{\alpha^s} \mathbb{R}} t_{\hat{\alpha}^s} = d^*(v_s)$ is independent of s. [Informally, a Kasparov product of Thom modules representing $d^*(t_\beta)$ and $t_{\hat{\beta}}$ gives a homotopy between a Kasparov product of Thom modules representing t_α and $t_{\hat{\alpha}}$ and a Kasparov product of Thom modules representing t_ι and $t_{\hat{\iota}}$.] □

LEMMA 19.3.8. *For any action α on a separable trivially graded C^*-algebra A, we have $u_\alpha = 1_A$.*

PROOF. In light of 19.3.7, we need only show that $u_\iota = 1_A$. But under the decomposition $A = A \otimes \mathbb{C}$, t_ι becomes $1_A \,\hat{\otimes}\, x$, $t_{\hat{\iota}}$ becomes $1_A \,\hat{\otimes}\, y$ by 19.3.4(a) and (b); thus $u_\iota = 1_A \,\hat{\otimes}\, xy = 1_A$ by 19.2.4.   □

PROOF OF 19.3.6.. From 19.3.8 t_α has a right inverse, namely $t_{\hat{\alpha}}$. But we may identify t_α with $t_{\hat{\alpha}}$ as in 19.3.5(c); $t_{\hat{\alpha}}$ has $t_{\hat{\alpha}}$ as left inverse.   □

Note that we have finished the proof of Bott Periodicity as a special case of 19.3.6; the proof of 19.3.6 does not use full Bott periodicity for KK or Bott periodicity of K-theory—it only uses 19.2.4.

One reason that the proof of the Thom isomorphism works so much more simply for KK than for K-theory, besides the power of the intersection product, is the fact that buried in the proof is use of the equality of $\sim_c$ and $\sim_h$: KK-elements are regarded as extensions, where the natural equivalence relation is $\sim_c$, but $\sim_h$ was used in an essential way in the proof of 19.3.7. The proof for KK could be adapted for K-theory, but one would essentially have to prove an analog of the agreement of the two equivalence relations.

We have the following corollary of the Thom isomorphism:

COROLLARY 19.3.9. *Let G be a simply connected solvable Lie group, α a continuous action on a separable (trivially graded) C^*-algebra A. Then $A \times_\alpha G$ is KK-equivalent to A if $\dim(G)$ is even, and to SA if $\dim(G)$ is odd.*

19.4. Mapping Cones and Puppe Sequences

We introduce some notions which are generalizations of some important standard constructions of topology.

DEFINITION 19.4.1. (a) If A is a (graded) C^*-algebra, the *cone* of A, denoted CA, is the (graded) C^*-algebra $A \hat{\otimes} C \cong C_0([0,1), A)$ (with the obvious grading).

(b) If $\phi : A \to B$ is a (graded) *-homomorphism, then the *mapping cone* of ϕ, denoted C_ϕ, is the (graded) C^*-subalgebra $\{(x, f) \mid \phi(x) = f(0)\}$ of $A \oplus CB$.

The mapping cone construction is an important example of a pullback (15.3).

There are maps $p : C_\phi \to A$ and $i : SB \to C_\phi$ given by $p(x, f) = x$, $i(f) = (0, f)$. (We identify SB with $C_0((0,1), B)$.) There is a standard short exact sequence

$$0 \longrightarrow SB \xrightarrow{\ i\ } C_\phi \xrightarrow{\ p\ } A \longrightarrow 0$$

associated to C_ϕ. The mapping cone construction is natural in A and B, i.e. if we have a commutative diagram

$$
\begin{array}{ccc}
A_1 & \xrightarrow{\ \phi\ } & B_1 \\
\downarrow{\scriptstyle f} & & \downarrow{\scriptstyle g} \\
A_2 & \xrightarrow{\ \psi\ } & B_2
\end{array}
$$

there is a map $\omega : C_\phi \to C_\psi$ making the following diagram commutative:

$$
\begin{array}{ccccccccc}
0 & \longrightarrow & SB_1 & \longrightarrow & C_\phi & \longrightarrow & A_1 & \longrightarrow & 0 \\
& & \downarrow{\scriptstyle Sg} & & \downarrow{\scriptstyle \omega} & & \downarrow{\scriptstyle f} & & \\
0 & \longrightarrow & SB_2 & \longrightarrow & C_\psi & \longrightarrow & A_2 & \longrightarrow & 0
\end{array}
$$

EXAMPLES 19.4.2. (a) $C_{1_A} \cong CA$; $C_{S\phi} \cong SC_\phi$ for any ϕ.

(b) If A, B, ϕ, C_ϕ, and p are as above, then C_p is isomorphic to the C^*-subalgebra $\{(f,g) \mid \phi(f(0)) = g(0)\}$ of $CA \oplus CB$. The map $j : SB \to C_p$ given by $j(g) = (0, g)$ is a homotopy equivalence, with homotopy inverse $\psi : C_p \to SB$ given by

$$\psi(f,g)(t) = \begin{cases} g(2t - 1) & \text{for } \frac{1}{2} \leq t < 1 \\ \phi(f(1 - 2t)) & \text{for } 0 < t \leq \frac{1}{2}. \end{cases}$$

(c) If A, B, ϕ, C_ϕ, and i are as above, then C_i is isomorphic to the subalgebra $\{(f,g) \mid \phi(f(t)) = g(0,t)\}$ of $SA \oplus C_0([0,1] \times (0,1), B)$. To show this, work through the intermediate algebra

$$\{(f,g) \mid \phi(f(t)) = g(0,t)\} \subseteq SA \oplus C_0([0,1) \times [0,1) \setminus \{0,0\}, B).$$

The map $q : C_i \to SA$ defined by $q(f,g) = f$ is a homotopy equivalence with homotopy inverse $\omega : SA \to C_i$ defined by $\omega(f) = (f,g)$, where $g(s,t) = \phi(f(s))$.

THEOREM 19.4.3 (PUPPE SEQUENCES). *Let A, B, D be graded C^*-algebras and $\phi : A \to B$ a graded $*$-homomorphism. Then the following sequences are exact:*

$$KK(D, SA) \xrightarrow{S\phi_*} KK(D, SB) \xrightarrow{i_*} KK(D, C_\phi) \xrightarrow{p_*} KK(D, A) \xrightarrow{\phi_*} KK(D, B)$$
$$KK(B, D) \xrightarrow{\phi^*} KK(A, D) \xrightarrow{p^*} KK(C_\phi, D) \xrightarrow{i^*} KK(SB, D) \xrightarrow{S\phi^*} KK(SA, D)$$

LEMMA 19.4.4. *Let A, B, D, ϕ be as in 19.4.3. Then the following short sequences are exact in the middle:*

$$KK(D, C_\phi) \xrightarrow{p_*} KK(D, A) \xrightarrow{\phi_*} KK(D, B)$$
$$KK(B, D) \xrightarrow{\phi^*} KK(A, D) \xrightarrow{p^*} KK(C_\phi, D)$$

PROOF. For the first sequence, let $(E_0, \psi_0, F_0) \in \mathbb{E}(D, A)$, and suppose that $\phi_*(E_0, \psi_0, F_0) = 0$ in $KK(D, B)$. Then there exists $(\bar{E}, \bar{\psi}, \bar{F}) \in \mathbb{E}(D, IB)$ with $f_{0*}(\bar{E}, \bar{\psi}, \bar{F}) = \phi_*(E_0, \psi_0, F_0)$ and $f_{1*}(\bar{E}, \bar{\psi}, \bar{F})$ is the 0-module. Then $(E_0 \oplus \bar{E}, \psi_0 \oplus \bar{\psi}, F_0 \oplus \bar{F})$ is a Kasparov (D, C_ϕ)-module in the obvious way, and its image under p_* is (E_0, ψ_0, F_0). Conversely, if l is the obvious projection of C_ϕ onto CB, and $(E, \psi, F) \in \mathbb{E}(D, C_\phi)$, then $l_*(E, \psi, F)$ gives a homotopy from $(\phi \circ p)_*(E, \psi, F)$ to the 0-module. Therefore the first sequence is exact.

The second sequence is harder. We first show $p^* \circ \phi^* = 0$. If $p_0 : CB \to B$ is evaluation at 0 and l is as above, we have $\phi \circ p = p_0 \circ l$. But CB is contractible, so $p_0^* = l^* = 0$, $p^* \circ \phi^* = (\phi_* p)^* = (p_0 \circ l)^* = l^* \circ p_0^* = 0$.

To show that $\ker p^* \subseteq \text{im } \phi^*$, let $(E, \psi, F) \in \mathbb{E}(A, D)$, and suppose that $p^*[(E, \psi, F)] = 0$ in $KK(C_\phi, D)$. Then there is a Kasparov (C_ϕ, CD)-module $(\bar{E}, \bar{\psi}, \bar{F})$ with $f_{0*}(\bar{E}, \bar{\psi}, \bar{F}) = p^*(E, \psi, F)$. Let $\tilde{E}$ be the kernel of the map $f_{0*} : \bar{E} \to p^*(E)$. ($\tilde{E}$ consists of elements of $\bar{E}$ which "vanish at 0".) $\tilde{E}$ is a submodule of $\bar{E}$, and the elements of $\mathbb{B}(\bar{E})$ leave $\tilde{E}$ fixed (since SD is an ideal in CD). Let $\tilde{F}$ be the restriction of $\bar{F}$ to $\tilde{E}$. Let $\omega = \psi \circ i : SB \to C_\phi \to \mathbb{B}(\bar{E}) \to \mathbb{B}(\tilde{E})$. Since $p \circ i = 0$, $\omega(SB) \cdot \bar{E} \subseteq \tilde{E}$. It follows that $(\tilde{E}, \omega, \tilde{F}) \in \mathbb{E}(SB, SD)$.

We define a Kasparov (SA, SD)-module $(\hat{E}, \hat{\psi}, \hat{F})$ as follows. Realize S as $C_0((-1,1))$, so $SA = C_0((-1,1), A)$ etc. Let $\hat{E} = \{(\xi, \eta) \mid \xi(0) = f_{0*}(\eta)\} \subseteq C_0((-1,0], E) \oplus \bar{E}$. $\hat{E}$ is a Hilbert $C_0((-1,1), D)$-module in the obvious way. Let $\pi : CA \to C_\phi$ be the obvious quotient map. If $g \in SA$, let $\hat{\psi}(g)$ be the operator on $\hat{E}$ defined by letting $g \mid_{(-1,0]}$ act on $C_0((-1,0], E)$ pointwise via ψ, and letting $g \mid_{[0,1)}$ act on $\bar{E}$ via $\bar{\psi} \circ \pi$. Let $\hat{F} = 1 \otimes F \oplus \bar{F}$. Then $(\hat{E}, \hat{\psi}, \hat{F}) \in \mathbb{E}(SA, SD)$.

Since the inclusions of $C_0((-1,0))$ and $C_0((0,1))$ into $C_0((-1,1))$ are homotopy equivalences, the Kasparov (SA, SD)-modules obtained by restricting $(\hat{E}, \hat{\psi}, \hat{F})$ to $C_0((-1,0), A)$ and $C_0((0,1), A)$ are homotopic. The restriction of $(\hat{E}, \hat{\psi}, \hat{F})$ to $C_0((-1,0), A)$ is $\tau_S(E, \psi, F)$, and the restriction to $C_0((0,1), A)$ is $(S\phi)^*(\tilde{E}, \omega, \tilde{F})$.

τ_S is an isomorphism by Bott periodicity; if $\boldsymbol{x}$ is the inverse image of the class of $(\tilde{E}, \omega, \tilde{F})$ in $KK(B, D)$, then the class of $(S\phi)^*(\tilde{E}, \omega, \tilde{F})$ is $(S\phi)^*(\tau_S(\boldsymbol{x}) = \tau_S(\phi^*(\boldsymbol{x}))$. Thus the class of $\tau_S(E, \psi, F)$ is $\tau_S(\boldsymbol{x})$, i.e. the class of (E, ϕ, F) is $\boldsymbol{x}$. $\qquad \square$

PROOF OF 19.4.3. Exactness at A follows from 19.4.4. To prove exactness at C_ϕ, apply 19.4.4 to the sequences $KK(D, C_p) \to KK(D, C_\phi) \to KK(D, A)$ and $KK(A, D) \to KK(C_\phi, D) \to KK(C_p, D)$ and note that the inclusion $j : SB \to C_p$ is a homotopy equivalence by 19.4.2(b). Similarly, to prove exactness at SB consider the sequences $KK(D, C_i) \to KK(D, SB) \to KK(D, C_\phi)$ and $KK(C_\phi, D) \to KK(SB, D) \to KK(C_i, D)$ and use 19.4.2(c). It is routine to check that the maps match up properly. The rest of the sequence follows from suspension and Bott periodicity. $\qquad \square$

19.5. Exact Sequences

We will now use the Puppe sequences to derive six-term cyclic exact sequences in each variable for the KK-groups corresponding to a short exact sequence of C^*-algebras.

Exact sequences in KK-theory do not hold in general (19.9.8), and the precise necessary and sufficient conditions for their existence are not known. We can only derive the sequences under the assumption that the quotient map in the exact sequence has a completely positive cross section.

DEFINITION 19.5.1. An exact sequence

$$0 \longrightarrow J \xrightarrow{j} A \xrightarrow{q} A/J \longrightarrow 0$$

of graded C^*-algebras is *semisplit* if there exists a completely positive, norm-decreasing, grading-preserving cross section for q. J is said to be a semisplit ideal in A.

Semisplit ideals are exactly the ideals corresponding to invertible extensions (15.7.1).

EXAMPLES 19.5.2. (a) If A is nuclear, then every ideal of A is semisplit (15.8.3).

(b) If $\phi : A \to B$ is a graded homomorphism, then the mapping cone sequence

$$0 \longrightarrow SB \longrightarrow C_\phi \longrightarrow A \longrightarrow 0$$

is semisplit: the map $\psi(a) = (a, (1-t)\phi(a))$ is a cross section.

LEMMA 19.5.3. *Let A be a separable graded C^*-algebra, J a graded ideal in A. If J is semisplit in A, then SJ is semisplit in CA.*

PROOF. Let ϕ be a completely positive grading-preserving cross section from $B = A/J$ to A for the quotient map $\pi : A \to B$, and let (u_t), for $0 < t < 1$ be a path in $J^{(0)}$ with $0 \le u_t \le 1$, $\lim_{t \to 1} u_t = 0$, such that (u_t) forms a quasicentral approximate identity for J as $t \to 0$. Set $u_0 = 1$, and $v_t = (1 - u_t^2)^{1/2}$.

Let $(a_0, (b_t)) \in CA/SJ \cong C_\pi$. Then $a_0 \in A$, b_t $(0 \le t < 1) \in B$, and $\pi(a_0) = b_0$. Define $\psi(a_0, (b_t)) \in CA$ by

$$[\psi(a_0, (b_t))](s) = u_s a_0 u_s + v_s \phi(b_s) v_s$$

The only nontrivial step in checking that ψ is a completely positive grading-preserving cross section for the quotient map from CA to CA/SJ is to show that $\psi(a_0, (b_t))$ is actually in CA, i.e. that $\psi(a_0, (b_t))$ is continuous at 0. To show this, note that $a_0 - \phi(b_0) = c \in J$, so $\lim_{s \to 0} a_0 - \phi(b_s) = c$. So

$$\lim_{s \to 0} [v_s(a_0 - \phi(b_s))v_s - v_s c v_s] = 0,$$

since (v_s) is bounded. We have $\lim_{s \to 0} v_s c v_s = 0$, so $\lim_{s \to 0} v_s(a_0 - \phi(b_s))v_s = 0$,

$$\lim_{s \to 0} ([\psi(a_0, (b_t))](s) - [u_s a_0 u_s + v_s a_0 v_s]) = 0$$

But $u_s a_0 u_s - a_0 u_s^2$ and $v_s a_0 v_s - a_0 v_s^2$ approach 0 as $s \to 0$ since (u_t) is quasicentral, so

$$\lim_{s \to 0} [\psi(a_0, (b_t))](s) = a_0 = [\psi(a_0, (b_t))](0). \qquad \square$$

REMARK 19.5.4. Using the Technical Theorem one can prove the more general result that if J and K are semisplit graded ideals in a separable graded C^*-algebra, then $J \cap K$ is also semisplit [Cuntz and Skandalis 1986, 2.2].

THEOREM 19.5.5. *Let A be a separable (graded) C^*-algebra, J a semisplit (graded) ideal of A, $q : A \to A/J$ the quotient map. Let $e : J \to C_q$ be defined by $e(x) = (x, 0)$. Then e (regarded as an element $\boldsymbol{e}$ of $KK(J, C_q)$) is a KK-equivalence.*

PROOF. The inverse of $\boldsymbol{e}$ is the element $\boldsymbol{u}$ of $KK(C_q, J) \cong KK^1(C_q, SJ)$ represented by the extension

$$0 \longrightarrow SJ \longrightarrow CA \xrightarrow{\ \pi\ } C_q \longrightarrow 0$$

as in 17.6.4. More specifically, if $\boldsymbol{v} \in KK^1(C_q, SJ)$ is the element represented by this extension, then $\boldsymbol{u} = \boldsymbol{v} \hat{\otimes}_S (\boldsymbol{1}_J \hat{\otimes} \boldsymbol{y})$, where $\boldsymbol{y}$ is as in 19.2.1.

LEMMA 19.5.6. $\boldsymbol{eu} = e^*(\boldsymbol{u}) = \boldsymbol{1}_J$.

PROOF. From the diagram

$$
\begin{array}{ccccccccc}
0 & \longrightarrow & SJ & \longrightarrow & CJ & \xrightarrow{\ f_0\ } & J & \longrightarrow & 0 \\
& & \| & & \downarrow & & \downarrow e & & \\
0 & \longrightarrow & SJ & \xrightarrow[p]{} & CA & \longrightarrow & C_q & \longrightarrow & 0
\end{array}
$$

we conclude that $ev = e^*(v) = 1_J \hat{\otimes} x$, where x is as in 19.2.1. Thus $eu = 1_J$. $\square$

PROOF OF 19.5.5 (CONT.). Let $\phi : CC_q \to C_q \to C(A/J)$ be evaluation at 0 followed by the obvious quotient map. Then $\ker \phi$ is semisplit (each of the quotient maps making up ϕ has an obvious cross section), and is isomorphic to C_e. Apply 19.5.6 to the exact sequence

$$
0 \longrightarrow C_e \to CC_q \xrightarrow{\ \phi\ } C(A/J) \longrightarrow 0
$$

to conclude that for any D, the map $KK(D, C_e) \to KK(D, C_\phi)$ is injective. However, by the first Puppe sequence we have that $KK(D, C_\phi) = 0$ since CC_ϕ and $SC(A/J)$ are contractible. Thus $KK(D, C_e) = 0$ for any D. Now apply the first Puppe sequence to e to conclude that for any $D e_* : KK(D, J) \to KK(D, C_q)$ is an isomorphism, i.e. e is an invertible element in $KK(J, C_q)$. Since u is a right inverse, it must be the inverse. $\square$

As a corollary, we obtain the general six-term exact sequences for KK:

THEOREM 19.5.7 (SIX-TERM EXACT SEQUENCE FOR KK). *Let*

$$
0 \longrightarrow J \xrightarrow{\ j\ } A \xrightarrow{\ q\ } A/J \longrightarrow 0
$$

be a semisplit short exact sequence of σ-unital graded C^*-algebras. Then, for any separable graded D, the following six-term sequence is exact:

$$
\begin{array}{ccccc}
KK(D, J) & \xrightarrow{\ j_*\ } & KK(D, A) & \xrightarrow{\ q_*\ } & KK(D, A/J) \\
\delta \uparrow & & & & \downarrow \delta \\
KK^1(D, A/J) & \xleftarrow{\ q_*\ } & KK^1(D, A) & \xleftarrow{\ j_*\ } & KK^1(D, J)
\end{array}
$$

If A is separable, then for any σ-unital graded C^*-algebra D the following six-term sequence is exact:

$$
\begin{array}{ccccc}
KK(J, D) & \xleftarrow{\ j^*\ } & KK(A, D) & \xleftarrow{\ q^*\ } & KK(A/J, D) \\
\delta \downarrow & & & & \uparrow \delta \\
KK^1(A/J, D) & \xrightarrow{\ q^*\ } & KK^1(A, D) & \xrightarrow{\ j^*\ } & KK^1(J, D)
\end{array}
$$

The map δ is multiplication by the element $\delta_q \in KK^1(A/J, J)$ corresponding to the extension. Under the identification of $KK^1(A/J, J)$ with $KK(S(A/J), J)$,

δ_q corresponds to $i^*(\boldsymbol{u})$, where i is the natural inclusion of $S(A/J)$ into C_q and $\boldsymbol{u} \in KK(C_q, J)$ is the inverse of $e : J \to C_q$ (19.5.6).

REMARK 19.5.8. The proof of 19.5.6 only requires Lemma 19.2.4 and the first (easier) short Puppe sequence, which does not depend on Bott periodicity. Thus one can obtain the six-term exact sequence in the second variable without full Bott periodicity.

If one applies this exact sequence to $0 \to \mathbb{K} \to C^*(v-1) \to S \to 0$, the extension of 19.2.1 which defines $\boldsymbol{y}$, and uses the fact that $C^*(v-1)$ is K-contractible (19.1.3(b)), one obtains an alternate proof of full Bott periodicity.

REMARK 19.5.9. Kasparov's original proof of the KK-theory exact sequences was rather different than the one given here. The argument used the isomorphism of $KK^1(A, B)$ and $Ext(A, B)^{-1}$, and thus only works if all algebras are trivially graded. There is one slight advantage to Kasparov's approach, however: if the first variable is a fixed separable nuclear C^*-algebra, and all algebras are trivially graded, then one obtains a six-term exact sequence in the second variable even for extensions which are not semisplit. (The exact sequence in the first variable requires semisplitness in any event.) A different proof of this more general result can be obtained from E-theory (25.5.13, 25.6.3). The existence of exact sequences in somewhat greater generality can be established using the notion of K-nuclearity (20.10.2).

19.6. Pimsner–Voiculescu Exact Sequences

By essentially the same procedure as in Section 10, we can obtain six-term exact sequences in each KK-variable separately corresponding to a crossed product by $\mathbb{Z}$.

If A is a trivially graded C^*-algebra and $\alpha \in \text{Aut}(A)$, we have an extension

$$0 \longrightarrow S(A \otimes \mathbb{K}) \longrightarrow (A \times_\alpha \mathbb{Z}) \times_{\hat{\alpha}} \mathbb{R} \longrightarrow A \otimes \mathbb{K} \longrightarrow 0$$

corresponding to the mapping torus construction. This exact sequence is always "locally split", hence semisplit. If we apply the exact sequences of KK-theory and the Thom isomorphism, we obtain

THEOREM 19.6.1. Let A be a trivially graded σ-unital C^*-algebra, $\alpha \in \text{Aut}(A)$. Then if D is any separable graded C^*-algebra, the following six-term sequence is exact:

$$
\begin{array}{ccccc}
KK(D, A) & \xrightarrow{\;1-\alpha_*\;} & KK(D, A) & \longrightarrow & KK(D, A \times_\alpha \mathbb{Z}) \\
\big\uparrow & & & & \big\downarrow \\
KK^1(D, A \times_\alpha \mathbb{Z}) & \longleftarrow & KK^1(D, A) & \xleftarrow{\;1-\alpha_*\;} & KK^1(D, A)
\end{array}
$$

If A is separable, then for any σ-unital graded C^*-algebra D, the following six-term sequence is exact:

$$
\begin{array}{ccccc}
KK(A,D) & \xleftarrow{\;1-\alpha^*\;} & KK(A,D) & \longleftarrow & KK(A \times_\alpha \mathbb{Z}, D) \\
\downarrow & & & & \uparrow \\
KK^1(A \times_\alpha \mathbb{Z},\, D) & \longrightarrow & KK^1(A,D) & \xrightarrow{\;1-\alpha^*\;} & KK^1(A,D)
\end{array}
$$

See 19.9.2 for an alternate derivation of this sequence.

REMARK 19.6.2. Similarly, we obtain analogs of the exact sequence of 10.6 for $\mathbb{T}$-actions and of 10.7.1 for $\mathbb{Z}_n$-actions.

19.7. Countable Additivity

We have seen (17.7) that KK is finitely additive in each variable. It is also countably additive in the first variable; the proof requires the fact that $\sim_h = \sim_{oh}$. This result is due to Rosenberg [1987, 1.12].

THEOREM 19.7.1. *Let $A = \bigoplus_i A_i$ be a countable C^*-direct sum of separable C^*-algebras. Then the coordinate inclusions $g_j : A_j \to A$ induce an isomorphism $KK(A,B) \cong \prod KK(A_i, B)$.*

PROOF. The g_j define a homomorphism $\theta = \prod g_j^* : KK(A,B) \to \prod KK(A_i, B)$. To show that θ is surjective, let $(E_i, \phi_i, F_i) \in \mathbb{E}(A_i, B)$. Let E be the "L^2-direct sum" of the E_i as Hilbert B-modules; then there is an evident *-homomorphism $\phi = \oplus \phi_i$ from A to $\mathbb{B}(E)$, and an operator $F = \oplus F_i$ on E; (E, ϕ, F) is a Kasparov (A,B)-bimodule since the intertwining relations only need to be checked for elements in the algebraic direct sum of the A_i, and for each such element everything happens inside a finite direct sum.

To prove injectivity, suppose $(E, \phi, F) \in \mathbb{E}(A, B)$ is in the kernel. By 18.6.1, we may assume (E, ϕ, F) extends to $\bigoplus A_i^+$, which is a split extension of A. Let p_i be the image of the identity of A_i^+ in $\mathbb{B}(E)$; then (E, ϕ, F) is a "compact perturbation" of $(\bigoplus p_i E, \phi, \bigoplus p_i F p_i) \oplus (\text{degenerate})$. By adding on another degenerate if necessary, we may assume that $(p_i E, \phi_i, p_i F p_i) \in \mathbb{E}(A_i, B)$ is operator homotopic to a degenerate element. These operator homotopies may be added in the strict topology of $\mathbb{B}(\bigoplus p_i E)$ to give a homotopy from $(\bigoplus p_i E, \phi, \bigoplus p_i F p_i)$ to a degenerate element. $\qquad\square$

Note the delicate interplay between the various notions of equivalence for Kasparov modules in this proof.

19.7.2. KK is not countably additive in the second variable in general. If $B = \bigoplus B_i$, then for any A there is a natural map ω from $\bigoplus KK(A, B_i)$ to $KK(A, B)$; however, ω is not surjective in general. As an example, let $A = B = c_0$, with $B_i = \mathbb{C}$. Then $1_A \in KK(A, A)$ is not in the image.

Using the Universal Coefficient Theorem, one can prove that if A is a "nice" C^*-algebra with $K_*(A)$ finitely generated, then $KK(A, \cdot)$ is countably additive (23.15.5).

19.8. Recapitulation

Let us once again summarize the principal facts we have proved about the Kasparov groups.

19.8.1. $KK^n(n = 0, 1)$ is a homotopy-invariant bifunctor from pairs (A, B) of graded C^*-algebras to abelian groups. This functor is well behaved only when A is separable and B is σ-unital; throughout the rest of this subsection we assume without mention that all C^*-algebras satisfy these size restrictions.

19.8.2. The elements of $KK^n(A, B)$ are equivalence classes of (A, B)-bimodules. The bimodules considered may be expressed in various standard forms, for example, in the Fredholm module picture (17.5), the Cuntz picture (17.6), the Baaj–Julg picture (17.9), or as extensions (19.2.6). The equivalence relation used is not at all critical; any equivalence relation weaker than $\sim_c$ and stronger than $\sim_h$ will do.

19.8.3. If B is trivially graded, then $KK^n(\mathbb{C}, B)$ is naturally isomorphic to $K_n(B)$ for $n = 0, 1$ (17.5.5, 17.5.6).

19.8.4. If A and B are trivially graded, then $KK^1(A, B)$ is naturally isomorphic to $Ext(A, B)^{-1}$. If A is nuclear, then $KK^1(A, B)$ is naturally isomorphic to $Ext(A, B)$.

19.8.5. KK^n is a stable invariant in each variable, i.e. there are natural isomorphisms

$$KK^n(A, B) \cong KK^n(A \hat{\otimes} \mathbb{K}, B) \cong KK^n(A, B \hat{\otimes} \mathbb{K}) \cong KK^n(A \hat{\otimes} \mathbb{K}, B \hat{\otimes} \mathbb{K}).$$

19.8.6. There is a product

$$\hat{\otimes}_D : KK^n(A_1, B_1 \hat{\otimes} D) \times KK^m(D \hat{\otimes} A_2, B_2) \to KK^{n+m}(A_1 \hat{\otimes} A_2, B_1 \hat{\otimes} B_2).$$

This product is associative and functorial in all possible senses. The product generalizes composition and tensor product of *-homomorphisms, cup and cap products, tensor product of elliptic pseudodifferential operators, and the pairing between K-theory and "K-homology".

19.8.7. There are natural isomorphisms

$$KK^n(A, B) \cong KK^n(SA, SB) \cong KK^{n+1}(A, SB) \cong KK^{n+1}(SA, B)$$

(addition mod 2) for any A and B.

19.8.8. If A is trivially graded and α is any action of $\mathbb{R}$ on A, then there are natural isomorphisms $KK^n(A \times_\alpha \mathbb{R}, B) \cong KK^{n+1}(A, B)$ and $KK^n(B, A \times_\alpha \mathbb{R}) \cong KK^{n+1}(B, A)$ for any B (addition mod 2).

The natural isomorphisms of 19.8.5, 19.8.7, and 19.8.8 are implemented by taking a product with a fixed KK-element.

19.8.9. For any semisplit extension of graded C^*-algebras, there are cyclic six-term exact sequences in each variable (19.5.7).

19.8.10. KK^n is countably additive in the first variable and finitely additive in the second.

19.9. EXERCISES AND PROBLEMS

19.9.1. Let

$$0 \longrightarrow B \xrightarrow{\quad j \quad} D \underset{q}{\overset{s}{\rightleftarrows}} A \longrightarrow 0$$

be a split short exact sequence of graded separable C^*-algebras. Show that the element $s \oplus j \in KK(A \oplus B, D)$ is a KK-equivalence with inverse $q \oplus \pi_s$ (17.8.2(d)). (Use the long exact sequence.)

19.9.2. Let A be a trivially graded unital C^*-algebra, $\alpha \in \mathrm{Aut}(A)$, and T the Toeplitz algebra generated by u (9.4.2). The *generalized Toeplitz algebra of* α, denoted T_α, is the C^*-subalgebra of $(A \times_\alpha \mathbb{Z}) \otimes T$ generated by $A \otimes 1$ and $v \otimes u$, where $v \in A \times_\alpha \mathbb{Z}$ is the unitary implementing α. There is a semisplit exact sequence

$$0 \to A \otimes \mathbb{K} \to T_\alpha \to A \times_\alpha \mathbb{Z} \to 0.$$

(a) Let $\phi^{(0)}$ be the embedding of T_α into $M^s(A)$ corresponding to this extension, and let $\phi^{(1)}(t) = (1 \otimes u)\phi^{(0)}(t)(1 \otimes u^*)$ for $t \in T_\alpha$. Then $(\phi^{(0)}, \phi^{(1)})$ is a quasihomomorphism from T_α to A; let x be its class in $KK(T_\alpha, A)$.

(b) Let $j : A \to T_\alpha$ be defined by $j(a) = a \otimes 1$, and $j \in KK(A, T_\alpha)$ the corresponding element. Show by an argument similar to 9.4.2 that j is invertible with inverse x [Cuntz 1984, 5.5].

(c) Let h be the inclusion $A \to A \otimes \mathbb{K} \to T_\alpha$ as in 17.8.2(b). Show that $hx = 1_A - [\alpha^{-1}] \in KK(A, A)$ [Cuntz 1984, 5.6].

(d) Apply the KK-theory exact sequences to the exact sequence above to obtain an alternate proof of the Pimsner–Voiculescu exact sequence.

This is a refined version (due to Cuntz) of the original argument of Pimsner and Voiculescu for the K-theory exact sequence.

19.9.3. Represent the x_n and y_n of 19.2 as follows [Kasparov 1980b, § 5].

(a) For x_n, regard $C_0(\mathbb{R}^n) \hat{\otimes} \mathbb{C}_n$ as the algebra of continuous functions from $\mathbb{R}^n$ to the Clifford algebra of $\mathbb{C}^n$. The function $f : \mathbb{R}^n \to \mathbb{C}^n$ defined by $f(x) = x(1 + \|x\|^2)^{-1/2}$ gives a multiplier F_n of $C_0(\mathbb{R}^n) \hat{\otimes} \mathbb{C}_n$, and the module $(C_0(\mathbb{R}^n) \hat{\otimes} \mathbb{C}_n, 1, F_n)$ represents $x_n \in KK(\mathbb{C}, C_0(\mathbb{R}^n) \hat{\otimes} \mathbb{C}_n)$. This representation of x_2 exactly gives the Bott element described in 19.2.5.

(b) For $\boldsymbol{y}_n$, let d be the operator of exterior differentiation on $\mathbb{H} = L^2({}_{\mathbb{C}}\mathbb{R}^n)$, δ its adjoint. Then $d + \delta$ defines an essentially self-adjoint unbounded operator of degree 1 on $\mathbb{H}$ (graded by degrees), and if $C_0(\mathbb{R}^n) \hat{\otimes} \mathbb{C}_n$ acts by Clifford multiplication μ, then $(\mathbb{H}, \mu, d+\delta)$ defines an unbounded Kasparov module (17.11). If an ordinary Kasparov module is desired, let $\Delta = d\delta + \delta d$ be the Laplace operator, and $\hat{F}_n = (d + \delta)(1 + \Delta)^{-1/2}$. $(\mathbb{H}, \mu, \hat{F}_n)$ represents $\boldsymbol{y}_n$.

(c) Calculate $\boldsymbol{x}_n \hat{\otimes}_{C_0(\mathbb{R}^n)} \boldsymbol{y}_n$ explicitly by computing the Fredholm index of a suitable operator obtained as in 18.10.1. If $n = 2$, the operator is "half" of the Euler characteristic operator of the 2-sphere.

(d) Show directly thet $\boldsymbol{y}_n \hat{\otimes}_{\mathbb{C}_n} \boldsymbol{x}_n = \boldsymbol{1}_{C_0(\mathbb{R}^n)}$ by finding a Kasparov product [Kasparov 1980b, § 5, Theorem 7].

(e) $\boldsymbol{y}_2$ is exactly the element $[\bar{\partial}_{\mathbb{R}}]$ of 17.1.2(f), where $T^*\mathbb{R}$ is identified with $\mathbb{R}^2$. (More generally, $\boldsymbol{y}_{2n} = [\bar{\partial}_{\mathbb{R}^n}]$.) As a corollary, we obtain that $[\bar{\partial}_{\mathbb{R}^n}]$ is invertible. More generally, $[\bar{\partial}_M]$ is invertible if M is any simply connected complete Riemannian manifold with non-positive sectional curvatures [Miščenko 1974; Kasparov 1995] (cf. 20.7.2).

19.9.4. TOPOLOGICAL THOM ISOMORPHISM. This is a "parametrized version of Bott Periodicity." Let X be a compact Hausdorff space and V a real vector bundle over X. Construct

$$\boldsymbol{x} \in KK(\Gamma(\mathrm{Cliff}(V)), C_0(V)) \quad \text{and} \quad \boldsymbol{y} \in KK(C_0(V), \Gamma(\mathrm{Cliff}(V))),$$

where $\Gamma(\mathrm{Cliff}(V))$ is the homogeneous C^*-algebra of sections of the (complex) Clifford bundle of V, in a manner analogous to the construction of 19.9.3.

Prove the following theorem [Kasparov 1980b, § 5, Theorem 8]:

THEOREM. *$\boldsymbol{x}$ and $\boldsymbol{y}$ are invertible elements which are inverses of each other.*

The proof is similar to the proof of Bott Periodicity outlined in 19.9.3.

If V has a spin^c-structure, then this structure gives a Morita equivalence (hence a KK-equivalence) between $\Gamma(\mathrm{Cliff}(V))$ and $C(X)$ if $\dim(V)$ is even, and between $\Gamma(\mathrm{Cliff}(V))$ and $C(X) \hat{\otimes} \mathbb{C}_1$ if $\dim(V)$ is odd. So the topological Thom isomorphism gives a KK-equivalence between $C_0(V)$ and $C(X)$ with degree shift $\dim(V) \bmod 2$. This is the "classical" Thom isomorphism for K-theory, expressed in KK language.

If X is a single point, we recover Bott Periodicity as done in 19.9.3. This is the justification for calling the topological Thom isomorphism a parametrized version of Bott periodicity, where the parameter runs over the space X.

19.9.5. Prove the following theorem:

THEOREM. *Let M, N, R be smooth manifolds, and $f : M \to N$, $g : N \to R$ be K-oriented maps. Let $f! \in KK(C_0(M), C_0(N))$ and $g! \in KK(C_0(N), C_0(R))$ be the corresponding shriek maps as defined in 17.1.2(g). Then $(g \circ f)! = f! \hat{\otimes}_{C_0(N)}$*

$g!$. *The same is true if M, N, R are foliated manifolds and f, g are foliation maps.*

This theorem, which is due to Kasparov in special cases and Connes and Skandalis [1984] (see [Hilsum and Skandalis 1987; Moore and Schochet 1988]) in full generality, is a fundamental fact; most of the applications of KK-theory to geometry and topology discussed in section 24 are (at least implicitly) based on this result.

19.9.3 (for n even) is a special case, where $M = R$ is a one-point space and $N = \mathbb{R}^n$; we have $\boldsymbol{x} = f!$ and $\boldsymbol{y} = g!$, where f and g are the obvious maps. 19.9.4 is also a special case of 19.9.5 if X is a spin^c-manifold and dim V is even: take $M = R = X$ and $N = V$.

19.9.6. CORRESPONDENCES. [Connes and Skandalis 1984] Let X be a locally compact (second countable) space and Y a (second countable) smooth manifold (not necessarily compact). A *correspondence* from X to Y is a 4-tuple (M, E, f_X, f_Y), where M is a smooth manifold, E is a vector bundle over M (so $\Gamma_0(E)$ is a finitely generated projective module over $C_0(M)$), $f_X : M \to X$ is a proper map, and $f_Y : M \to Y$ is a K-oriented map (17.1.2(g)).

(a) E defines modules

$$(\Gamma_0(E), 1, 0) \in \mathbb{E}(\mathbb{C}, C_0(M))$$

and $(\Gamma_0(E), \mu, 0) \in \mathbb{E}(C_0(M), C_0(M))$, where μ is the action of $C_0(M)$ on $\Gamma_0(E)$ by multiplication. Let $[E] \in KK(\mathbb{C}, C_0(M))$ and $[[E]] \in KK(C_0(M), C_0(M))$ be the corresponding equivalence classes. If $\Delta : M \to M \times M$ is the diagonal map (a proper map), and $[\Delta] \in KK(C_0(M) \otimes C_0(M), C_0(M))$ is its class, then we have $[[E]] = [E] \hat{\otimes}_{C_0(M)} [\Delta]$.

(b) A correspondence (M, E, f_X, f_Y) from X to Y defines the element

$$f_X^*([[E]] \hat{\otimes}_{C_0(M)} f_Y!) \in KK(C_0(X), C_0(Y)).$$

Show that every element of $KK(C_0(X), C_0(Y))$ comes from a correspondence.

(c) Let (M_1, E_1, f_X, f_Y) be a correspondence from X to Y and (M_2, E_2, g_Y, g_Z) a correspondence from Y to Z. Define the *composition* to be the correspondence $(M_1 \times_Y M_2, \pi_1^*(E_1) \oplus \pi_2^*(E_2), f_X \circ \pi_1, g_Z \circ \pi_2)$ from X to Z, where $\pi_i : M_1 \times_Y M_2 \to M_i$ are the coordinate projections. (It may be necessary to perturb f_Y and g_Y to smooth transverse maps via a homotopy.)

(d) Show that if $\boldsymbol{x} \in KK(C_0(X), C_0(Y))$ and $\boldsymbol{y} \in KK(C_0(Y), C_0(Z))$ are represented by the correspondences (M_1, E_1, f_X, f_Y) and (M_2, E_2, g_Y, g_Z) respectively, then $\boldsymbol{x} \hat{\otimes}_{C_0(Y)} \boldsymbol{y}$ is represented by the composition of the two correspondences. Thus at least for commutative C^*-algebras the intersection product can be constructed on a purely geometric level.

19.9.7. Work out the details of Kasparov theory for real and "real" C^*-algebras (a "real" C^*-algebra is the complexification of a real C^*-algebra). Kasparov treats these cases simultaneously with the complex case. The major part of the theory is identical except that more systematic use of Clifford algebras is necessary. See also [Rosenberg 1986a] for results such as Connes' Thom isomorphism in the real case. The theory of real C^*-algebras is outlined in [Rosenberg 1986a] and described in more detail in [Goodearl 1982; Madsen and Rosenberg 1988; Schröder 1993].

19.9.8. [Skandalis 1988] Let Γ be a lattice in a simply connected Lie group locally isomorphic to $\mathrm{Sp}(n,1)$, and let $A = C^*(\Gamma)$, J the kernel of the regular representation, $q : A \to A/J = B = C_r^*(\Gamma)$. ($\Gamma$ has property T, and J is not semisplit in A.) If $e : J \to C_q$ is as in 19.5.5, then $e \in KK(J, C_q)$ is not invertible, specifically there is no $\boldsymbol{u} \in KK(C_q, J)$ such that $\boldsymbol{u}e = \mathbf{1}_{C_q}$ in $KK(C_q, C_q)$. So the sequence $KK(C_q, J) \xrightarrow{e_*} KK(C_q, C_q) \to KK(C_q, CB) = 0$ corresponding to $0 \longrightarrow J \xrightarrow{e} C_q \longrightarrow CB \longrightarrow 0$ is not exact in the middle; therefore for arbitrary fixed first coordinate one does not have a six-term exact sequence in the second variable for arbitrary extensions.

20. Equivariant KK-Theory

In this section, we indicate how the construction of the KK-groups can be modified to give equivariant KK-groups for actions of locally compact groups, and give a survey of the most important properties.

This section was based on Kasparov's conspectus [Kasparov 1995]. This conspectus contains only bare outlines of proofs; a more complete version appeared in [Kasparov 1988]. Some of the details were filled in by Fack [1983] and Rosenberg [1986a]. A complete account of equivariant E-theory has recently appeared [Guentner et al. 1997]. It was my original hope to give a complete treatment in these notes, but the project unfortunately proved to be impractical; thus we must content ourselves with a survey.

The results here for compact groups are a rather simple extension of the non-equivariant case, and were treated in Kasparov's original paper; the substance does not differ markedly from the results of Section 17. The noncompact case, however, presents some considerable additional technical difficulties. Not surprisingly, it is the noncompact case which is of the greatest interest, not only because it provides a good framework for noncompact equivariant K-theory, but also because some beautiful applications have been made to the Novikov conjecture.

Kasparov's approach for noncompact groups is rather different than the Baum–Connes–Phillips approach to equivariant K-theory; Kasparov's theory does not agree with the ordinary KK-theory of the crossed products in general. For this reason, Kasparov's theory might more properly be called "G-continuous" KK-theory.

Throughout this section, all topological groups considered will be *locally compact* but not necessarily compact. To avoid potential difficulties, we will assume (except in 20.1.5) that all groups considered are second countable and all C^*-algebras separable, although much of the work can be done in greater generality (the reader who is so inclined may amuse himself by working out the possibilities). A C^*-algebra with a continuous action of G is called a *G-algebra*.

20.1. Preliminaries

We must first introduce some terminology. The next definition is similar to 11.2.1.

DEFINITION 20.1.1. Let (B, G, β) be a covariant system, and E a Hilbert B-module. A *continuous action* of G on E is a homomorphism from G into the invertible bounded linear transformations on E (not necessarily module homomorphisms) which is continuous in the strong operator topology, i.e. $g \to \|\langle g \cdot x, g \cdot x \rangle\|$ is continuous for all $x \in E$, and for which

$$g \cdot (xb) = (g \cdot x)\beta_g(b) \quad \text{for} \quad g \in G, \ x \in E, \ b \in B$$

A Hilbert B-module with a continuous action of G is called a *Hilbert (B, G, β)-module*.

If E_1 and E_2 are Hilbert (B, G, β)-modules, then there is a natural induced action of G on $\mathbb{B}(E_1, E_2)$ and $\mathbb{K}(E_1, E_2)$ as in 11.3.

If $T \in \mathbb{B}(E_1, E_2)$, then it is not true in general that the function $g \to g \cdot T$ is norm-continuous (it will only be strong-$*$-operator continuous in general).

DEFINITION 20.1.2. $T \in \mathbb{B}(E_1, E_2)$ is *G-continuous* if $g \to g \cdot T$ is norm-continuous.

The set of G-continuous elements of $\mathbb{B}(E)$ form a C^*-subalgebra containing $\mathbb{K}(E)$.

Every G-equivariant map is clearly G-continuous. If G is compact, any G-continuous map can be averaged over G to give a canonically associated G-equivariant map.

DEFINITION 20.1.3. A *graded G-algebra* is a G-algebra A with a grading $A = A^{(0)} \oplus A^{(1)}$, where the $A^{(n)}$ are (globally) invariant under the action of G. A *graded covariant system* is a covariant system (A, G, α) in which A is a graded G-algebra.

Just as in the non-equivariant case, there is a corresponding notion of graded Hilbert modules.

One can form (graded) tensor products of G-algebras; the action of G on the tensor product is the diagonal action. If E_1 is a (graded) Hilbert A-module, E_2 a (graded) Hilbert B-module, each with a continuous action of G, and ϕ is an equivariant homomorphism from A to $\mathbb{B}(E_2)$, then $E_1 \hat{\otimes}_\phi E_2$ can be given a continuous action of G in the obvious way. An important special case is $\hat{\mathbb{H}}_B^G = L^2(G) \hat{\otimes}_{\mathbb{C}} \hat{\mathbb{H}}_B$.

The equivariant stabilization theorem says that $\hat{\mathbb{H}}_B^G$ is the "universal" graded Hilbert (B, G, β)-module:

THEOREM 20.1.4. [Kasparov 1995, §2; Mingo and Phillips 1984] *Let (B, G, β) be a graded covariant system, and E a countably generated Hilbert (B, G, β)-module. Then there is a G-continuous isometric module isomorphism of degree 0 from $E \oplus \hat{\mathbb{H}}_B^G$ onto $\hat{\mathbb{H}}_B^G$.*

The isomorphism cannot be chosen to be G-equivariant in general unless G is compact.

The proof in [Mingo and Phillips 1984] is quite simple. The result follows immediately from the ordinary stabilization theorem and the following two facts:

(1) If E is a countably generated graded Hilbert (B, G, β)-module, then there is a G-continuous isomorphism $E \oplus L^2(G, E^\infty) \cong L^2(G, E^\infty)$ of graded Hilbert B-modules.

(2) If E_1 and E_2 are countably generated graded Hilbert (B, G, β)-modules, and $E_1 \cong E_2$ as graded Hilbert B-modules, then there is a G-equivariant isomorphism $L^2(G, E_1) \cong L^2(G, E_2)$ of graded Hilbert B-modules.

In both cases, there are straightforward explicit formulas for the isomorphisms.

The final preliminary result is the equivariant version of the Kasparov Technical Theorem (14.6.2). The version we give is not quite as general as the one given in [Kasparov 1995], but is sufficient for applications.

THEOREM 20.1.5. *Let J be a σ-unital graded G-algebra. Let A_1 and A_2 be σ-unital C^*-subalgebras of $M(J)$, and Δ a separable graded subspace of $M(J)$. Suppose A_1, A_2, Δ consist of G-continuous elements, that $A_1 \cdot A_2 \subseteq J$, and that Δ derives A_1. Then there are G-continuous elements $M, N \in M(J)$ of degree 0 such that $0 \leq M \leq 1$, $N = 1 - M$, $M \cdot A_1 \subseteq J$, $N \cdot A_2 \subseteq J$, $[M, \Delta] \subseteq J$, and $g \cdot M - M \in J$ for all $g \in G$.*

If G is compact, then M and N may be chosen to be G-invariant.

20.2. The Equivariant KK-Groups

The definitions of this subsection are exact analogs of the ones of Section 17.

DEFINITION 20.2.1. Let A and B be graded G-algebras. $\mathbb{E}_G(A, B)$, the set of *Kasparov G-modules* for (A, B), is the set of triples (E, ϕ, F), where E is a countably generated Hilbert B-module with a continuous action of G, $\phi :$ $A \to \mathbb{B}(E)$ is an equivariant graded *-homomorphism, and F is a G-continuous operator in $\mathbb{B}(E)$ of degree 1, such that $[F, \phi(a)]$, $(F^2 - 1)\phi(a)$, $(F - F^*)\phi(a)$, and $(g \cdot F - F)\phi(a)$ are all in $\mathbb{K}(E)$ for all $a \in A$ and $g \in G$. The set $\mathbb{D}_G(A, B)$ of degenerate Kasparov G-modules is defined correspondingly.

The equivalence relations $\sim_h$, $\sim_{oh}$, $\sim_c$ are defined exactly as in the nonequivariant case.

Direct sum makes $\mathbb{E}_G(A, B)$ into an abelian semigroup.

DEFINITION 20.2.2. $KK_G(A, B)$ is the quotient of $\mathbb{E}_G(A, B)$ by $\sim_h$.

Equivariant KK_{oh}- and KK_c-groups may be similarly defined.

PROPOSITION 20.2.3. $KK_G(A, B)$ is an abelian group. KK_G is a bifunctor from pairs of G-algebras to abelian groups, contravariant in the first variable and covariant in the second.

If G is compact, by a simple averaging we may reduce to the case where the operator F is G-invariant:

PROPOSITION 20.2.4. If G is compact, then any Kasparov G-module is a "compact perturbation" of a Kasparov G-module in which the F is G-invariant. Homotopies and operator homotopies may be taken to lie within this class. So if G is compact, it suffices to consider Kasparov G-modules in which F is invariant under the action of G.

This result can fail for G noncompact.

Kasparov G-modules can be visualized in various ways as in Section 17. If $f : A \to B$ is an equivariant *-homomorphism, then f can be viewed as the $KK_G(A, B)$-element $\boldsymbol{f} = [(B, f, 0)]$.

We may define $KK_G^1(A, B)$ to be $KK_G(A, B \hat{\otimes} \mathbb{C}_1)$, where G acts trivially on $\mathbb{C}_1$. We then have formal Bott Periodicity:

PROPOSITION 20.2.5. There are natural isomorphisms
$KK_G^1(A, B) \cong KK_G(A \hat{\otimes} \mathbb{C}_1, B)$
and
$KK_G(A, B) \cong KK_G^1(A, B \hat{\otimes} \mathbb{C}_1) \cong KK_G^1(A \hat{\otimes} \mathbb{C}_1, B) \cong KK_G(A \hat{\otimes} \mathbb{C}_1, B \hat{\otimes} \mathbb{C}_1)$

We sometimes write $KK_G^0(A, B)$ for $KK_G(A, B)$. We can define $KK_G^n(A, B)$ as before; these groups are periodic mod 2.

DEFINITION 20.2.6. $K_i^G(B) = KK_G^i(\mathbb{C}, B)$. $K_G^i(A) = KK_G^i(A, \mathbb{C})$.

Caution: We have used superscripts and subscripts according to our usual convention (which should be the universal convention). However, some authors such as Kasparov [1980b; 1995] use the opposite convention.
Note also that if A and B are not evenly graded, then $K_i^G(B)$ and $K_G^i(A)$ do not coincide with the ordinary equivariant K-groups (with the usual indexing).

We have the Green–Julg theorem, and its dual:

THEOREM 20.2.7. (a) If G is compact, then $K_i^G(B) \cong K_i(B \times_\beta G)$.
(b) If G is discrete, then $K_G^i(A) \cong K^i(A \times_\alpha G)$.

20.3. The Intersection Product

The intersection product has the same form and properties as in the nonequivariant case. The proofs are similar, using the equivariant version of the Technical Theorem (20.1.5).

THEOREM 20.3.1. *Let A_1, A_2, B_1, B_2, D be G-algebras. Then there is a bilinear pairing*

$$\hat{\otimes}_D : KK_G^m(A_1, B_1 \hat{\otimes} D) \times KK_G^n(D \hat{\otimes} A_2, B_2) \to KK_G^{m+n}(A_1 \hat{\otimes} A_2, B_1 \hat{\otimes} B_2)$$

which is associative and functorial in all possible senses.

As a consequence, one has Bott periodicity. If G is compact, there are six-term cyclic exact sequences in both variables for semisplit extensions, just as in the non-equivariant case. The conditions for existence of exact sequences in the noncompact case are not well understood.

If G is compact, we have a more general equivariant version of Bott Periodicity:

THEOREM 20.3.2. [Kasparov 1980b] *Let G be compact, and let V be a finite-dimensional real vector space with a continuous linear G-action. Then G has an induced action on the complex Clifford algebra $\mathbb{C}_V$, and there are invertible elements $x \in KK_G(\mathbb{C}_V, C_0(V))$ and $y \in KK_G(C_0(V), \mathbb{C}_V)$ (defined as in 19.9.3), which are inverses of each other. If the action of G on V is spinor (in particular, if V is a complex vector space and the action of G is $\mathbb{C}$-linear), then $C_0(V)$ with the induced action is KK_G-equivalent to $C_0(V)$ with the trivial action.*

As a corollary, one obtains Atiyah's equivariant Bott periodicity theorem for K-theory (11.9.5). This result has been generalized to the case where G is noncompact and the space is infinite-dimensional [Higson et al. 1997].

20.4. The Representation Ring

We have an immediate corollary of 20.3.1:

COROLLARY 20.4.1. *If A is a G-algebra, then $KK_G(A, A)$ is an associative ring with unit $\mathbf{1}_A$. $KK_G^*(A, A) = KK_G(A, A) \oplus KK_G^1(A, A)$ is an associative graded ring with unit.*

DEFINITION 20.4.2. *The representation ring of G is the graded ring $R_*(G) = KK_G^*(\mathbb{C}, \mathbb{C})$. $R_0(G)$ is the ring $KK_G(\mathbb{C}, \mathbb{C})$.*

PROPOSITION 20.4.3. *$R_*(G)$ is a commutative ring in the graded sense, i.e. $xy = (-1)^{\partial x \cdot \partial y} yx$; $R_0(G)$ is a commutative ring.*

PROPOSITION 20.4.4. *If G is compact, $R_*(G) = R_0(G)$ is the usual representation ring $R(G)$ defined in 11.1.3, and $R_1(G) = 0$.*

THEOREM 20.4.5. *For any A and B, $KK_G(A, B)$ is an $R_0(G)$-module via intersection product. Functoriality of KK_G and the bilinearity of the product respect the module structure, i.e. KK_G is a bifunctor from pairs of G-algebras to $R_0(G)$-modules. Similarly, KK_G^* is a bifunctor from pairs of G-algebras to graded $R_*(G)$-modules. $KK_G(A, A)$ is an $R_0(G)$-algebra; $KK_G^*(A, A)$ is a graded $R_*(G)$-algebra.*

20.5. Restriction and Induction

We now consider to what extent KK_G is functorial in G. There is one obvious morphism:

DEFINITION 20.5.1. Let $f : H \to G$ be a continuous homomorphism, and let A and B be G-algebras. A and B may be regarded as H-algebras via f, and the map $\rho_f : KK_G(A, B) \to KK_H(A, B)$ defined in the obvious way is called the *restriction homomorphism*. If f is the inclusion of a subgroup, we sometimes write ρ_H or $\rho_{G \downarrow H}$ for ρ_f.

The name "restriction homomorphism" is perhaps not appropriate in the general case; for example, one important instance arises when f is a quotient map. However, the terminology is established.

ρ_f is functorial in A and B and compatible with intersection products. If $f_1 : \Gamma \to H$, $f_2 : H \to G$, then $\rho_{f_2 \circ f_1} = \rho_{f_2} \circ \rho_{f_1}$. If $A = B$, $\rho_f(1_A) = 1_A$.

To describe the induction morphism, we first need the following definition, which is natural in light of the ordinary process of induction of representations:

DEFINITION 20.5.2. Let H be a subgroup of G, and B a H-algebra. $\mathrm{Ind}_{H \uparrow G}(B)$ is the C^*-algebra of all continuous functions f from G to B such that $f(gh) = h^{-1} \cdot (f(g))$ for all $g \in G$, $h \in H$, and such that $\|f\|$, regarded as a function on G/H, vanishes at infinity.

G acts on $\mathrm{Ind}_{H \uparrow G}(B)$ by left translation.

EXAMPLES 20.5.3. (a) If $\alpha \in \mathrm{Aut}(A)$, then the mapping torus (10.3.1) M_α is $\mathrm{Ind}_{\mathbb{Z} \uparrow \mathbb{R}}(A)$.

(b) If B is a G-algebra viewed as an H-algebra by restriction of the action, then $\mathrm{Ind}_{H \uparrow G}(B)$ is equivariantly isomorphic to $C_0(G/H)) \otimes B$ via Φ, where $[\Phi(f)](g) = g \cdot f(g)$.

If E is a Hilbert B-module with continuous action of H, then the same formulas as above define a Hilbert $\mathrm{Ind}_{H \uparrow G}(B)$-module $\mathrm{Ind}_{H \uparrow G}(E)$ with continuous action of G.

THEOREM 20.5.4. *Let H be a subgroup of G, and A and B H-algebras. Then there is an induction homomorphism*

$$\iota_{H \uparrow G} : KK_H(A, B) \to KK_G(\mathrm{Ind}_{H \uparrow G}(A), \mathrm{Ind}_{H \uparrow G}(B))$$

which is functorial in A and B and compatible with the intersection products. If $\Gamma \subseteq H \subseteq G$, then $\iota_{H \uparrow G} \circ \iota_{\Gamma \uparrow H} = \iota_{\Gamma \uparrow G}$. If $A = B$, then $\iota_{H \uparrow G}(1_A) = 1_{\mathrm{Ind}_{H \uparrow G}(A)}$.

$\iota_{H \uparrow G}$ is defined as follows. Let $(E, \phi, F) \in \mathbb{E}_G(A, B)$. There is an obvious homomorphism $\psi : \mathrm{Ind}_{H \uparrow G}(A) \to \mathbb{B}(\mathrm{Ind}_{H \uparrow G}(E))$. Let $\tilde{F} \in \mathbb{B}(\mathrm{Ind}_{H \uparrow G}(E))$ be defined by

$$\tilde{F}(g) = \int_H h \cdot F\alpha(gh)\, dh$$

where α is a nonnegative scalar-valued function on G with $\int_H \alpha(gh)\, dh = 1$ for all g and such that $\int_H |\alpha(gh) - \alpha(g_0 h)|\, dh \to 0$ as $g \to g_0$. $\iota_{H\uparrow G}[(E, \phi, F)] = [(\mathrm{Ind}_{H\uparrow G}(E), \psi, \tilde{F})]$.

We have the following "Frobenius Reciprocity Theorem", due to A. Wassermann [1983]:

THEOREM 20.5.5. *If H is a closed subgroup of a compact group G, and if A is a G-algebra and B an H-algebra, then there is an isomorphism $KK_H(A, B) \to KK_G(A, \mathrm{Ind}_{H\uparrow G}(B))$ given by $x \to \psi^*(\iota_{H\uparrow G}(x))$, where ψ is the inclusion $A \cong A \otimes 1 \subseteq A \otimes C(G/H) \cong \mathrm{Ind}_{H\uparrow G}(A)$. The inverse is given by $y \to \eta_*(\rho_{G\downarrow H}(y))$, where $\eta : \mathrm{Ind}_{H\uparrow G}(B) \to B$ is evaluation at the identity.*

20.6. Relation with Crossed Products

In this subsection, we give the fundamental relationship between equivariant KK-theory and the K-theory of crossed products.

Let (B, G, β) be a covariant system, and let E be a Hilbert B-module. The algebra $C_c(G, B)$ acts on $C_c(G, E)$ by

$$(xf)(t) = \int_G x(s)\beta_s(f(s^{-1}t))\, dt$$

Define a $C_c(G, B)$-valued inner product on $C_c(G, E)$ by

$$\langle x, y \rangle(t) = \int_G \beta_{s^{-1}}(\langle x(s), y(st) \rangle_B)\, ds$$

DEFINITION 20.6.1. *The completion of $C_c(G, E)$ with this inner product is a Hilbert $B \times_\beta G$-module, denoted $E \times_\beta G$.*

(Note that E is not assumed to have a G-action.)

Now suppose $(E, \phi, F) \in \mathbb{E}_G(A, B)$. ϕ induces $\psi : A \times_\alpha G \to \mathbb{B}(E \times_\beta G)$ by

$$(\psi(a)x)(t) = \int_G \phi(a(s)) \cdot [s \cdot x(s^{-1}t)]\, ds \text{ for } a \in C_c(G, A), x \in C_c(G, E)$$

Define $\tilde{F} \in \mathbb{B}(E \times_\beta G)$ by $(\tilde{F}x)(t) = F(x(t))$ for $x \in C_c(G, E)$.

THEOREM 20.6.2. *$(E \times_\beta G, \psi, \tilde{F}) \in \mathbb{E}(A \times_\alpha G, B \times_\beta G)$. The map sending (E, ϕ, F) to $(E \times_\beta G, \psi, \tilde{F})$ gives a homomorphism*

$$j_G : KK_G(A, B) \to KK(A \times_\alpha G, B \times_\beta G)$$

which is functorial in A and B and compatible with the intersection product. If $A = B$ then $j_G(1_A) = 1_{A \times_\alpha G}$.

20.7. Connected Groups

20.7.1. In this subsection, we obtain more specific information about the maps of 20.5 and 20.6 for connected groups. In good cases, the restrictions to the maximal compact subgroup are isomorphisms; in general, this restriction picks out the "interesting part" of the KK-theory.

Throughout this subsection, G will denote a *connected* locally compact group and H a maximal compact subgroup of G (recall that all maximal compact subgroups are conjugate).

The results use some *canonical elements* associated to G and H. If M is a complete Riemannian manifold with continuous G-action, we may form the canonical element $y \in KK_G(\Gamma_0(\text{Cliff}(T^*M)), \mathbb{C})$ corresponding to the "Dirac operator" (analogous to 19.9.3). If M is simply connected and has nonpositive sectional curvature, another canonical element $x \in KK_G(\mathbb{C}, \Gamma_0(\text{Cliff}(T^*M)))$ can be constructed as the "dual Dirac" element as in 19.9.3. In the case of G and H, G has a compact normal subgroup K with G/K a Lie group, and $M = G/H$ is a homogeneous space of G/K. Thus the element y can be constructed for M as above. If G/K is semisimple with finite center, then M (which is simply connected in any case) has nonpositive sectional curvature, and x may also be constructed as above. In the general case x must be constructed using an inductive argument on the dimension of G/K.

If V is the tangent space to $M = G/H$ at the identity coset, then the C^*-algebra $\Gamma_0(\text{Cliff}(T^*M))$ is isomorphic to $\text{Ind}_{H\uparrow G}(\mathbb{C}_V)$, where $\mathbb{C}_V$ is the complex Clifford algebra of V. Thus we can use the KK_H-equivalence of $\mathbb{C}_V$ and $C_0(V)$ (20.3.2) to transfer x and y to elements of $K_0^G(C_0(T^*M))$ and $K_G^0(C_0(T^*M))$ respectively. (Note that $\mathbb{C}_V$ is graded, so if $\dim M$ is odd, then x and y are really "KK_G^1-elements".)

If the canonical action of H on V is spinor, we may further reduce to $x \in K_n^G(C_0(M))$, $y \in K_G^n(C_0(M))$, where $n = \dim M \mod 2$.

THEOREM 20.7.2. (a) $y \hat{\otimes}_{\mathbb{C}} x = \mathbf{1}_{C_0(T^*M)}$.

(b) $z_G = x \hat{\otimes}_{C_0(T^*M)} y$ is an idempotent in $R_0(G)$; if Γ is a connected subgroup of G, then $\rho_{G\downarrow\Gamma}(z_G) = z_\Gamma$.

(c) If Γ is an amenable subgroup of G, then $\rho_{G\downarrow\Gamma}(z_G) = \mathbf{1}$ (the identity of $R_0(\Gamma)$). In particular, if G is amenable, then $z_G = \mathbf{1}$.

The Theorem remains true in the case where M is a simply connected G-manifold with nonpositive sectional curvature. The proof in both cases is along the lines of 19.9.3.

x, y, z are called β, α, γ respectively in [Kasparov 1995]; x and y are called $\hat{\delta}$ and δ respectively in [Connes 1994].

If G has Kazhdan's Property (T), then $z_G \neq \mathbf{1}$. See 20.9 for further discussion of the circumstances under which $z_G = \mathbf{1}$.

The idempotent z_G may be thought of as the support of the "well-behaved" part of $R_0(G)$. The next few results and conjectures make this precise.

THEOREM 20.7.3. *The restriction map $\rho_{G\downarrow H}$ is surjective; its kernel is exactly the kernel of module multiplication by z_G, denoted $\ker z_G$. $KK_G(A, B) \cong$ $(\ker z_G) \oplus (\ker(1 - z_G))$, and $\ker(1 - z_G) \cong KK_H(A, B)$. If G is amenable, $\rho_{G\downarrow H} : KK_G(A, B) \to KK_H(A, B)$ is an isomorphism.*

Set $n = \dim M$. If (A, G, α) is a covariant system, then G has a diagonal action δ on $A \hat{\otimes} \mathbb{C}_n$, where the action on $\mathbb{C}_n$ comes from identifying it with the Clifford algebra of V. By sending $\boldsymbol{x} \hat{\otimes} 1_A$ and $\boldsymbol{y} \hat{\otimes} 1_A$ over with j_G, and using the Morita equivalence of $(\operatorname{Ind}_{H\uparrow G}(A\hat{\otimes}\mathbb{C}_n)) \times_\tau G$ with $(A\hat{\otimes}\mathbb{C}_n) \times_\delta H$ [Green 1977], we obtain elements $\tilde{\boldsymbol{x}} \in KK(A \times_\alpha G, (A \hat{\otimes} \mathbb{C}_n) \times_\delta H)$ and $\tilde{\boldsymbol{y}} \in KK((A\hat{\otimes}\mathbb{C}_n) \times_\delta H, A \times_\alpha G)$.

If the representation of H on V is spinor, then $\tilde{\boldsymbol{x}}$ and $\tilde{\boldsymbol{y}}$ may be regarded as elements of $KK^n(A \times_\alpha G, A \times_\alpha H)$ and $KK^n(A \times_\alpha H, A \times_\alpha G)$ respectively.

From the functoriality of j_G, we have that $\tilde{\boldsymbol{y}} \hat{\otimes}_{A\times_\alpha G} \tilde{\boldsymbol{x}} = 1_{(A\times\mathbb{C}_n)\times_\delta H}$, and $\tilde{\boldsymbol{x}} \hat{\otimes}_{(A\times\mathbb{C}_n)\times_\delta H} \tilde{\boldsymbol{y}} = \tilde{z} = j_G(z_G) \hat{\otimes} 1_A$. $\tilde{z}$ is an idempotent in the ring $KK(A \times_\alpha G, A \times_\alpha G)$; if G is amenable, then $\tilde{z} = 1_{A\times_\alpha G}$.

We summarize the results of this construction in a theorem.

THEOREM 20.7.4. *For any A and B the homomorphism from $KK(A \times_\alpha G, B)$ to $KK((A \hat{\otimes} \mathbb{C}_n) \times_\delta H, B)$ given by left multiplication by $\tilde{\boldsymbol{x}}$ is surjective. Its kernel is the same as the kernel of left multiplication by $\tilde{z}$. A similar statement holds for right multiplication by $\tilde{\boldsymbol{y}}$.*

COROLLARY 20.7.5. *If G is amenable, then $A \times_\alpha G$ and $(A \hat{\otimes} \mathbb{C}_n) \times_\delta H$ are KK-equivalent. If the action of H on V is spinor, then $A \times_\alpha G$ and $A \times_\alpha H$ are KK-equivalent, with a shift in parity equal to $\dim G/H \mod 2$.*

The Thom isomorphism (19.3) is a special case of this theorem, with $G = \mathbb{R}$.

One feature of the canonical element $\boldsymbol{x} \in K_G^0(C_0(T^*M))$ is that the natural homomorphism from $C^*(G)$ to $\mathbb{B}(E \times_\beta G)$ coming from the transition from KK_G to KK of the crossed product factors through $C_r^*(G)$, i.e. if $\pi : C^*(G) \to C_r^*(G)$ is the quotient map, there is an element $\tilde{\boldsymbol{x}}_r \in KK(C_r^*(G), \mathbb{C}_n \times_\delta H)$ with $\tilde{\boldsymbol{x}} = \pi^*(\tilde{\boldsymbol{x}}_r)$. Set $\tilde{z}_r = \tilde{\boldsymbol{x}}_r \hat{\otimes}_{\mathbb{C}_n \times_\delta H} \tilde{\boldsymbol{y}}$. Then $\tilde{z}_r \in KK(C_r^*(G), C^*(G))$ and $\tilde{z} = \pi^*(\tilde{z}_r)$. We may push $\tilde{z}_r$ via π_* to obtain $\tilde{w} = \pi_*(\tilde{z}_r) \in KK(C_r^*(G), C_r^*(G))$.

The following conjecture is called the *Connes–Kasparov–Rosenberg Conjecture* [Connes and Moscovici 1982a; 1982b; Rosenberg 1984, 4.1; Kasparov 1995, § 6]:

CONJECTURE 20.7.6. $\tilde{w} = 1_{C_r^*(G)}$.

The conjecture can also be stated for crossed products.

If this conjecture is true for a group G, it has the following consequence:

CONSEQUENCE 20.7.7. *$\tilde{z}_r \in KK(C_r^*(G), C^*(G))$ defines a cross section (splitting) for the K-theory exact sequence corresponding to the extension*

$$0 \to \ker \pi \to C^*(G) \to C_r^*(G) \to 0$$

so there is a canonical splitting

$$K_*(C^*(G)) \cong K_*(C_r^*(G)) \oplus K_*(\ker \pi)$$

The conjecture is true for many classes of groups. However, the conjecture fails for $G = \Gamma$ of 19.9.8 [Skandalis 1988].

The original form of the Connes–Kasparov Conjecture was:

CONJECTURE 20.7.8. *If $\tilde{\boldsymbol{y}}_r = \pi^*(\tilde{\boldsymbol{y}}) \in KK(\mathbb{C}_n \otimes_\delta H, C_r^*(G))$, then right multiplication by $\tilde{\boldsymbol{y}}_r$ gives an isomorphism between $K_*(\mathbb{C}_n \otimes_\delta H)$ and $K_*(C_r^*(G))$, and similarly for crossed products.*

There are two differences between 20.7.8 and 20.7.6: (1) under 20.7.8, $\tilde{\boldsymbol{y}}_r$ could be a KK-equivalence whose inverse is not $\tilde{\boldsymbol{x}}_r$; or (2) $\tilde{\boldsymbol{y}}_r$ might not be a KK-equivalence but might nonetheless give an isomorphism on K-theory. (2) cannot happen in the presence of the UCT by 23.10.1; the failure of the UCT in the example may exactly account for the failure of 20.7.6 in this case.

We will discuss applications of the results of this section to the Novikov conjecture in Section 24.

20.8. Discrete Groups

[Kasparov 1995] also contains a number of results about discrete subgroups of connected groups. One example is the following, which is useful in computing the ring $R_*(\Pi)$ for discrete Π.

THEOREM 20.8.1. *Let G be a connected Lie group, H a maximal compact subgroup, and Π a discrete subgroup of G. Then the subgroup $\{x \in R_i(\Pi) \mid \rho_{G\downarrow\Pi}(z_G)x = x\}$ of the group $R_i(\Pi)$ is isomorphic to $K_i^H(\Gamma(\mathit{Cliff}(G/\Pi)))$.*

Other results give ways of calculating the KK-groups of $A \times_\alpha \Pi$, including a spectral sequence, whenever Π is a torsion-free discrete subgroup of a connected group. As a very special case one recovers the Pimsner–Voiculescu exact sequence.

20.9. K-Theoretic Amenability for Groups

If G is a locally compact group, the study of the full group C^*-algebra $C^*(G)$ and the reduced group C^*-algebra $C_r^*(G)$ is important in many contexts. The structure of these C^*-algebras is, of course, of fundamental importance in the theory of group representations. In addition, natural examples of interesting C^*-algebras arise as group C^*-algebras (cf. 6.10.4). When full and reduced crossed product algebras are also considered, the range of interesting examples and applications is greatly expanded.

Since group C^*-algebras and crossed products often seem to have a rather intractable structure, it is very desirable to have techniques for calculating K-groups, *Ext*-groups, and more generally KK-groups for such algebras. The results of chapter V are basic examples of such results.

It is frequently much simpler to deal with $C^*(G)$ than with $C_r^*(G)$, since $C^*(G)$ has a readily identified universal property for representations. In order to apply results on $C^*(G)$ to $C_r^*(G)$, one would like general results identifying the

K-theories, ideally a KK-equivalence. Cuntz [1983b] was thus led to introduce the notion of K-theoretic amenability for discrete groups. Julg and Valette [1983, 1984, 1985] subsequently extended the definition to nondiscrete groups.

DEFINITION 20.9.1. A locally compact group G is K-*amenable* if the element $1_{\mathbb{C}} \in KK_G(\mathbb{C}, \mathbb{C})$ is represented by a module $(\mathbb{H}_0 \oplus \mathbb{H}_1^{op}, 1, F)$, where G acts on $\mathbb{H}_0$ and $\mathbb{H}_1$ by representations weakly contained in the regular representation.

At least for discrete groups, the definition may be rephrased in Cuntz' original form:

PROPOSITION 20.9.2. (a) *If G is K-amenable, then the quotient map λ : $C^*(G) \to C_r^*(G)$ is a KK-equivalence. More generally, if A is a G-algebra, then $\lambda : C^*(G, A) \to C_r^*(G, A)$ is a KK-equivalence.*
(b) *If G is discrete and λ is a KK-equivalence, then G is K-amenable.*

Any amenable group is obviously K-amenable, but there are also many nonamenable groups which are K-amenable, as the next theorem shows.

THEOREM 20.9.3. [Cuntz 1983b] *The class of K-amenable discrete groups is closed under extensions, direct limits, direct products, and free products. A closed subgroup of a K-amenable group is K-amenable.*

Thus any free group, or more generally any free product of discrete abelian groups, is K-amenable.

There are groups which are not K-amenable: for example, any group with Kazhdan's property (T) has a direct summand of $\mathbb{C}$ in its full group C^*-algebra which is in the kernel of λ, and therefore cannot be K-amenable. It follows that a quotient of a K-amenable group need not be K-amenable (any discrete group is a quotient of a free group).

Pimsner [1986] has shown the following extension of a result of Julg and Valette:

THEOREM 20.9.4. *If G is a locally compact group acting without involution on a tree, and if all the stability groups are K-amenable, then G is K-amenable.*

It follows that $SL_2(\mathbb{Q}_p)$ is K-amenable.

Kasparov [1984b] has shown that any group locally isomorphic to $SO(n, 1)$ is K-amenable. It is also conjectured that groups locally isomorphic to $SU(n, 1)$ are K-amenable. If this is true, then a connected Lie group would be K-amenable if and only if it has a composition series consisting of $SO(n, 1)$'s, $SU(n, 1)$'s, and compact groups (any other simple Lie group has property (T)).

If G is connected and the element z_G of 20.7.2 is 1, then G is K-amenable. The converse is conjectured but still open, but is true for Lie groups [Julg and Kasparov 1995]. If this is true in general, then throughout 20.7 "amenable" could be replaced by "K-amenable".

See [Julg and Valette 1985] for a further discussion of K-amenability.

20.10. EXERCISES AND PROBLEMS

20.10.1. Work out the details of all of the results outlined in this section and the additional results in [Kasparov 1995]. Develop structure theorems for $R_*(G)$ for standard classes of G.

20.10.2. K-THEORETIC NUCLEARITY. [Skandalis 1988]

(a) A Kasparov (A, B)-module (E, ϕ, F) is called *nuclear* if $\phi : A \to \mathbb{B}(E)$ can be approximated in the topology of pointwise strong operator convergence by completely positive finite-rank contractions (cf. 15.8.1).

(b) In analogy with 20.9.1, a separable (graded) C^*-algebra A is K-*nuclear* if $1_A \in KK(A, A)$ is represented by a nuclear Kasparov module. Any nuclear or K-contractible C^*-algebra is K-nuclear.

(c) A Kasparov product of a nuclear Kasparov module with any Kasparov module (in either order) is nuclear. Thus the class of K-nuclear C^*-algebras is closed under KK-equivalence.

(d) The class of K-nuclear C^*-algebras is closed under minimal and maximal tensor products, and under (full or reduced) crossed products by K-amenable groups. In particular, if G is K-amenable, then $C^*(G)$ and $C_r^*(G)$ are K-nuclear. If $0 \to J \to A \to A/J \to 0$ is a semisplit exact sequence, and two of the algebras are K-nuclear, then so is the third.

(e) If A is K-nuclear, then for every (separable) B the quotient map from $A \hat{\otimes}_{\max} B$ to $A \hat{\otimes} B$ is a KK-equivalence, and the kernel is K-contractible.

(f) If A is K-nuclear, then $KK(A, \cdot)$ is a homology theory on the category of separable C^*-algebras, i.e. there is a cyclic 6-term exact sequence in the second variable for arbitrary (not necessarily semisplit) graded extensions.

(g) If $0 \to J \to A \to A/J \to 0$ is an exact sequence (not necessarily semisplit) of K-nuclear graded C^*-algebras, then for any separable B there is a cyclic 6-term exact sequence for $KK(\cdot, B)$ corresponding to this extension.

(h) Is every K-nuclear C^*-algebra KK-equivalent to a nuclear C^*-algebra?

(i) Formulate an equivariant version of K-nuclearity and examine the relationship with K-amenability.

(j) The C^*-algebra $C_r^*(\Gamma)$ is not K-nuclear, where Γ is as in 19.9.8. (Use (f).)

CHAPTER IX

FURTHER TOPICS

21. Homology and Cohomology Theories on C^*-Algebras

In this section, we define and derive some basic properties of abstract homology and cohomology theories for C^*-algebras. We consider only *trivially graded* C^*-algebras.

The results and ideas of this section are taken primarily from [Schochet 1984a], although many of them are at least based on folklore and some are straightforward translations of standard results of topology.

21.1. Basic Definitions

Let **S** be a subcategory of the category of all C^*-algebras, which is closed under quotients, extensions, and closed under suspension in the sense that if $A \in \mathrm{Ob}(\mathbf{S})$, then $SA \in \mathrm{Ob}(\mathbf{S})$, that $\phi \in \mathrm{Hom}_\mathbf{S}(A, B)$ implies $S\phi \in \mathrm{Hom}_\mathbf{S}(SA, SB)$, and that $C_0(\mathbb{R}) \in \mathrm{Ob}(\mathbf{S})$ and every *-homomorphism from $C_0(\mathbb{R})$ to A is in $\mathrm{Hom}_\mathbf{S}(C_0(\mathbb{R}), A)$.

We will consider covariant and contravariant functors F from **S** to **Ab**, the category of abelian groups, although we could more generally work with functors to any abelian category with arbitrary limits (for some results less is required). We will consider only functors which satisfy the following *homotopy* axiom:

(H) If $f_0, f_1 : A \to B$ are homotopic, then $f_{0*} = f_{1*} : F(A) \to F(B)$ [or $f_0^* = f_1^* : F(B) \to F(A)$ in the contravariant case].

DEFINITION 21.1.1. A *homology theory* on **S** is a sequence $\{h_n\}$ of covariant functors from **S** to **Ab**, satisfying (H) and

(LX) If $0 \longrightarrow J \overset{i}{\longrightarrow} A \overset{q}{\longrightarrow} B \longrightarrow 0$ is a short exact sequence in **S**, then for each n there is a connecting map $\partial : h_n(B) \to h_{n-1}(J)$ making exact the long sequence

$$\cdots \overset{\partial}{\longrightarrow} h_n(J) \overset{i_*}{\longrightarrow} h_n(A) \overset{q_*}{\longrightarrow} h_n(B) \overset{\partial}{\longrightarrow} h_{n-1}(J) \overset{i_*}{\longrightarrow} \cdots$$

∂ is natural with respect to morphisms of short exact sequences.

A *cohomology theory* on **S** is a sequence $\{h^n\}$ of contravariant functors from **S** to **Ab**, satisfying (H) and

(LX′) If $0 \longrightarrow J \xrightarrow{\ i\ } A \xrightarrow{\ q\ } B \longrightarrow 0$ is a short exact sequence in **S**, then for each n there is a connecting map $\partial : h^n(J) \to h^{n+1}(B)$ making exact the long sequence

$$\cdots \xrightarrow{\ \partial\ } h^n(B) \xrightarrow{\ q^*\ } h^n(A) \xrightarrow{\ i^*\ } h^n(J) \xrightarrow{\ \partial\ } h^{n+1}(B) \xrightarrow{\ q^*\ } \cdots$$

∂ is natural with respect to morphisms of short exact sequences.

∂ must be 0 for a split exact sequence by naturality. It follows that the functors in a homology or cohomology theory are automatically additive.

We will allow homology or cohomology theories where the index set is all of $\mathbb{Z}$, or a set of the form $\{n : n \geq n_0\}$ or $\{n : n \leq n_0\}$.

If **S** is closed under stable isomorphism, then a homology or cohomology theory on **S** is called *stable* if each functor satisfies

(S) If $\eta : A \to A \otimes \mathbb{K}$ sends a to $a \otimes p$ for a one-dimensional projection p, then η_* is an isomorphism.

If **S** is closed under finite direct sums and (countable) inductive limits, it is closed also under (countable) direct sums (where a direct sum of C^*-algebras is understood to be the c_0-direct sum). A homology theory on such an **S** is called $\sigma - additive$ if it satisfies

(A) If $A = \bigoplus A_i$ is a countable direct sum, then for each n the canonical maps $h_n(A_i) \to h_n(A)$ induce an isomorphism $\bigoplus_i h_n(A_i) \to h_n(A)$.

A cohomology theory on **S** is called σ-additive if it satisfies

(A′) If $A = \oplus A_i$ is a countable direct sum, then for each n the canonical maps $h^n(A) \to h^n(A_i)$ induce an isomorphism $h^n(A) \to \prod_i h^n(A_i)$.

A theory is called *completely additive* if the same is true for arbitrary direct sums.

The notion of σ-additivity (or complete additivity) is called "additivity" in some references such as [Rosenberg and Schochet 1987]. We have not used this terminology since it is easily confused with (finite) additivity of functors.

EXAMPLES 21.1.2. (a) $\{K_n\}$ is a stable completely additive homology theory on the category of all C^*-algebras.

(b) For a fixed separable A, $\{KK^n(A, \cdot)\}$ is a stable homology theory on the category of σ-unital nuclear C^*-algebras (19.5.7). It is not σ-additive in general (19.7.2). If A is separable and nuclear, then $\{KK^n(A, \cdot)\}$ is a stable homology theory on the category of all σ-unital C^*-algebras (19.5.6).

(c) For a fixed σ-unital B, $\{KK^n(\cdot, B)\}$ is a stable σ-additive cohomology theory on the category of separable nuclear C^*-algebras (19.5.7, 19.7.1).

(d) Any homology or cohomology theory (in the usual sense) on locally compact Hausdorff spaces gives respectively a cohomology or homology theory on the category of commutative C^*-algebras. Some, but not all, of these theories are

σ-additive or completely additive. For example, Čech cohomology with compact supports gives a completely additive homology theory on the category of all commutative C^*-algebras.

21.2. Mayer–Vietoris Sequence

We now obtain a noncommutative analog of the Mayer–Vietoris exact sequence of topology, which tells how to compute the homology of a pushout. More specifically, if a space X is the union of two closed subspaces X_1 and X_2, the Mayer–Vietoris sequence relates the homology of X to that of X_1, X_2, and $X_1 \cap X_2$.

We begin with one simple general property of homology and cohomology theories.

PROPOSITION 21.2.1. *Let $\{h_n\}$ [resp. $\{h^n\}$] be a homology [resp. cohomology] theory on* **S**. *Then, for any A, we have $h_{n+1}(A) \cong h_n(SA)$ [resp. $h^{n+1}(A) \cong h^n(SA)$].*

PROOF. Apply the long exact sequence to $0 \to SA \to CA \to A \to 0$ and note that CA is contractible, so $h_n(CA) = 0$.   □

THEOREM 21.2.2 (MAYER–VIETORIS SEQUENCE FOR HOMOLOGY). *Let $\{h_n\}$ be a homology theory on* **S**, *and let the following commutative "pullback" diagram (15.3) of C^*-algebras in* **S** *be given:*

$$
\begin{array}{ccc}
P & \xrightarrow{\ g_1\ } & A_1 \\
{\scriptstyle g_2}\big\downarrow & & \big\downarrow{\scriptstyle f_1} \\
A_2 & \xrightarrow{\ f_2\ } & B
\end{array}
$$

with $P = \{(a_1, a_2) \mid f_1(a_1) = f_2(a_2)\} \subseteq A_1 \oplus A_2$, and f_1 and f_2 surjective. Then there is a long exact sequence

$$
\cdots \xrightarrow{\ \gamma\ } h_n(P) \xrightarrow{\ (g_{1*}, g_{2*})\ } h_n(A_1) \oplus h_n(A_2) \xrightarrow{\ f_{2*} - f_{1*}\ } h_n(B) \xrightarrow{\ \gamma\ } h_{n-1}(P) \longrightarrow \cdots
$$

which is natural with respect to morphisms of pullback diagrams.

PROOF. Let $g = g_1 \oplus g_2 : P \to A_1 \oplus A_2$. Then

$$
C_g \cong \{(h_1, h_2) \mid f_1(h_1(0)) = f_2(h_2(0))\} \subseteq CA_1 \oplus CA_2.
$$

There is a canonical map $\psi : C_g \to SB$ defined by

$$
[\psi(h_1, h_2)](t) = \begin{cases} f_1(h_1(1 - 2t)) & \text{for } t \leq \frac{1}{2}, \\ f_2(h_2(2t - 1)) & \text{for } t \geq \frac{1}{2}. \end{cases}
$$

Since the f_i are surjective, ψ is surjective. Thus we have a natural short exact sequence

$$
0 \longrightarrow CJ_1 \oplus CJ_2 \longrightarrow C_g \xrightarrow{\ \psi\ } SB \longrightarrow 0,
$$

where $J_i = \ker f_i$. Since $\ker \psi$ is contractible, ψ_* is an isomorphism.

Let $k : SA_1 \oplus SA_2 \to SB$ be the restriction of ψ to $SA_1 \oplus SA_2 \subseteq C_g$. Then k is homotopic to the function $\omega : SA_1 \oplus SA_2 \to SB$ defined by $\omega(h_1, h_2)(t) = f_2(h_2(t)) - f_1(h_1(t))$, so $k_* : h_n(SA_1) \oplus h_n(SA_2)$ is given by $k_*(x, y) = f_{2*}(y) - f_{1*}(x)$.

The long exact sequence corresponding to $0 \to SA_1 \oplus SA_2 \to C_g \to P \to 0$ is

$$\cdots \longrightarrow h_{n+1}(P) \xrightarrow{\ \partial\ } h_n(SA_1 \oplus SA_2) \longrightarrow h_n(C_g) \longrightarrow h_n(P) \longrightarrow \cdots$$

$$\cong \Big\downarrow \qquad\qquad \cong \Big\downarrow \psi_*$$

$$h_n(SA_1) \oplus h_n(SA_2) \xrightarrow{\ k_*\ } h_n(SB)$$

Applying the suspension isomorphism $h_{n+1}(D) \cong h_n(SD)$ (21.2.1), we obtain the desired sequence. It is easy to check that the map corresponding to ∂ under this identification is g_*. $\qquad\qquad\qquad\qquad\qquad\qquad\qquad\qquad\qquad\qquad\square$

The hypothesis that f_1 and f_2 be surjective may be removed [Schochet 1984a, 4.5].

The statement and proof of the Mayer–Vietoris Theorem for Cohomology is nearly identical:

THEOREM 21.2.3 (MAYER–VIETORIS SEQUENCE FOR COHOMOLOGY). *Let $\{h^n\}$ be a cohomology theory on* **S**. *Then, given a pullback diagram as in 21.2.2, there is a natural long exact sequence*

$$\cdots \longrightarrow h^n(B) \xrightarrow{(-f_1^*, f_2^*)} h^n(A_1) \oplus h^n(A_2) \xrightarrow{g_1^* + g_2^*} h^n(P) \longrightarrow h^{n+1}(B) \longrightarrow \cdots$$

21.3. Continuity

There is another axiom which turns out to be equivalent to σ-additivity. A covariant functor F from **S** to **Ab** is called *continuous* if it satisfies

(C) If $A = \varinjlim A_i$ is a countable inductive limit, then $F(A) = \varinjlim F(A_i)$ via the canonical maps.

A homology theory on **S** is continuous if each h_n is continuous.

THEOREM 21.3.1. *A homology theory on* **S** *is σ-additive if and only if it is continuous.*

PROOF. Continuity clearly implies σ-additivity. To prove the converse, we use a "mapping telescope" construction due to L. Brown. Let $A = \varinjlim A_k$, and define $T = \{f \mid f(t) \in A_k \text{ for } t \leq k\} \subseteq C_0((0, \infty], A)$. Then T is contractible, so $h_*(T) = 0$. The kernel J of the evaluation map $p_\infty : T \to A$ $(p_t(f) = f(t))$ satisfies $h_n(J) \cong h_{n+1}(A)$ from the long exact sequence. Set

$$B_k = \{f \in C([k, k+1], A_{k+1}) \mid f(k) \in A_k\}.$$

Then the inclusion of A_k into B_k via constant functions is a homotopy equivalence with inverse p_k. Set $D_1 = \bigoplus_{k \text{ odd}} B_k$, $D_2 = \bigoplus_{k \text{ even}} B_k$. $D_1 \oplus D_2$

is homotopy equivalent to $\bigoplus A_k$. There are obvious maps $r_i : J \to D_i$ and $q_i : D_i \to \bigoplus A_k$, and it is easily checked that the following diagram is a pullback diagram:

$$
\begin{array}{ccc}
J & \xrightarrow{\ r_1\ } & D_1 \\
{\scriptstyle r_2}\downarrow & & \downarrow{\scriptstyle q_1} \\
D_2 & \xrightarrow{\ q_2\ } & \bigoplus A_k
\end{array}
$$

We apply Mayer–Vietoris to obtain

$$\cdots \longrightarrow h_n(J) \longrightarrow h_n(D_1) \oplus h_n(D_2) \xrightarrow{\ \phi\ } h_n\!\left(\bigoplus A_k\right) \longrightarrow h_{n-1}(J) \longrightarrow \cdots,$$

where $\phi = q_{2*} - q_{1*}$. We have $h_n(J) \cong h_{n+1}(A)$ and $h_n\!\left(\bigoplus A_k\right) \cong h_n(D_1) \oplus h_n(D_2)$, and $h_n\!\left(\bigoplus A_k\right) \cong \bigoplus h_n(A_k)$ by assumption, so by substituting the long exact sequence becomes

$$\cdots \longrightarrow h_{n+1}(A) \longrightarrow \bigoplus_k h_n(A_k) \xrightarrow{\ \Phi_n\ } \bigoplus_k h_n(A_k) \longrightarrow h_n(A) \longrightarrow \cdots$$

with $\Phi_n\!\left(\bigoplus x_k\right) = \bigoplus(x_k - f_{k-1*}(x_{k-1}))$, where f_k is the embedding of A_k into A_{k+1}. Unsplicing yields short exact sequences

$$0 \longrightarrow \operatorname{coker} \Phi_n \longrightarrow h_n(A) \longrightarrow \ker \Phi_{n+1} \longrightarrow 0.$$

We have $\operatorname{coker} \Phi_n \cong \varinjlim h_n(A_k)$ and $\ker \Phi_{n+1} = 0$, establishing the theorem. $\square$

There is no obvious corresponding notion of continuity for contravariant functors. However, the next theorem gives a property of σ-additive cohomology theories which is analogous to continuity:

THEOREM 21.3.2. *If h^* is a σ-additive cohomology theory on $\mathbf{S}$, and $A = \varinjlim A_i$ in $\mathbf{S}$, then for each n there is a natural Milnor $\lim^1$-sequence*

$$0 \longrightarrow \varprojlim{}^1 h^{n-1}(A_i) \longrightarrow h^n(A) \longrightarrow \varprojlim h^n(A_i) \longrightarrow 0$$

PROOF. The proof is identical to that of 21.3.1 (with appropriate indices raised and arrows reversed) until the very end. We get a short exact sequence

$$0 \longrightarrow \operatorname{coker} \Phi^{n-1} \longrightarrow h^n(A) \longrightarrow \ker \Phi^n \longrightarrow 0.$$

The map $\Phi^n : \prod_k h^n(A_k) \to \prod_k h^n(A_k)$ is given by

$$\Phi^n\!\left(\prod x_k\right) = \prod(x_k - f_k^*(x_{k-1})),$$

so $\ker \Phi^n \cong \varprojlim h^n(A_k)$ and $\operatorname{coker} \Phi^{n-1} \cong \varprojlim{}^1 h^{n-1}(A_k)$. $\square$

21.4. Half-Exact Functors

The final result of this section is a sort of converse to 21.2.1, giving a way of generating a homology or cohomology theory from a single functor.

We say that a covariant [resp. contravariant] functor F from $\mathbf{S}$ to $\mathbf{Ab}$ is *half-exact* if it satisfies

(HX) If $0 \longrightarrow J \xrightarrow{i} A \xrightarrow{q} B \longrightarrow 0$ is a short exact sequence, then $F(J) \xrightarrow{i_*} F(A) \xrightarrow{q_*} F(B)$ [or $F(B) \xrightarrow{q^*} F(A) \xrightarrow{i^*} F(J)$, as the case may be] is exact in the middle.

A functor F from $\mathbf{S}$ to an additive category $\mathbf{A}$ is *half-exact* if $\operatorname{Hom}_{\mathbf{A}}(F(D), F(\,\cdot\,))$ and $\operatorname{Hom}_{\mathbf{A}}(F(\,\cdot\,), F(D))$ are half-exact for every $D \in \mathbf{S}$.

PROPOSITION 21.4.1. *Let* $0 \longrightarrow J \xrightarrow{j} A \xrightarrow{q} B \longrightarrow 0$ *be a short exact sequence in* $\mathbf{S}$, *and let* F *be a half-exact homotopy-invariant additive functor (covariant or contravariant) on* $\mathbf{S}$. *If* B *is contractible, then* j *induces an isomorphism between* $F(J)$ *and* $F(A)$.

PROOF. We first prove the result for F a covariant functor to $\mathbf{Ab}$; the contravariant case is essentially identical. We have that $F(B) = 0$, so that $j_* : F(J) \to F(A)$ is surjective. Applying half-exactness to the exact sequence $0 \longrightarrow SB \xrightarrow{i} C_q \xrightarrow{p} A \longrightarrow 0$, we also obtain that $p_* : F(C_q) \to F(A)$ is injective, since SB is contractible and hence $F(SB) = 0$. Finally, set $Z = \{f : [0,1] \to A \mid f(1) \in J\}$. Then the inclusion k of J into Z as constant functions is a homotopy equivalence. We have an exact sequence $0 \longrightarrow CJ \longrightarrow Z \xrightarrow{\pi} C_q \longrightarrow 0$; since CJ is contractible, we have that $\pi_* : F(Z) \to F(C_q)$ is injective. Since $j = p \circ \pi \circ k$, the map $j_* = p_* \circ \pi_* \circ k_*$ is injective.

Now suppose F is a (covariant) functor to an additive category $\mathbf{A}$. By the first part of the proof, there is an $h \in \operatorname{Hom}_{\mathbf{A}}(F(A), F(J))$ such that $F(j) \circ h = id_{F(A)}$ (using the fact that $id_{F(A)}$ is in the image of $j_* : \operatorname{Hom}_{\mathbf{A}}(F(A), F(J)) \to \operatorname{Hom}_{\mathbf{A}}(F(A), F(A))$). Similarly, using surjectivity of $j^* : \operatorname{Hom}_{\mathbf{A}}(F(A), F(J)) \to \operatorname{Hom}_{\mathbf{A}}(F(J), F(J))$, there is a $k \in \operatorname{Hom}_{\mathbf{A}}(F(A), F(J))$ with $k \circ F(j) = id_{F(J)}$. Then $F(j) \circ h \circ F(j) = F(j) \circ k \circ F(j) = F(j)$, i.e. $j_*(h \circ F(j)) = j_*(k \circ F(j))$, so $h \circ F(j) = k \circ F(j)$ since j_* is injective. So $j^*(h) = j^*(k)$, and by injectivity of j^* we get $h = k$ as the inverse of j. $\qquad\square$

We have the following analog of 19.5.5. Note that no semisplitting is assumed.

COROLLARY 21.4.2. *Let* $0 \longrightarrow J \xrightarrow{j} A \xrightarrow{q} B \longrightarrow 0$ *be a short exact sequence in* $\mathbf{S}$, *and let* F *be a half-exact homotopy-invariant additive functor on* $\mathbf{S}$. *Then the map* $e : J \to C_q$ *defined in 19.5.5 induces an isomorphism between* $F(J)$ *and* $F(C_q)$.

PROOF. Apply 21.4.1 to the exact sequence $0 \longrightarrow J \xrightarrow{e} C_q \longrightarrow CB \longrightarrow 0$ and note that CB is contractible. $\qquad\square$

THEOREM 21.4.3. *Let F be an additive covariant functor from $\mathbf{S}$ to $\mathbf{Ab}$ satisfying (H) and (HX). Set $F_n(A) = F(S^n A)$ and $\partial = e_*^{-1} \circ i_* : F(SB) \to C_q \to F(J)$. Then $\{F_n\}$ is a homology theory on $\mathbf{S}$.*

Let F be an additive contravariant functor from $\mathbf{S}$ to $\mathbf{Ab}$ satisfying (H) and (HX). Set $F^{-n}(A) = F(S^n A)$, and $\partial = i^* \circ e^{*-1}$. Then $\{F^n\}$ is a cohomology theory on $\mathbf{S}$.

PROOF. We prove the covariant statement; the other is analogous. If $0 \longrightarrow J \xrightarrow{j} A \xrightarrow{q} B \longrightarrow 0$ is a short exact sequence, it suffices to prove that the sequence

$$\cdots \xrightarrow{j_*} F(SA) \xrightarrow{q_*} F(SB) \xrightarrow{\partial} F(J) \xrightarrow{j_*} F(A) \xrightarrow{q_*} F(B)$$

is exact at $F(SB)$ and at $F(J)$. Exactness at $F(J)$ comes from half-exactness applied to $0 \to SB \to C_q \to A \to 0$ and the isomorphism of $F(J)$ and $F(C_q)$. For exactness at $F(SB)$, define

$$T = \{(f,g) : g(0) = q(f(0))\} \subseteq C_0((-1,0], A) \oplus C_0([0,1), B).$$

Then the canonical embedding $k : SB \to T$ gives an isomorphism $k_* : F(SB) \to F(T)$ since the quotient is contractible. Also, the map $SA \cong C_0((-1,0), A) \to T$ is homotopic to the composite $SA \xrightarrow{Sq} SB \xrightarrow{k} T$. So exactness at SB follows from half-exactness applied to

$$0 \longrightarrow C_0((-1,0), A) \longrightarrow T \longrightarrow C_q \longrightarrow 0.$$

It is easy to check that the maps match up correctly. □

Analogously, we have:

THEOREM 21.4.4. *Let F be an additive, homotopy-invariant, exact functor from $\mathbf{S}$ to an additive category $\mathbf{A}$. Then F has long exact sequences: if $0 \longrightarrow J \xrightarrow{j} A \xrightarrow{q} B \longrightarrow 0$ is a short exact sequence in $\mathbf{S}$, then there is a natural long exact sequence*

$$\cdots \xrightarrow{j_*} F(SA) \xrightarrow{q_*} F(SB) \xrightarrow{\partial} F(J) \xrightarrow{j_*} F(A) \xrightarrow{q_*} F(B).$$

21.5. Applications to KK-Theory

We finish by explicitly stating the principal results of this section in the special case of the KK-groups.

THEOREM 21.5.1 (MAYER–VIETORIS). (a) *Let*

$$
\begin{array}{ccc}
P & \xrightarrow{\ g_1\ } & A_1 \\
\downarrow{\scriptstyle g_2} & & \downarrow{\scriptstyle f_1} \\
A_2 & \xrightarrow{\ f_2\ } & D
\end{array}
$$

be a pullback diagram of separable nuclear C^*-algebras. Then, for any σ-unital C^*-algebra B, the following six-term sequence is exact:

$$KK(D,B) \xrightarrow{(-f_1^*, f_2^*)} KK(A_1,B) \oplus KK(A_2,B) \xrightarrow{g_1^* + g_2^*} KK(P,B)$$

$$\uparrow \qquad\qquad\qquad\qquad\qquad\qquad\qquad\qquad\qquad\qquad \downarrow$$

$$KK^1(P,B) \xleftarrow{g_1^* + g_2^*} KK^1(A_1,B) \oplus KK^1(A_2,B) \xleftarrow{(-f_1^*, f_2^*)} KK^1(D,B)$$

(b) Let

$$
\begin{array}{ccc}
P & \xrightarrow{\ g_1\ } & B_1 \\
{\scriptstyle g_2}\downarrow & & \downarrow{\scriptstyle f_1} \\
B_2 & \xrightarrow{\ f_2\ } & D
\end{array}
$$

be a pullback diagram of σ-unital nuclear C^*-algebras. Then for any separable C^*-algebra A, the following six-term sequence is exact:

$$KK(A,P) \xrightarrow{(g_{1*}, g_{2*})} KK(A,B_1) \oplus KK(A,B_2) \xrightarrow{f_{2*} - f_{1*}} KK(A,D)$$

$$\uparrow \qquad\qquad\qquad\qquad\qquad\qquad\qquad\qquad\qquad\qquad \downarrow$$

$$KK^1(A,D) \xleftarrow{f_{2*} - f_{1*}} KK^1(A,B_1) \oplus KK^1(A,B_2) \xleftarrow{(g_{1*}, g_{2*})} KK^1(A,P).$$

Both sequences are natural with respect to morphisms of pullback diagrams.

THEOREM 21.5.2. *Let B be a σ-unital C^*-algebra, and let $A = \varinjlim A_n$ be a countable inductive limit of separable C^*-algebras. Then there is a natural Milnor $\lim^1$-sequence*

$$0 \longrightarrow \varprojlim^1 KK^1(A_n, B) \longrightarrow KK(A,B) \longrightarrow \varprojlim KK(A_n,B) \longrightarrow 0$$

There is no reasonable relationship between $KK(A, \varinjlim B_n)$ and $\varinjlim KK(A, B_n)$ in general (19.7.2).

See [Rosenberg 1982b; Schochet 1984a] for more detailed analysis of the properties of homology and cohomology theories on C^*-algebras. See [Nagy 1995; Dădărlat 1994] for nonstable homology theories.

22. Axiomatic K-Theory

In this section, we will give a universal categorical construction of KK-theory, and show that K-theory can be characterized by a simple set of axioms (at least for a suitably well-behaved class of C^*-algebras). The ideas of this section are primarily due to Cuntz [1984] and Higson [1983], based in part on earlier results of Rosenberg [1982b].

Throughout this section, we will only consider *separable trivially graded C^*-* algebras. For unexplained terminology and results from category theory, the reader may consult [MacLane 1971].

22.1. KK as a Category

We begin with a categorical description of KK. Let **KK** be the category whose objects are separable (trivially graded) C^*-algebras, and for which the morphisms from A to B are the elements of $KK(A, B)$. Composition of morphisms is via the intersection product. **KK** is an additive category. There is an obvious functor from the category **SC*** of all separable C^*-algebras and *-homomorphisms to **KK**, which has the coproduct property:

DEFINITION 22.1.1. Let F be a functor (covariant or contravariant) from **SC*** to a category **A**. Then F has the *coproduct property* (CP) if for every split short exact sequence

$$0 \longrightarrow A \xrightarrow{\quad j \quad} D \overset{s}{\underset{p}{\longleftarrow\!\!\!\longrightarrow}} B \longrightarrow 0$$

in **SC***, the morphisms Fj and Fs make $F(D)$ a coproduct of $F(A)$ and $F(B)$ in **A**. [If F is contravariant, $F(D)$ should be a product of $F(A)$ and $F(B)$.]

If **A** is an additive category, then F is automatically additive, and there is a **A**-morphism $\pi : F(D) \to F(A)$ [$\pi : F(A) \to F(D)$ in the contravariant case] making $F(D)$ the biproduct of $F(A)$ and $F(B)$. So if **A** is additive, F has the coproduct property if and only if F sends split exact sequences to split exact sequences.

In **KK**, the morphism π is π_s, defined in 17.1.2(b).

PROPOSITION 22.1.2.   *A functor satisfying (H) and (HX) has the coproduct property.*

PROOF. Follows immediately from 21.4.3. $\square$

We will construct a universal enveloping additive category with the coproduct property, and show it is isomorphic to **KK**. First, let **S** be the category whose objects are C^*-algebras for which the morphisms from A to B are the homotopy classes of *-homomorphisms from A to $B \otimes \mathbb{K}$, denoted $[A, B \otimes \mathbb{K}]$. Morphisms are composed as follows: if $\phi : A \to B \otimes \mathbb{K}$ and $\psi : B \to D \otimes \mathbb{K}$, then $\psi\phi$ is $(\psi \otimes 1_\mathbb{K}) \circ \phi : A \to B \otimes \mathbb{K} \to D \otimes \mathbb{K} \otimes \mathbb{K} \cong D \otimes \mathbb{K}$. Since there is only one *-isomorphism from $\mathbb{K}$ to $\mathbb{K}$ up to homotopy, composition is well defined. The morphism $[\eta_A] : A \to A$, where $\eta_A : A \to A \otimes \mathbb{K}$ sends a to $a \otimes p$, p a one-dimensional projection, is the identity morphism on A; the same map, regarded as a morphism κ_A from A to $A \otimes \mathbb{K}$, is an isomorphism in **S**. So the morphisms in **S** from A to B could have been defined to be $[A \otimes \mathbb{K}, B \otimes \mathbb{K}]$ with ordinary composition. (The construction of **S** is an example of the construction of the Kleisli category of a monad [Higson 1983, §3; MacLane 1971, VI.5].)

There is an obvious functor from the category **SC*** to **S**.

Let F be a functor from **SC*** to any other category, which satisfies (H) and (S) of Section 21. Then F factors in an obvious way through **S**, so that F may be regarded as a functor on **S**. In particular, K_0, K_1, *Ext*, and more generally KK

(for a fixed first or second argument) may be regarded as functors on **S**. In fact, there is an obvious functor from **S** to **KK** implementing these other functors.

There is an additive structure on the Hom-sets of **S**, defined as in 15.6 (cf. [Rosenberg 1982b]). The canonical functor from **S** to **KK** is additive and has the coproduct property.

22.2. Universal Enveloping Categories

If **A** is any category with an additive structure, then there is a standard construction of a universal enveloping additive category **L**, with the property that any additive functor from **A** to another category, with the coproduct property, factors through **L**. The only restriction on **A** is that it be small enough to avoid logical difficulties; it is enough that **A** have a small skeletal subcategory **A**$_0$. This assumption implies that there is a *set* $\{\mathbf{B}_\gamma\}$ of small additive categories and additive covariant functors F_γ from **A**$_0$ to **B**$_\gamma$ with the coproduct property, such that if F is any additive covariant [resp. contravariant] functor with the coproduct property from **A**$_0$ to a category **B**, there is a subcategory of **B** isomorphic [resp. anti-isomorphic] to one of the **B**$_\gamma$, with F isomorphic [resp. anti-isomorphic] to F_γ. Let **L**$'$ be the smallest additive subcategory of $\prod \mathbf{B}_\gamma$ with objects $\{\prod F_\gamma a : a \in \mathrm{Ob}(\mathbf{A})\}$ and containing the morphisms $\{\prod F_\gamma \phi : \phi \in \mathrm{Hom}(\mathbf{A})\}$ and $\{\prod \pi_\gamma\}$ for each split short exact sequence, where π_γ is the splitting morphism in **B**$_\gamma$. Now define **L** to be the category whose objects are the objects of **A** and for which $\mathbf{L}(a,b) = \mathbf{L}'(\prod F_\gamma a_0, \prod F_\gamma b_0)$, where a_0, b_0 are the objects of **A**$_0$ isomorphic to a and b respectively.

We now fix **L** to be the category obtained by applying the above construction to **S**. **L** is an additive category whose objects are separable C^*-algebras. Any functor with (CP) from **S** to any additive category factors through **L**; hence any functor from **SC*** satisfying (H), (S), and (CP) factors through **L**. In particular, there is a functor F from **L** to **KK** induced from the canonical functor from **S** to **KK**.

THEOREM 22.2.1. *F is an isomorphism of categories.*

PROOF. The objects of **L** and **KK** are the same, and F induces the identity map on objects. Thus to prove the theorem it is enough to prove that for each fixed A the natural transformation β, defined by F between the functors $B \to \mathbf{L}(A,B)$ and $B \to KK(A,B)$ from **SC*** to **Ab**, the category of abelian groups, has a (natural) inverse.

To define the inverse α, fix B (as well as A), and let $\boldsymbol{x} \in KK(A,B)$. Let $[\phi^{(0)}, \phi^{(1)}]$ be a quasihomomorphism representing $\boldsymbol{x}$. Then there is a canonically associated split exact sequence as in 17.8.3, such that $\boldsymbol{x}$ is the composition of an ordinary homomorphism and the splitting morphism. Let $\alpha(\boldsymbol{x})$ be the morphism of **L** which is the composition of the corresponding homomorphism and splitting morphism. This α is obviously the inverse of β; we need only show that α is well defined.

If $[\phi^{(0)}, \phi^{(1)}]$ is degenerate, i.e. $\phi^{(0)} = \phi^{(1)}$, then the corresponding morphism of **L** is zero by the coproduct property. And if $[\phi_0^{(0)}, \phi_0^{(1)}]$ and $[\phi_1^{(0)}, \phi_1^{(1)}]$ are homotopic in the sense of 17.6.3, i.e. there is a quasihomomorphism $[\phi^{(0)}, \phi^{(1)}]$ from A to IB implementing the homotopy, then $\alpha([\phi_i^{(0)}, \phi_i^{(1)}]) = \tilde{f}_{i\circ}\alpha([\phi^{(0)}, \phi^{(1)}])$, where $\tilde{f}_i$, for $i = 0, 1$, is the **L**-morphism corresponding to the homomorphism $f_i : IB \to B$ of evaluation at i. But f_0 and f_1 are homotopic, so $\tilde{f}_0 = \tilde{f}_1$. The equivalence relation on $KK(A, B)$ is generated by homotopy and addition of degenerates; so α is well defined.   □

22.3. Functors on KK and Axiomatic K-Theory

The following is an immediate corollary of 22.2.1:

COROLLARY 22.3.1. *Let F be a covariant [resp. contravariant] functor from* **SC*** *to any additive category, satisfying (H), (S), and (CP). Then F factors canonically through* **KK**. *So for any such functor, there is a pairing $F(A) \times KK(A, B) \to F(B)$ [resp. $KK(A, B) \times F(B) \to F(A)$] which is compatible (associative) with the intersection product. Any such functor satisfies Bott periodicity: $F(A) \cong F(S^2 A)$ for any A.*

An elegant direct proof of the last two statements can be found in [Cuntz 1984] (cf. 9.4.2), based on the fact that the Toeplitz algebra is KK-equivalent to $\mathbb{C}$. This proof is unnecessary for us because of the categorical approach we have taken above. (Actually the result here is slightly more general since the functors of [Cuntz 1984] are covariant functors to the category **Ab**; but the general result can be established with the same proof.)

COROLLARY 22.3.2. *Let F be an additive functor from* **S** *to* **Ab** *satisfying (H), (S), and (HX). Then the associated homology or cohomology theory satisfies Bott periodicity, and the long exact sequence becomes a cyclic six-term exact sequence.*

We now wish to describe K-theory by a set of axioms. To do this, we consider covariant functors F from **SC*** to **Ab** satisfying (H), (S), and (CP). The "canonical" examples (maybe the universal examples) of such functors are the functors $K_{A,B}$, where $K_{A,B}(D) = KK(A, D \otimes B)$, with A, B fixed. We have $K_{\mathbb{C},\mathbb{C}} = K_0$.

There are two additional axioms which K_0 satisfies, continuity and the *dimension axiom*

(D) $F(\mathbb{C}) = \mathbb{Z}$, $F(C_0(\mathbb{R})) = 0$.

In addition, K_0 satisfies (HX), which, in the presence of (H), is stronger than (CP) (22.1.2).

PROPOSITION 22.3.3. *Let F be a covariant functor from* **SC*** *to* **Ab** *satisfying (H), (S), (CP), and (D). Then there is a natural transformation from K_0 to F. If F satisfies (HX), then this natural transformation extends to a natural transformation from K_1 to F_1; the natural transformations are compatible with* *suspensions and, if F satisfies (HX), with the connecting maps in the six-term exact sequences.*

PROOF. Let f be a generator of $F(\mathbb{C})$. Associate to $x \in K_0(A) = KK(\mathbb{C}, A)$ the element $fx \in F(A)$ coming from the pairing $F(\mathbb{C}) \times KK(\mathbb{C}, A) \to F(A)$. It is easy to verify that this association gives a natural transformation.   □

If F is a covariant functor from **SC**** to **Ab** satisfying (H), (S), (HX) [or just (CP)], (C), and (D), it is plausible (but unknown) that F agrees with K_0. It is not clear how to begin to approach the general problem. However, it is possible to prove by "bootstrap" methods that F must agree with K_0 on at least a substantial subcategory, one which seems to contain all C^*-algebras of interest.

DEFINITION 22.3.4. Let N be the smallest class of separable nuclear C^*-algebras with the following properties:

(N1) N contains $\mathbb{C}$.

(N2) N is closed under countable inductive limits.

(N3) If $0 \to A \to D \to B \to 0$ is an exact sequence, and two of the terms are in N, then so is the third.

(N4) N is closed under KK-equivalence.

Let H be the smallest class of separable C^*-algebras closed under the same operations (H1)–(H4). Let N_0 (resp. H_0) be the smallest class of separable nuclear (resp. separable) C^*-algebras closed under (N2), (N4), and

(N1′) N_0 contains $\mathbb{C}$ and $C_0(\mathbb{R})$.

(N3′) If $0 \to A \to D \to B \to 0$ is a *split* exact sequence, and two of the terms are in N_0, then so is the third.

Let **NC*** (resp. **HC***, **nC***, **hC***) be the full subcategory of **SC*** whose objects are the C^*-algebras in N (resp. H, N_0, H_0).

We will show later (23.10.3, 23.10.4) that N is just the smallest class of nuclear C^*-algebras containing the commutative C^*-algebras and closed under KK-equivalence, and that N_0 exactly consists of the algebras in N with torsion-free K-theory. H appears to be larger than just the class of C^*-algebras KK-equivalent to algebras in N, because of being closed under property (H3). It will be necessary to distinguish between H and N in section 23.

N is often called the *bootstrap category*. However, this terminology is not uniform: some authors use the term "bootstrap category" to mean a (possibly) smaller category closed under a different set of properties such as (N1)–(N3) and crossed products by $\mathbb{Z}$. N is also precisely the category of all separable nuclear C^*-algebras A which satisfy the conclusion of the Universal Coefficient Theorem (23.1.1) for all B. The C^*-algebras in H do not appear to always satisfy the UCT for every B, since $KK(A, -)$ is (probably) not exact in general for $A \in H$; but the analogous class with KK-equivalence replaced by E-equivalence gives the class of A for which the E-theory UCT holds for every B (25.7.5).

22.3.5. All the algebras in N are nuclear; however, the class N contains most C^*-algebras which arise "naturally":

(a) All classes are closed under stable isomorphism by (N4).

(b) N_0 is closed under split extensions by (N3$'$), and in particular under direct sums. Thus all finite-dimensional C^*-algebras are in N_0. So N_0 contains all AF algebras.

(c) N_0 contains $C([0,1]^n)$, which is homotopy-equivalent (and thus KK-equivalent) to $\mathbb{C}$. N_0 also contains $C_0(\mathbb{R}^n)$, which is KK-equivalent to $\mathbb{C}$ or $C_0(\mathbb{R})$. So N_0 also contains $C(S^n)$ by (N3$'$).

(d) N contains $C(X)$ for every finite simplicial complex X. Since every compact space is an inverse limit of simplicial complexes, $C(X) \in N$ for every compact X. By (N3), N also contains $C_0(X)$ for every locally compact X. Thus N contains all (separable) commutative C^*-algebras.

(e) N contains all type I C^*-algebras; hence also all inductive limits of type I C^*-algebras (any stable type I C^*-algebra can be built up by extensions and inductive limits from C^*-algebras of the form $C_0(X) \otimes \mathbb{K}$).

(f) N and N_0 are closed under tensor products, since if B is built up from $\mathbb{C}$ by the operations (N2)–(N4), then $A \otimes B$ can be built up from $A \otimes \mathbb{C}$ by the same sequence of operations. It is not clear that H and H_0 are closed under minimal tensor products, since the minimal tensor product is not exact in general; but they are closed under maximal tensor products (maximal tensor products preserve KK-equivalence by [Skandalis 1988].)

(g) All these classes are closed under crossed products by $\mathbb{R}$ by 19.3.6. H and N are also closed under crossed products by $\mathbb{Z}$ because of the Toeplitz extension. So N contains the Cuntz algebras O_n (10.11.8), since $O_n \otimes \mathbb{K}$ can be written as a crossed product of an AF algebra by $\mathbb{Z}$. More generally, the Cuntz–Krieger algebras O_A (10.11.9) are all in N.

(h) All the classes are closed under crossed products by simply connected solvable Lie groups by successive applications of (f).

(i) It is, however, not clear that any of these classes are closed under crossed products by finite cyclic groups or by $\mathbb{T}$ (23.15.12).

H contains many nonnuclear C^*-algebras such as $C^*(\mathbb{F}_n)$ and $C_r^*(\mathbb{F}_n)$, where $\mathbb{F}_n$ is the free group on n generators, since these C^*-algebras are KK-equivalent to commutative C^*-algebras. In fact, it may be that H contains all separable C^*-algebras. See 23.13 for further comments on this question.

THEOREM 22.3.6. *Let F be a covariant functor from* $\mathbf{HC^*}$ *to* $\mathbf{Ab}$, *satisfying* (H), (S), (HX), (C), *and* (D). *Then* $F = K_0$. *If F is a covariant functor from* $\mathbf{hC^*}$ *to* $\mathbf{Ab}$, *satisfying* (H), (S), (CP), (C), *and* (D), *then* $F = K_0$.

PROOF. Let H' be the class of C^*-algebras for which the natural transformation of 22.3.3 is an isomorphism. By (D), H' contains $\mathbb{C}$ and $C_0(\mathbb{R})$, and by (C) it is closed under inductive limits. Since F factors through KK by 22.3.1, H' is closed under KK-equivalence. If F satisfies (HX), H' is closed under the procedure of (H3); in F satisfies (CP) it is closed under (H3'). Thus H' contains H_0, and contains H if F satisfies (HX).   □

Thus, at least on the class H, K_0 is completely characterized by homotopy invariance, stability, half-exactness, continuity, and normalization. On the smaller class H_0, half-exactness may be weakened to split exactness.

If (D) is replaced by a similar axiom stating that $F(\mathbb{C}) = 0$ and $F(C_0(\mathbb{R})) = \mathbb{Z}$, one obtains an axiomatic characterization of K_1. One can similarly characterize the functors $K_{A,B}$, as long as A and B are in N and have free abelian K-groups. The Universal Coefficient Theorem explains what axioms must be taken in the general case (the continuity axiom may fail, for example).

It is more difficult to axiomatize the contravariant functor K^0, since there is no satisfactory analog of the continuity axiom (C). There is a natural $\lim^1$-sequence which may be used as a substitute (21.3.2). The difficulty is analogous to the difficulty of axiomatizing homology on general compact spaces, as opposed to cohomology. (There is a theory of "the contravariant functor dual to a homology theory" which can be used, due to D. Andersson, P. Kainen, and Z. Yosimura, which is relevant to this problem; cf. [Madsen and Rosenberg 1988].)

22.4. EXERCISES AND PROBLEMS

22.4.1. Let F be a functor on the category of all C^*-algebras (or a suitable subcategory) satisfying (S), (CP), and

(U) If α is an inner automorphism of A, then $\alpha_* : F(A) \to F(A)$ is the identity.

Then F satisfies (H) [Higson 1988]. Conversely, a functor satisfying (S) and (H) satisfies (U) (use 3.4.1).

22.4.2. Ordinary Eilenberg–Steenrod cohomology with compact supports defines a continuous homology theory $\{h_n\}$ on the category of commutative C^*-algebras by $h_n(C_0(X)) = H_c^n(X)$. This theory satisfies the (Eilenberg–Steenrod) dimension axiom

(ESD) $h_0(\mathbb{C}) = \mathbb{Z}$, $h_n(\mathbb{C}) = 0$ for $n > 0$.

Show from 22.3.6 and 21.2.1 that this homology theory cannot extend to a stable continuous homology theory on $\mathbf{HC}^*$. Thus to define "ordinary cohomology" for any reasonable class of noncommutative C^*-algebras one of the axioms must be discarded. The most reasonable ones to discard are (S) and (HX). The next problem considers a weakening of (HX); see [Dădărlat 1994] for theories not satisfying (S). [Property (S) is a very natural axiom when dealing with arbitrary C^*-algebras because of the fact that stable isomorphism coincides with strong Morita equivalence for σ-unital C^*-algebras (13.7.1).]

22.4.3. A *relative homology theory* on a (suitable) category $\mathbf{S}$ of C^*-algebras is a sequence $\{h_n\}$ of covariant functors from pairs (A, J) to abelian groups, where

A is a C^*-algebra in $\mathbf{S}$ and J is a closed two-sided ideal in A [a morphism from (A, J) to (B, K) is a *-homomorphism $\phi : A \to B$ with $\phi(J) \subseteq K$], satisfying (H) (appropriately modified in the obvious way) and

(LX) For any pair (A, J) there is a long exact sequence

$$\cdots \xrightarrow{\partial} h_n(A, J) \xrightarrow{j_*} h_n(A, A) \xrightarrow{q_*} h_n(A/J, A/J) \xrightarrow{\partial} h_{n-1}(A, J) \longrightarrow$$
$$\cdots \longrightarrow h_0(A/J, A/J).$$

The long exact sequence is natural.

(WEX) (weak excision) If K is an ideal of A orthogonal to J, then the map $A \to A/K$ induces an isomorphism $h_n(A, J) \cong h_n(A/K, J/K)$ for all n.

A relative homology theory is *stable* [resp. *continuous*] if it satisfies the appropriately modified version of (S) [resp. (C)].

The *dimension axiom* is

(ESD) $h_0(\mathbb{C}, \mathbb{C}) = \mathbb{Z}$; $h_n(\mathbb{C}, \mathbb{C}) = 0$ for $n > 0$.

(a) Show that a relative homology theory on the category of unital commutative C^*-algebras is exactly a cohomology theory on compact pairs (X, Y), in the sense of Eilenberg–Steenrod. (X, Y) corresponds to (A, J), where $A = C(X)$ and J is the ideal of all functions vanishing on Y. So if $\{h_n\}$ is a continuous relative homology theory satisfying (ESD) on a category containing all unital separable commutative C^*-algebras, then for any pair $(C(X), J)$ as above, $h_n(C(X), J) = H^n(X, Y)$.

(b) If $\{h_n\}$ is a continuous stable relative homology theory satisfying (ESD) on a category containing all unital AF algebras, then for any pair (A, J) with A unital AF, $h_0(A, J) = K_0(J)$, $h_n(A, J) = 0$ for $n > 0$.

(c) There is a unital AF algebra A and a unital embedding $\phi : C(\mathbb{T}^2) \to A$ such that $\phi_* : K_0(C(\mathbb{T}^2)) \to K_0(A)$ is injective [Loring 1986]. (Similar embeddings for irrational rotation algebras were obtained in [Pimsner and Voiculescu 1980b].)

(d) If $\mathbf{S}$ is a full subcategory of $\mathbf{SC}^*$ containing all separable commutative unital C^*-algebras and closed under stable isomorphism and countable inductive limits, then there is no continuous stable relative homology theory on $\mathbf{S}$ satisfying (ESD). So ordinary cohomology (or even rational cohomology) on spaces cannot be extended to a stable theory on the most restricted natural class of noncommutative C^*-algebras. (Use (a), (b), (c), and the Chern character (1.6.6).)

22.4.4. (a) If $\{h_n\}$ is a continuous relative homology theory and (A, J) is a pair with A unital and $\mathrm{Prim}(J)$ (regarded as an open subset of $\mathrm{Prim}(A)$) a union of closed sets (in particular, if $\mathrm{Prim}(A)$ is Hausdorff), then $\{h_n\}$ satisfies *strong excision* for (A, J): the inclusion map $(J^+, J) \to (A, J)$ induces an isomorphism on h_*. (Use weak excision and continuity.)

(b) If $\{h_n\}$ is a relative homology theory satisfying strong excision for all pairs, then $h_n(A, J)$ depends only on J, and $\{h_n\}$ defines a homology theory by $h_n(J) = h_n(J^+, J)$. Conversely, any homology theory defines a relative homology theory satisfying strong excision by $h_n(A, J) = h_n(J)$ (cf. 5.4).

(c) Let T be the Toeplitz algebra (9.4.2). If $\{h_n\}$ is a relative homology theory satisfying (ESD) (on a suitably large category of C^*-algebras) which has a Chern character, i.e. $\bigoplus_{n \, \text{even}} h_n(B, B) \otimes \mathbb{Q} \cong K_0(B) \otimes \mathbb{Q}$, $\bigoplus_{n \, \text{odd}} h_n(B, B) \otimes \mathbb{Q} \cong K_1(B) \otimes \mathbb{Q}$ for all B, then $h_0(T, \mathbb{K}) = 0$, $h_2(T, \mathbb{K}) \cong \mathbb{Z}$, so $\{h_n\}$ does not satisfy strong excision for $(T, \mathbb{K})$.

(d) Does there exist a continuous stable relative homology theory on noncommutative C^*-algebras which does not satisfy strong excision?

23. Universal Coefficient Theorems and Künneth Theorems

In this section, we establish several results which allow computation of KK-groups (and, therefore, K-groups) of "nice" C^*-algebras. Among other things, we prove that KK-equivalence is completely determined by isomorphism of the K_0 and K_1 groups for C^*-algebras in the class N defined in 22.3.4, and consequently that any C^*-algebra in N is KK-equivalent to a commutative C^*-algebra. The results of this section are due to Rosenberg and Schochet [1987; Schochet 1982], based on earlier work by L. Brown [1984]. Cuntz had also previously obtained the Universal Coefficient sequence for the O_A.

All C^*-algebras in this section will be *separable* and *trivially graded*. We will consider K-theory and KK-theory to be $\mathbb{Z}_2$-graded theories in order to simplify notation: $KK^*(A, B)$ will denote $KK(A, B) \oplus KK^1(A, B)$, and similarly $K_*(A) = K_0(A) \oplus K_1(A)$, $K^*(A) = K^0(A) \oplus K^1(A)$. If G and H are graded abelian groups, then $\text{Hom}(G, H)$ is also graded by degree-preserving/degree-reversing maps. A tensor product of abelian groups is, of course, the ordinary tensor product over $\mathbb{Z}$. A tensor product of graded groups has a natural grading defined as in the C^*-algebra case (14.4).

23.1. Statements of the Theorems

For any separable C^*-algebras A and B, there are maps

$$\alpha : K_*(A) \otimes K_*(B) \to K_*(A \otimes B),$$
$$\beta : K^*(A) \otimes K_*(B) \to KK^*(A, B),$$
$$\gamma : KK^*(A, B) \to \text{Hom}(K_*(A), K_*(B))$$

which are natural in A and B.

These maps are defined using the intersection product. α comes from the pairing $KK^*(\mathbb{C}, A) \times KK^*(\mathbb{C}, B) \to KK^*(\mathbb{C}, A \otimes B)$. More specifically, α is induced by the four pairings

$$KK(\mathbb{C}, A) \times KK(\mathbb{C}, B) \to KK(\mathbb{C}, A \otimes B),$$
$$KK(C_0(\mathbb{R}), A) \times KK(\mathbb{C}, B) \to KK(C_0(\mathbb{R}), A \otimes B)$$
$$KK(\mathbb{C}, A) \times KK(C_0(\mathbb{R}), B) \to KK(C_0(\mathbb{R}), A \otimes B)$$
$$KK(C_0(\mathbb{R}), A) \times KK(C_0(\mathbb{R}), B) \to KK(C_0(\mathbb{R}^2), A \otimes B) \cong KK(\mathbb{C}, A \otimes B)$$

The map β comes from the pairing $KK^*(A, \mathbb{C}) \times KK^*(\mathbb{C}, B) \to KK^*(A, B)$. Finally, γ is the adjoint of the pairing $KK^*(\mathbb{C}, A) \times KK^*(A, B) \to KK^*(\mathbb{C}, B)$.

When $A = B$, $KK^*(A, A)$ is a graded ring, and γ is a ring-homomorphism (if the intersection product is written composition-style).

The surjectivity of β measures to what extent a general KK-element factors through $\mathbb{C}$ or $C_0(\mathbb{R})$ (or through $C(S^1)$, which is KK-equivalent to $\mathbb{C} \oplus C_0(\mathbb{R})$).

It is not necessary to use KK-theory to define α, β, or γ. α has a straightforward K-theoretic definition (which actually defines a map from $K_*(A) \otimes K_*(B)$ to $K_*(A \otimes_{\max} B)$), and for β everything may be rephrased in terms of the pairing between K-theory and "K-homology" (16.3.2, 18.10.3). γ has a nice interpretation in terms of extensions, which could be taken as an alternate definition. If $\tau \in KK^1(A, B)$ is represented by the extension

$$0 \longrightarrow B \longrightarrow D \longrightarrow A \longrightarrow 0$$

then $\gamma(\tau)$ is given by the connecting maps in the associated six-term exact sequence of K-theory. (If $B = \mathbb{K}$, then γ is the map γ_∞ of 16.3.2.)

One might hope that α, β, and γ would be isomorphisms, but they cannot be in general for (essentially) homological algebra reasons. If the sequences are modified in the appropriate way to incorporate the homological algebra obstructions, sequences are obtained which are valid at least for C^*-algebras in N.

The additional ingredient is easiest to describe in the case of γ. If $\gamma(\tau) = 0$ for an extension τ, then the six-term K-theory sequence degenerates into two short exact sequences of the form

$$0 \longrightarrow K_i(B) \longrightarrow K_i(D) \longrightarrow K_i(A) \longrightarrow 0$$

and thus determines an element $\kappa(\tau) \in \text{Ext}^1_{\mathbb{Z}}(K_*(A), K_*(B))$ [this is the $\text{Ext}^1_{\mathbb{Z}}$-group of homological algebra, the derived functor of the Hom functor, *not* the *Ext*-group of Chapter VII.] Note that κ reverses degree. The maps γ and κ are generalizations of the Adams d and e operations in topological K-theory.

The obstruction for α and β is an element of $\text{Tor}^{\mathbb{Z}}_1$, the derived functor of the tensor product functor. It is natural to expect a $\text{Tor}^{\mathbb{Z}}_1$ obstruction, since $\text{Tor}^{\mathbb{Z}}_1$ measures the deviation from exactness of the tensor product functor on groups.

The statements of the theorems are as follows.

THEOREM 23.1.1 (UNIVERSAL COEFFICIENT THEOREM (UCT)). [Rosenberg and Schochet 1987] *Let A and B be separable C^*-algebras, with $A \in N$. Then* *there is a short exact sequence*

$$0 \longrightarrow \operatorname{Ext}^1_{\mathbb{Z}}(K_*(A), K_*(B)) \xrightarrow{\delta} KK^*(A, B) \xrightarrow{\gamma} \operatorname{Hom}(K_*(A), K_*(B)) \longrightarrow 0$$

The map γ has degree 0 and δ has degree 1. The sequence is natural in each variable, and splits unnaturally. So if $K_*(A)$ is free or $K_*(B)$ is divisible, then γ is an isomorphism.

THEOREM 23.1.2 (KÜNNETH THEOREM (KT)). *[Rosenberg and Schochet 1987] Let A and B be separable C^*-algebras, with $A \in N$, and suppose $K_*(A)$ or $K_*(B)$ is finitely generated. Then there is a short exact sequence*

$$0 \longrightarrow K^*(A) \otimes K_*(B) \xrightarrow{\beta} KK^*(A, B) \xrightarrow{\rho} \operatorname{Tor}^{\mathbb{Z}}_1(K^*(A), K_*(B)) \longrightarrow 0.$$

The map β has degree 0 and ρ has degree 1. The sequence is natural in each variable, and splits unnaturally. So if $K^*(A)$ or $K_*(B)$ is torsion-free, β is an isomorphism.

THEOREM 23.1.3 (KÜNNETH THEOREM FOR TENSOR PRODUCTS (KTP)). *[Schochet 1982] Let A and B be C^*-algebras, with $A \in N$. Then there is a short exact sequence*

$$0 \longrightarrow K_*(A) \otimes K_*(B) \xrightarrow{\alpha} K_*(A \otimes B) \xrightarrow{\sigma} \operatorname{Tor}^{\mathbb{Z}}_1(K_*(A), K_*(B)) \longrightarrow 0.$$

The map α has degree 0 and σ has degree 1. The sequence is natural in each variable, and splits unnaturally. So if $K_*(A)$ or $K_*(B)$ is torsion-free, α is an isomorphism.

The names given to these theorems reflect the fact that they are analogs of the ordinary Universal Coefficient Theorem and Künneth Theorem of algebraic topology. The UCT can also be regarded as a theorem about K-theory with coefficients; see 23.15.6 and 23.15.7.

The KT and especially the KTP can be stated and proved without reference to KK-theory. The proofs, however, are similar to and require some of the same machinery as the proof of the UCT, so it is most efficient to consider all three together.

The strategy of proof for all three theorems will be the same. We will first prove the theorems by bootstrap methods for arbitrary A, with fixed B of a form making α, β, γ isomorphisms. Then we will deduce the general results by "abstract nonsense", using an appropriate exact sequence giving a resolution of a general B into ones of the special form.

23.2. Proof of the Special UCT

We first prove the UCT in the case where $K_*(B)$ is divisible. We must show that if $A \in N$ and B is separable with divisible K-groups, then γ is an isomorphism. We will fix B, and vary A according to the defining properties of N.

PROPOSITION 23.2.1. *If $K_*(B)$ is divisible, then $\operatorname{Hom}(K_*(\cdot), K_*(B))$ is a σ-additive (in fact completely additive) cohomology theory on C^*-algebras (21.1.1).*

PROOF. $K_*(\,\cdot\,)$ is a σ-additive homology theory by 21.1.2(a). If $K_*(B)$ is divisible, then $\mathrm{Hom}(\,\cdot\,,K_*(B))$ is an exact functor, so $\mathrm{Hom}(K_*(\,\cdot\,),K_*(B))$ is a cohomology theory; it is σ-additive since Hom transforms direct sums in the first variable into direct products.   □

$KK^*(\,\cdot\,,B)$ is also a σ-additive cohomology theory on separable nuclear C^*-algebras by 21.1.2(c) (with no restriction on B except that it be σ-unital). γ is a natural transformation between the theories.

Let N' be the smallest class of separable C^*-algebras A for which $\gamma(A,B)$ is an isomorphism. N' contains $\mathbb{C}$ since $KK^i(\mathbb{C},B) = K_i(B) = \mathrm{Hom}(\mathbb{Z},K_i(B))$. A simple application of the Five Lemma to the long exact cohomology sequences shows that if two of A, J, and A/J are in N', then so is the third, so N' is closed under $(N3)$. A similar application of the Five Lemma to the $\lim^1$-sequences, which exist by 21.3.2, yields that N' is closed under inductive limits. And the naturality of γ with respect to the intersection product shows that N' is closed under KK-equivalence. Thus N' contains N, and the special case of the UCT is proved.

23.3. Proof of the Special KT

We now prove the special case of the Künneth Theorem. We will show that if $A \in N$ and $K_*(B)$ is finitely generated and free, then β is an isomorphism.

PROPOSITION 23.3.1. *If $K_*(B)$ is torsion-free, then $K^*(\,\cdot\,) \otimes K_*(B)$ is a cohomology theory, and $\beta(\,\cdot\,,B)$ is a natural transformation of cohomology theories. If $K_*(B)$ is finitely generated, then $K^*(\,\cdot\,) \otimes K_*(B)$ is a σ-additive cohomology theory.*

PROOF. Tensoring with a torsion-free abelian group G is exact, and preserves products if G is finitely generated.   □

Fix B with $K_*(B)$ finitely generated and free. Let N' be the smallest class of separable C^*-algebras A for which $\beta(A,B)$ is an isomorphism. N' contains $\mathbb{C}$ as for the UCT. Applying the Five Lemma to the long exact cohomology sequences and the $\lim^1$-sequences shows that N' is closed under $(N2)$ and $(N3)$, and the naturality of β shows that N' is closed under KK-equivalence. Thus N' contains N, and the special case of the KT is proved.

23.4. Proof of the Special KTP

Finally, we prove the special case of the Künneth Theorem for tensor products. Let B be a C^*-algebra with torsion-free K-theory, and let N' be the class of all C^*-algebras A for which $\alpha(A,B)$ is an isomorphism. N' obviously contains $\mathbb{C}$. If $A = \varinjlim A_n$, then $A \otimes B = \varinjlim (A_n \otimes B)$, so if each A_n is in N', then A is too since K-theory commutes with inductive limits. So N' is closed under $(N2)$. If two of A, J, A/J are in N', tensor the six-term K-theory exact sequence for A and J with $K_*(B)$; the resulting sequence remains exact since $K_*(B)$ is torsion-free. Apply the Five Lemma (using the naturality of α) to conclude that the remaining algebra is in N', so N' is closed under $(N3)$. N' is obviously closed under KK-equivalence. So N' contains N.

23.5. Geometric Resolutions of C^*-Algebras

In order to prove the general theorems, we must have a way of realizing a projective or injective resolution of the K-groups of a C^*-algebra by means of maps on the C^*-level. Given a C^*-algebra B, we seek a short exact sequence of C^*-algebras, containing B, with the property that the associated K-theory six-term exact sequence separates into two short exact sequences giving a (projective or injective) resolution of $K_*(B)$.

If B is a general separable C^*-algebra, it is not difficult to construct geometric projective and injective resolutions for $S^2 B \otimes \mathbb{K}$. This will be sufficient for our purposes, since all of the groups considered in the theorems are invariant under KK-equivalence.

The first step in constructing a geometric projective resolution is the following proposition.

PROPOSITION 23.5.1. *Let B be a separable C^*-algebra. Then there is a separable commutative C^*-algebra F, whose spectrum consists of a disjoint union of lines and planes, and a homomorphism $\phi : F \to SB \otimes \mathbb{K}$, such that $\phi_* : K_*(F) \to K_*(SB)$ is surjective. If $K_*(B)$ is finitely generated, we may require that $K_*(F)$ be finitely generated.*

PROOF. Choose a set of generators for $K_1(B)$. Each generator can be represented by a unitary in $M_n(B)^+$ for some n, and hence by a *-homomorphism from $C_0(\mathbb{R})$ into $M_n(B)$. Thus, if $F_0 = C_0(X_0)$, where X_0 is a disjoint union of lines, one for each generator, there is a *-homomorphism ϕ_0 from F_0 to $B \otimes \mathbb{K}$ such that $\phi_{0*} : K_1(F_0) \to K_1(B)$ is surjective. We have $K_0(F_0) = 0$. Do the same construction starting with SB to obtain an F_1 whose spectrum is a disjoint union of lines, and a *-homomorphism $\phi_1 : F_1 \to SB \otimes \mathbb{K}$ with $\phi_{1*} : K_1(F_1) \to K_1(SB)$ surjective. Set $F = SF_0 \oplus F_1$. Then $\phi = S\phi_0 \oplus \phi_1 : F \to SB \otimes \mathbb{K} \oplus SB \otimes \mathbb{K} \subseteq M_2(SB \otimes \mathbb{K}) \cong SB \otimes \mathbb{K}$ has the required properties. $\square$

Note that the K-groups of F are free abelian groups.

Now construct the geometric projective resolution for B as follows. Replace B by $S^2 B \otimes \mathbb{K}$ if necessary, and find an F as in 23.5.1 and a map $\phi : F \to B$ with ϕ_* surjective. The mapping cone sequence

$$0 \longrightarrow SB \longrightarrow C_\phi \overset{s}{\longrightarrow} F \longrightarrow 0$$

(which is semisplit) has associated K-theory sequence which degenerates to

$$0 \longrightarrow K_i(C_\phi) \overset{s_*}{\longrightarrow} K_i(F) \longrightarrow K_{i-1}(SB) \longrightarrow 0$$

since ϕ_* is surjective. This last sequence gives a projective (free) resolution of $K_*(B)$.

To obtain an injective resolution, we first need an injective version of 23.5.1:

PROPOSITION 23.5.2. *Let B be a separable C^*-algebra. Then there exists a separable C^*-algebra D whose K-groups are divisible, and a $*$-homomorphism $\psi : SB \to D$ such that $\psi_* : K_*(B) \to K_*(D)$ is injective.*

PROOF. Let F and ϕ be as in 23.5.1, and let $s : C_\phi \to F$ be the standard map. Let R be the UHF algebra whose dimension group is $\mathbb{Q}$ (7.5), $RF = F \otimes R$, and $t : F \to RF$ given by $t(x) = x \otimes 1$. Then $K_*(RF) \cong K_*(F) \otimes \mathbb{Q}$ (RF is an inductive limit of matrix algebras over F), and $t_* : K_*(F) \to K_*(RF)$ is injective. The mapping cone sequence for $ts : C_\phi \to F \otimes R$ has associated degenerate K-theory sequence

$$0 \longrightarrow K_*(C_\phi) \overset{(ts)_*}{\longrightarrow} K_*(RF) \longrightarrow K_*(C_{ts}) \longrightarrow 0$$

since $(ts)_*$ is injective. This implies that the K-groups of $D = C_{ts}$ are divisible. The naturality of the cone construction implies that there is a map of mapping cone sequences

$$
\begin{array}{ccccccccc}
SC_\phi & \longrightarrow & SF & \longrightarrow & C_s & \longrightarrow & C_\phi & \overset{s}{\longrightarrow} & F \\
\downarrow & & \downarrow & & \downarrow u & & \downarrow & & \downarrow u \\
SC_\phi & \longrightarrow & SRF & \longrightarrow & C_{ts} & \longrightarrow & C_\phi & \longrightarrow & RF
\end{array}
$$

and hence a commuting diagram of short exact sequences

$$
\begin{array}{ccccccccc}
 & & & & 0 & & & & \\
 & & & & \downarrow & & & & \\
0 & \longrightarrow & K_*(C_\phi) & \longrightarrow & K_*(F) & \longrightarrow & K_*(C_s) & \longrightarrow & 0 \\
 & & \| & & \downarrow t_* & & \downarrow u_* & & \\
0 & \longrightarrow & K_*(C_\phi) & \longrightarrow & K_*(RF) & \longrightarrow & K_*(C_{ts}) & \longrightarrow & 0
\end{array}
$$

A simple diagram chase shows that $u_* : K_*(C_s) \to K_*(C_{ts})$ is injective.

Since $SB \otimes \mathbb{K}$ is the kernel of s, we have that v_* is an isomorphism, where $v : SB \otimes \mathbb{K} \to C_s$ is the natural map (19.4.1). Define ψ to be the composite

$$SB \overset{w}{\longrightarrow} SB \otimes \mathbb{K} \overset{v}{\longrightarrow} C_s \overset{u}{\longrightarrow} C_{ts} = D.$$

Then ψ_* is injective and $K_*(D)$ is divisible. $\qquad\square$

We now finish the construction of the injective resolution by again using a mapping cone construction. The mapping cone sequence of ψ is

$$0 \longrightarrow SD \longrightarrow C_\psi \longrightarrow SB \longrightarrow 0$$

Since ψ_* is injective, the associated K-theory sequence degenerates to two short exact sequences

$$0 \longrightarrow K_i(SB) \longrightarrow K_{1-i}(SD) \longrightarrow K_{1-i}(C_\psi) \longrightarrow 0$$

The groups $K_*(C_\psi)$ are quotients of divisible groups, hence divisible; so we have an injective resolution of $K_*(SB)$.

It should be noted that the construction of projective and injective resolutions for C^*-algebras is much simpler than the construction in topology, which involves Grassmanians. The reason for the simplification is that it suffices to work stably for KK-theory, and there is "much more room" (i.e. many more morphisms) when one deals with stable C^*-algebras.

23.6. Proof of the General KTP

We now prove the exact sequences of the three theorems in the general case, beginning with the KTP.

PROPOSITION 23.6.1. *Let A be a nuclear C^*-algebra. Suppose the KTP holds for A and B whenever B is separable with torsion-free K-groups. Then the KTP holds for A and B, for every separable B.*

PROOF. Let B be a separable C^*-algebra, which we may take to be stable without loss of generality. Construct a geometric resolution

$$0 \longrightarrow SB \longrightarrow C \xrightarrow{\nu} F \longrightarrow 0$$

Since A is nuclear, the sequence

$$0 \longrightarrow A \otimes SB \longrightarrow A \otimes C \xrightarrow{\mu} A \otimes F \longrightarrow 0$$

is also exact, where $\mu = 1 \otimes \nu$. The associated K-theory exact sequence is

$$\cdots \longrightarrow K_i(A \otimes C) \xrightarrow{\mu_*} K_i(A \otimes F) \longrightarrow K_i(A \otimes B)$$
$$\longrightarrow K_{i-1}(A \otimes C) \xrightarrow{\mu_*} K_{i-1}(A \otimes F) \longrightarrow \cdots.$$

From this we obtain the short exact sequences

$$0 \longrightarrow \mathrm{Coker}(\mu_*)_i \xrightarrow{\omega} K_i(A \otimes B) \longrightarrow \mathrm{Ker}(\mu_*)_{i-1} \longrightarrow 0.$$

Since we are using a free resolution of $K_*(B)$, the sequence

$$0 \longrightarrow \mathrm{Tor}_1^{\mathbb{Z}}(K_*(A), K_*(B)) \longrightarrow K_*(A) \otimes K_*(C) \xrightarrow{\eta} K_*(A) \otimes K_*(F)$$
$$\longrightarrow K_*(A) \otimes K_*(B) \longrightarrow 0$$

is exact, where $\eta = 1 \otimes \nu_*$. Since C and F have torsion-free K-theory, we may replace $K_*(A) \otimes K_*(C)$ and $K_*(A) \otimes K_*(F)$ by $K_*(A \otimes C)$ and $K_*(A \otimes F)$ respectively; this identification replaces η by μ_*. Thus $\mathrm{Coker}(\mu_*) = K_*(A) \otimes K_*(B)$, and $\mathrm{Ker}(\mu_*) = \mathrm{Tor}_1^{\mathbb{Z}}(K_*(A), K_*(B))$. It is easy to check that ω corresponds to α under these identifications. $\square$

23.7. Proof of the General KT

The proof of the general exact sequence in the KT is quite similar to the proof of the KTP.

PROPOSITION 23.7.1. *Let A be a separable C^*-algebra. Suppose that for all separable B with $K_*(B)$ finitely generated and free, $\beta(A, B)$ is an isomorphism. Then for every separable B with finitely generated K-theory, the exact sequence in the KT holds for A and B.*

PROOF. Let B be a separable C^*-algebra, which may be assumed stable without loss of generality. Construct a geometric projective resolution

$$0 \longrightarrow SB \longrightarrow C \xrightarrow{\ g\ } F \longrightarrow 0$$

With C and F separable C^*-algebras with finitely generated free K-groups. The associated six-term exact KK-sequence becomes

$$\cdots \longrightarrow KK^i(A, C) \xrightarrow{\ \omega\ } KK^i(A, F) \longrightarrow KK^i(A, B) \longrightarrow KK^{i-1}(A, C) \longrightarrow \cdots,$$

which unsplices to yield a short exact sequence

$$0 \longrightarrow \mathrm{Coker}(\omega) \longrightarrow KK^*(A, B) \longrightarrow \mathrm{Ker}(\omega) \longrightarrow 0.$$

We may replace $KK^*(A, C)$ and $KK^*(A, F)$ by $K^*(A) \otimes K_*(C)$ and $K^*(A) \otimes K_*(F)$ respectively; with this identification ω becomes $1 \otimes g_*$. So $\mathrm{Coker}(\omega) = \mathrm{Coker}(1 \otimes g_*) \cong K^*(A) \otimes K_*(B)$ and $\mathrm{Ker}(\omega) = \mathrm{Ker}(1 \otimes g_*) \cong \mathrm{Tor}_1^{\mathbb{Z}}(K^*(A), K_*(B))$ (since we are working with a projective resolution of $K_*(B)$). Again, it is easy to see that the maps match up properly. $\square$

23.8. Proof of the General UCT

Finally, we use a geometric injective resolution to establish the general form of the UCT exact sequence.

PROPOSITION 23.8.1. *Let A be a separable C^*-algebra. Suppose that for all separable B with divisible K-groups, $\gamma(A, B)$ is an isomorphism. Then for all separable B, the exact sequence of the UCT holds for A and B.*

PROOF. Let B be separable and stable. Form a geometric injective resolution

$$0 \longrightarrow D \xrightarrow{\ g\ } C \longrightarrow SB \longrightarrow 0$$

as in 23.5. The associated K-theory sequence degenerates to

$$0 \longrightarrow K_i(B) \longrightarrow K_i(D) \xrightarrow{\ g_*\ } K_i(C) \longrightarrow 0,$$

which is an injective resolution of $K_*(B)$. The six-term exact sequence of KK-theory is

$$\cdots \longrightarrow KK^i(A, D) \xrightarrow{\ \omega\ } KK^i(A, C) \longrightarrow KK^i(A, B) \longrightarrow KK^{i-1}(A, D) \longrightarrow \cdots,$$

which unsplices to two short exact sequences

$$0 \longrightarrow \operatorname{Coker}(\omega) \longrightarrow KK^i(A, B) \longrightarrow \operatorname{Ker}(\omega) \longrightarrow 0.$$

By hypothesis we are allowed to replace $KK^*(A, D)$ by $\operatorname{Hom}(K_*(A), K_*(D))$ and $KK^*(A, C)$ by $\operatorname{Hom}(K_*(A), K_*(C))$; the map ω is replaced by $\operatorname{Hom}(1, g_*)$ under this identification. Thus we have

$$\operatorname{Ker}(\omega) \cong \operatorname{Ker}(\operatorname{Hom}(1, g_*)) \cong \operatorname{Hom}(K_*(A), K_*(B))$$

and $\operatorname{Coker}(\omega) \cong \operatorname{Coker}(\operatorname{Hom}(1, g_*)) \cong \operatorname{Ext}^1_{\mathbb{Z}}(K_*(A), K_*(B))$. It is easy to check that the map from $KK^*(A, B)$ to $\operatorname{Hom}(K_*(A), K_*(B))$ is the map γ. And although it is not necessary for the proof, it is also easy to check that the map δ from $\operatorname{Ext}^1_{\mathbb{Z}}(K_*(A), K_*(B))$ is the inverse of the map κ defined in 23.1. $\square$

23.9. Naturality

Although the maps α, β, γ in the theorems are clearly natural, there appears to be a slight problem with the naturality of the other maps since the construction of geometric resolutions is not functorial. However, a careful inspection of the proofs shows that the induced maps between the kernels or cokernels (as the case may be), which are natural, agree with the (obviously natural) maps between $\operatorname{Ext}^1_{\mathbb{Z}}$ or $\operatorname{Tor}^{\mathbb{Z}}_1$.

At least in the case of the UCT, the naturality can be seen another way: the proof shows that the natural map $\kappa : \ker \gamma \to \operatorname{Ext}^1_{\mathbb{Z}}$ is an isomorphism, and its inverse map δ is therefore also natural.

23.10. Some Corollaries

Before completing the proofs of the theorems, we obtain some corollaries of the results proved so far. The first result is the rather surprising fact that the K-groups of a C^*-algebra in N completely determine its KK-theory. Let N' be the class of all C^*-algebras A for which the UCT holds for (A, D) for every D.

PROPOSITION 23.10.1. *Let A and B be in N', and let $\boldsymbol{x} \in KK(A, B)$ have the property that $\gamma(\boldsymbol{x}) \in \operatorname{Hom}(K_*(A), K_*(B))$ is an isomorphism. Then $\boldsymbol{x}$ is a KK-equivalence.*

PROOF. For any separable D the induced maps $\eta : \operatorname{Hom}(K_*(B), K_*(D)) \to \operatorname{Hom}(K_*(A), K_*(D))$ and $\theta : \operatorname{Ext}^1_{\mathbb{Z}}(K_*(B), K_*(D)) \to \operatorname{Ext}^1_{\mathbb{Z}}(K_*(A), K_*(D))$ are isomorphisms, and by naturality we have a commutative diagram

$$
\begin{array}{ccccccccc}
0 & \longrightarrow & \operatorname{Ext}^1_{\mathbb{Z}}(K_*(B), K_*(D)) & \longrightarrow & KK^*(B, D) & \longrightarrow & \operatorname{Hom}(K_*(B), K_*(D)) & \longrightarrow & 0 \\
 & & \downarrow{\scriptstyle \theta} & & \downarrow{\scriptstyle (\cdot) \circ \boldsymbol{x}} & & \downarrow{\scriptstyle \eta} & & \\
0 & \longrightarrow & \operatorname{Ext}^1_{\mathbb{Z}}(K_*(A), K_*(D)) & \longrightarrow & KK^*(A, D) & \longrightarrow & \operatorname{Hom}(K_*(A), K_*(D)) & \longrightarrow & 0
\end{array}
$$

Apply the Five Lemma to conclude that

$$\boldsymbol{x} \otimes_B (\cdot) : KK^*(B, D) \to KK^*(A, D)$$

is an isomorphism. Taking $D = A$, we conclude that $\boldsymbol{x}$ has a right inverse. A similar argument using naturality in the second argument shows that

$$(\,\cdot\,) \otimes_A \boldsymbol{x} : KK^*(D, A) \to KK^*(D, B)$$

is an isomorphism for all D. Taking $D = B$, we conclude that $\boldsymbol{x}$ also has a left inverse. $\qquad\square$

COROLLARY 23.10.2. *Let A and B be C^*-algebras in N'. If $K_*(A) \cong K_*(B)$, then A and B are KK-equivalent.*

So, for "nice" C^*-algebras, every isomorphism of the K-groups is implemented by a KK-equivalence, and hence comes from a natural transformation.

COROLLARY 23.10.3. *Let B be any (separable) C^*-algebra. Then there is a (separable) commutative C^*-algebra C, whose spectrum has dimension at most 3, and an element $\boldsymbol{x} \in KK(C, B)$ for which $\gamma(\boldsymbol{x}) : K_*(C) \to K_*(B)$ is an isomorphism. We have that $(\,\cdot\,) \otimes_C \boldsymbol{x} : KK^*(A, C) \to KK^*(A, B)$ is an isomorphism for all $A \in N'$. If $B \in N'$, $\boldsymbol{x}$ is a KK-equivalence. We may choose C to be of the form $C_0 \oplus C_1$, where $K_1(C_0) = K_0(C_1) = 0$. If $K_*(B)$ is finitely generated, we may also choose C so that its spectrum is a finite complex of dimension at most 3.*

PROOF. Since the map $\gamma : KK^*(C, B) \to \text{Hom}(K_*(C), K_*(B))$ is surjective, it is only necessary to find a commutative C^*-algebra of the specified form whose K-groups are isomorphic to those of B. There is a standard way to construct a C_0 with specified K^0 and trivial K^1: choose a free resolution

$$0 \longrightarrow F_1 \overset{f}{\longrightarrow} F_2 \longrightarrow K_0(B) \longrightarrow 0;$$

let D_1 and D_2 each be a c_0-direct sum of copies of S and $\phi : D_1 \to D_2$ with $\phi_* = f$, and let C_0 be the mapping cone of ϕ. The direct sum of one such algebra with the suspension of another gives an algebra of the required form with arbitrary K-theory. The proof that $(\,\cdot\,) \otimes_C \boldsymbol{x}$ is an isomorphism is identical to the proof of 23.10.1, using naturality in the second variable. $\qquad\square$

Thus, from the point of view of KK-theory, every C^*-algebra in N might as well be commutative. So without additional structure K-theory and KK-theory give only a very weak invariant.

COROLLARY 23.10.4. *Let A be a C^*-algebra in N with torsion-free K-theory. Then there are simple AF algebras A_0 and A_1 such that A is KK-equivalent to $A_0 \oplus SA_1$. So A is in the class N_0 (22.4.5).*

PROOF. Any countable torsion-free abelian group G can be embedded as an additive subgroup of $\mathbb{R}$; the image is dense unless $G = \mathbb{Z}$. If $K_i(A) = \mathbb{Z}$, set $A_i = \mathbb{C}$; otherwise embed $K_i(A)$ into $\mathbb{R}$ and let A_i be an AF algebra with the image as dimension group (7.5). $\qquad\square$

The next result was observed by Skandalis [Skandalis 1988]; it is implicitly contained in [Rosenberg and Schochet 1987]. The proof is quite similar to that of 23.10.1.

THEOREM 23.10.5. *Let B be a C^*-algebra. The following conditions are equivalent*:

(i) $B \in N'$.
(ii) B *is KK-equivalent to a C^*-algebra in N.*
(iii) B *is KK-equivalent to a commutative C^*-algebra.*
(iv) *If D is any C^*-algebra with $K_*(D) = 0$, then $KK^*(B, D) = 0$.*

PROOF. (i) $\implies$ (iv) is trivial, (iii) $\implies$ (ii) is 22.3.5(d), and (ii) $\implies$ (i) is the UCT. To prove (iv) $\implies$ (iii), let A be a commutative C^*-algebra and $\boldsymbol{x} \in KK(A, B)$ with $\gamma(\boldsymbol{x}) : K_*(A) \to K_*(B)$ an isomorphism (23.10.3). Represent $\boldsymbol{x}$ by a semisplit extension

$$0 \longrightarrow SB \otimes \mathbb{K} \longrightarrow D \longrightarrow A \longrightarrow 0.$$

By the K-theory exact sequence, $K_*(D) = 0$, so $KK^*(B, D) = 0$. Apply the KK-theory exact sequence in the second variable to conclude that right multiplication by $\boldsymbol{x}$ is an isomorphism from $KK(B, A)$ to $KK(B, B)$. If $\boldsymbol{y}$ is a left inverse for $\boldsymbol{x}$, by the same argument applied to A instead of B, and using the implication (iii) $\implies$ (iv), we conclude $\boldsymbol{y}$ is left invertible, so $\boldsymbol{x}$ is invertible. □

This result has an interesting and important consequence. We first need a definition:

DEFINITION 23.10.6. If A and B are separable C^*-algebras, then A KK-*dominates* B if there are $\boldsymbol{x} \in KK(A, B)$, $\boldsymbol{y} \in KK(B, A)$ such that $\boldsymbol{yx} = \boldsymbol{1}_B$ in $KK(B, B)$.

EXAMPLES 23.10.7. (a) If A homotopy-dominates B (or shape-dominates B) then A KK-dominates B.

(b) $A \oplus B$ KK-dominates both A and B.

(c) If $0 \to J \to A \to B \to 0$ is a split exact sequence of separable C^*-algebras, then A KK-dominates both J and B.

COROLLARY 23.10.8. *If $A \in N'$, and A KK-dominates B, then $B \in N'$.*

PROOF. Let D be a (separable) C^*-algebra with $K_*(D) = 0$, and let $\boldsymbol{z} \in KK(B, D)$. Then, if $\boldsymbol{x}$, $\boldsymbol{y}$ are as in 23.10.6, we have $\boldsymbol{xz} = 0$ in $KK(A, D) = 0$. Thus $\boldsymbol{yxz} = \boldsymbol{z} = 0$, i.e. $KK(B, D) = 0$. □

A simple consequence, which can be easily proved directly, is that $A_1 \oplus A_2 \in N'$ if and only if $A_1, A_2 \in N'$. A more interesting application is the proof in [Tu 1998] that the C^*-algebra of an amenable groupoid is always in N.

23.11. Splitting

We can now use the results of 23.10 to conclude that the exact sequences in the three theorems split (unnaturally). 23.10.3 allows us to reduce the splitting problem to the case where A and B are commutative, $A = A_0 \oplus A_1$, $B = B_0 \oplus B_1$, $K_i(A_j) = K_i(B_j) = 0$ for $i \neq j$. Then

$$KK(A, B) \cong KK(A_0, B_0) \oplus KK(A_0, B_1) \oplus KK(A_1, B_0) \oplus KK(A_1, B_1),$$

$$K_*(A \otimes B) \cong K_*(A_0 \otimes B_0) \oplus K_*(A_0 \otimes B_1) \oplus K_*(A_1 \otimes B_0) \oplus K_*(A_1 \otimes B_1).$$

Apply the theorems to each term separately, and note that in each case one of the maps is an isomorphism because the other end term vanishes. This proves that all the sequences split.

It is known [Rosenberg and Schochet 1987] that the exact sequences in these theorems cannot have *natural* splittings even in the commutative case (an "explanation" is given in 23.15.11).

PROPOSITION 23.11.1. *Let $A \in N$. Then the UCT sequence for $KK^*(A, A)$ is a split exact sequence of graded rings. So $KK^*(A, A)$ is isomorphic to*

$$\bigoplus \mathrm{Ext}_{\mathbb{Z}}^n(K_i(A), K_j(A)), \quad \text{where } i, j, n = 0, 1,$$

with the following ring structure: the product of any two $\mathrm{Ext}_{\mathbb{Z}}^1$-terms is 0, and $\mathrm{Ext}_{\mathbb{Z}}^0 = \mathrm{Hom}$ acts on Hom and $\mathrm{Ext}_{\mathbb{Z}}^1$ as usual. So the $\mathrm{Ext}_{\mathbb{Z}}^1$-terms in $KK^*(A, A)$ form an ideal with square 0.

PROOF. As above, we may assume that $A = A_0 \oplus A_1$. $KK^*(A_0, A_0)$ and $KK^*(A_1, A_1)$ are graded subrings of $KK^*(A, A)$, and if $\boldsymbol{x} \in KK^*(A, A_i) \subseteq KK^*(A, A)$, $\boldsymbol{y} \in KK^*(A_j, A) \subseteq KK^*(A, A)$ for $i \neq j$, then $\boldsymbol{x} \otimes_A \boldsymbol{y} = 0$. It is routine to check all the various cases of products from the different summands to verify that the ring structure is as described. $\square$

23.12. The General KT

The proofs of all the theorems except the KT are now complete. We have, however, only proved that the KT holds if $K_*(B)$ is finitely generated. To show that it also holds if $K_*(A)$ is finitely generated, we first reduce to the case where A and B are commutative, say $A = C_0(X)$, $B = C_0(Y)$. Then $KK^*(A, B) \cong \tilde{K}^{*-1}(F(X^+) \wedge Y^+)$ (16.4.4). One must then replace $F(X^+)$ by a finite complex and use the Künneth Theorem of algebraic topology.

23.13. Extensions of the Theorems

23.10.5 shows that the UCT holds for all B if and only if A is KK-equivalent to a commutative C^*-algebra. The class N' of A for which the UCT holds for every B does not consist of all separable C^*-algebras: if $A \in N'$, then there are six-term exact sequences in $KK(A, \cdot)$ for arbitrary extensions. Thus $C_r^*(\Gamma)$, where Γ is as in 19.9.8, is not in N'.

Let N'' be the set of all A for which the KT holds for all B. 23.10.5(iv) shows that $N'' \subseteq N'$, and N'' at least contains all algebras in N' with finitely generated K-theory. Some type of finite generation hypothesis is necessary in the KT (23.15.4). It should be possible (and interesting) to characterize the algebras in N'' in terms of their K-groups. In this connection, the following result is relevant:

PROPOSITION 23.13.1. [Rosenberg and Schochet 1987, 7.14] *Let* $A_i \in N''$, *and* $A = \varinjlim A_i$. *Then* $A \in N''$ *if and only if the Milnor* $\varprojlim^1$-*sequence holds*:

$$0 \longrightarrow \varprojlim{}^1 K^*(A_i) \longrightarrow K^*(A) \longrightarrow \varprojlim K^*(A_i) \longrightarrow 0$$

To characterize N'', one would need to write a general $A \in N$ in some canonical way as an inductive limit of algebras in N with finitely generated K-theory.

23.13.2. It is less clear which A satisfy the KTP for all B. An argument that the KTP should not hold in general is that (minimal) tensor product with a fixed C^*-algebra is not always an exact functor. The class of exact C^*-algebras is strictly larger than the class of nuclear C^*-algebras; it is the same as the class of subnuclear C^*-algebras. See [Archbold 1982], [Lance 1982], or [Wassermann 1994] for a discussion of exact C^*-algebras.

The KTP appears to have a better chance of holding in general if maximal instead of minimal tensor products are considered, since $\otimes_{\max}$ is always exact. If A_1 and A_2 are KK-equivalent, then $A_1 \otimes_{\max} B$ and $A_2 \otimes_{\max} B$ are KK-equivalent [Skandalis 1988]. So the class of A for which the $\otimes_{\max}$-KTP holds for all B is closed under KK-equivalence.

If Γ is as in 19.9.8, then the quotient map from $C_r^*(\Gamma) \otimes_{\max} C_r^*(\Gamma)$ to $C_r^*(\Gamma) \otimes C_r^*(\Gamma)$ does not induce an isomorphism on K-theory (there is a nontrivial kernel), so either the KTP or the $\otimes_{\max}$-KTP (probably the KTP or possibly both) fails for $A = B = C_r^*(\Gamma)$.

23.13.3. There is also some "fine structure" in the KK-groups related to the UCT. See [Elliott and Gong 1994; Gong 1994] for the action of the product in relation to the UCT. There is also a natural topology on $KK(A, B)$; the subgroup $\mathrm{Pext}(A, B)$ of $\mathrm{Ext}^1_{\mathbb{Z}}(A, B)$ consisting of pure extensions is the closure of 0 in this topology. Pext has played a natural role in classification problems [Rørdam 1993]. There is a "universal multicoefficient theorem" [Dădărlat and Loring 1996b] involving Pext and mod p K-theory. See [Schochet 1996] for more on the fine structure of the KK-groups.

23.14. Equivariant Theorems

There are equivariant versions of the UCT and KT, but even the correct statements are much more complicated to formulate. See [Rosenberg and Schochet 1986]. See [Madsen and Rosenberg 1988] for more equivariant UCT's and a UCT for real C^*-algebras.

23.15. EXERCISES AND PROBLEMS

23.15.1. If $A \in N$, a special case of the UCT (with $B = \mathbb{C}$) gives a short exact sequence

$$0 \longrightarrow \mathrm{Ext}^1_{\mathbb{Z}}(K_0(A), \mathbb{Z}) \longrightarrow Ext(A) \longrightarrow \mathrm{Hom}(K_1(A), \mathbb{Z}) \longrightarrow 0$$

which is natural in A and splits unnaturally. This is Brown's original Universal Coefficient Theorem [Brown 1984] (16.3.3). Show that this sequence must split (unnaturally) for purely algebraic reasons [Kahn et al. 1977].

23.15.2. If A and B are AF algebras, then the UCT says that $KK(A, B) \cong \mathrm{Hom}(K_0(A), K_0(B))$. In particular, A and B are KK-equivalent if and only if their dimension groups are isomorphic as groups (ignoring the order structure completely). Find a direct proof of this fact.

23.15.3. If A and B are AF algebras, then the UCT says that $Ext(A, B) \cong \mathrm{Ext}^1_{\mathbb{Z}}(K_0(A), K_0(B))$ (cf. 16.4.7). The correspondence is given by the map κ of 23.1. This result (at least for A and B simple) is originally due to Handelman [Handelman 1982], with the equivalence relation clarified by Brown and Elliott [Brown and Elliott 1982]. The case $B = \mathbb{C}$ comes from Brown's UCT (23.15.1). Recall that any extension of AF algebras is again AF.

23.15.4. Let R be the UHF algebra with dimension group $\mathbb{Q}$ (7.5). From the UCT we have that $KK(R, R) = \mathbb{Q}$ and $K^*(R) = 0$. Thus the Künneth theorem fails for $A = B = R$, showing that the finite generation hypothesis in the KT is necessary. This example is due to G. Elliott.

The difficulty stems from the fact that KK is not σ-additive in the second variable (19.7).

23.15.5. Show that if $A \in N$ and $K_*(A)$ is finitely generated, then $KK(A, \cdot)$ is σ-additive (19.7.2) [Rosenberg and Schochet 1987, 7.13].

23.15.6. K-THEORY WITH RATIONAL COEFFICIENTS. (a) Let $D \in N$ with $K_0(D) = \mathbb{Q}$, $K_1(D) = 0$ (e.g. $D = R$ of 23.15.4). Define $K_*(B; \mathbb{Q}) = K_*(B \otimes D)$, and more generally $KK(A, B; \mathbb{Q}) = KK(A \otimes D, B \otimes D) \cong KK(A, B \otimes D)$. Then $KK(\cdot, \cdot; \mathbb{Q})$ is independent of D and is a bifunctor from pairs of C^*-algebras to rational vector spaces, with the same basic properties as the ordinary KK-groups (homotopy invariance, stability, Bott periodicity, long exact sequences for semisplit extensions, etc.); the UCT and KTP hold with appropriate (simplifying) modifications.

(b) It is not true in general that $KK(A, B; \mathbb{Q}) \cong KK(A, B) \otimes_{\mathbb{Z}} \mathbb{Q}$: for example, $KK(D, \mathbb{C}; \mathbb{Q}) = \mathbb{Q}$, but $KK(D, \mathbb{C}) = 0$. It is true that $K_*(B; \mathbb{Q}) \cong K_*(B) \otimes_{\mathbb{Z}} \mathbb{Q}$ for any B.

(c) Show that the functor $K_0(\cdot; \mathbb{Q})$ is characterized on the class N by homotopy invariance, stability, continuity, *split* exactness, and the "rationalized" dimension axiom ($F(\mathbb{C}) = \mathbb{Q}$, $F(C_0(\mathbb{R})) = 0$).

23.15.7. K-THEORY mod p. (a) As in 23.15.6, let D be a C^*-algebra in N with $K_0(D) = \mathbb{Z}_p$, $K_1(D) = 0$. For example, let D be the Cuntz algebra O_{p+1} (or a suitable commutative C^*-algebra). Define $K_*(B; \mathbb{Z}_p) = K_*(B \otimes D)$. $K_*(B; \mathbb{Z}_p)$ is a $\mathbb{Z}_p$-module.

(b) Show that these modules are independent of the choice of D. It is not in general true that $K_*(B; \mathbb{Z}_p) \cong K_*(B) \otimes_{\mathbb{Z}} \mathbb{Z}_p$: $K_1(D; \mathbb{Z}_p)$ is nontrivial since $\text{Ext}^1_{\mathbb{Z}}(\mathbb{Z}_p, \mathbb{Z}_p)$ is nontrivial.

(c) There is for any B an exact sequence

$$
\begin{array}{ccc}
K_0(B) & \xrightarrow{\ p\ } K_0(B) \longrightarrow K_0(B; \mathbb{Z}_p) \\
\uparrow & \qquad\qquad\qquad\downarrow \\
K_1(B; \mathbb{Z}_p) \longleftarrow K_1(B) & \xleftarrow{\ p\ } K_1(B)
\end{array}
$$

where p denotes multiplication by p, which may be used to calculate $K_*(B; \mathbb{Z}_p)$ [Cuntz 1981c; Schochet 1984b].

(d) Show from the UCT that $K^0(D) = 0$, $K^1(D) = \mathbb{Z}_p$ (cf. 16.4.5).

(e) Define $KK(A, B; \mathbb{Z}_p) = KK(A, B \otimes D)$. Show from (d) that $KK(A, B; \mathbb{Z}_p)$ is isomorphic to $KK^1(A \otimes D, B)$. Determine which properties of ordinary KK carry over to the mod p case.

Rosenberg and Schochet [Rosenberg and Schochet 1987] have obtained some very interesting results about mod p K-theory. Cuntz and Schochet [Schochet 1984b] had done earlier work on the subject. The topological mod p theory was developed earlier [Araki and Toda 1965, 1966].

More recently, mod p K-theory has been used in the classification of non-simple approximately homogeneous C^*-algebras of real rank zero [Dădărlat and Loring 1996a]. The invariant is $\mathbf{K}(A) = K_*(A) \oplus \bigoplus_p K_*(A; \mathbb{Z}_p)$ with a natural order structure and Bockstein operations. There is a universal multicoefficient theorem relating KK, Pext, and homomorphisms of this invariant [Dădărlat and Loring 1996b].

23.15.8. Let **KKN** be the full subcategory of **KK** with objects in N. Show that **KKN** is an abelian category by using a mapping cone construction to yield kernels and cokernels. Find the projective and injective objects in this category.

23.15.9. (a) Let A and B be graded C^*-algebras. Define a map $\sigma : \mathbb{E}(A, B) \to \mathbb{E}(A, B)$ by $\sigma(E, \phi, F) = (E, \phi, -F)$. σ drops to an involution of $KK(A, B)$, also denoted σ, and hence to an involution of $KK^*(A, B)$. If A and B are evenly graded, then $\sigma = 1$, whereas if one is evenly graded and the other has an odd grading, then $\sigma = -1$. We have $\sigma(\boldsymbol{x} \hat{\otimes}_D \boldsymbol{y}) = \sigma(\boldsymbol{x}) \hat{\otimes}_D \sigma(\boldsymbol{y})$.

(b) Regard $KK^*(A, B)$ as a graded abelian group using the involution σ, i.e. $KK^*(A, B)^{(0)} = \{\boldsymbol{x} : \sigma(\boldsymbol{x}) = \boldsymbol{x}\}$, $KK^*(A, B)^{(1)} = \{\boldsymbol{x} : \sigma(\boldsymbol{x}) = -\boldsymbol{x}\}$. If A and B are trivially graded (or evenly graded), then $KK^*(A, B)^{(0)} = KK(A, B)$,

$KK^*(A,B)^{(1)} = KK^1(A,B)$, so this grading on KK^* extends the grading considered in Section 23.

(c) Formulate and prove graded versions of the UCT, KT, and KTP. $K_*(A)$ and $K^*(A)$ are replaced by $KK^*(\mathbb{C},A)$ and $KK^*(A,\mathbb{C})$, with grading defined as in (b); Hom , Ext, Tor are interpreted as respecting the grading in the appropriate sense.

23.15.10. Is every graded C^*-algebra KK-equivalent to a trivially graded C^*-algebra? 23.10.3 suggests it is possible that a general graded C^*-algebra is KK-equivalent to a C^*-algebra $A \oplus B$, where A has a standard even grading and B a standard odd grading. Such a C^*-algebra is KK-equivalent to $A \oplus (SB \,\hat{\otimes}\, \mathbb{C}_1)$ with standard even grading; this is in turn KK-equivalent to a trivially graded C^*-algebra.

23.15.11. (a) Let A and B be (separable) C^*-algebras, with $A \in N$. By identifying $KK(A,B)$ with $Ext(A,SB)$, show that to each $\boldsymbol{x} \in KK(A,B)$ there is a six-term cyclic exact sequence

$$
\begin{array}{ccccc}
E_1 & \longleftarrow & K_1(B) & \xleftarrow{\ \phi_1\ } & K_1(A) \\
\downarrow & & & & \uparrow \\
K_0(A) & \xrightarrow{\ \phi_0\ } & K_0(B) & \longrightarrow & E_0
\end{array}
$$

which is well defined up to the usual notion of equivalence.

(b) Such an exact sequence is completely determined by the elements of

$$\mathrm{Ext}^1_{\mathbb{Z}}(\ker \phi_0, \mathrm{coker}\, \phi_1) \quad \text{and} \quad \mathrm{Ext}^1_{\mathbb{Z}}(\ker \phi_1, \mathrm{coker}\, \phi_0)$$

at E_1 and E_0 respectively.

(c) Suppose $A \in N$. Use the fact that the natural map

$$\mathrm{Ext}^1_{\mathbb{Z}}(K_i(A), K_{1-i}(B)) \to \mathrm{Ext}^1_{\mathbb{Z}}(\ker \phi_i, \mathrm{coker}\, \phi_{1-i})$$

is surjective (because $\mathrm{Ext}^2_{\mathbb{Z}} = 0$) to show that every such exact sequence comes from a KK-element.

(d) The natural maps in (c) are not injective in general, so different KK-elements can yield the same six-term exact sequences. For example, if $A = B = O_3 \otimes O_3$, then $|KK(A,B)| = 16$, but there are only 7 such exact sequences.

   The fact that there is no natural way of associating a KK-element to a six-term exact sequence is closely related to the fact that the UCT exact sequence has no natural splitting.

23.15.12. One of the outstanding open questions of C^*-algebra theory is whether the bootstrap class N of 22.3.4, which by the results of this section consists precisely of the separable nuclear C^*-algebras A for which the UCT holds for all B, contains all nuclear C^*-algebras. By 22.3.5(g), if $A \in N$ and $\alpha \in Aut(A)$, then $A \times_\alpha \mathbb{Z} \in N$.

(a) Let B be the stable AF algebra with dimension group $\mathbb{Z} + \mathbb{Z}\gamma$, where γ is the golden ratio $\frac{1}{2}(1 + \sqrt{5})$. Then B is simple and KK-equivalent to $\mathbb{C} \oplus \mathbb{C}$, and there is an automorphism β of B with β_* multiplication by γ. $B \times_\beta \mathbb{Z}$ is purely infinite, and its K-theory is trivial by the Pimsner–Voiculescu sequence, so $B \times_\beta \mathbb{Z} \cong O_2 \otimes \mathbb{K}$ by [Kirchberg 1998].

(b) Let A be a separable nuclear C^*-algebra. Let δ be the automorphism $1 \otimes \beta$ of $D = A \otimes B$. Then $D \times_\delta \mathbb{Z} \cong A \otimes O_2 \in N$ because it is K-contractible. Also, D is KK-equivalent to $A \oplus A$.

(c) Suppose $A \times_\alpha \mathbb{Z} \in N \implies A \in N$. (By Takai duality, this is equivalent to saying that N is closed under crossed products by $\mathbb{T}$.) Then, if A is as in (b), we can conclude that $A \oplus A \in N$ and thus $A \in N$ by 23.13.8.

(d) A similar argument might show that if N is closed under crossed products by finite cyclic groups (even by $\mathbb{Z}_2$), then N contains all separable nuclear C^*-algebras.

24. Survey of Applications to Geometry and Topology

In this section, we give a brief survey of the principal applications of KK-theory so far in geometry and topology. This section is intended only as an introduction to an already formidable array of deep mathematics, and it is hoped that readers will further pursue some of these topics by referring to the original works. Other detailed surveys of this material can be found in [Higson 1990; Connes 1994].

24.1. Index Theorems

The Atiyah–Singer Index Theorem, which asserts that the Fredholm index of an elliptic pseudodifferential operator can be calculated from purely topological data associated with the operator and the underlying spaces, is regarded as one of the great achievements of modern mathematics. In this paragraph we will give a brief description of the index theorem, how noncommutative topology is relevant, and some of the ways the theorem can be generalized using noncommutative topology. The interested reader can learn much more about index theorems in [Atiyah and Singer 1968a; 1968b; 1971a; 1971b; Atiyah and Segal 1968; Baum and Douglas 1982a; Fack 1983; Kasparov 1984a; Moore and Schochet 1988].

We will not attempt to even summarize the theory of pseudodifferential operators here; [Trèves 1980] is one good reference for the general theory.

If E is a smooth vector bundle over a smooth (C^∞) manifold M, we will write $\Gamma^\infty(E)$ [resp. $\Gamma_c^\infty(E)$, $\Gamma_0^\infty(E)$] to denote the smooth sections [resp. with compact support, vanishing at ∞] of E. If D is a pseudodifferential operator from $E^{(0)}$ to $E^{(1)}$, where the $E^{(i)}$ are smooth vector bundles over M (i.e. $D : \Gamma^\infty(E^{(0)}) \to \Gamma^\infty(E^{(1)}))$, then D defines a *symbol* $\sigma_D : \pi^* E^{(0)} \to \pi^* E^{(1)}$, where $\pi^* E^{(i)}$ is the pullback of $E^{(i)}$ to a vector bundle over the cotangent bundle $T^* M$ of M via the projection map π. D is *elliptic* if $\sigma_D(\xi)$ is an isomorphism for every nonzero cotangent vector ξ.

If D is elliptic and M is compact, then the kernel and cokernel of D are finite-dimensional subspaces of $\Gamma^\infty(E^{(0)})$ and $\Gamma^\infty(E^{(1)})$ respectively, and the *analytic index* of D is defined to be

$$\mathrm{Ind}^a(D) = \dim \ker D - \dim \mathrm{coker}\, D$$

If M is oriented, the topological index of D is defined as

$$\mathrm{Ind}^t(D) = \langle \tau_!(\mathrm{ch}(\sigma_D)) \cup \mathrm{Td}(T^*M \otimes_{\mathbb{R}} \mathbb{C}), [M]\rangle$$

where $\mathrm{ch}(\sigma_D)$ is the Chern character of $\sigma_D \in H_c^*(T^*M), \mathbb{Q})$, $\mathrm{Td}(T^*M \otimes_{\mathbb{R}} \mathbb{C}) \in H^*(M; \mathbb{Q})$ is the Todd class of the complexified cotangent bundle, $\cup$ is the cup product in cohomology, $\tau_! : H_c^*(T^*M) \to H_c^*(M)$ is the inverse of the Thom isomorphism, $[M] \in H_*(M)$ is the fundamental class, and $\langle \cdot, \cdot \rangle$ is the standard pairing of cohomology and homology.

THEOREM 24.1.1 (ATIYAH–SINGER INDEX THEOREM). *If M is compact and oriented, and D is an elliptic pseudodifferential operator as above, then*

$$\mathrm{Ind}^a(D) = \mathrm{Ind}^t(D).$$

Two special cases illustrate the typical content of the Index Theorem in the odd- and even-dimensional cases respectively:

EXAMPLES 24.1.2. (a) Let $M = S^1$, $E^{(0)} = E^{(1)} = $ one-dimensional trivial bundle, so $\Gamma^\infty(E^{(i)}) \cong C^\infty(S^1)$. Let $f \in C^\infty(S^1)$, f never 0. Write $C^\infty(S^1) = C^\infty(S^1)_+ \oplus C^\infty(S^1)_-$, where $C^\infty(S^1)_+$ is the Hardy space; let T_f be the compression of the multiplication operator M_f to $C^\infty(S^1)_+$, and set $D_f = T_f \oplus 1$. The Index Theorem then says that $\mathrm{Ind}^a(D_f) = -$(winding number of f).

(b) Let M be a compact Riemann surface of genus g, and L a holomorphic line bundle. Let $\bar{\partial} : \Gamma^\infty(L) \to \Gamma^\infty(L \otimes \bar{T}^*)$ be the standard $\bar{\partial}$-operator ($\bar{\partial}$ really gives the Cauchy–Riemann equations; the solutions of $\bar{\partial}$ are the holomorphic sections.) The Index Theorem in this case gives the *Riemann–Roch Theorem:* $\mathrm{Ind}^a(\bar{\partial}) = d - g + 1$, where d is the *degree* of L, defined via intersection theory.

There are many possible directions in which the Index Theorem has potential generalizations. In all cases, the first difficulty is to make sense of the analytic index, since the operators considered are not Fredholm operators in the usual sense. Sometimes, as in the L^2 index theorems, a numerical index can be defined using dimension theory in von Neumann algebras instead of ordinary dimension theory; but usually the analytic index is instead interpreted as an element of a certain K-group. The topological index is most cleanly expressed in terms of the intersection product. A general index theorem is then a result that a K-group element defined analytically from a pseudodifferential operator is the same as another K-group element defined in purely topological terms from the symbol of the operator and the topology of the underlying spaces.

The original Index Theorem can be rephrased in this manner: the topological index may be regarded as defining a homomorphism from $K^0(T^*M)$ to $\mathbb{Z}$ (actually to $\mathbb{Q}$), which can be alternately described without any explicit mention of cohomology (see [Atiyah and Singer 1968a]). The symbol of an elliptic operator defines an element of $K^0(T^*M)$. The analytic index is the composite of the homomorphism $K^0(T^*M) \to K_0(M)$ (given by a certain intersection product) which sends the class of a symbol to the class of the associated elliptic operator, with $p_* : K_0(M) \to \mathbb{Z}$ induced by $p : M \to \{\text{pt}\}$.

24.1.3. The first generalization we consider is the index theorem for families, also due to Atiyah and Singer. Suppose Y is a locally compact space (called the parameter space), $D_y(y \in Y)$ a continuous family of elliptic pseudodifferential operators on (vector bundles over) a compact smooth manifold M, which are invertible except for a compact set of y. This time the index (both analytic and topological) takes values in $K^0(Y)$. The analytic index is easy to (approximately) describe: $y \to \ker D_y$ and $y \to \operatorname{coker} D_y$ (essentially) define vector bundles over Y which have compact support; the analytic index is the difference of the equivalence classes of these vector bundles. (Actually the definition is a little more complicated since these "vector bundles" need not be locally trivial.) The topological index is given by a formula similar to the one in the original index theorem.

24.1.4. The next index theorem we consider, which is a generalization of the previous ones, is the result of Miščenko and Fomenko [1979]. If B is a unital C^*-algebra, then a *B-vector bundle* over a space X is a locally trivial bundle of Banach spaces over X whose fibers have the structure of finitely generated projective B-modules. We write $K^0(X; B)$ for the Grothendieck group of the semigroup of stable isomorphism classes of B-vector bundles over X (if X is not compact, we take the kernel of the map $K^0(X^+, B) \to K^0(+, B)$ as in ordinary K-theory.) We have $K^0(X; B) \cong K_0(C_0(X) \otimes B)$.

A theory of elliptic pseudodifferential B-operators between smooth B-vector bundles over a smooth compact manifold M can be developed in complete analogy with the ordinary theory; such an operator D has an analytic B-index: after perturbing D by a suitable "B-compact" operator, the kernel and cokernel of D are finitely generated projective B-modules, so we can let $\operatorname{Ind}_B^a(D)$ be the difference of these modules in $K_0(B)$.

There is a topological formula for the index in this case which is formally identical with the formula in the original index theorem. The Chern character, however, must be suitably interpreted: from the Künneth Theorem for tensor products in K-theory (23.1.3) we have

$$K^0(X; B) \otimes \mathbb{Q} \cong (K^0(X) \otimes K_0(B) \otimes \mathbb{Q}) \oplus (K^1(X) \otimes K_1(B) \otimes \mathbb{Q})$$

and composing this isomorphism with the ordinary Chern character we get a map

$$\text{ch} : K^0(X;B) \otimes \mathbb{Q} \to (H^{\text{even}}(X;\mathbb{Q}) \otimes K_0(B)) \oplus (H^{\text{odd}}(X;\mathbb{Q}) \otimes K_1(B))$$

which is a rational isomorphism. This is the Chern character used in the Miščenko–Fomenko index formula.

The Atiyah–Singer Index Theorem is the case $B = \mathbb{C}$, and the Index Theorem for Families is the case $B = C_0(Y)$.

24.1.5. Another generalization comes from considering a G-equivariant elliptic operator on a proper G-space M, where G is a locally compact group. If G is compact, then the original index theorem applies to give an equality between two numbers; but a finer index theorem is possible in this situation. As first noted by Bott and then developed by Atiyah and Singer, one should use G-equivariant K-theory (section 11); then everything can be developed in the identical manner to the original situation, except that the indexes take values in the representation ring $R(G) = K_0^G(\mathbb{C}) = K_G^0(+)$ instead of $\mathbb{Z} = K_0(\mathbb{C}) = K^0(+)$.

Whether or not G is compact, if M/G and all the stability subgroups are compact one obtains an analytic and topological index taking values in $K_0(C_r^*(G))$; Kasparov [1984b] proved that these two indexes coincide.

An important special case is when $M = G/H$, where G is a connected Lie group and H is a compact subgroup. One can identify a G-invariant elliptic operator on G/H with a G-invariant operator on G, so one can obtain an index in $K_0(C_r^*(G))$ for such an operator. If G is unimodular, the Plancherel dimension of the kernel and cokernel (defined, if you like, using the canonical trace on $\lambda(G)''$) are finite, and the difference gives a real-valued analytic index for the operator. This index may be calculated by a topological formula as in the original index theorem. Several important results on the existence or non-existence of L^2-solutions to elliptic partial differential equations have been proved using this theorem, which is due to Connes and Moscovici [1982a; 1982b].

One can also consider the equivariant version of the Miščenko–Fomenko index theorem: if M is a proper G-space and B is a G-algebra, one can obtain indexes in $K_0^G(B)$ or in $K_0(C_r^*(G,B))$ for a G-invariant elliptic operator between B-vector bundles over M.

24.1.6. The final example we discuss is a generalization of Kasparov's index theorem of 24.1.5: the longitudinal index theorem for foliations of Connes and Skandalis [1984]. If V is a smooth manifold with foliation F, then there is an index theorem for pseudodifferential operators which are elliptic in the longitudinal direction. In this case the indexes take values in $K_0(C^*(V/F))$, where $C^*(V/F)$ is the C^*-algebra of the foliation, i.e. the groupoid C^*-algebra of the holonomy groupoid of V/F [Connes 1982]. See [Moore and Schochet 1988] for a detailed treatment of this result. There is a version of this theorem for real C^*-algebras [Schröder 1993].

There are other index theorems; for example, Teleman's index theorem [Teleman 1984], which can be used to prove the homotopy invariance of the rational Pontrjagin classes. See [Hilsum 1985] for a proof using KK-theory. Baum, Douglas, and Taylor [Baum et al. 1989; Baum and Douglas 1991] have developed a relative KK-theory with the goal of studying index theorems on manifolds with boundary.

24.1.7. The general principle in all these index theorems is that if (D_y) is a family of elliptic operators parametrized by a space Y, then the index should be an element of $K^0(Y)$. In several of the settings, Y is a "singular space" (e.g. the dual space of a group or the leaf space of a foliation) for which there is a noncommutative C^*-algebra B which plays the role of $C_0(Y)$ in a natural way. The general principle is then that the index should take values in $K_0(B)$.

24.1.8. To place the index theorems more in the language of the main body of work in these notes, we consider the following construction. Let M be a compact manifold and P_M the algebra of pseudodifferential operators of degree 0 on M. Then the symbol map defines a homomorphism $\phi_M : P_M \to C(S^*M)$, where S^*M is the cosphere bundle on M (defined using a Riemannian metric on M.) We thus get an extension

$$0 \to \mathbb{K}(L^2(M)) \to^i \bar{P}_M \xrightarrow{\phi_M} C(S^*M) \to 0$$

($\bar{P}_M$ is the closure of P_M.) If D is elliptic, then σ_D is invertible in $C(S^*M)$; the analytic index is the image of its class in $K_1(C(S^*M))$ under the connecting map $K_1(C(S^*M)) \to K_0(\mathbb{K}) \cong \mathbb{Z}$ in the corresponding K-theory exact sequence. The Atiyah–Singer Index Theorem essentially gives a topological formula for this image element.

In the other index theorems, a similar extension can be obtained, with the terms $\mathbb{K}(L^2(M))$ and $C(S^*M)$ replaced by other suitable C^*-algebras. For example, in the longitudinal index theorem for foliations, $\mathbb{K}(L^2(M))$ is replaced by $C^*(V/F)$. In each case the analytic index can be interpreted nicely as the image of the symbol element under the K-theory connecting map of the exact sequence.

Although these index theorems are extremely powerful and useful as they stand, they are a bit crude in the sense that they really describe the extension of C^*-algebras defined by taking the closure of the set of operators of interest. By taking closures, the smooth structure of the operators is lost and only the underlying topology remains. If the differential structure is retained, a more delicate analysis is possible: for example, in the original setting, we actually have an extension

$$0 \longrightarrow \mathbb{K}_p(L^2(M)) \longrightarrow P_M \longrightarrow C^\infty(S^*M) \longrightarrow 0$$

where $\mathbb{K}_p$ is the Schatten p-class, for suitably large p. The analysis in this case requires the study of "p-summable Fredholm modules" and is one of the motivations for the development of cyclic cohomology theory [Connes 1994]. Such analysis can properly be called *noncommutative differential geometry*.

24.2. Homotopy Invariance of Higher Signatures

A rather surprising application of noncommutative topology is to the (so far partial) solution to the Novikov Conjecture on the homotopy invariance of higher signatures on manifolds.

Let M be an oriented connected compact smooth n-manifold. Then M has a fundamental class $[M] \in H_n(M)$. A number obtained by evaluating a characteristic class in $H^*(M; \mathbb{Q})$ on $[M]$ is called a *characteristic number*. The question then arises which characteristic numbers are (oriented) homotopy invariant.

One particularly important characteristic class is the $\mathbb{L}$-class $\mathbb{L}(M)$. We denote the characteristic number $\langle \mathbb{L}(M), [M] \rangle$ by $\mathrm{Sign}(M)$, the *signature* of M. If $n = 4k$, the Hirzebruch signature theorem says that $\mathrm{Sign}(M)$ is equal to the signature (in the ordinary linear algebra sense) of the symmetric nondegenerate bilinear form on $H^{2k}(M; \mathbb{Q})$ induced by the cup product. This signature is obviously homotopy invariant.

If M is simply connected, then Novikov proved that the signature is the only homotopy-invariant characteristic number. However, if M is not simply connected, there are generally others also.

Let Π be the fundamental group of M, and let $\tilde{M}$ be the universal cover of M. Then there is a continuous map f (well defined up to homotopy), called the *classifying map* of the covering $\tilde{M} \to M$, $f : M \to B\Pi$, where $B\Pi$ is the classifying space of Π (also well defined up to homotopy equivalence). f induces an isomorphism of fundamental groups. f induces a map f^* from $H^*(B\Pi)$ to $H^*(M)$. If $x \in H^*(B\Pi; \mathbb{Q})$, then the cup product $\mathbb{L}(M) \cup f^*(x)$ is a characteristic class of M; the corresponding characteristic number is called a *higher signature*. It can be shown using bordism theory that any homotopy-invariant characteristic number is a higher signature; the question then becomes whether all higher signatures are homotopy invariant.

CONJECTURE 24.2.1 (NOVIKOV CONJECTURE). *All higher signatures are invariant under (oriented) homotopy equivalence. Alternately, $f_*(\mathbb{L}(M) \cap [M])$ is an (oriented) homotopy invariant of M in $H_*(\Pi; \mathbb{Q})$.*

The original Novikov conjecture only applied to elements of $H^*(B\Pi)$ which are products of one-dimensional classes; but all evidence is that the more general conjecture holds.

If Π has torsion, the conjecture would definitely be false if we did not tensor with $\mathbb{Q}$.

The homotopy invariance of the signature has had important consequences in topology; for example, the existence of exotic spheres and partial classification of homotopy $\mathbb{RP}^{4k-1}$'s. The Novikov conjecture (to the extent to which it is true) gives important information about obstructions for maps being homotopy equivalences and homotopy equivalences being homotopic to homeomorphisms (if the fundamental group is "large", there aren't many obstructions.) For example, there is a conjecture of Borel, which is only slightly stronger than the Novikov conjecture, that any two aspherical closed manifolds of the same dimension and the same fundamental group are homeomorphic (and furthermore that any homotopy equivalence is homotopic to a homeomorphism.)

For future reference there is a technical strengthening of the Novikov conjecture. Kasparov defined maps $\alpha : K^*(C^*(\Pi)) \to RK^*(B\Pi)$ and $\beta : RK_*(B\Pi) \to K_*(C^*(\Pi))$ (RK^* is representable K-theory), which are adjoints of each other. The Strong Novikov Conjecture (SNC) is:

CONJECTURE 24.2.2 (STRONG NOVIKOV CONJECTURE). *β is rationally injective (injective after tensoring with $\mathbb{Q}$), or, equivalently, the image of α (after tensoring with $\mathbb{Q}$) is dense in the projective limit topology.*

The fact that the SNC implies the Novikov Conjecture follows from a theorem of Miščenko and Kasparov that the *generalized signature* $\mathrm{Sign}_{C^*(\Pi)}(M) = \langle \mathbb{L}(M) \cup \mathrm{ch}[V], [M] \rangle \in K_0(C^*(\Pi)) \otimes \mathbb{Q}$, where $V = \tilde{M} \times_\Pi C^*(\Pi)$ is the "universal flat $C^*(\Pi)$-vector bundle" over M, is a homotopy invariant (this is proved using surgery theory.) $\mathrm{Sign}_{C^*(\Pi)}(M)$ may be thought of as the "Π-equivariant signature of M." If the SNC holds, to prove the Novikov conjecture it suffices to consider only those elements of $H^*(B\Pi; \mathbb{Q})$ which can be pulled back to $K^*(C^*(\Pi))$; the higher signature corresponding to such an element y can be calculated as $\langle \mathrm{Sign}_{C^*(\Pi)}(M), y \rangle$ using the pairing $K_*(C^*(\Pi)) \times K^*(C^*(\Pi)) \to \mathbb{Z}$ of K-theory and "K-homology" (actually a special case of the intersection product.) The result is obviously homotopy invariant.

The Novikov Conjecture and the SNC may really be thought of as questions about the group Π, since for a fixed Π the conjectures concern homotopy invariance under maps between arbitrary closed manifolds with fundamental group Π. We say "the conjecture holds for Π" if it holds for manifolds with fundamental group Π.

The best results known so far are due to Miščenko, Kasparov, Connes, Gromov, Moscovici, Skandalis, Higson, and Rosenberg:

THEOREM 24.2.3. *Let Π be a finitely presented (discrete) group. Then the SNC holds for Π if any of the following are satisfied:*

(a) [Miščenko 1974] *There is a closed orientable $K(\Pi, 1)$-manifold admitting a metric with non-positive sectional curvatures. (Kasparov [1995] has shown that "closed" may be replaced by "complete").*

(b) [Kasparov 1995] *Π can be embedded as a discrete subgroup of a connected Lie group.*

(c) [Kasparov and Skandalis 1994] *Π is "bolic" (this contains (a) and (b) as special cases, as well as hyperbolic groups [Connes 1994]).*

(d) [Higson and Kasparov 1997] Π *admits an affine, isometric and metrically proper action on a Euclidean space (in the language of Gromov, Π is a-T-menable); in particular, all amenable groups satisfy the SNC.*

(e) [Rosenberg 1983] *The class of Π for which the SNC holds contains all countable solvable groups with torsion-free abelian composition factors, and is closed under extensions by finite groups (in either order) and under free products.*

The proofs of all parts are somewhat similar, using equivariant KK-theory in a fundamental way.

There are versions of this theorem for real K-theory due to Rosenberg [1986a].

C. Ogle has announced a proof of the Novikov conjecture in full generality, but his argument has not been checked at this writing.

[Ferry et al. 1995] and its references contain a much more detailed account of the Novikov Conjecture and its history. There is also a Novikov Conjecture Home Page (www.math.umd.edu/~jmr/NC.html).

24.3. Positive Scalar Curvature

A problem from differential topology which appears on the surface to have even less to do with operator algebras than the Novikov Conjecture is the question of characterizing which closed manifolds admit a Riemannian metric with everywhere positive scalar curvature. To avoid verbosity, we will say by slight abuse of language that such a manifold *has positive scalar curvature.*

If M is a closed 2-manifold, then the Gauss–Bonnet theorem says that M can have positive scalar curvature only if its Euler characteristic is positive, i.e. S^2 and $\mathbb{RP}^2$ are the only possibilities (and both do, in fact, have positive scalar curvature.) The situation becomes far more complicated even in dimension 3.

To simplify the discussion, from now on we will consider only smooth manifolds which are closed and have a spin structure (and are in particular oriented.) Most results can be generalized to the case where M is only oriented and its universal cover $\tilde{M}$ has a spin structure. Such a spin manifold M has a canonically defined Dirac operator D on the spinor bundle; if M is even-dimensional, then the Dirac operator decomposes into D^+ and D^- mapping the positive [resp. negative] half-spinors to the negative [resp. positive] ones. These are Fredholm operators. More generally, if $\Pi = \pi_1(M)$, and B is a unital C^*-algebra, we may define Dirac operators D_V^+ and D_V^- for any flat B-vector bundle V over M (flat means pulled back from $B\Pi$); such a Dirac operator has an index $\operatorname{Ind}_B(D_V^+) \in K_0(B) \otimes \mathbb{Q}$.

There is one necessary condition for positive scalar curvature:

THEOREM 24.3.1. [Rosenberg 1983, 1.1] *Let M be even-dimensional, and let B be any unital C^*-algebra. If M admits a metric with positive scalar curvature, then $\operatorname{Ind}_B(D_V^+) = 0$ for any flat B-vector bundle V. In particular, $\operatorname{Ind}(D^+) = 0$.*

OUTLINE OF PROOF. The key to the proof is to write $D_V^+ D_V^-$ and $D_V^- D_V^+$ as $\nabla^*\nabla + \kappa/4$, where $\nabla^*\nabla$ is a "Laplacian" and κ is the scalar curvature. Since the scalar curvature is everywhere positive, $D_V^+ D_V^-$ and $D_V^- D_V^+$ are one-to-one with dense ranges and bounded inverses. Hence $\mathrm{Ind}_B(D_V^+) = 0$. $\qquad\square$

$\mathrm{Ind}_B(D_V^+)$ can be rewritten using the Miščenko–Fomenko index theorem as $\langle \hat{\mathbf{A}}(M) \cup \mathrm{ch}[V], [M] \rangle$, where $\hat{\mathbf{A}}(M) \in H^*(M; \mathbb{Q})$ is the total $\hat{A}$-class of M (a certain polynomial in the rational Pontrjagin classes.) So the theorem may be rephrased:

COROLLARY 24.3.2. *Under the hypotheses of* 24.3.1, *we have*

$$\langle \hat{\mathbf{A}}(M) \cup \mathrm{ch}[V], [M] \rangle = 0.$$

As a special case, one obtains a theorem of Lichnerowicz: if M has positive scalar curvature, then the $\hat{A}$-genus $\hat{A}(M) = \langle \hat{\mathbf{A}}(M), [M] \rangle$ is 0.

The analogy of this result with that of Miščenko–Kasparov on the generalized signature (24.2) suggests the following version of a conjecture of Gromov and Lawson:

CONJECTURE 24.3.3. *If M has positive scalar curvature, then for all $x \in H^*(B\Pi; \mathbb{Q})$, the higher $\hat{A}$-genus $\langle \hat{\mathbf{A}}(M) \cup f^*(x), [M] \rangle$ is 0 (where $f : M \to B\Pi$ is the classifying map.) Equivalently, $f_*(\hat{\mathbf{A}}(M) \cap [M]) = 0$.*

Just as in the case of the Novikov Conjecture, 24.3.3 would follow from 24.3.2 and the fact that Kasparov's map β is rationally injective. Thus the SNC implies 24.3.3. So we have the following corollary of 24.2.3 and 24.3.2:

THEOREM 24.3.4. *Let Π be a finitely presented group. Then all higher $\hat{A}$-genera for M vanish, for any M with a spin structure, positive scalar curvature, and fundamental group Π, if*

(a) *there exists an orientable $K(\Pi, 1)$-manifold admitting a complete metric with non-positive sectional curvatures*

or

(b) *Π can be embedded as a discrete subgroup in a connected Lie group.*

One important consequence is the following:

COROLLARY 24.3.5. *Let Π be a group for which the SNC holds, and let N be a closed spin manifold for which there is a map $g : N \to B\Pi$ with $g_*[N] \neq 0$ in $H_*(B\Pi; \mathbb{Q})$. Then N does not have positive scalar curvature. In particular, no $K(\Pi, 1)$-manifold has positive scalar curvature.*

The finer version of the conjecture of Gromov and Lawson says that a spin manifold M has positive scalar curvature if and only if the image of the spin bordism class of M in the *real* K-theory group $KO_n(B\Pi)$ ($n = \dim(M)$) vanishes. Conjecture 24.3.3 is roughly the Gromov–Lawson conjecture modulo torsion. Gromov–Lawson, Rosenberg, and Miyazaki have obtained partial results on the

G-L conjecture; Rosenberg has shown that the conjecture must be modified if Π has torsion. Some of Rosenberg's work requires use of real KK-theory.

The reader interested in a more detailed account of the positive scalar curvature problem is urged to read the papers of Rosenberg [1983; 1986b; 1986a]. There is a recent survey article [Rosenberg and Stolz 1998].

24.4. The Baum–Connes Conjecture

Let G be a Lie group with countably many components (e.g. a connected Lie group or a countable discrete group), and let M be a smooth manifold (not necessarily connected) on which G acts smoothly. Then the analytic equivariant K-theory of M can be defined as $K_*(C^*(G, C_0(M)))$ or $K_*(C_r^*(G, C_0(M)))$. It is desirable to construct a corresponding topological equivariant K-theory $K^*(G, M)$. For compact G this was done in Section 11.

An approach by Baum and Connes [1982] reduces the study of general actions to proper actions. The cocycles in the Baum–Connes theory are triples (Z, σ, f), where Z is a smooth manifold with a proper G-action, $f : Z \to M$ a G-equivariant smooth submersion, and σ is a G-equivariant symbol along the fibers of f. There is a map $\mu : K^i(G, M) \to K_i(C_r^*(G, C_0(M))$, called the *(reduced) assembly map* : on each fiber σ gives an elliptic operator D_x, and $\mu(Z, \sigma, f)$ is the index of the family (D_x). In KK-terms, σ gives an element of $KK(\mathbb{C}, C^*(G, C_0(V))$, where V is the subbundle of TZ of vectors tangent to the fibers of f. The map $\phi : V \to M$ is K-oriented, so there is an element $\phi! \in KK_G(C_0(V), C_0(M))$. $\mu(Z, \sigma, f) = [\sigma] \hat{\otimes}_{C^*(G, C_0(V))} j_G(\phi!)$.

CONJECTURE 24.4.1. $\mu : K^i(G, M) \to K_i(C_r^*(G, C_0(M))$ *is always an isomorphism.*

Most of the previous index theorems are special cases. For example, Atiyah–Singer is (essentially) the case $G = \{e\}$, $M = \{\text{pt}\}$.

$K^i(G, M)$ is usually easier to compute than $K_i(C_r^*(G, C_0(M)))$; there is machinery from algebraic topology which can be applied.

Consider the case where G is discrete and M is a point. Let $\{\gamma_1, \gamma_2, \ldots\}$ be a complete set of representatives for the conjugacy classes of elements of G of finite order. For each γ_i, let $Z(\gamma_i)$ be the centralizer in G. Then the Chern character gives a map ch : $K^*(G, \cdot) \to \oplus_i H_*(Z(\gamma_i); \mathbb{C})$.

THEOREM 24.4.2. [Baum and Connes 1982]

$$\text{ch} : K^*(G, \cdot) \otimes_{\mathbb{Z}} \mathbb{C} \to \oplus_i H_*(Z(\gamma_i); \mathbb{C})$$

is an isomorphism.

So in this case, the conjecture becomes:

CONJECTURE 24.4.3. $\mu : \oplus_i H_*(Z(\gamma_i); \mathbb{C}) \to K_*(C_r^*(G)) \otimes_{\mathbb{Z}} \mathbb{C}$ *is an isomorphism.*

The identity of G is one of the γ_i, and the centralizer in this case is all of G. μ restricted to this summand is (essentially) Kasparov's map β (24.2). So injectivity of μ implies the Strong Novikov Conjecture (and is equivalent if G is torsion-free.) Injectivity of μ is known if G can be embedded as a discrete subgroup of a connected Lie group. In general, injectivity of μ is relevant to problems in topology, and surjectivity is relevant to problems in C^*-algebras.

The Baum–Connes Conjecture implies the *Generalized Kadison Conjecture* that $C_r^*(G)$ is projectionless whenever G is a torsion-free discrete group (cf. 10.8.2.) There are many other structure questions about $C_r^*(G)$ which would be answered as a consequence of the conjecture.

The conjecture is known to be true in many cases, e.g. for a-T-menable groups (24.2.3(d)) [Higson and Kasparov 1997]; see also [Julg and Kasparov 1995; Guentner et al. 1997]. The conjecture can be stated for groupoids, where similar results hold [Tu 1998].

24.5. KK-Theoretic Proofs

There is one general line of argument using KK-theory which, with relatively minor variations, can be used to prove most of the results described above. The argument in this (approximate) form is due to Kasparov; the specific exposition given here is based on lecture notes of J. Rosenberg.

Throughout this section, we let M be a closed smooth even-dimensional manifold (if $n = \dim M$ is odd, M can be replaced by $M \times \mathbb{T}^n$.) Fix a Riemannian metric and a smooth measure on M. The cotangent bundle T^*M has a canonical almost-complex structure. Let $\pi : T^*M \to M$ be the bundle projection, $\omega : M \to +$ the collapse map ($+$ is a one-point space), and $\Delta : M \to M \times M$ the diagonal map. We will also fix a separable unital C^*-algebra B, called the *auxiliary algebra*.

As a special case of the topological Thom isomorphism (19.9.3) there are elements $\boldsymbol{x} \in KK(\Gamma(\mathrm{Cliff}(T^*M)), C_0(T^*M))$ and $\boldsymbol{y} \in KK(C_0(T^*M), \Gamma(\mathrm{Cliff}(T^*M)))$ which define a KK-equivalence. If M has a spin^c-structure, $\boldsymbol{x}$ and $\boldsymbol{y}$ may be regarded as elements of $KK(C(M), C_0(T^*M))$ and $KK(C_0(T^*M), C(M))$ respectively.

We now introduce the Dirac operator and the $\bar{\partial}$-operator. The $\bar{\partial}$-operator is defined as in 17.1.2(f), giving an element $[\bar{\partial}] = [\bar{\partial}_M] \in KK(C_0(T^*M), \mathbb{C})$. If M has a spin^c-structure, then there is a true Dirac operator D on M which defines a class in $KK(C(M), \mathbb{C})$. In the general situation, we may still define a "Dirac operator" as an elliptic pseudodifferential operator T with principal symbol $\sigma_D(x, \xi) = i\rho(\xi/\|\xi\|)$, with ρ right Clifford multiplication on $\mathrm{Cliff}(T^*M)$. Then $(L^2(\mathrm{Cliff}^{\mathrm{even}}(T^*M)) \oplus L^2(\mathrm{Cliff}^{\mathrm{odd}}(T^*M)), \lambda, T)$ is a Kasparov $(\Gamma(\mathrm{Cliff}(T^*M)), \mathbb{C})$-module whose class in $KK(\Gamma(\mathrm{Cliff}(T^*M)), \mathbb{C})$ is denoted $[D]$. (D is actually the unbounded operator associated to T as in 17.11.4. D is only determined up to operator homotopy.)

LEMMA 24.5.1. $\boldsymbol{x} \,\hat{\otimes}_{C_0(T^*M)}\, [\bar{\partial}] = [D]$ *and* $\boldsymbol{y} \,\hat{\otimes}_{\Gamma(\mathrm{Cliff}(T^*M))}\, [D] = [\bar{\partial}]$.

The proof is a straightforward calculation.

Thus it is sufficient to define either the $\bar{\partial}$-operator or the Dirac operator; in fact, both constructions are really the same thing from different points of view.

If P is an elliptic pseudodifferential operator between smooth B-vector bundles over M, then P defines a class $[P] \in KK(C(M), B)$ by $(L^2(E^{(0)}) \oplus L^2(E^{(1)})$, $\mu, P)$, where μ is the action of $C(M)$ by pointwise multiplication. We define $\mathrm{Ind}_B^a(P)$ to be $\omega_*([P]) \in KK(\mathbb{C}, B) \cong K_0(B)$. $\mathrm{Ind}_B^a(P)$ is given by the same module as $[P]$ but with the action of $C(M)$ suppressed.

The symbol σ_P defines an element $[[\sigma_P]] \in KK(C(M), C_0(T^*M) \otimes B)$ by $(\Gamma_0(\pi^*E^{(0)}) \oplus \Gamma_0(\pi^*E^{(1)}), \mu, \sigma_P)$; we also have an element $[\sigma_P] = \omega_*([[\sigma_P]]) \in KK(\mathbb{C}, C_0(T^*M) \otimes B)$ obtained by forgetting the action of $C(M)$.

We also have an element $z \in KK(C(M) \otimes C_0(T^*M), C_0(T^*M))$ induced by $\pi \times 1 : T^*M \to M \times T^*M$.

LEMMA 24.5.2. $[[\sigma_P]] = [\sigma_P] \hat{\otimes}_{C_0(T^*M)} z$.

In the special case that P is the "Dirac operator" D_E with coefficients in a B-vector bundle E, defined in the usual way using a connection, we write $[[E]]$ for the element of $KK(C(M), C(M) \otimes B)$ defined by the module $(\Gamma(E) \oplus 0, \mu, 0)$, and $[E] = \omega_*([[E]])$ the same module without $C(M)$-action. Then, just as in Lemma 24.5.2, we have

LEMMA 24.5.3. (a) $[[E]] = [E] \hat{\otimes}_{C(M)} w$, where w is the class of Δ in the group $KK(C(M) \otimes C(M), C(M))$.
(b) $[D_E] = [[E]] \hat{\otimes}_{C(M)} [D]$ in $KK(C(M), B)$.

LEMMA 24.5.4. $[D_E] = [[\sigma_{D_E}]] \hat{\otimes}_{C_0(T^*M)} [\bar{\partial}]$.

PROOF. We have $[[\sigma_{D_E}]] \hat{\otimes}_{C_0(T^*M)} y = [[E]]$, so the result reduces to 24.5.3. $\square$

The following theorem can be regarded as the "Index Theorem":

THEOREM 24.5.5. *If P is an elliptic pseudodifferential operator between smooth B-vector bundles over M, then $[P] = [\sigma_P] \hat{\otimes}_{C_0(T^*M)} [\bar{\partial}]$.*

SKETCH OF PROOF. Reduce from a general pseudodifferential operator to Dirac operators, using the fact that there is a Thom isomorphism between $K^*(T^*M)$ and $K^*(M)$, and thus the classes $[\sigma_{D_E}]$ generate all of $K_0(C_0(T^*M) \otimes B)$. Then apply 24.5.4. $\square$

To prove the Miščenko–Fomenko index theorem, apply ω_* to both sides and use the fact that $\mathrm{Td}(T^*M)$ is the image under Poincaré duality and Chern character of $[\bar{\partial}]$ (this could be taken as the definition).

For the case of an elliptic operator invariant under a group G, as in 24.1.5, one replaces KK by KK_G everywhere. The index is then an element of $K_0^G(B)$. One can push this index into $K_0(C_r^*(G, B))$ using j_G of 20.6.2 and the canonical element $[c] \in K_0(C_0(M) \rtimes_\alpha G)$ defined using a cutoff function [Kasparov 1984a, §4].

To prove the Strong Novikov Conjecture in the case that Π is a discrete subgroup of a connected Lie group G, first reduce by standard arguments to the case where Π is torsion-free. In this case, we can let $B\Pi = \Pi \backslash G / H$, where H is a maximal compact subgroup of G. Then $M = G/H$ is a universal cover for $B\Pi$. If there is a complete orientable $K(\Pi, 1)$-manifold admitting a metric with nonpositive sectional curvatures, we may take this manifold for $B\Pi$; then let M be a universal cover for $B\Pi$. In either case the statement that β is rationally injective can be rephrased using Poincaré duality as the statement that right intersection product by $j_\Pi([\bar{\partial}_M])$ is rationally injective. This follows from 20.7.2, which says that $[\bar{\partial}_M] \in KK_\Pi(\mathbb{C}, C_0(M))$ has a right inverse. So in these cases β is actually injective. (If Π has torsion in the first case, then β may only be rationally injective.)

Some of the natural approaches to the Baum–Connes conjecture involve cyclic cohomology. But that is a subject for another book [Connes 1994].

25. E-Theory

While KK-theory is enormously useful, it has at least two defects: it is technically difficult and does not have exact sequences in general. Both features are remedied by E-theory. E-theory was developed abstractly in categorical terms by Higson as a universal enveloping category of $\mathbf{SC}^*$ with the same abstract properties as KK and for which the "excision" map e of 19.5.5 is an isomorphism for every exact sequence (it follows then that $E(A, \cdot)$ has six-term exact sequences for every separable A). Connes and Higson then found a concrete realization of E-theory, which has the additional advantage that the technical problems inherent in KK are greatly reduced, in addition to being a natural realization which has immediate applicability.

In this section, we cover the basics of E-theory. There has so far been no definitive treatment; the original article [Connes and Higson 1990] and the book [Connes 1994] are sketchy on some details. The most detailed treatment so far is [Samuel 1997], but its coverage of some of the technicalities is also inadequate. [The reader comparing our exposition to [Connes 1994] should keep in mind that our definition of the mapping cone of a homomorphism (19.4.1) differs in that the algebra A is placed at 0 rather than 1. We have retained our definition from earlier chapters since I believe it makes the arguments slightly simpler and more natural.] Recently, a detailed exposition including equivariant E-theory has become available [Guentner et al. 1997], using a different approach to some technical parts of the theory including construction of the product (cf. 25.7.4).

We will not attempt to describe applications of E-theory, not because they are insignificant—in fact, E-theory has made possible some of the most important applications of operator K-theory—but because most of the applications are nicely discussed in [Connes 1994]. Perhaps the best application so far, though, is in Kirchberg's remarkable classification theorem for purely infinite separable nuclear simple C^*-algebras [Kirchberg 1998] (in part proved independently by N. C. Phillips). The tools of asymptotic morphisms and E-theory play an essential role in the proof.

25.1. Asymptotic Morphisms

DEFINITION 25.1.1. Let A and B be C^*-algebras. An *asymptotic morphism* from A to B is a family $\langle \phi_t \rangle (t \in [1, \infty))$ of maps from A to B with the following properties:

(i) $t \to \phi_t(a)$ is continuous for every $a \in A$

(ii) The set $\langle \phi_t \rangle$ is asymptotically *-linear and multiplicative:

$$\lim_{t \to \infty} \| \phi_t(a + b) - (\phi_t(a) + \phi_t(b)) \| = 0$$

for all $a, b \in A$, etc.

Two asymptotic morphisms $\langle \phi_t \rangle$ and $\langle \psi_t \rangle$ are (*strictly*) *equivalent* if

$$\lim_{t \to \infty} \| \phi_t(a) - \psi_t(a) \| = 0$$

for all $a \in A$.

A *homotopy* between asymptotic morphisms $\langle \phi_t^{(0)} \rangle$ and $\langle \phi_t^{(1)} \rangle$ from A to B is an asymptotic morphism $\langle \phi_t \rangle$ from A to $C([0, 1], B)$ such that $[\phi_t(a)](s) = \phi_t^{(s)}(a)$ for $s = 0, 1$, all t, and all a.

Denote the set of homotopy classes of asymptotic morphisms from A to B by $[[A, B]]$.

EXAMPLES/REMARKS 25.1.2. (a) A *-homomorphism ϕ from A to B defines an asymptotic morphism by setting $\phi_t = \phi$ for all t. We will, by slight abuse of terminology, consider ordinary *-homomorphisms to be asymptotic morphisms in this way. A homotopy of homomorphisms gives a homotopy of the corresponding asymptotic morphisms, i.e. there is a natural map from the set $[A, B]$ of homotopy classes of *-homomorphisms from A to B, to $[[A, B]]$. This map is far from surjective in general, and can also fail to be injective. More generally, a (point-norm continuous) path $\langle \phi_t \rangle$ of *-homomorphisms from A to B defines an asymptotic morphism from A to B, which is homotopic to the homomorphism ϕ_1 (or to ϕ_{t_0} for any t_0).

(b) The choice of the interval $[1, \infty)$ as the domain of t is arbitrary. We could have just as easily taken $[t_0, \infty)$ for any t_0. We could also instead (or in addition) require that $\phi_1(a) = 0$ for all a; then by replacing t by t^{-1} we may take the index set to be $(0, 1]$ with the requirement that $\phi_1 = 0$ and that ϕ_t becomes asymptotically *-linear and multiplicative as $t \to 0$.

(c) A *deformation* from A to B is a continuous field $\langle B(t) \rangle$ of C^*-algebras over $[0, 1]$ such that $B(0) \cong A$, $B(t) \cong B$ for $0 < t \leq 1$, and such that the field restricted to $(0, 1]$ is isomorphic to the constant field. [Some references such as [Connes 1994] place the A at 1 rather than at 0; we have put it at 0 since

it seems more natural and is consistent with our definition of mapping cones.]
Every deformation from A to B defines an asymptotic morphism from A to B
as in (b). However, not every asymptotic morphism from A to B comes from a
deformation, even a deformation of a quotient of A to B.

(d) If A and B are unital, then an asymptotic morphism $\langle \phi_t \rangle$ from A to B is
unital if $\phi_t(1_A) = 1_B$ for all t. If A and B are arbitrary and $\langle \phi_t \rangle$ is any asymptotic morphism from A to B, then ϕ extends canonically to a unital asymptotic
morphism $\langle \tilde{\phi}_t \rangle$ from A^+ to $\tilde{B}$ by the formula $\tilde{\phi}_t(x + \lambda 1) = \phi_t(x) + \lambda 1$. (It is easy
to verify that this gives an asymptotic morphism.)

(e) An asymptotic morphism ϕ from A to B applied coordinatewise gives an
asymptotic morphism $\phi^{(n)}$ from $M_n(A)$ to $M_n(B)$, for any n. [This is a special
case of a tensor product, discussed in 25.2.]

(f) If ϕ is an asymptotic morphism from A to B, and p is a projection in A,
then for sufficiently large t, $\phi_t(p)$ is close to a projection q in B as in 4.5.1; the
class of q in $\mathrm{Proj}(B)$ is independent of the choices made, and depends only on
the class of p in $\mathrm{Proj}(A)$. Passing to matrix algebras as in (e), we get an induced
homomorphism $\phi_* : V(A) \to V(B)$. With some additional preliminaries, we can
extend this to get maps $\phi_* : K_i(A) \to K_i(B)$ for $i = 0, 1$ (25.1.6).

(g) Equivalent asymptotic morphisms are homotopic via the "straight line" in
between, i.e. $\phi_t^{(s)}(a) = s\phi_t(a) + (1 - s)\psi_t(a)$ for $0 \le s \le 1$.

(h) A *reparametrization* of an asymptotic morphism $\langle \phi_t \rangle$ is an asymptotic morphism $\langle \psi_t \rangle$, where $\psi_t = \phi_{r(t)}$ for some continuous function $r : [1, \infty) \to [1, \infty)$
with $\lim_{t \to \infty} r(t) = \infty$. (We usually only consider reparametrizations where r is
increasing, but this is not necessary.) An asymptotic morphism is homotopic to
any reparametrization via the reparametrizations $r_s(t) = st + (1 - s)r(t)$.

PROPOSITION 25.1.3. *If A and B are C^*-algebras and $\langle \phi_t \rangle$ is an asymptotic
morphism from A to B, then for each $a \in A$ the function $t \to \phi_t(a)$ is norm-
bounded; in fact, $\limsup_t \|\phi_t(a)\| \le \|a\|$.*

PROOF. By 25.1.2(d) we may assume A, B, and ϕ are unital. Since, for any
$\lambda \in \mathbb{C}$, $\phi_t(\lambda a) \approx \lambda\phi_t(a)$ for t large, we may assume $\|a\| = 1$. Then $1 - a^*a = x^*x$
for some $x \in A$, so for any $\varepsilon > 0$ we have for all sufficiently large t an element
$b_t \in B$ with $b_t = b_t^*$ and $\|b_t\| < \varepsilon$, such that $\phi_t(a)^*\phi_t(a) = 1_B - \phi_t(x)^*\phi_t(x) + b_t \le
(1 + \varepsilon)1_B$. Thus $\limsup_t \|\phi_t(a)\| \le 1$.   □

25.1.4. (a) In light of 25.1.3, an asymptotic morphism from A to B defines a
*-homomorphism from A to $B_\infty = C_b([1, \infty), B)/C_0([1, \infty), B)$. Two asymptotic
morphisms define the same *-homomorphism if and only if they are equivalent.
Conversely, if $\phi : A \to B_\infty$ is a *-homomorphism, any set-theoretic cross section
for ϕ is an asymptotic morphism from A to B, and any two such are equivalent.

(b) With this observation, we may view an asymptotic morphism as a generalized mapping cone. An asymptotic morphism ϕ, indexed by $(0, 1]$ as in 25.1.2(b)

with $\phi_1 = 0$, may be thought of as an extension $0 \to SB \to E \to A \to 0$. If ϕ comes from an actual homomorphism ψ (i.e. $\lim_{t\to 0} \phi_t(a) = \psi(a)$ for all a), then E is just the mapping cone of ψ and the exact sequence is the same as in 19.4.1. [Conversely, if B is unital, then any extension of A by SB which is "concentrated at 0" gives an asymptotic morphism from A to B via the Busby invariant, since in this case $M(SB) = C_b((0,1), B)$ by 12.1.1(c).]

(c) *Caution:* If $\phi^{(0)}$ and $\phi^{(1)}$ are homotopic homomorphisms from A to B_∞ via a homotopy $\phi^{(s)}$, i.e. $\phi^{(s)}$ is given by a *-homomorphism from A into $C([0,1], B_\infty)$, then the $\phi^{(s)}$ define a homotopy between the corresponding asymptotic morphisms (or more properly, between representatives of the corresponding equivalence classes of asymptotic morphisms). But a homotopy between asymptotic morphisms does not in general induce a homotopy between the corresponding *-homomorphisms into B_∞: there is a natural inclusion of $C([0,1], B_\infty)$ into $C([0,1], B)_\infty$ induced by the obvious inclusion of $C([0,1], C_b([1, \infty), B))$ into $C_b([1, \infty), C([0,1], B))$, but these inclusions are never surjective (even if $B = \mathbb{C}$). Said another way, if f is a continuous function from $[1, \infty)$ to $C([0,1], B)$, the corresponding function $\tilde{f} : [1, \infty) \times [0,1] \to B$ is only uniformly continuous on compact subsets of $[1, \infty) \times [0,1]$ in general, not on the whole set. Thus homotopy of asymptotic morphisms is a more general notion.

25.1.5. 25.1.4 also allows us to replace a given asymptotic morphism with an equivalent one with special properties in a number of ways:

(a) By elementary linear algebra, a *-homomorphism from A to B_∞ always has a *-linear (not necessarily bounded) lifting to $C_b([1, \infty), B)$. Thus every asymptotic morphism from A to B is equivalent to an asymptotic morphism $\langle \phi_t \rangle$, where each ϕ_t is *-linear (but not necessarily bounded).

(b) By the Bartle–Graves Selection Theorem [Bessaga and Pełczyński 1975], a *-homomorphism from A to B_∞ always has a continuous (not necessarily linear) lifting to $C_b([1, \infty), B)$. Thus every asymptotic morphism from A to B is equivalent to an asymptotic morphism $\langle \phi_t \rangle$, where each ϕ_t is continuous (but not necessarily linear), in fact $(t, a) \to \phi_t(a)$ is a continuous function from $[1, \infty) \times A$ to B, and such that ϕ_t becomes asymptotically *-linear and multiplicative uniformly on compact sets as $t \to \infty$, i.e. for every $\varepsilon > 0$ and compact $K \subseteq A$ there is a t_0 such that the following inequalities are satisfied for all $t \geq t_0$ and for all $x, y \in K$ and all $\lambda \in \mathbb{C}$ with $|\lambda| \leq 1$:

$$\|\phi_t(x) + \lambda\phi_t(y) - \phi_t(x + \lambda y)\| < \varepsilon,$$
$$\|\phi_t(x)\phi_t(y) - \phi_t(xy)\| < \varepsilon,$$
$$\|\phi_t(x)^* - \phi_t(x^*)\| < \varepsilon,$$
$$\|\phi_t(x)\| < \|x\| + \varepsilon.$$

It is a routine exercise to show that an asymptotic morphism coming from a continuous map from A to $C_b([1, \infty), B))$ satisfies these conditions.

An asymptotic morphism satisfying all these conditions (i.e. coming from a continuous map from A to $C_b([1, \infty), B)$) will be called *uniform*.

Two equivalent uniform asymptotic morphisms $\langle \phi_t \rangle$ and $\langle \psi_t \rangle$ are uniformly equivalent in the sense that for any $\varepsilon > 0$ and compact $K \subseteq A$ there is a t_0 such that $\|\phi_t(x) - \psi_t(x)\| < \varepsilon$ for all $t \geq t_0$ and all $x \in K$.

By the same argument applied to $C([0, 1], B)_\infty$, if $\langle \phi_t^{(0)} \rangle$ and $\langle \phi_t^{(1)} \rangle$ are homotopic asymptotic morphisms, then the $\phi_t^{(i)}$ are equivalent to uniform asymptotic morphisms $\psi_t^{(i)}$ which are homotopic via a path $\psi_t^{(s)} (0 \leq s \leq 1)$ which are all uniform and which even vary uniformly in s on compact sets.

(c) If A is nuclear, then a $*$-homomorphism from A to B_∞ has a completely positive contractive lifting to $C_b([1, \infty), B)$, for any B. Thus every asymptotic morphism from A to any B is equivalent to an asymptotic morphism $\langle \phi_t \rangle$, where each ϕ_t is a completely positive linear contraction (such an asymptotic morphism is called a *completely positive asymptotic morphism*). A completely positive asymptotic morphism is automatically uniform.

25.1.6. Using 25.1.5(b), we may extend the idea of 25.1.2(f). If ϕ is a unital asymptotic morphism from A to B and x is invertible in A, then $\phi_t(x)$ is invertible in B for sufficiently large t, and its (path) component in $GL_1(B)$ is independent of t and depends only on the (path) component of x in $GL_1(A)$. If ϕ is an arbitrary asymptotic morphism between not necessarily unital A and B, then by 25.1.2(d) and (e) ϕ induces a homomorphism $\phi_* : K_1(A) \to K_1(B)$. By suspension, a map $\phi_* : K_0(A) \cong K_1(SA) \to K_1(SB) \cong K_0(B)$ is obtained. It is routine to check that this map agrees with the map induced from the $\phi_* : V(A) \to V(B)$ of 25.1.2(f) if A and B are unital.

There is an even closer connection between asymptotic morphisms and K-theory (25.1.8, 25.4.5).

PROPOSITION 25.1.7. *Let A and B be C^*-algebras, with A semiprojective (4.7.1). Then every asymptotic morphism from A to B is homotopic to a $*$-homomorphism. The natural map from $[A, B]$ to $[[A, B]]$ is a bijection.*

PROOF. Let J_n be the ideal of elements of $C_b([1, \infty), B)$ which vanish on $[n, \infty)$. Then $[\cup J_n]^- = C_0([1, \infty), B)$. Given an asymptotic morphism $\langle \phi_t \rangle$ from A to B, the corresponding $*$-homomorphism from A to $C_b([1, \infty), B)/C_0([1, \infty), B)$ lifts to a $*$-homomorphism from A to $C_b([1, \infty), B)/J_n$, which can be naturally identified with $C_b([n, \infty), B)$, for some n, i.e. with an asymptotic morphism $\langle \psi_t \rangle$ equivalent to $\langle \phi_t \rangle$, where each ψ_t is a $*$-homomorphism from A to B. (Set $\psi_t = \psi_n$ for $t < n$.) But then $\langle \psi_t \rangle$ is homotopic to the constant asymptotic morphism $\langle \psi_n \rangle$ in an evident way. This shows that the map from $[A, B]$ to $[[A, B]]$ is surjective. But the same procedure can be used on maps from A to $C([0, 1], B)$ to show that if two homomorphisms are homotopic as asymptotic morphisms, they are homotopic as homomorphisms, also proving injectivity. $\square$

COROLLARY 25.1.8. *For any C^*-algebra B, $[[\mathbb{C}, B \otimes \mathbb{K}]] \cong V(B)$ (at least as a set); $[[S, B \otimes \mathbb{K}]] \cong K_1(B)$, and hence $[[S, SB \otimes \mathbb{K}]] \cong K_0(B)$.*

PROOF. Combine 25.1.7 with 4.7.1(e) and 9.4.4. □

25.2. Tensor Products and Suspensions

25.2.1. One must be slightly careful in defining the tensor product of two asymptotic morphisms. Suppose $\langle \phi_t \rangle$ is an asymptotic morphism from A to B, and $\langle \psi_t \rangle$ from C to D. Then $\langle \phi_t \otimes 1 \rangle$ and $\langle 1 \otimes \psi_t \rangle$ define asymptotic morphisms from A and B, respectively, to $\tilde{C} \otimes_{\max} \tilde{D}$ and hence homomorphisms to $(\tilde{C} \otimes_{\max} \tilde{D})_\infty$. The images commute, so there is an induced homomorphism from $A \otimes_{\max} B$ to $(\tilde{C} \otimes_{\max} \tilde{D})_\infty$. The image is easily seen to lie in $(C \otimes_{\max} D)_\infty$, which sits naturally as an ideal in $(\tilde{C} \otimes_{\max} \tilde{D})_\infty$; hence there is a well-defined (up to equivalence) asymptotic morphism $\phi \otimes \psi$ from $A \otimes_{\max} B$ to $C \otimes_{\max} D$. The class of $\phi \otimes \psi$ in $[[A \otimes_{\max} B, C \otimes_{\max} D]]$ depends only on the classes of ϕ and ψ in $[[A, C]]$ and $[[B, D]]$.

However, there is a difficulty in doing the same construction for the minimal cross norm. One obtains in the same way commuting homomorphisms from $A \otimes_{\max} B$ into $(\tilde{C} \otimes \tilde{D})_\infty$ and hence an asymptotic morphism from $A \otimes_{\max} B$ to $C \otimes D$, but it is not obvious in general that it factors through $A \otimes B$. Of course, if A or B is nuclear, there is no problem.

25.2.2. In particular, if ϕ is an asymptotic morphism from A to B, there is a well-defined suspension $S\phi = id_S \otimes \phi$ from $SA = S \otimes A$ to $SB = S \otimes B$; so there is a suspension map from $[[A, B]]$ to $[[SA, SB]]$.

25.2.3. In fact, one can define tensor product morphisms more generally. If $\langle \phi_t \rangle$ and $\langle \psi_t \rangle$ are asymptotic morphisms from A and B, respectively, into C, and if $\lim_{t \to \infty}[\phi_t(a), \psi_t(b)] = 0$ for all $a \in A$, $b \in B$, there are induced commuting homomorphisms from A and B to C_∞ which induce an asymptotic morphism from $A \otimes_{\max} B$ to C. The same construction works if $\langle \phi_t \rangle$ and $\langle \psi_t \rangle$ are asymptotic morphisms into $M(C)$ such that $\phi_t(a)\psi_t(b) \in C$ for all a, b, t. Even more generality is possible: see, for example, the asymptotic morphism constructed in 25.5.1.

Caution: If $\phi^{(0)}$ and $\phi^{(1)}$ are asymptotic morphisms from A to C which both asymptotically commute with the asymptotic morphism ψ from B to C, so that $\phi^{(0)} \otimes \psi$ and $\phi^{(1)} \otimes \psi$ are defined, and $[\phi^{(0)}] = [\phi^{(1)}]$ in $[[A, C]]$, it is not necessarily true that $[\phi^{(0)} \otimes \psi] = [\phi^{(1)} \otimes \psi]$ in $[[A \otimes_{\max} B, C]]$, because there may not be a path $\phi^{(s)}$ all asymptotically commuting with ψ.

25.3. Composition

We want a way of composing asymptotic morphisms which generalizes composition of ordinary morphisms. Unfortunately, if $\langle \phi_t \rangle$ and $\langle \psi_t \rangle$ are asymptotic morphisms from A to B and B to C, respectively, then $\langle \psi_t \circ \phi_t \rangle$ is not in general an asymptotic morphism from A to C (although it is if one of the asymptotic morphisms is an actual homomorphism). But the problem can be circumvented by reparametrization:

THEOREM 25.3.1. (a) *If A, B, and C are separable C^*-algebras, and $\langle \phi_t \rangle$ and $\langle \psi_t \rangle$ are uniform asymptotic morphisms from A to B and B to C, respectively, then for any increasing r growing sufficiently quickly the family $\langle \psi_{r(t)} \circ \phi_t \rangle$ is an asymptotic morphism from A to C.*

(b) *The resulting asymptotic morphism depends up to homotopy only on the homotopy classes of $\langle \phi_t \rangle$ and $\langle \psi_t \rangle$, and thus defines a "composition" $[[A, B]] \times [[B, C]] \to [[A, C]]$.*

(c) *Composition is associative, commutes with tensor products, and agrees with ordinary composition for homomorphisms.*

PROOF. (a) Let A_0 be a dense σ-compact *-subalgebra of A (e.g. the polynomials in a countable generating set and their adjoints). Write $A_0 = \cup K_n$, where K_n is compact, $K_n + K_n \subseteq K_{n+1}$, $K_n K_n \subseteq K_{n+1}$, $K_n^* = K_n$, and $\lambda K_n \subseteq K_{n+1}$ for $|\lambda| \leq n$. Inductively choose $t_n \geq t_{n-1}$, $t_n \geq n$, such that ϕ_t satisfies the conditions of 25.1.4(b) with $\varepsilon = 1/n$ and $K = K_n$, for all $t \geq t_n$. Let $K_n' = \{\phi_t(a) : a \in K_n, t \leq t_{n+1}\}$. K_n' is a compact subset of B. Let $K_1'' = K_1'$, and inductively let $K_{n+1}'' = K_{n+1}' \cup (K_n'' + K_n'') \cup K_n'' K_n'' \cup \{\lambda K_n'' : |\lambda| \leq n\}$. Choose r_n so that ψ_t satisfies the conditions of 25.1.4(b) with $\varepsilon = 1/n$ and $K = K_{n+2}''$. Then let $r(t)$ be any increasing function with $r(t_n) \geq r_n$ for all n. If $x, y \in A_0$, then $x, y \in K_n$ for some n. Then, for $m > n$ and t between t_m and t_{m+1}, we have

$$\left\| \psi_{r(t)}(\phi_t(x+y)) - [\psi_{r(t)}(\phi_t(x)) + \psi_{r(t)}(\phi_t(y))] \right\|$$

$$\leq \left\| \psi_{r(t)}(\phi_t(x+y)) - \psi_{r(t)}(\phi_t(x) + \phi_t(y)) \right\|$$

$$+ \left\| \psi_{r(t)}(\phi_t(x) + \phi_t(y)) - [\psi_{r(t)}(\phi_t(x)) + \psi_{r(t)}(\phi_t(y))] \right\|$$

$$\leq \left\| \psi_{r(t)}([\phi_t(x) + \phi_t(y)] + [\phi_t(x+y) - \phi_t(x) - \phi_t(y)]) \right.$$

$$- [\psi_{r(t)}(\phi_t(x) + \phi_t(y)) + \psi_{r(t)}(\phi_t(x+y) - \phi_t(x) - \phi_t(y))]$$

$$+ [\psi_{r(t)}(\phi_t(x) + \phi_t(y)) + \psi_{r(t)}(\phi_t(x+y) - \phi_t(x) - \phi_t(y))]$$

$$\left. - \psi_{r(t)}(\phi_t(x) + \phi_t(y)) \right\| + \tfrac{1}{m}$$

$$\leq \left\| \psi_{r(t)}([\phi_t(x) + \phi_t(y)] + [\phi_t(x+y) - \phi_t(x) - \phi_t(y)]) \right.$$

$$\left. - [\psi_{r(t)}(\phi_t(x) + \phi_t(y)) + \psi_{r(t)}(\phi_t(x+y) - \phi_t(x) - \phi_t(y))] \right\|$$

$$+ \left\| \psi_{r(t)}(\phi_t(x+y) - \phi_t(x) - \phi_t(y)) \right\| + \tfrac{1}{m}.$$

Since $x + y \in K_m$, $\phi_t(x) + \phi_t(y)$ and $\phi_t(x+y) - \phi_t(x) - \phi_t(y)$ are in K_{n+2}'', and so the first term is at most $\tfrac{1}{m}$. Also, $\|\phi_t(x+y) - \phi_t(x) - \phi_t(y)\| < \tfrac{1}{m}$, so the second term is $\leq \tfrac{1}{m} + \tfrac{1}{m}$. Thus we get

$$\left\| \psi_{r(t)}(\phi_t(x+y)) - [\psi_{r(t)}(\phi_t(x)) + \psi_{r(t)}(\phi_t(y))] \right\| < \tfrac{4}{m}.$$

The other conditions are similarly verified, including that $\|\psi_{r(t)}(\phi_t(x))\| < \|x\| + \frac{2}{m}$ for $x \in K_m$ and $t_m \leq t \leq t_{m+1}$.

Thus $\langle \psi_{r(t)} \circ \phi_t \rangle$ defines a *-homomorphism from A_0 to C_∞ which is norm-decreasing, hence extends to a homomorphism from A to C_∞. The homomorphisms defined for different r (increasing and satisfying $r(t_n) \geq r_n$ for all n) are obviously homotopic, and so a well-defined class in $[[A, C]]$ is obtained independent of the choice of r.

Note that in fact the asymptotic morphism $\langle \psi_{r(t)} \circ \phi_t \rangle$ is uniform on the sets K_n (we cannot assert that it is uniform on all compact subsets of A!) Also note that for the construction it suffices to assume that $\langle \phi_t \rangle$ is uniform on the K_n and $\langle \psi_t \rangle$ is uniform on the K_n''.

(b) If a different collection $\{\tilde{K}_n\}$ is used with $A_0 = \bigcup \tilde{K}_n$, then the rate at which r must increase for the $\tilde{K}_n$ may be different; but if r increases fast enough to work for both the K_n and $\tilde{K}_n$, the same asymptotic morphism is obtained. Thus the class in $[[A, C]]$ is independent of the choice of the K_n.

If a different σ-compact *-subalgebra A_1 is used, then the *-subalgebra A_2 generated by A_0 and A_1 is also σ-compact, and an r which grows fast enough for A_2 (with respect to a suitably chosen increasing sequence of compact sets) will work also for both A_0 and A_1; thus the class in $[[A, C]]$ is independent of the choice of A_0.

If ϕ is replaced by an equivalent ϕ' which is also uniform on the K_n, then $\|\phi_t(a) - \phi_t'(a)\| \to 0$ uniformly on the K_n. It follows easily that if r grows sufficiently fast, then $\langle \psi_{r(t)} \circ \phi_t \rangle$ and $\langle \psi_{r(t)} \circ \phi_t' \rangle$ are equivalent. A more elementary argument shows that if ψ is replaced by an equivalent ψ' which is also uniform on the K_n'', and r grows sufficiently rapidly, then $\langle \psi_{r(t)} \circ \phi_t \rangle$ and $\langle \psi_{r(t)}' \circ \phi_t \rangle$ are equivalent.

If $\langle \phi_t^{(0)} \rangle$ and $\langle \phi_t^{(1)} \rangle$ are homotopic uniform morphisms, then by the previous paragraph and the last part of 25.1.4(b) we may assume that they are connected by a homotopy $\langle \phi_t^{(s)} \rangle$ of uniform morphisms uniform in s. Then in the construction take K_n' to be $\{\phi_t^{(s)}(x) : x \in K_n, t \leq t_n, 0 \leq s \leq 1\}$. The set K_n' is compact. The r thus obtained works simultaneously for all $\psi_{r(t)} \circ \phi_t^{(s)}$, and the asymptotic morphisms thus obtained are obviously homotopic. Thus the class in $[[A, C]]$ depends only on the class of ϕ in $[[A, B]]$. Similarly, if $\psi_t^{(s)}$ is a uniform homotopy, then the numbers r_n in the above proof can be chosen to work simultaneously for all s; thus a single r can be chosen to work for all $\psi_{r(t)}^{(s)} \circ \phi_t$, so the class in $[[A, C]]$ depends only on the class of ψ in $[[B, C]]$.

A cleaner way of phrasing these arguments is: if ψ is a uniform asymptotic morphism from B to $C([0, 1], C)$ implementing a (uniform) homotopy from $\psi^{(0)}$ to $\psi^{(1)}$, and ϕ a uniform asymptotic morphism from A to B, and r grows sufficiently rapidly, then $\langle \psi_{r(t)} \circ \phi_t \rangle$ gives a homotopy between $\langle \psi_{r(t)}^{(0)} \circ \phi_t \rangle$ and $\langle \psi_{r(t)}^{(1)} \circ \phi_t \rangle$. Similarly, if ϕ is a uniform asymptotic morphism from A to $C([0, 1], B)$ giving a homotopy between $\phi^{(0)}$ and $\phi^{(1)}$, and ψ is a uniform asymptotic morphism from B to C, form the asymptotic morphism $\omega = \psi \otimes id_{C[0,1]}$ from $C([0,1], B)$ to $C([0,1], C)$, and choose a uniform representative $\langle \omega_t \rangle$; then for a sufficiently fast-growing r the composition $\langle \omega_{r(t)} \circ \phi_t \rangle$ gives a homotopy between asymptotic morphisms equivalent to

$$\langle \psi_{r(t)} \circ \phi_t^{(0)} \rangle \quad \text{and} \quad \langle \psi_{r(t)} \circ \phi_t^{(1)} \rangle$$

respectively.

(c) The fact that the composition agrees with ordinary composition for homomorphisms is obvious. For associativity, suppose $\langle \phi_t \rangle$, $\langle \psi_t \rangle$, and $\langle \omega_t \rangle$ are uniform asymptotic morphisms from A to B, B to C, and C to D respectively. Choose A_0 and K_n as before, and let $B_0 = \cup K_n''$. Then B_0 is a dense σ-compact *-subalgebra of B. In constructing $\omega \circ \psi$, use B_0 and K_n''; then the construction of $(\omega \circ \psi) \circ \phi$ works as before [even though $\omega \circ \psi$ is not a uniform asymptotic morphism, it is uniform on the K_n'', which is all that is needed for the construction], and yields an asymptotic morphism $\langle \omega_{s(t)} \circ \psi_{r(t)} \circ \phi_t \rangle$ which, for suitable choice of r and s, is the same as is obtained from the construction for $\omega \circ (\psi \circ \phi)$ using the K_n for both steps.

The fact that composition commutes with tensor products is easier. Suppose ϕ, ψ, ω, θ are uniform asymptotic morphisms from A to B, B to C, D to E, and E to F respectively. Choose dense σ-compact *-subalgebras A_0 of A and D_0 of D. Then $A_0 \odot D_0$ is a dense σ-compact *-subalgebra of $A \otimes_{\max} D$. Choose r growing fast enough that $\langle \psi_{r(t)} \circ \phi_t \rangle$, $\langle \theta_{r(t)} \circ \omega_t \rangle$, and $\langle (\psi \otimes \theta)_{r(t)} \circ (\phi \otimes \omega)_t \rangle$ are all asymptotic morphisms. Then $\langle (\psi_{r(t)} \circ \phi_t) \otimes (\theta_{r(t)} \circ \omega_t) \rangle$ is equivalent to $\langle (\psi \otimes \theta)_{r(t)} \circ (\phi \otimes \omega)_t \rangle$. $\qquad \square$

25.3.2. So one can form a category **AM** whose objects are separable C^*-algebras and in which the morphisms from A to B are $[[A, B]]$. Similarly, we can take the morphisms to be $[[SA, SB]]$, obtaining a category **SAM**. There are obvious functors $\mathbf{SC^*} \to \mathbf{AM} \to \mathbf{SAM}$.

25.4. Additive structure

25.4.1. Next we want to obtain an additive structure on asymptotic morphisms. We proceed just as in *Ext* and *KK*: if we consider asymptotic morphisms from A to $B \otimes \mathbb{K}$, we can define addition by fixing an isomorphism from $M_2(\mathbb{K})$ to $\mathbb{K}$ and taking the orthogonal sum. The resulting sum is well defined up to homotopy and makes $[[A, B \otimes \mathbb{K}]]$ into an abelian semigroup. The addition is functorial in all senses, and composition distributes over addition.

There is a natural map from $[[A, B \otimes \mathbb{K}]]$ to $[[A \otimes \mathbb{K}, B \otimes \mathbb{K} \otimes \mathbb{K}]] \cong [[A \otimes \mathbb{K}, B \otimes \mathbb{K}]]$ given by tensoring on the identity map on $\mathbb{K}$. This is a bijection since the map from $\mathbb{K}$ to $\mathbb{K} \otimes \mathbb{K}$ given by $x \to x \otimes e_{11}$ is homotopic to an isomorphism. (The proof of this in [Samuel 1997] is incorrect and incomplete; however, by a judicious choice of bases it suffices to find a strongly continuous path of isometries on a separable Hilbert space linking the identity to an isometry with infinite codimension. The operators V_s on $L^2[0, 1]$, where $[V_s f](t) = s^{-1/2} f(st)$ for, say, $1 \leq s \leq 2$, do the trick. The argument of 7.7.5 can also be used.) This fact also shows that the (class of the) zero homomorphism is an additive identity.

25.4.2. There is also an "addition" on $[[A, SB]]$ for any A and B. The interval $(0, 1)$ can be homotopically "squeezed" down to any subinterval, so any asymptotic morphism from A to SB is homotopic to an asymptotic morphism whose range is supported on (a, b) for any $0 \leq a < b \leq 1$ (in the sense that the composition with evaluation at s is the zero map from A to B for $s \notin (a, b)$). If ϕ and ψ are asymptotic morphisms from A to SB, move ϕ and ψ homotopically to asymptotic morphisms ϕ', ψ' supported on $(0, \frac{1}{2})$ and $(\frac{1}{2}, 1)$ respectively; then $\langle \phi'_t + \psi'_t \rangle$ is an asymptotic morphism from A to SB whose class depends only on $[\phi]$ and $[\psi]$ in $[[A, SB]]$. We denote this class by $[\phi] + [\psi]$, and sometimes by slight abuse of notation we write $\phi + \psi$ for the actual asymptotic morphism defined this way.

Warning: This "addition" on $[[A, SB]]$ is not commutative in general. (It is the C^*-analog of concatenation of loops.)

PROPOSITION 25.4.3. (a) *Addition on $[[A, SB]]$ is associative.*
(b) *The two definitions of addition (of 25.4.1 and 25.4.2) agree on $[[A, SB \otimes \mathbb{K}]]$.*
(c) *$[[A, SB]]$ is a group; the inverse of an asymptotic morphism ϕ is $\phi \circ (\rho \otimes id)$, where $\rho : S \to S$ is "reversal", i.e. $(\rho(f))(s) = f(1 - s)$.*

PROOF. The proofs of (a) and (b) are straightforward exercises left to the reader. They are good practice in working with homotopies. Proof of (c): it suffices to show that the homomorphism $id_S + \rho : S \to S$ is homotopic to the zero homomorphism. Let $h \in S$ be defined by

$$h(s) = \begin{cases} 2s & \text{if } 0 \leq s \leq \frac{1}{2}, \\ 1 - 2s & \text{if } \frac{1}{2} \leq s \leq 1. \end{cases}$$

Then $(id + \rho)(f) = f \circ h = f(h)$ for $f \in S$, where $f(h)$ denotes functional calculus. Then $\phi^{(r)} : S \to S$ defined by $\phi^{(r)}(f) = f(rh)(0 \leq r \leq 1)$ gives a path of homomorphisms from $id + \rho$ to the zero homomorphism. (Alternately, from a topological point of view, the concatenation of the identity loop on S^1 with a loop of winding number -1 is homotopic to a constant loop.) $\square$

In a similar way, if ϕ and ψ are asymptotic morphisms from A to SC and B to SC respectively, we can define an asymptotic morphism $\phi + \psi$ from $A \oplus B$ to SC.

Although $[[A, B \otimes \mathbb{K}]]$ is not in general a group, it is in certain special cases (e.g. [Dădărlat and Loring 1994]).

DEFINITION 25.4.4. If A and B are (separable) C^*-algebras, then $E(A, B) = [[SA, SB \otimes \mathbb{K}]]$.

Using the natural map from $[[SB, SC \otimes \mathbb{K}]]$ to $[[SB \otimes \mathbb{K}, SC \otimes \mathbb{K}]]$ one may obtain a composition $E(A, B) \times E(B, C) \to E(A, C)$. Thus there is an additive category $\mathbf{E}$ whose objects are C^*-algebras and with $E(A, B)$ the morphisms from A to B. There is an obvious functor from $\mathbf{SAM}$ to $\mathbf{E}$.

PROPOSITION 25.4.5. *The correspondences of 25.1.8 are (semi)group isomorphisms.*

25.5. Exact Sequences

Perhaps the most important example of an asymptotic morphism is the "connecting morphism" associated to an exact sequence of C^*-algebras:

PROPOSITION 25.5.1. (a) *Let* $0 \longrightarrow J \longrightarrow A \overset{q}{\longrightarrow} B \longrightarrow 0$ *be an exact sequence of* C^*-*algebras. Suppose* J *has a continuous approximate unit* $\langle u_t \rangle$ *which is quasicentral for* A *(this will always be true if* A *is separable [Voiculescu 1976]). Choose a cross section* σ *for* q. *Then the function* ϕ_t *defined by* $\phi_t(f \otimes b) = f(u_t)\sigma(b)$ *defines an asymptotic morphism from* $C_0((0, 1)) \odot B$ *to* J *which extends to an asymptotic morphism from* SB *to* J.
(b) *The class* ε_q *of this asymptotic morphism in* $[[SB, J]]$ *is independent of the choice of* σ *and* $\langle u_t \rangle$.

It is difficult to motivate the definition of ε_q, but from knowing what properties it should have one can work backwards. Some of the key properties the connecting morphism should have are given in 25.7.1, along with an outline of a direct proof that the desired properties hold; the reader is urged to work through this problem both for justification of the definition and for practice in working with the types of homotopies used in this section. The examples of 25.5.4 give additional motivation for the relevance and importance of the definition.

Compare 25.5.1 to the association of an element of $KK(SB, J) \cong Ext(B, J)^{-1}$ to the extension if it is semisplit. The fact that this connecting morphism is defined even for extensions which are not semisplit is the crucial difference between E-theory and KK-theory.

LEMMA 25.5.2. *Let* D *be a* C^*-*algebra and* $\langle u_\lambda \rangle$ *a net of positive elements of* D *of norm* ≤ 1. *Let* $x \in D$ *and* $f \in C_0((0, 1))$.

(a) *If* $\lim_{\lambda \to \infty} [u_\lambda, x] = 0$, *then* $\lim_{\lambda \to \infty} [f(u_\lambda), x] = 0$
(b) *If* $\lim_{\lambda \to \infty} u_\lambda x = x$, *then* $\lim_{\lambda \to \infty} f(u_\lambda)x = 0$.

PROOF. For both parts, we may assume f is a polynomial vanishing at 0 and 1, since such polynomials are dense in $C_0((0, 1))$ by the Stone–Weierstrass theorem. For polynomials, (a) follows immediately from the fact that $\|[y^k, x]\| \leq k \|y\|^{k-1} \|[y, x]\|$ for any y and k. [Write

$$[y^k, x] = (y^k x - y^{k-1} x y) + (y^{k-1} x y - y^{k-2} x y^2) + \cdots + (y x y^{k-1} - x y^k)$$
$$= y^{k-1}[y, x] + y^{k-2}[y, x]y + \cdots + [y, x]y^{k-1}. \ ]$$

Similarly, writing $y^k x - x = (y^k x - y^{k-1} x) + \cdots + (yx - x)$, we obtain $\|y^k x - x\| \leq k \|yx - x\|$ for $\|y\| \leq 1$. For part (b), write $f(z) = \sum_{k=1}^{n} \alpha_k z^k$; then $\sum_{k=1}^{n} \alpha_k = f(1) = 0$. Then

$$\sum_{k=1}^{n} \alpha_k u_\lambda^k x = \sum_{k=1}^{n} \alpha_k u_\lambda^k x - \sum_{k=1}^{n} \alpha_k x = \sum_{k=1}^{n} \alpha_k (u_\lambda^k x - x).$$

By the previous observation, each term approaches 0 as $\lambda \to \infty$.   □

PROOF OF 25.5.1. The maps $\theta_t : (f, x) \to f(u_t)\sigma(x)$ clearly give a map from $C_0((0,1)) \times B$ to $C_b([1, \infty), J))$. (Continuity in t follows from continuity of functional calculus.) If τ is another cross section for q, then $\sigma(x) - \tau(x) \in J$ for any $x \in B$, and so by 25.5.2 $\lim_{t \to \infty} \|f(u_t)\sigma(x) - f(u_t)\tau(x)\| = 0$ for any f; so modulo $C_0([1, \infty), J)$ θ does not depend on the choice of σ. For fixed x, $\theta_t(\cdot, x)$ is linear in f for each t. The fact that $\langle \theta_t(f, \cdot) \rangle$ is asymptotically linear in x for fixed f follows easily from 25.5.2, since $\sigma(x + y) - \sigma(x) - \sigma(y)$, $\sigma(x^*) - \sigma(x)^*$, and $\sigma(\lambda x) - \lambda \sigma(x)$ are all in J for any $x, y \in B$. It also follows easily from 25.5.2 that $\|f(u_t)\sigma(x)g(u_t)\sigma(y) - (fg)(u_t)\sigma(xy)\| \to 0$ as $t \to \infty$ for any f, g, x, y, since $\sigma(xy) - \sigma(x)\sigma(y) \in J$, $(fg)(u_t) = f(u_t)g(u_t)$, and $\langle u_t \rangle$ is quasicentral. $\|f(u_t)\sigma(x)\| \leq \|f\| \|\sigma(x)\|$. Thus θ defines a *-homomorphism ψ from $C_0((0,1)) \odot B$ to J_∞; since ψ is independent of the choice of σ, and for any $x \in B$ we can choose a σ with $\|\sigma(x)\| = \|x\|$, we have $\|\psi(f \otimes x)\| \leq \|f\| \|x\|$. Thus ψ extends to a *-homomorphism from SB to J_∞, i.e. an asymptotic morphism. The set of quasicentral continuous approximate units is convex, and by continuity of functional calculus the "straight-line" homotopy between two continuous quasicentral approximate units induces a homotopy between the corresponding asymptotic morphisms. Thus the class of ψ in $[[SB, J]]$ is independent of the choice of $\langle u_t \rangle$.   □

REMARK 25.5.3. While this argument is very much in the spirit of 25.2.3, it would be misleading to say that 25.5.1 is merely an application of 25.2.3. For, although there is an obvious related asymptotic morphism $\langle \psi_t \rangle$ from S to J given by $\psi_t(f) = f(u_t)$ (in fact, this is a path of actual *-homomorphisms), it is important to realize that there is generally *no* asymptotic morphism ω from B to J such that $\phi = \psi \otimes \omega$ in the sense of 25.2.3. In fact, if there is such an ω, then (modulo the caution of 25.2.3) the argument of 25.5.5(b) below shows that the tensor product morphism is homotopic to 0.

EXAMPLES 25.5.4. (a) Let ϕ be an asymptotic morphism from A to B. As in 25.1.4(b), we get a naturally associated extension $0 \to SB \to E \to A \to 0$, with $E \subseteq A \oplus CB$. Choose the approximate unit for $C_0((0,1))$ to be $\langle h_t \rangle$, where $h_t (t \geq 3)$ is the function which is 0 on $[0, 1/(t+1)]$ and $[1 - 1/(t+1), 1]$, 1 on $[1/t, 1 - 1/t]$ and linear on $[1/(t+1), 1/t]$ and $[1 - 1/t, 1 - 1/(t+1)]$. Thus, if $f \in C_0((0,1))$, then $f(h_t)$ consists of two copies of f, one transferred to $[1/(t+1), 1/t]$ (more precisely, $[f(h_t)](\frac{1-s}{t+1} + \frac{s}{t}) = f(s)$) and the other, with orientation reversed, transferred to $[1-1/t, \, 1-1/(t+1)]$, extended by 0 on the rest of $[0,1]$. By choosing a cross section σ for the quotient map $q : E \to A$ which vanishes on $[\frac{1}{2}, 1]$, the copy of f on $[1-1/t, \, 1-1/(t+1)]$ has no effect on the asymptotic morphism ψ from SA to SB defined by the exact sequence. Expanding the interval $[1/(t+1), \, 1/t]$ to $[0, \frac{1}{2}]$ via a homotopy converts ψ into an asymptotic morphism homotopic to the suspension $S\phi$.

(b) As an important special case of (a), let $0 \to S \to C \to \mathbb{C} \to 0$ be the standard extension of $\mathbb{C}$ by S (19.2). The asymptotic morphism associated to the exact sequence is homotopic to the identity map on S.

(c) If

$$ 0 \longrightarrow J \xrightarrow{\ i\ } A \underset{q}{\overset{\sigma}{\rightleftarrows}} B \longrightarrow 0 $$

is a split exact sequence, there is an associated exact sequence $0 \to SJ \to E \to A \to 0$, where E is the C^*-subalgebra of $C([0,1], A)$ generated by SJ and $\{\tau(x) : x \in A\}$, where $[\tau(x)](s) = (1-s)x + s[\sigma \circ q](x)$. The class of the corresponding asymptotic morphism from SA to SJ is called the *splitting morphism* of the exact sequence, denoted η_q (it should properly be $\eta_{q,\sigma}$ since it depends on the choice of σ). η_q is the exact analog in this context of the splitting morphism of KK-theory defined in 17.1.2(b), which is exactly the KK-element defined by this extension under the standard identification of $Ext(A, SJ)^{-1}$ with $KK(A, J)$.

The next simple proposition is good practice in working with homotopies of asymptotic morphisms:

PROPOSITION 25.5.5. *Let*

$$ 0 \longrightarrow J \longrightarrow A \xrightarrow{\ q\ } B \longrightarrow 0 $$

be an exact sequence of separable C^*-algebras.

(a) *If* $0 \to SJ \to SA \xrightarrow{Sq} SB \to 0$ *is the suspended exact sequence, then*
   $\varepsilon_{Sq} = S\varepsilon_q$ *in* $[[S^2B, SJ]]$
(b) *If* $0 \longrightarrow J \longrightarrow A \xrightarrow{\ q\ } B \longrightarrow 0$ *splits, then* $\varepsilon_q = [0]$ *in* $[[SB, J]]$.

[*Caution:* The associated element η_q of $[[SA, SJ]]$ of 25.5.4(c) is not 0 in general!]

PROOF. (a) The asymptotic morphism associated to Sq is given as follows: let $\langle u_t \rangle$ be a quasicentral continuous unit for J in A and σ a cross section for q, and let h_t be as in 25.5.4(a); then $\phi_t(f \otimes (g \otimes x)) = f(h_t \otimes u_t)(g \otimes \sigma(x))$, that is, if it is thought of as a function from $(0, 1)$ to J (for fixed t, f, g, x) its value at s is $f(h_t(s)u_t)g(s)\sigma(x)$. For $0 \le r < 1$ define $\phi_t^{(r)}(f \otimes (g \otimes x))$ to be the function whose value at s is $f(h_{t+r/(1-r)}(s)u_t)g(s)\sigma(x)$, and $\phi^{(1)}(f \otimes (g \otimes x))$ to be the function whose value at s is $f(u_t)g(s)\sigma(x)$. Then the $\phi^{(r)}$ define an asymptotic morphism from S^2B to $C([0,1], SJ)$ (this is proved exactly like 25.5.1; the continuity in r

as $r \to 1$ is the only additional thing to be checked), which gives a homotopy between $\phi^{(0)}$ and $\phi^{(1)}$. We have $[\phi^{(0)}] = \varepsilon_{Sq}$ and $[\phi^{(1)}] = S\varepsilon_q$.

(b) Let $\phi^{(1)}$ as above be the constructed representative for ε_q using a splitting σ, and define $\phi_t^{(s)}(f \otimes x) = f(su_t)\sigma(x)$. Then the $\phi_t^{(s)}$ define an asymptotic morphism from SB to $C([0,1], J)$ (the proof of this is almost identical to a subset of the proof of 25.5.1) which gives a homotopy between $\phi^{(0)} = 0$ and $\phi^{(1)}$. □

PROPOSITION 25.5.6. *If*

$$0 \longrightarrow J \xrightarrow[i]{} A \underset{q}{\overset{\sigma}{\rightleftarrows}} B \longrightarrow 0$$

is a split exact sequence of separable C^*-algebras and η_q is the splitting morphism of 25.5.4(c), then

(a) $\eta_q \circ [Si] = [id_{SJ}]$ *in* $[[SJ, SJ]]$, *and*
(b) $[Si] \circ \eta_q = [id_{SA}] - [S(\sigma \circ q)]$ *in* $[[SA, SA]]$.

PROOF. The two proofs are nearly identical. Choose a cross section ω for the map from E to A. While we can take $\omega = \tau$ of 25.5.4(c), it is more convenient to choose ω so that $[\omega(x)](s) = x$ for $0 \le s \le \frac{1}{3}$ and $[\omega(x)](s) = [\sigma \circ q](x)$ for $\frac{2}{3} \le s \le 1$. If $\langle u_t \rangle$ is a continuous approximate unit for J which is quasicentral for A, then a simple compactness argument shows that $\langle h_t \otimes u_t \rangle$ is a continuous approximate unit for SJ which is quasicentral for E, where h_t is as in 25.5.4(a). Thus a representative for η_q is ϕ, where $\phi_t(f \otimes x) = f(h_t \otimes u_t)\omega(x)$. Thought of as a function from $(0,1)$ to J (for fixed t, f, x), the value at s is $f(h_t(s)u_t)[\omega(x)](s)$. If we compose with Si on the right [resp. left], we get an asymptotic morphism from SJ to SJ [resp. from SA to SA] whose value at s (for fixed t, f, x) is given by the same formula. For $0 \le r \le 1$ let $v_t \in C([0,1], \tilde{A})$ be defined by $v_t(r) = r1_{\tilde{A}} + (1-r)u_t$. Then $\langle v_t \rangle$ is a net of positive elements of norm ≤ 1 in $C([0,1], \tilde{A})$, continuous in t, and a simple compactness argument shows that, with $D = C([0,1], \tilde{A})$, $\langle v_t \rangle$ satisfies the hypotheses of 25.5.2(a) for $x \in C([0,1], A)$ and of 25.5.2(b) if $x \in C([0,1], J)$. Thus, if we define $\phi_t^{(r)}(f \otimes x)$ to be the function whose value at s is $f(h_t(s)v_t(r))[\omega(x)](s)$, an argument identical to the proof of 25.5.1 shows that $\langle \phi_t^{(r)} \rangle$ defines an asymptotic morphism from SJ to $C([0,1], SJ)$ [resp. from SA to $C([0,1], SA)$] giving a homotopy between $\langle \phi_t \rangle$ and the asymptotic morphism $\langle \psi_t \rangle$ from SA to SA which, as a function of s for fixed t, f, x, takes value $f(h_t(s))[\omega(x)](s)$. As in 25.5.4(a), $s \to f(h_t(s))$ consists of two "copies" of f, one supported on $[1/(t+1), 1/t]$ and the other, with orientation reversed, on $[1-1/t, 1-1/(t+1)]$. Since $[\omega(x)](s) = x$ on $[0, \frac{1}{3}]$ and $[\omega(x)](s) = [\sigma \circ q](x)(= 0$ if $x \in J)$ on $[\frac{2}{3}, 1]$, we get that ψ is homotopic to id_{SJ} [resp. that ψ is homotopic to $id_{SA} + (\rho \otimes (\sigma \circ q))]$. □

COROLLARY 25.5.7. *If*

$$0 \longrightarrow J \xrightarrow{\ i\ } A \underset{q}{\overset{\sigma}{\rightleftarrows}} B \longrightarrow 0$$

is a split exact sequence of separable C^*-algebras and η_q is the splitting morphism of 25.5.4(c), then $\eta_q \oplus [Sq]$ is an isomorphism in **SAM** *from A to $J \oplus B$, with inverse $[Si] + [S\sigma]$.*

Compare this with the fact that the splitting morphism of KK-theory defines a KK-equivalence between A and $J \oplus B$ in exactly the same manner (19.9.1).

We get the following crucial consequence of this by applying 22.3.1:

COROLLARY 25.5.8. *The canonical functor from* **SC*** *to* **E** *factors through* **KK**, *i.e. there is a functor $F : $ **KK** $\to$ **E** *sending $KK(A, B)$ to $E(A, B)$ for every separable A, B. F respects addition and tensor products (hence suspensions).*

An explicit formula for the map is given in 25.5.1, identifying $KK(SB, J)$ with $Ext(B, J)^{-1}$. (Of course, it is not *a priori* obvious that this formula gives a well-defined map.)

COROLLARY 25.5.9 (BOTT PERIODICITY). *For any A and B, there are canonical isomorphisms $E(A, B) \cong E(S^2 A, B) \cong E(A, S^2 B) \cong E(SA, SB)$, which are natural in A and B. Specifically, if ϕ is the homomorphism from S to $M_2(C_0(\mathbb{R}^3))$ corresponding to the generator of $K_1(C_0(\mathbb{R}^3))$, then the map $[\phi] \in E(\mathbb{C}, C_0(\mathbb{R}^2))$ is an isomorphism in* **E**, *whose inverse is $S\varepsilon_q$ for the Toeplitz extension $0 \longrightarrow \mathbb{K} \longrightarrow T_0 \xrightarrow{\ q\ } S \longrightarrow 0$ (9.4.2); the isomorphism from $E(A, B)$ to $E(S^2 A, B)$ is given by tensoring with ε_q (or composing on the left with $\varepsilon_q \otimes [id_{SA}]$). Similarly the isomorphism $E(A, B) \cong E(A, S^2 B)$ is given by tensoring with $[\phi]$. The isomorphism from $E(A, B)$ to $E(SA, SB)$ is given by tensoring with id_S.*

A direct proof of 25.5.9 can be given; in fact, it is virtually identical to the argument in 9.4.2.

We can thus define $E^0(A, B) = E(A, B)$ and $E^1(A, B) = E(A, SB) \cong E(SA, B)$; then $E^1(SA, B) \cong E^1(A, SB) \cong E(A, B)$.

We now turn to general exact sequences.

THEOREM 25.5.10. *E is half-exact: for any extension $0 \longrightarrow J \xrightarrow{\ i\ } A \xrightarrow{\ q\ } B \longrightarrow$ 0 of separable C^*-algebras, and every separable C^*-algebra D, the sequences*

$$E(D, J) \xrightarrow{\ i_*\ } E(D, A) \xrightarrow{\ q_*\ } E(D, B),$$

$$E(B, D) \xrightarrow{\ q^*\ } E(A, D) \xrightarrow{\ i^*\ } E(J, D)$$

are exact in the middle, where i_* [resp. i^*] is composition on the right [resp. on the left] with $[i] \in E(J, A)$, etc.

Note that no semisplitting is assumed.

We need two lemmas for the proof of 25.5.10. If $0 \longrightarrow J \overset{i}{\longrightarrow} A \overset{q}{\longrightarrow} B \longrightarrow 0$ is an exact sequence as above, let $0 \longrightarrow SJ \longrightarrow CA \overset{p}{\longrightarrow} C_q \longrightarrow 0$ be the associated mapping cone sequence. Let α be the quotient map from C_q to A.

LEMMA 25.5.11. $[Si \circ \varepsilon_p] = [S\alpha]$ in $[[SC_q, SA]]$.

PROOF. We proceed as in the proof of 25.5.6. Choose a continuous quasicentral approximate unit $\langle u_t \rangle$ for J in A; if h_t is as in 25.5.4(a), then $\langle h_t \otimes u_t \rangle$ is a continuous approximate unit for SJ which is quasicentral for CA. Choose a cross section σ for p. Then for $f \in S$, $x \in C_q$, the constructed representative for ε_p is the function $\phi_t(f \otimes x) = f(h_t \otimes u_t)\sigma(x)$. Thought of as a function from $(0, 1)$ to J (for fixed t, f, x), the value at s is $f(h_t(s)u_t)[\sigma(x)](s)$. If we compose with Si, we get an asymptotic morphism from SC_q to $C_0((0, 1), A)$ whose value at s (for fixed t, f, x) is given by the same formula. Now for $0 \leq r \leq 1$ let $v_t \in C([0, 1], \tilde{A})$ be defined by $v_t(r) = r1_{\tilde{A}} + (1 - r)u_t$. Then as in 25.5.6, if we define $\phi_t^{(r)}(f \otimes x)$ to be the function whose value at s is $f(h_t(s)v_t(r))[\sigma(x)](s)$, $\langle \phi_t^{(r)} \rangle$ defines an asymptotic morphism from SC_p to $C([0, 1], SA)$ giving a homotopy between $\langle \phi_t \rangle$ and the asymptotic morphism $\langle \psi_t \rangle$ from SC_q to SA which, as a function of s for fixed t, f, x, takes value $f(h_t(s))[\sigma(x)](s)$. As in 25.5.4(a), $s \to f(h_t(s))$ consists of two "copies" of f, one supported on $[1/(t+1), 1/t]$ and the other, with orientation reversed, on $[1-1/t, 1-1/(t+1)]$. Since $[\sigma(x)](s) \to 0$ as $s \to 1$, for large t the second copy of f has negligible effect (for fixed f and x). Thus $\langle \psi_t \rangle$ is equivalent to the asymptotic morphism $\langle \omega_t \rangle$, where $[\omega_t(f \otimes x)](s) = \tilde{f}_t(s)[\sigma(x)](s)$, where $\tilde{f}_t$ is the function equal to $f \circ h_t$ on $[1/(t+1), 1/t]$ and 0 elsewhere. Next define $[\omega_t^{(r)}(f \otimes x)](s) = \tilde{f}_t(s)[\sigma(x)](s - r)$ $([\sigma(x)](s - r) = [\sigma(x)](0) = x(0)$ if $s - r < 0$); $\langle \omega_t^{(r)} \rangle$ defines a homotopy between $\omega = \omega^{(0)}$ and $\omega^{(1)}$, which is given by the formula $[\omega^{(1)}(f \otimes x)](s) = \tilde{f}_t(s)x(0)$. We can then homotopically expand $[1/(t+1), 1/t]$ to all of $[0, 1]$ to convert $\omega^{(1)}$ to $S\alpha$. □

LEMMA 25.5.12. Let $0 \longrightarrow J \overset{i}{\longrightarrow} A \overset{q}{\longrightarrow} B \longrightarrow 0$ be an extension of separable C^*-algebras, and D a separable C^*-algebra.

(a) If h is an asymptotic morphism from D to A, and $[q \circ h] = [0]$ in $[[D, B]]$, then there is an asymptotic morphism k from SD to SJ such that $[Si \circ k] = [Sh]$ in $[[SD, SA]]$.

(b) If h is an asymptotic morphism from A to D, and $[h \circ i] = [0]$ in $[[J, D]]$, then there is an asymptotic morphism k from S^2B to S^2D such that $[k \circ S^2q] = [S^2h]$ in $[[S^2A, S^2D]]$.

PROOF. (a) Let $\langle \phi_t^{(s)} \rangle$ be a homotopy from $q \circ h$ to 0. Then $\phi^{(s)}$ is an asymptotic morphism from D to $C_0([0, 1), B)$, and yields an asymptotic morphism $\psi = h \oplus \phi^{(s)}$ from D to C_q with $h = \alpha \circ \psi$. Then by 25.5.11 we have $[Sh] = [S\alpha] \circ [S\psi] = [Si] \circ (\varepsilon_p \circ [S\psi])$, so k may be taken to be any representative of $\varepsilon_p \circ [S\psi]$.

(b) First suppose $[h \circ \alpha] = [0]$ in $[[C_q, D]]$, and let $\phi^{(s)}$ be a homotopy from $h \circ \alpha$ to 0. The map $\phi^{(s)}$ is an asymptotic morphism from C_q to $C_0([0,1), D)$; and since $\alpha = 0$ on the ideal SB of C_q, restricting $\phi^{(s)}$ to SB (or composing with the inclusion from SB into C_q) gives an asymptotic morphism k from SB to $C_0((0,1), D) = SD$.

We claim that $[k \circ Sq] = [Sh]$ in $[[SA, SD]]$. For $0 \le r < 1$ let β_r be the homeomorphism of $[r, 1]$ to $[0, 1]$ given by $\beta_r(\lambda) = (1 - r)^{-1}(\lambda - r)$, and γ_r the inverse homeomorphism. Then for $f \in S$, $x \in A$, $0 \le r \le 1$, $0 < s < 1$, $t \ge 1$, define

$$[\psi_t^{(r)}(f \otimes x)](s) = \begin{cases} f(s)h_t(x) & \text{if } s \le r, \\ \phi_t^{(\beta_r(s))}(f(r)x \oplus [(f \circ \gamma_r) \otimes q(x)]) & \text{if } s > r. \end{cases}$$

One must check continuity in s at $s = r$ for fixed r. Continuity in r as $r \to 1$ follows from the fact that for fixed f, $f \circ \gamma_r \to 0$ as $r \to 1$. $\langle \psi_t^{(r)} \rangle$ thus defines an asymptotic morphism from SA to $C([0,1], SD)$ giving a homotopy from $\psi^{(0)} = k \circ Sq$ to $\psi^{(1)} = Sh$.

Now under the hypotheses of (b), we have that $[Sh \circ Si] = 0$, so $0 = [Sh] \circ [Si] \circ \varepsilon_p = [Sh \circ S\alpha]$ by 25.5.11, so by the first part of the argument we get k from S^2B to S^2D with the desired property. $\square$

PROOF.   Proof of 25.5.10 It follows immediately from 25.5.12 that these sequences are exact in the middle:

$$[[SD, SJ]] \xrightarrow{Si_*} [[SD, SA]] \xrightarrow{Sq_*} [[SD, SB]],$$
$$[[S^2B, S^2D]] \xrightarrow{S^2q^*} [[S^2A, S^2D]] \xrightarrow{S^2i_*} [[S^2J, S^2D]]$$

(we have im $Si_* \subseteq \ker Sq_*$ by functoriality). These, combined with Bott periodicity, give the desired exact sequences for E. $\square$

COROLLARY 25.5.13.   *E-Theory has six-term exact sequences in each variable for arbitrary extensions of separable C^*-algebras. Specifically, if $0 \longrightarrow J \xrightarrow{j} A \xrightarrow{q} B \longrightarrow 0$ is a short exact sequence of separable C^*-algebras, and D is any separable C^*-algebra, then the following sequences are exact:*

$$\begin{array}{ccccc}
E^0(D, J) & \xrightarrow{i_*} & E^0(D, A) & \xrightarrow{q_*} & E^0(D, B) \\
{\scriptstyle \varepsilon_{q*}} \uparrow & & & & \downarrow {\scriptstyle \varepsilon_{q*}} \\
E^1(D, B) & \xleftarrow{q_*} & E^1(D, A) & \xleftarrow{i_*} & E^1(D, J)
\end{array}$$

$$\begin{array}{ccccc}
E^0(B, D) & \xrightarrow{q^*} & E^0(A, D) & \xrightarrow{i^*} & E^0(J, D) \\
{\scriptstyle \varepsilon_q^*} \uparrow & & & & \downarrow {\scriptstyle \varepsilon_q^*} \\
E^1(J, D) & \xleftarrow{i^*} & E^1(A, D) & \xleftarrow{q^*} & E^1(B, D)
\end{array}$$

PROOF. This mostly follows immediately from 21.4.4 and 25.5.9. The only thing remaining to check is that the connecting maps, which are of the form $i_* \circ e_*^{-1}$ (or $e^{*\,-1} \circ i^*$), are equal to ε_{q*}; this is an immediate consequence of 25.7.1(c). □

25.6. Axiomatic E-Theory

We begin by recalling from earlier sections some of the properties that half-exact homotopy-invariant functors enjoy. If F is a half-exact homotopy-invariant functor from $\mathbf{SC}^*$ to an additive category $\mathbf{A}$, then by suspension F has long exact sequences (21.4.3, 21.4.4); in particular, if $0 \to J \to A \to B \to 0$ is a short exact sequence of separable C^*-algebras, and one of J, A, B is contractible, then the induced map between the other two is an isomorphism (with a degree shift if A is contractible). In addition, if F is stable, then F factors through $\mathbf{KK}$ (22.1.2, 22.3.1), and hence satisfies Bott Periodicity in the sense of 25.5.9.

The main result of this section is the universal property of $\mathbf{E}$:

THEOREM 25.6.1. *Any functor from* $\mathbf{SC}^*$ *to an additive category* $\mathbf{A}$, *which is homotopy-invariant and half-exact, factors uniquely through* $\mathbf{AM}$; *if the functor is also stable, it factors uniquely through* $\mathbf{E}$. *In other words, if F is a homotopy-invariant, stable, half-exact covariant [resp. contravariant] functor from* $\mathbf{SC}^*$ *to an additive category* $\mathbf{A}$, *then there is a pairing* $F(A) \times E(A, B) \to F(B)$ *[resp.* $E(A, B) \times F(B) \to F(A)$] *which is compatible (associative) with respect to composition in* $\mathbf{E}$, *and which agrees with usual functoriality for homomorphisms.*

To prove this theorem, we first show that every element of $E(A, B)$ can be canonically written as an ordinary homomorphism composed with the inverse of another. This fact is closely related in spirit to the canonical factorization of KK-elements (17.8.3) used to prove the universality of KK-theory.

If ϕ is an asymptotic morphism from A to B, then there is an associated extension $0 \to C_0([1, \infty), B) \to D \to A \to 0$: if $\pi : C_b([1, \infty), B) \to B_\infty$ is the quotient map, set $D = \pi^{-1}(A)$ (or the pullback of π and id_A if ϕ is not faithful). We denote also by π the quotient map from D to A; $x \to \langle \phi_t(x) \rangle$ gives a cross section for π. Since $C_0([1, \infty), B)$ is contractible, $[\pi]$ is an isomorphism in $E(D, A)$. Denote by ρ_t the map from E to B given by evaluation at t.

PROPOSITION 25.6.2. $[\rho_1] \circ [\pi]^{-1} = [\phi]$ *in* $E(A, B)$.

PROOF. It suffices to show that $[\phi] \circ [\pi] = [\rho_1]$, which is almost obvious: the asymptotic morphism $\langle \phi_t \circ \pi \rangle$ from D to B is equivalent to the asymptotic morphism $\langle \rho_t \rangle$, which is obviously homotopic to the constant morphism ρ_1. □

PROOF OF 25.6.1. We prove the result for F covariant; the contravariant case is essentially identical. If ϕ is an asymptotic morphism from A to B, let $0 \to C_0([1, \infty), B) \to D \xrightarrow{\pi} A \to 0$ be the associated exact sequence as above. Then $F(\pi)$ is an isomorphism in $\mathbf{A}$ from $F(D)$ to $F(A)$ by the long exact sequence for F and the contractibility of $C_0([1, \infty), B)$. We then set $F([\phi]) = F(\rho_1) \circ F(\pi)^{-1}$. To show that this is well defined, suppose $\phi^{(0)}$ and

$\phi^{(1)}$ are asymptotic morphisms from A to B which are homotopic via $\phi^{(s)}$. Then we have corresponding extensions $0 \longrightarrow C_0([1,\infty), B) \longrightarrow D^{(i)} \xrightarrow{\pi^{(i)}} A \longrightarrow 0$ for $i = 0, 1$, and $0 \to C_0([1,\infty), IB) \to D \xrightarrow{\pi} A \to 0$, and for each i a commutative diagram

$$
\begin{array}{ccccccccc}
0 & \longrightarrow & C_0([1,\infty), IB) & \longrightarrow & D & \xrightarrow{\pi} & A & \longrightarrow & 0 \\
 & & \downarrow{\gamma^{(i)}} & & \downarrow{\gamma^{(i)}} & & \downarrow{id_A} & & \\
0 & \longrightarrow & C_0([1,\infty), B) & \longrightarrow & D^{(i)} & \xrightarrow{\pi^{(i)}} & A & \longrightarrow & 0
\end{array}
$$

where $\gamma^{(i)} : C_b([1,\infty), C([0,1], B)) \to C_b([1,\infty), B)$ is evaluation at $i = 0, 1 \in [0,1]$. By functoriality, $F(\gamma^{(i)})$ is an isomorphism from $F(D)$ to $F(D^{(i)})$, and that $F(\phi^{(i)})$, as defined above as $F(\rho_1^{(i)}) \circ F(\pi^{(i)})^{-1}$, is equal to

$$
F(\beta^{(i)}) \circ F(\rho_1) \circ F(\gamma^{(i)})^{-1} \circ F(\pi^{(i)})^{-1} = F(\beta^{(i)}) \circ F(\rho_1) \circ F(\pi)^{-1},
$$

where $\beta^{(i)} : C([0,1], B) \to B$ is evaluation at i. But $\beta^{(0)}$ and $\beta^{(1)}$ are homotopic, so $F(\beta^{(0)}) = F(\beta^{(1)})$, i.e. $F(\phi^{(0)}) = F(\phi^{(1)})$, and

$$
F([\phi]) \in \mathrm{Hom}_{\mathbf{A}}(F(A), F(B))
$$

is well defined.

It is obvious that this definition of $F([\phi])$ agrees with the usual functorial definition of $F(\phi)$ if ϕ is an actual *-homomorphism; and $\phi \to F([\phi])$ respects composition since F respects composition of homomorphisms and their inverses by functoriality. Thus F factors through $\mathbf{AM}$.

By composing F with the suspension functor, by the same argument we can associate to an element of $[[A, B]]$ an element of $\mathrm{Hom}_{\mathbf{A}}(F(SA), F(SB))$.

If F is also stable, it satisfies Bott Periodicity, and there are standard isomorphisms between $F(A)$, $F(S^2A)$, $F(A \otimes \mathbb{K})$, and $F(S^2A \otimes \mathbb{K})$, for any A. Using these identifications, for any element of $E(A, B)$ we get a well-defined element of $\mathrm{Hom}_{\mathbf{A}}(F(S^2A \otimes \mathbb{K}), F(S^2B \otimes \mathbb{K})) \cong \mathrm{Hom}_{\mathbf{A}}(F(A), F(B))$, giving the desired factorization of F. $\qquad\square$

THEOREM 25.6.3. *Let A be a separable C^*-algebra for which $KK(A, \cdot)$ is half-exact. Then $E(A, B)$ is naturally isomorphic to $KK(A, B)$ for every separable B. In particular, if A is separable nuclear (or K-nuclear (20.10.2)), then $E(A, B) \cong KK(A, B)$ for every separable B.*

PROOF. By 25.6.1 there is a pairing $KK(A, D) \times E(D, B) \to KK(A, B)$ for every D, B. Setting $D = A$, one can apply this pairing to $\mathbf{1}_A$ to get a homomorphism from $E(A, B)$ to $KK(A, B)$ for any B. It is easy to check that this homomorphism is an inverse to the canonical homomorphism from $KK(A, B)$ to $E(A, B)$ from 25.5.8. $\qquad\square$

Cuntz [1997] has given an elegant unified approach to KK-theory, E-theory, and cyclic homology. Houghton-Larsen and Thomsen [1996] also showed how to obtain KK-theory via asymptotic morphisms (25.7.3).

25.7. EXERCISES AND PROBLEMS

25.7.1. Let $0 \longrightarrow J \xrightarrow{\ j\ } A \xrightarrow{\ q\ } B \longrightarrow 0$ be an exact sequence of separable C^*-algebras, and ε_q the associated asymptotic morphism from SB to J as in 25.5.1.

(a) Show directly that $[j] \circ \varepsilon_q = [0]$ in $[[SB, A]]$ by using the homotopy $\phi_t^{(s)}(f \otimes x) = f((1-s)u_t + s1_{\tilde{A}})\sigma(x)$.

(b) Show directly that $\varepsilon_q \circ [Sq] = [0]$ in $[[SA, J]]$ by using the homotopy $\phi_t^{(s)}(f \otimes y) = f(u_t)[(1-s)\sigma(q(y)) + sy]$ followed by the homotopy of 25.5.5(b).

(c) Let $0 \longrightarrow J \xrightarrow{\ e\ } C_q \longrightarrow CB \longrightarrow 0$ be the associated "excision" sequence. Use the following homotopy to show directly that $[e] \circ \varepsilon_q = [j]$ in $[[SB, C_q]]$, where $j : SB \to C_q$ is the natural inclusion: let $\phi_t^{(r)}(f \otimes x)$ be the element of C_q whose A-coordinate is $f((1-r)u_t + r1_{\tilde{A}})\sigma(x)$ and whose CB-coordinate is the function whose value at s is $f(s+1-r)x$ (where $f(s+1-r) = 0$ if $s+1-r \geq 1$).

(d) Show directly that the induced maps $\varepsilon_{q*} : K_i(SB) \cong K_{1-i}(B) \to K_i(J)$ (25.1.6) are exactly the connecting maps in the six-term exact sequence of K-theory associated to the extension.

25.7.2. This problem explores other equivalence relations on asymptotic morphisms which have been important in applications.

(a) Two asymptotic morphisms $\langle \phi_t \rangle$ and $\langle \psi_t \rangle$ are *unitarily equivalent* [resp. *multiplier unitarily equivalent*] if there is a unitary u in $\tilde{B}$ [resp. $M(B)$] such that $u\phi_t(a)u^* = \psi_t(a)$ for all $a \in A$. Two asymptotic morphisms $\langle \phi_t \rangle$ and $\langle \psi_t \rangle$ are *asymptotically unitarily equivalent* [resp. *asymptotically multiplier unitarily equivalent*] if there is a path $\langle u_t \rangle$ of unitaries in $\tilde{B}$ [resp. $M(B)$] such that $\lim_{t\to\infty} \|\psi_t(a) - u_t\phi_t(a)u_t^*\| = 0$ for all $a \in A$.

(b) [Multiplier] unitarily equivalent asymptotic morphisms are not in general homotopic, unless the implementing unitary is in $U(M(B))_0$. If $\langle \phi_t \rangle$ and $\langle \psi_t \rangle$ are [multiplier] asymptotically unitarily equivalent, then ψ is homotopic to an asymptotic morphism [multiplier] unitarily equivalent to ϕ.

(c) Asymptotic unitary equivalence of asymptotic morphisms is a useful notion, but it has a defect: an asymptotic morphism is rarely multiplier asymptotically unitarily equivalent to its reparametrizations. There is a stronger notion: two asymptotic morphisms $\langle \phi_t \rangle$ and $\langle \psi_t \rangle$ from A to B are *decoupled asymptotically unitarily equivalent* if there is a continuous function $(s, t) \to u_{s,t}$ from $[1, \infty) \times [1, \infty)$ to $U(M(B))$ such that $\lim_{s,t\to\infty} \|\psi_t(a) - u_{s,t}\phi_s(a)u_{s,t}^*\| = 0$ for all $a \in A$. Decoupled unitary equivalence is a very strong condition. For example, we have the following fact, which is crucial in Phillips' approach to the classification problem for purely infinite algebras.

PROPOSITION.   *Let $\langle \phi_t \rangle$ be an asymptotic morphism from A to B. If A is separable and $\langle \phi_t \rangle$ is decoupled asymptotically unitarily equivalent to itself, then $\langle \phi_t \rangle$ is homotopic to a homomorphism from A to B.*

SKETCH OF PROOF.   Let $\langle u_{s,t} \rangle$ be the implementing family of unitaries. Replacing $u_{s,t}$ by $u_{t,t}^* u_{s,t}$, we may assume $u_{t,t} = 1$ for all t. Choose an increasing sequence $\langle F_n \rangle$ of finite subsets of A with dense union, and find an increasing sequence $1 < t_1 < t_2 < \cdots$ with $t_n \to \infty$, such that ϕ_{t_n} is approximately *-linear and multiplicative within 2^{-n} on F_n, and such that $\|u_{t_n,t} \phi_{t_{n+1}}(a) u_{t_n,t}^* - \phi_{t_n}(a)\| < 2^{-n}$ for $a \in F_n$ and $t \geq t_n$. For $t_n \leq t \leq t_{n+1}$, define $v_t = u_{t_n,t} u_{t_{n-1},t_n} \cdots u_{t_1,1}$, and let $\psi_t(a) = v_t^* \phi_t(a) v_t$. Then $\langle \psi_t \rangle$ is an asymptotic morphism which is asymptotically unitarily equivalent to ϕ, hence homotopic since $v_1 = 1$. For $a \in F_n$, $\omega(a) = \lim_{t \to \infty} \psi_t(a)$ exists in B, and ω is a homomorphism. The constant asymptotic morphism $\langle \omega \rangle$ is equivalent to ψ. $\square$

25.7.3. [Houghton-Larsen and Thomsen 1996] If A and B are C^*-algebras, let $[[A, B]]_{cp}$ be the set of homotopy classes of completely positive asymptotic morphisms from A to B (where the asymptotic morphisms in the homotopy are also required to be completely positive, i.e. the homotopy is in $[[A, C([0,1], B)]]_{cp}$).

(a) Show that all constructions and results from 25.1-25.4 have exact analogs in this setting. Define $E_{cp}(A, B) = [[SA, SB \otimes \mathbb{K}]]_{cp}$.

(b) Suppose $0 \to J \to A \to B \to 0$ is a *semisplit* exact sequence of separable C^*-algebras. The formula of 25.5.1 defines an asymptotic morphism which is equivalent to a completely positive asymptotic morphism, if the section σ is chosen completely positive. Results analogous to 25.5.1-25.5.14 hold.

(c) As in 25.5.10, E_{cp} has six-term exact sequences in each variable for *semisplit* extensions.

(d) If $\mathbf{E}_{cp}$ is the corresponding category with separable C^*-algebras as objects and Hom-sets $E_{cp}(A, B)$, then a proof analogous to that of 25.6.1 shows that E_{cp} is the universal homotopy-invariant, stable functor with exact sequences for semisplit extensions. Thus the natural map from $\mathbf{KK}$ to $\mathbf{E}_{cp}$ is an isomorphism, i.e. $KK(A, B) \cong E_{cp}(A, B)$ for any separable A, B.

(e) Since $E_{cp}(A, B) = E(A, B)$ for any B if A is nuclear, we get an alternate proof of 25.6.3 for A nuclear.

25.7.4. [Guentner et al. 1997] Here is an alternate approach to constructing the product.

(a) If B is a C^*-algebra, let $\alpha B = B_\infty$ (25.1.4(a)), and inductively let $\alpha^n B = \alpha(\alpha^{n-1} B)$. There is a natural inclusion of $\alpha^n B$ into $\alpha^{n+1} B$ as "constant" functions. α is a functor: a *-homomorphism $\phi : A \to B$ defines a natural *-homomorphism $\alpha\phi : \alpha A \to \alpha B$. Two *-homomorphisms $\phi^{(0)}, \phi^{(1)}$ from A to $\alpha^n B$ are *n-homotopic* if there is a *-homomorphism $\phi : A \to \alpha^n C([0,1], B)$ with $(\alpha^n \pi_s) \circ \phi = \phi^{(s)}$ for $s = 0, 1$. Denote by $[[A, B]]_n$ the set of n-homotopy classes of *-homomorphisms from A to $\alpha^n B$. There are natural maps from $[[A, B]]_n$ to $[[A, B]]_{n+1}$ (corresponding to the inclusion $\alpha^n B \subseteq \alpha^{n+1} B$) and from $[[A, B]]_n$ to $[[\alpha A, \alpha B]]_{n+1}$ (given by $[\phi] \to [\alpha\phi]$) for each n.

(b) There is a natural product $[[A, B]]_n \times [[B, C]]_m \to [[A, C]]_{n+m}$ given by $[\phi] \times [\psi] = [(\alpha^n \psi) \circ \phi]$. Thus, if $[[A, B]]_\infty = \varinjlim [[A, B]]_n$, there is a natural product $[[A, B]]_\infty \times [[B, C]]_\infty \to [[A, C]]_\infty$.

(c) Show that if A is separable, then the natural map $[[A, B]]_1 \to [[A, B]]_n$ is a bijection for any n and any C^*-algebra B [Guentner et al. 1997, Theorem 2.16]. Thus $[[A, B]]_\infty \cong [[A, B]]$ as defined in 25.1.1, and one obtains a product on $[[\,\cdot\,,\,\cdot\,]]$. Show that this agrees with the product as defined in 25.3.1.

Compare the constructions and results of this problem with those of 18.13.3.

25.7.5. Let $\mathscr{B}$ be the smallest class of separable C^*-algebras closed under $(N1)-$ $(N3)$ of 22.3.4, and closed under E-equivalence (isomorphism in $\mathbf{E}$). Show by arguments essentially identical to those in Section 23 that if A and B are separable C^*-algebras, with $A \in \mathscr{B}$, then (A, B) satisfies the E-theory versions of the UCT and (if $K_*(B)$ is finitely generated) the KTP: there are natural exact sequences

$$0 \longrightarrow \mathrm{Ext}^1_{\mathbb{Z}}(K_*(A), K_*(B)) \xrightarrow{\delta} E^*(A, B) \xrightarrow{\gamma} \mathrm{Hom}(K_*(A), K_*(B)) \longrightarrow 0,$$

$$0 \longrightarrow K^*(A) \otimes K_*(B) \xrightarrow{\beta} E^*(A, B) \xrightarrow{\rho} \mathrm{Tor}^{\mathbb{Z}}_1(K^*(A), K_*(B)) \to 0.$$

Is $\mathscr{B}$ the class of all separable C^*-algebras?

25.7.6. [Guentner et al. 1997] Work out the details of equivariant E-theory and compare them with the equivariant KK-theory of Section 20.

REFERENCES

[Akemann et al. 1973] C. A. Akemann, G. K. Pedersen, and J. Tomiyama, "Multipliers of C^*-algebras", *J. Functional Analysis* **13** (1973), 277–301.

[Anderson 1978] J. Anderson, "A C^*-algebra $\mathscr{A}$ for which $\mathrm{Ext}(\mathscr{A})$ is not a group", *Ann. of Math.* (2) **107**:3 (1978), 455–458.

[Anderson et al. 1991] J. Anderson, B. Blackadar, and U. Haagerup, "Minimal projections in the reduced group C^*-algebra of $Z_n * Z_m$", *J. Operator Theory* **26**:1 (1991), 3–23.

[Araki and Toda 1965] S. Araki and H. Toda, "Multiplicative structures in $\mathrm{mod}\, q$ cohomology theories. I", *Osaka J. Math.* **2** (1965), 71–115.

[Araki and Toda 1966] S. Araki and H. Toda, "Multiplicative structures in mod_q cohomology theories, II", *Osaka J. Math.* **3** (1966), 81–120.

[Araki et al. 1960] S. Araki, I. M. James, and E. Thomas, "Homotopy-abelian Lie groups", *Bull. Amer. Math. Soc.* **66** (1960), 324–326.

[Archbold 1982] R. J. Archbold, "Approximating maps and exact C^*-algebras", *Math. Proc. Cambridge Philos. Soc.* **91**:2 (1982), 285–289.

[Arveson 1977] W. Arveson, "Notes on extensions of C^*-algebras", *Duke Math. J.* **44**:2 (1977), 329–355.

[Atiyah 1967] M. F. Atiyah, *K-theory*, W. A. Benjamin, New York and Amsterdam, 1967. Notes by D. W. Anderson. Second edition: Addison-Wesley, Reading, MA, 1989.

[Atiyah 1968] M. F. Atiyah, "Bott periodicity and the index of elliptic operators", *Quart. J. Math. Oxford Ser.* (2) **19** (1968), 113–140.

[Atiyah and Hirzebruch 1961] M. F. Atiyah and F. Hirzebruch, "Vector bundles and homogeneous spaces", pp. 7–38 in *Differential Geometry*, edited by C. B. Allendoerfer, Proc. Sympos. Pure Math. **3**, Amer. Math. Soc., Providence, 1961.

[Atiyah and Segal 1968] M. F. Atiyah and G. B. Segal, "The index of elliptic operators, II", *Ann. of Math.* (2) **87** (1968), 531–545.

[Atiyah and Singer 1968a] M. F. Atiyah and I. M. Singer, "The index of elliptic operators, I", *Ann. of Math.* (2) **87** (1968), 484–530.

[Atiyah and Singer 1968b] M. F. Atiyah and I. M. Singer, "The index of elliptic operators, III", *Ann. of Math.* (2) **87** (1968), 546–604.

[Atiyah and Singer 1971a] M. F. Atiyah and I. M. Singer, "The index of elliptic operators, IV", *Ann. of Math.* (2) **93** (1971), 119–138.

[Atiyah and Singer 1971b] M. F. Atiyah and I. M. Singer, "The index of elliptic operators, V", *Ann. of Math.* (2) **93** (1971), 139–149.

[Baaj and Julg 1983] S. Baaj and P. Julg, "Théorie bivariante de Kasparov et opérateurs non bornés dans les C^*-modules hilbertiens", *C. R. Acad. Sci. Paris Sér. I Math.* **296**:21 (1983), 875–878.

[Bass 1968] H. Bass, *Algebraic K-theory*, W. A. Benjamin, New York and Amsterdam, 1968.

[Baum and Connes 1982] P. Baum and A. Connes, "Geometric K-theory for Lie groups and foliations", Preprint, 1982.

[Baum and Douglas 1982a] P. Baum and R. G. Douglas, "Index theory, bordism, and K-homology", pp. 1–31 in *Operator algebras and K-theory* (San Francisco, 1981), edited by R. Douglas and C. Schochet, Contemp. Math. **10**, Amer. Math. Soc., Providence, 1982.

[Baum and Douglas 1982b] P. Baum and R. G. Douglas, "K homology and index theory", pp. 117–173 in *Operator Algebras and Applications*, edited by R. V. Kadison, Proc. Sympos. Pure Math. **38**, Amer. Math. Soc., Providence, 1982.

[Baum and Douglas 1991] P. Baum and R. G. Douglas, "Relative K homology and C^* algebras", *K-Theory* **5**:1 (1991), 1–46.

[Baum et al. 1989] P. Baum, R. G. Douglas, and M. E. Taylor, "Cycles and relative cycles in analytic K-homology", *J. Differential Geom.* **30**:3 (1989), 761–804.

[Behncke and Cuntz 1976] H. Behncke and J. Cuntz, "Local completeness of operator algebras", *Proc. Amer. Math. Soc.* **62**:1 (1976), 95–100.

[Berberian 1972] S. K. Berberian, *Baer *-rings*, Grundlehren der mathematischen Wissenschaften **195**, Springer, New York, 1972.

[Bessaga and Pełczyński 1975] C. Bessaga and A. Pełczyński, *Selected topics in infinite-dimensional topology*, Monografie Matematyczne **58**, PWN (Polish Scientific Publishers), Warsaw, 1975.

[Blackadar 1978] B. E. Blackadar, "Weak expectations and nuclear C^*-algebras", *Indiana Univ. Math. J.* **27**:6 (1978), 1021–1026.

[Blackadar 1980a] B. E. Blackadar, "A simple C^*-algebra with no nontrivial projections", *Proc. Amer. Math. Soc.* **78**:4 (1980), 504–508.

[Blackadar 1980b] B. E. Blackadar, "Traces on simple AF C^*-algebras", *J. Funct. Anal.* **38**:2 (1980), 156–168.

[Blackadar 1981] B. E. Blackadar, "A simple unital projectionless C^*-algebra", *J. Operator Theory* **5**:1 (1981), 63–71.

[Blackadar 1983a] B. Blackadar, "Notes on the structure of projections in simple C*-algebras (Semesterbericht Funktionalanalysis)", Technical Report W82, Universität Tübingen, March 1983.

[Blackadar 1983b] B. Blackadar, "A stable cancellation theorem for simple C^*-algebras", *Proc. London Math. Soc.* (3) **47**:2 (1983), 303–305.

[Blackadar 1985a] B. Blackadar, "Nonnuclear subalgebras of C^*-algebras", *J. Operator Theory* **14**:2 (1985), 347–350.

[Blackadar 1985b] B. Blackadar, "Shape theory for C^*-algebras", *Math. Scand.* **56**:2 (1985), 249–275.

[Blackadar 1990] B. Blackadar, "Symmetries of the CAR algebra", *Ann. of Math.* (2) **131**:3 (1990), 589–623.

[Blackadar 1998] B. Blackadar, "Semiprojectivity in simple C^*-algebras", Preprint, 1998.

[Blackadar and Handelman 1982] B. Blackadar and D. Handelman, "Dimension functions and traces on C^*-algebras", *J. Funct. Anal.* **45**:3 (1982), 297–340.

[Blackadar and Kumjian 1985] B. Blackadar and A. Kumjian, "Skew products of relations and the structure of simple C^*-algebras", *Math. Z.* **189**:1 (1985), 55–63.

[Blackadar and Rørdam 1992] B. Blackadar and M. Rørdam, "Extending states on preordered semigroups and the existence of quasitraces on C^*-algebras", *J. Algebra* **152**:1 (1992), 240–247.

[Bratteli 1972] O. Bratteli, "Inductive limits of finite dimensional C^*-algebras", *Trans. Amer. Math. Soc.* **171** (1972), 195–234.

[Brown 1976] L. G. Brown, "Extensions and the structure of C^*-algebras", pp. 539–566 in *Convegno sulle Algebre C^* e loro Applicazioni in Fisica Teorica, Convegno sulla Teoria degli Operatori Indice e Teoria K* (Rome, 1975), Symposia Mathematica **20**, Academic Press, London, 1976.

[Brown 1977] L. G. Brown, "Stable isomorphism of hereditary subalgebras of C^*-algebras", *Pacific J. Math.* **71**:2 (1977), 335–348.

[Brown 1981] L. G. Brown, "Ext of certain free product C^*-algebras", *J. Operator Theory* **6**:1 (1981), 135–141.

[Brown 1984] L. G. Brown, "The universal coefficient theorem for Ext and quasidiagonality", pp. 60–64 in *Operator algebras and group representations, I* (Neptun, Romania, 1980), edited by G. Arsene et al., Monographs Stud. Math. **17**, Pitman, Boston, Mass., 1984.

[Brown and Elliott 1982] L. G. Brown and G. A. Elliott, "Extensions of AF-algebras are determined by K_0", *C. R. Math. Rep. Acad. Sci. Canada* **4**:1 (1982), 15–19.

[Brown and Pedersen 1991] L. G. Brown and G. K. Pedersen, "C^*-algebras of real rank zero", *J. Funct. Anal.* **99**:1 (1991), 131–149.

[Brown et al. 1973] L. G. Brown, R. G. Douglas, and P. A. Fillmore, "Unitary equivalence modulo the compact operators and extensions of C^*-algebras", pp. 58–128 in *Proceedings of a Conference on Operator Theory* (Halifax, N.S., 1973), edited by P. A. Fillmore, Lecture Notes in Math. **345**, Springer, Berlin, 1973.

[Brown et al. 1977a] L. G. Brown, R. G. Douglas, and P. A. Fillmore, "Extensions of C^*-algebras and K-homology", *Ann. of Math.* (2) **105**:2 (1977), 265–324.

[Brown et al. 1977b] L. G. Brown, P. Green, and M. A. Rieffel, "Stable isomorphism and strong Morita equivalence of C^*-algebras", *Pacific J. Math.* **71**:2 (1977), 349–363.

[Bunce and Deddens 1975] J. W. Bunce and J. A. Deddens, "A family of simple C^*-algebras related to weighted shift operators", *J. Functional Analysis* **19** (1975), 13–24.

[Busby 1968] R. C. Busby, "Double centralizers and extensions of C^*-algebras", *Trans. Amer. Math. Soc.* **132** (1968), 79–99.

[Choi 1979] M. D. Choi, "A simple C^*-algebra generated by two finite-order unitaries", *Canad. J. Math.* **31**:4 (1979), 867–880.

[Choi and Christensen 1983] M. D. Choi and E. Christensen, "Completely order isomorphic and close C^*-algebras need not be *-isomorphic", *Bull. London Math. Soc.* **15**:6 (1983), 604–610.

[Choi and Effros 1976] M. D. Choi and E. G. Effros, "The completely positive lifting problem for C^*-algebras", *Ann. of Math.* (2) **104**:3 (1976), 585–609.

[Clarke 1986] N. P. Clarke, "A finite but not stably finite C^*-algebra", *Proc. Amer. Math. Soc.* **96**:1 (1986), 85–88.

[Coburn 1973/74] L. A. Coburn, "Singular integral operators and Toeplitz operators on odd spheres", *Indiana Univ. Math. J.* **23** (1973/74), 433–439.

[Connes 1981] A. Connes, "An analogue of the Thom isomorphism for crossed products of a C^*-algebra by an action of **R**", *Adv. in Math.* **39**:1 (1981), 31–55.

[Connes 1982] A. Connes, "A survey of foliations and operator algebras", pp. 521–628 in *Operator algebras and applications* (Kingston, Ont., 1980), vol. 1, edited by R. V. Kadison, Proc. Sympos. Pure Math. **38**, Amer. Math. Soc., Providence, 1982.

[Connes 1994] A. Connes, *Noncommutative geometry*, Academic Press Inc., San Diego, CA, 1994.

[Connes and Higson 1990] A. Connes and N. Higson, "Déformations, morphismes asymptotiques et K-théorie bivariante", *C. R. Acad. Sci. Paris Sér. I Math.* **311**:2 (1990), 101–106.

[Connes and Moscovici 1982a] A. Connes and H. Moscovici, "The L^2-index theorem for homogeneous spaces of Lie groups", *Ann. of Math.* (2) **115**:2 (1982), 291–330.

[Connes and Moscovici 1982b] A. Connes and H. Moscovici, "L^2-index theory on homogeneous spaces and discrete series representations", pp. 419–433 in *Operator algebras and applications*, edited by R. V. Kadison, Proc. Sympos. Pure Math. **38**, Amer. Math. Soc., Providence, 1982.

[Connes and Skandalis 1984] A. Connes and G. Skandalis, "The longitudinal index theorem for foliations", *Publ. Res. Inst. Math. Sci.* **20**:6 (1984), 1139–1183.

[Cuntz 1977a] J. Cuntz, "Simple C^*-algebras generated by isometries", *Comm. Math. Phys.* **57**:2 (1977), 173–185.

[Cuntz 1977b] J. Cuntz, "The structure of addition and multiplication in simple C^*-algebras", *Math. Scand.* **40** (1977), 215–233.

[Cuntz 1978] J. Cuntz, "Dimension functions on simple C^*-algebras", *Math. Ann.* **233**:2 (1978), 145–153.

[Cuntz 1981a] J. Cuntz, "A class of C^*-algebras and topological Markov chains. II. Reducible chains and the Ext-functor for C^*-algebras", *Invent. Math.* **63**:1 (1981), 25–40.

[Cuntz 1981b] J. Cuntz, "K-theory for certain C^*-algebras", *Ann. of Math.* (2) **113**:1 (1981), 181–197.

[Cuntz 1981c] J. Cuntz, "K-theory for certain C^*-algebras, II", *J. Operator Theory* **5**:1 (1981), 101–108.

[Cuntz 1982a] J. Cuntz, "The internal structure of simple C^*-algebras", pp. 85–115 in *Operator Algebras and Applications, 1*, edited by R. V. Kadison, Proc. Sympos. Pure Math. **38**, Amer. Math. Soc., Providence, 1982.

[Cuntz 1982b] J. Cuntz, "The K-groups for free products of C^*-algebras", pp. 81–84 in *Operator Algebras and Applications, 1*, edited by R. V. Kadison, Proc. Sympos. Pure Math. **38**, Amer. Math. Soc., Providence, 1982.

[Cuntz 1983a] J. Cuntz, "Generalized homomorphisms between C^*-algebras and KK-theory", pp. 31–45 in *Dynamics and processes* (Bielefeld, 1981), edited by edited by Ph. Blanchard and L. Streit, Lecture Notes in Math. **1031**, Springer, Berlin, 1983.

[Cuntz 1983b] J. Cuntz, "K-theoretic amenability for discrete groups", *J. Reine Angew. Math.* **344** (1983), 180–195.

[Cuntz 1984] J. Cuntz, "K-theory and C^*-algebras", pp. 55–79 in *Algebraic K-theory, number theory, geometry and analysis* (Bielefeld, 1982), edited by A. Bak, Lecture Notes in Math. **1046**, Springer, Berlin, 1984.

[Cuntz 1987] J. Cuntz, "A new look at KK-theory", *K-Theory* **1**:1 (1987), 31–51.

[Cuntz 1997] J. Cuntz, "Bivariante K-Theorie für lokalkonvexe Algebren und der Chern-Connes-Charakter", *Doc. Math.* **2** (1997), 139–182.

[Cuntz and Higson 1987] J. Cuntz and N. Higson, "Kuiper's theorem for Hilbert modules", pp. 429–435 in *Operator algebras and mathematical physics* (Iowa City, 1985), edited by P. E. T. Jorgensen and P. S. Muhly, Contemp. Math. **62**, Amer. Math. Soc., Providence, 1987.

[Cuntz and Krieger 1980] J. Cuntz and W. Krieger, "A class of C^*-algebras and topological Markov chains", *Invent. Math.* **56**:3 (1980), 251–268.

[Cuntz and Skandalis 1986] J. Cuntz and G. Skandalis, "Mapping cones and exact sequences in KK-theory", *J. Operator Theory* **15**:1 (1986), 163–180.

[Dădărlat 1994] M. Dădărlat, "A note on asymptotic homomorphisms", *K-Theory* **8**:5 (1994), 465–482.

[Dădărlat and Loring 1994] M. Dădărlat and T. A. Loring, "K-homology, asymptotic representations, and unsuspended E-theory", *J. Funct. Anal.* **126**:2 (1994), 367–383.

[Dădărlat and Loring 1996a] M. Dădărlat and T. A. Loring, "Classifying C^*-algebras via ordered, mod-p K-theory", *Math. Ann.* **305**:4 (1996), 601–616.

[Dădărlat and Loring 1996b] M. Dădărlat and T. A. Loring, "A universal multicoefficient theorem for the Kasparov groups", *Duke Math. J.* **84**:2 (1996), 355–377.

[Davidson 1982] K. R. Davidson, "Lifting commuting pairs of C^* algebras", *J. Funct. Anal.* **48**:1 (1982), 20–42.

[Davidson 1996] K. R. Davidson, *C^*-algebras by example*, Fields Institute Monographs **6**, Amer. Math. Soc., Providence, RI, 1996.

[Diep 1974] D. N. Z'ep, "The structure of the group C^*-algebra of the group of affine transformations of the line", *Funkcional. Anal. i Priložen.* **9**:1 (1974), 63–64. In Russian.

[Dixmier 1967] J. Dixmier, "On some C^*-algebras considered by Glimm", *J. Functional Analysis* **1** (1967), 182–203.

[Dixmier 1969] J. Dixmier, *Les C^*-algèbres et leurs représentations*, 2nd ed., Cahiers Scientifiques **29**, Gauthier-Villars, Paris, 1969. Reprinted by Éditions Jacques Gabay, Paris, 1996. Translated as C^*-*algebras*, North-Holland, Amsterdam, 1977.

[Douglas 1980] R. G. Douglas, C^*-*algebra extensions and K-homology*, Annals of Mathematics Studies **95**, Princeton University Press, Princeton, N.J., 1980.

[Dykema and Rørdam 1998] K. J. Dykema and M. Rørdam, "Projections in free product C^*-algebras", *Geom. Funct. Anal.* **8**:1 (1998), 1–16.

[Effros 1981] E. G. Effros, *Dimensions and C^*-algebras*, CBMS Regional Conference Series in Mathematics **46**, Amer. Math. Soc., Providence, 1981.

[Effros and Haagerup 1985] E. G. Effros and U. Haagerup, "Lifting problems and local reflexivity for C^*-algebras", *Duke Math. J.* **52**:1 (1985), 103–128.

[Effros and Hahn 1967] E. G. Effros and F. Hahn, *Locally compact transformation groups and C^*- algebras*, Memoirs of the Amer. Math. Soc. **75**, Amer. Math. Soc., Providence, 1967.

[Effros and Kaminker 1986] E. G. Effros and J. Kaminker, "Homotopy continuity and shape theory for C^*-algebras", pp. 152–180 in *Geometric methods in operator algebras* (Kyoto, 1983), edited by H. Araki and E. Effros, Pitman Res. Notes Math. Ser. **123**, Longman Sci. Tech., Harlow, 1986.

[Effros and Lance 1977] E. G. Effros and E. C. Lance, "Tensor products of operator algebras", *Adv. Math.* **25**:1 (1977), 1–34.

[Effros et al. 1980] E. G. Effros, D. E. Handelman, and C. L. Shen, "Dimension groups and their affine representations", *Amer. J. Math.* **102**:2 (1980), 385–407.

[Eilenberg and Steenrod 1952] S. Eilenberg and N. Steenrod, *Foundations of algebraic topology*, Princeton University Press, Princeton, New Jersey, 1952.

[Elliott 1976] G. A. Elliott, "On the classification of inductive limits of sequences of semisimple finite-dimensional algebras", *J. Algebra* **38**:1 (1976), 29–44.

[Elliott 1979] G. A. Elliott, "On totally ordered groups, and K_0", pp. 1–49 in *Ring theory* (Waterloo, 1978), edited by D. Handelman and J. Lawrence, Lecture Notes in Math. **734**, Springer, Berlin, 1979.

[Elliott 1984] G. A. Elliott, "On the K-theory of the C^*-algebra generated by a projective representation of a torsion-free discrete abelian group", pp. 157–184 in *Operator algebras and group representations, I* (Neptun, Romania, 1980), edited by G. Arsene et al., Monographs Stud. Math. **17**, Pitman, Boston, 1984.

[Elliott 1996] G. A. Elliott, "An invariant for simple C^*-algebras", pp. 61–90 in *Canadian Mathematical Society 1945–1995, vol. 3: Invited papers*, edited by J. B. Carrell and R. Murty, Canadian Math. Soc., Ottawa, 1996.

[Elliott and Evans 1993] G. A. Elliott and D. E. Evans, "The structure of the irrational rotation C^*-algebra", *Ann. of Math.* (2) **138**:3 (1993), 477–501.

[Elliott and Gong 1994] G. Elliott and G. Gong, "On the classification of C^*-algebras of real rank zero", Preprint, 1994.

[Exel 1985] R. Exel, *Rotation numbers for automorphisms of C^*-algebras*, Ph.D. thesis, University of California, Berkeley, 1985.

[Fack 1983] T. Fack, "K-théorie bivariante de Kasparov", pp. 149–166 in *Séminaire Bourbaki*, 1982/83, Astérisque **105–106**, Soc. Math. France, Paris, 1983.

[Fack and Maréchal 1979] T. Fack and O. Maréchal, "Sur la classification des symétries des C^*-algèbres UHF", *Canad. J. Math.* **31**:3 (1979), 496–523.

[Fack and Maréchal 1981] T. Fack and O. Maréchal, "Sur la classification des automorphismes périodiques des C^*-algèbres UHF", *J. Funct. Anal.* **40**:3 (1981), 267–301.

[Fack and Skandalis 1981] T. Fack and G. Skandalis, "Connes' analogue of the Thom isomorphism for the Kasparov groups", *Invent. Math.* **64**:1 (1981), 7–14.

[Farrell and Hsiang 1970] F. T. Farrell and W.-C. Hsiang, "A formula for $K_1 R_\alpha [T]$", pp. 192–218 in *Applications of Categorical Algebra* (New York, 1968), edited by A. Heller, Proc. Sympos. Pure Math. **17**, Amer. Math. Soc., Providence, 1970.

[Fathi and Herman 1977] A. Fathi and M. R. Herman, "Existence de difféomorphismes minimaux", pp. 37–59 in *Dynamical systems* (Warsaw, 1977), vol. I, Astérisque **49**, Soc. Math. France, Paris, 1977.

[Ferry et al. 1995] S. C. Ferry, A. Ranicki, and J. Rosenberg, "A history and survey of the Novikov conjecture", pp. 7–66 in *Novikov conjectures, index theorems and rigidity* (Oberwolfach, 1993), vol. 1, edited by S. C. Ferry et al., London Math. Soc. Lecture Note Ser. **226**, Cambridge Univ. Press, Cambridge, 1995.

[Fillmore 1996] P. A. Fillmore, *A user's guide to operator algebras*, Canadian Mathematical Society Series of Monographs and Advanced Texts, John Wiley & Sons Inc., New York, 1996.

[Germain 1997] E. Germain, "Amalgamated free product C^*-algebras and KK-theory", pp. 89–103 in *Free probability theory* (Waterloo, ON, 1995), edited by D.-V. Voiculescu, Fields Inst. Commun. **12**, Amer. Math. Soc., Providence, RI, 1997.

[Glimm 1960] J. G. Glimm, "On a certain class of operator algebras", *Trans. Amer. Math. Soc.* **95** (1960), 318–340.

[Gong 1994] G. Gong, "Classification of C^*-algebras of real rank zero and unsuspended E-equivalent types", Preprint, 1994.

[Goodearl 1977/78] K. R. Goodearl, "Algebraic representations of Choquet simplexes", *J. Pure Appl. Algebra* **11**:1–3 (1977/78), 111–130.

[Goodearl 1982] K. R. Goodearl, *Notes on real and complex C^*-algebras*, Shiva Mathematics Series, Shiva Publishing Ltd., Nantwich, 1982.

[Goodearl and Handelman 1976] K. R. Goodearl and D. Handelman, "Rank functions and K_0 of regular rings", *J. Pure Appl. Algebra* **7**:2 (1976), 195–216.

[Grätzer 1968] G. Grätzer, *Universal algebra*, Van Nostrand, Princeton, NJ, and Toronto, 1968. Second edition: Springer, New York, 1979.

[Green 1977] P. Green, "C^*-algebras of transformation groups with smooth orbit space", *Pacific J. Math.* **72**:1 (1977), 71–97.

[Green 1982] P. Green, "Equivariant K-theory and crossed product C^*-algebras", pp. 337–338 in *Operator algebras and applications* (Kingston, Ont., 1980), vol. 1, edited by R. V. Kadison, Proc. Sympos. Pure Math. **38**, Amer. Math. Soc., Providence, 1982.

[Guentner et al. 1997] E. Guentner, N. Higson, and J. Trout, "Equivariant E-theory for C^*-algebras", Preprint, 1997.

[Haagerup 1992] U. Haagerup, "Quasitraces on exact C^*-algebras are traces", Preprint, 1992.

[Halmos 1967] P. R. Halmos, *A Hilbert space problem book*, Van Nostrand, Princeton and Toronto, 1967. Second edition, revised and enlarged: Graduate Texts in Mathematics **19**, Springer, New York, 1982.

[Handelman 1982] D. Handelman, "Extensions for AF C^* algebras and dimension groups", *Trans. Amer. Math. Soc.* **271**:2 (1982), 537–573.

[Handelman and Rossmann 1984] D. Handelman and W. Rossmann, "Product type actions of finite and compact groups", *Indiana Univ. Math. J.* **33**:4 (1984), 479–509.

[Harpe and Skandalis 1984] P. de la Harpe and G. Skandalis, "Déterminant associé à une trace sur une algébre de Banach", *Ann. Inst. Fourier (Grenoble)* **34**:1 (1984), 241–260.

[Herman and Rosenberg 1981] R. H. Herman and J. Rosenberg, "Norm-close group actions on C^*-algebras", *J. Operator Theory* **6**:1 (1981), 25–37.

[Herman and Vaserstein 1984] R. H. Herman and L. N. Vaserstein, "The stable range of C^*-algebras", *Invent. Math.* **77**:3 (1984), 553–555.

[Higson 1983] N. Higson, *On Kasparov Theory*, M.a. thesis, Dalhousie University, Halifax, 1983.

[Higson 1988] N. Higson, "Algebraic K-theory of stable C^*-algebras", *Adv. in Math.* **67**:1 (1988), 140.

[Higson 1990] N. Higson, "A primer on KK-theory", pp. 239–283 in *Operator theory: operator algebras and applications* (Durham, NH, 1988), vol. 1, edited by W. B. Arveson and R. G. Douglas, Proc. Sympos. Pure Math. **51**, Amer. Math. Soc., Providence, RI, 1990.

[Higson and Kasparov 1997] N. Higson and G. Kasparov, "Operator K-theory for groups which act properly and isometrically on Hilbert space", Preprint, 1997.

[Higson et al. 1997] N. Higson, G. Kasparov, and J. Trout, "A Bott periodicity theorem for infinite dimensional Euclidean space", Preprint, 1997.

[Hilsum 1985] M. Hilsum, "The signature operator on Lipschitz manifolds and unbounded Kasparov modules", pp. 254–288 in *Operator algebras and their connections with topology and ergodic theory* (Buşteni, Romenia, 1983), edited by H. Araki et al., Lecture Notes in Math. **1132**, Springer, Berlin, 1985.

[Hilsum and Skandalis 1987] M. Hilsum and G. Skandalis, "Morphismes K-orientés d'espaces de feuilles et fonctorialité en théorie de Kasparov (d'après une conjecture d'A. Connes)", *Ann. Sci. École Norm. Sup.* (4) **20**:3 (1987), 325–390.

[Hirsch 1976] M. W. Hirsch, *Differential topology*, Graduate Texts in Mathematics **33**, Springer, New York, 1976. Corrected reprint, 1994.

[Houghton-Larsen and Thomsen 1996] T. Houghton-Larsen and K. Thomsen, "Universal (co)homology theories", Preprint, 1996.

[Husemoller 1966] D. Husemoller, *Fibre bundles*, McGraw-Hill, New York, 1966. Second edition, Graduate Texts in Mathematics, **20**, Springer, New York, 1975; third edition, 1994.

[Jiang and Su 1997] X. Jiang and H. Su, "On a unital simple projectionless C^*-algebra", Preprint, 1997.

[Julg 1981]   P. Julg, "K-théorie équivariante et produits croisés", *C. R. Acad. Sci. Paris Sér. I Math.* **292**:13 (1981), 629–632.

[Julg and Kasparov 1995]   P. Julg and G. Kasparov, "Operator K-theory for the group SU$(n, 1)$", *J. Reine Angew. Math.* **463** (1995), 99–152.

[Julg and Valette 1983]   P. Julg and A. Valette, "K-moyennabilité pour les groupes opérant sur les arbres", *C. R. Acad. Sci. Paris Sér. I Math.* **296**:23 (1983), 977–980.

[Julg and Valette 1984]   P. Julg and A. Valette, "K-theoretic amenability for SL$_2(\mathbf{Q}_p)$, and the action on the associated tree", *J. Funct. Anal.* **58**:2 (1984), 194–215.

[Julg and Valette 1985]   P. Julg and A. Valette, "Group actions on trees and K-amenability", pp. 289–296 in *Operator algebras and their connections with topology and ergodic theory* (Buşteni, Romenia, 1983), edited by H. Araki et al., Lecture Notes in Math. **1132**, Springer, Berlin, 1985.

[Kahn et al. 1977]   D. S. Kahn, J. Kaminker, and C. Schochet, "Generalized homology theories on compact metric spaces", *Michigan Math. J.* **24**:2 (1977), 203–224.

[Kaplansky 1968]   I. Kaplansky, *Rings of operators*, W. A. Benjamin, New York and Amsterdam, 1968.

[Karoubi 1978]   M. Karoubi, *K-theory: An introduction*, Grundlehren der Mathematischen Wissenschaften, Springer, Berlin, 1978.

[Kasparov 1975]   G. G. Kasparov, "Topological invariants of elliptic operators, I: K-homology", *Izv. Akad. Nauk SSSR Ser. Mat.* **39**:4 (1975), 796–838. In Russian; translated in *Math. USSR Izv.* **9**:4 (1975), 751–792.

[Kasparov 1980a]   G. G. Kasparov, "Hilbert C^*-modules: theorems of Stinespring and Voiculescu", *J. Operator Theory* **4**:1 (1980), 133–150.

[Kasparov 1980b]   G. G. Kasparov, "The operator K-functor and extensions of C^*-algebras", *Izv. Akad. Nauk SSSR Ser. Mat.* **44**:3 (1980), 571–636, 719. In Russian; translation in *Math. USSR Izvestija* **16** (1981), 513–572.

[Kasparov 1984a]   G. Kasparov, "Operator K-theory and its applications: elliptic operators, group representations, higher signatures, C^*-extensions", pp. 987–1000 in *Proc. International Congress of Mathematicians* (Warsaw, 1983), edited by Z. Ciesielski and C. Olech, PWN-Polish Scientific Publishers, Warsaw, 1984.

[Kasparov 1984b]   G. G. Kasparov, "Lorentz groups: K-theory of unitary representations and crossed products", *Dokl. Akad. Nauk SSSR* **275**:3 (1984), 541–545.

[Kasparov 1988]   G. G. Kasparov, "Equivariant KK-theory and the Novikov conjecture", *Invent. Math.* **91**:1 (1988), 147–201.

[Kasparov 1995]   G. G. Kasparov, "K-theory, group C^*-algebras, and higher signatures (conspectus)", pp. 101–146 in *Novikov conjectures, index theorems and rigidity* (Oberwolfach, 1993), vol. 1, edited by S. C. Ferry et al., London Math. Soc. Lecture Note Ser. **226**, Cambridge Univ. Press, Cambridge, 1995.

[Kasparov and Skandalis 1994]   G. Kasparov and G. Skandalis, "Groupes "boliques" et conjecture de Novikov", *C. R. Acad. Sci. Paris Sér. I Math.* **319**:8 (1994), 815–820.

[Kirchberg 1998]   E. Kirchberg, "The classification of purely infinite C^*-algebras using Kasparov's theory", in *Lectures in Operator Algebras*, Fields Institute Monographs, Amer. Math. Soc., 1998.

[Kishimoto and Kumjian 1998] A. Kishimoto and A. Kumjian, "The Ext class of an approximately inner automorphism II", Preprint, 1998.

[Lance 1982] C. Lance, "Tensor products and nuclear C^*-algebras", pp. 379–399 in *Operator Algebras and Applications, 1*, edited by R. V. Kadison, Proc. Sympos. Pure Math. **38**, Amer. Math. Soc., Providence, 1982.

[Lance 1983] E. C. Lance, "K-theory for certain group C^*-algebras", *Acta Math.* **151**:3-4 (1983), 209–230.

[Lance 1985] C. Lance, "Some problems and results on reflexive algebras", pp. 324–330 in *Operator algebras and their connections with topology and ergodic theory* (Buşteni, Romenia, 1983), edited by H. Araki et al., Lecture Notes in Math. **1132**, Springer, Berlin, 1985.

[Loday 1992] J.-L. Loday, *Cyclic homology*, Grundlehren der Mathematischen Wissenschaften **301**, Springer, Berlin, 1992.

[Loring 1986] T. Loring, *The torus and noncommutative topology*, Ph.D. thesis, Univ. California, Berkeley, 1986.

[Loring 1997] T. A. Loring, *Lifting solutions to perturbing problems in C^*-algebras*, Fields Institute Monographs **8**, Amer. Math. Soc., Providence, RI, 1997.

[MacLane 1971] S. MacLane, *Categories for the working mathematician*, Graduate Texts in Mathematics **5**, Springer, New York, 1971.

[Madsen and Rosenberg 1988] I. Madsen and J. Rosenberg, "The universal coefficient theorem for equivariant K-theory of real and complex C^*-algebras", pp. 145–173 in *Index theory of elliptic operators, foliations, and operator algebras* (New Orleans and Indianapolis, 1986), edited by J. Kaminker et al., Contemp. Math. **70**, Amer. Math. Soc., Providence, RI, 1988.

[Magurn 1985] B. Magurn (editor), *Reviews in K-Theory*, Amer. Math. Soc., Providence, 1985.

[McGovern et al. 1981] R. McGovern, V. Paulsen, and N. Salinas, "A classification theorem for essentially binormal operators", *J. Funct. Anal.* **41**:2 (1981), 213–235.

[Mingo 1987] J. A. Mingo, "K-theory and multipliers of stable C^*-algebras", *Trans. Amer. Math. Soc.* **299**:1 (1987), 397–411.

[Mingo and Phillips 1984] J. A. Mingo and W. J. Phillips, "Equivariant triviality theorems for Hilbert C^*-modules", *Proc. Amer. Math. Soc.* **91**:2 (1984), 225–230.

[Miščenko 1974] A. S. Miščenko, *Infinite-dimensional representations of discrete groups, and higher signatures*, 1974.

[Miščenko and Fomenko 1979] A. Miščenko and A. Fomenko, "The index of elliptic operators over C^*-algebras", *Izv. Akad. Nauk SSSR Ser. Mat.* **43** (1979), 831–859.

[Moore and Schochet 1988] C. C. Moore and C. Schochet, *Global analysis on foliated spaces*, Mathematical Sciences Research Institute Publications, Springer, New York, 1988.

[Murphy 1990] G. J. Murphy, *C^*-algebras and operator theory*, Academic Press Inc., Boston, MA, 1990.

[Nagy 1995] G. Nagy, "Bivariant K-theories for C^*-algebras", Preprint, 1995.

[Natsume 1985] T. Natsume, "On $K_*(C^*(\mathrm{SL}_2(\mathbf{Z})))$", *J. Operator Theory* **13**:1 (1985), 103–118. Appendix to [Lance 1983].

[O'Donovan 1977] D. P. O'Donovan, "Quasidiagonality in the Brown–Douglas–Fillmore theory", *Duke Math. J.* **44**:4 (1977), 767–776.

[Olesen and Pedersen 1978] D. Olesen and G. K. Pedersen, "Applications of the Connes spectrum to C^*-dynamical systems", *J. Funct. Anal.* **30**:2 (1978), 179–197.

[Olesen and Pedersen 1986] D. Olesen and G. K. Pedersen, "Partially inner C^*-dynamical systems", *J. Funct. Anal.* **66**:2 (1986), 262–281.

[Paschke 1973] W. L. Paschke, "Inner product modules over B^*-algebras", *Trans. Amer. Math. Soc.* **182** (1973), 443–468.

[Paschke 1981] W. L. Paschke, "K-theory for commutants in the Calkin algebra", *Pacific J. Math.* **95**:2 (1981), 427–434.

[Paschke and Salinas 1979] W. L. Paschke and N. Salinas, "Matrix algebras over O_n", *Michigan Math. J.* **26**:1 (1979), 3–12.

[Pedersen 1979] G. K. Pedersen, *C^*-algebras and their automorphism groups*, London Mathematical Society Monographs **14**, Academic Press, London, 1979.

[Pedersen 1980] G. K. Pedersen, "The linear span of projections in simple C^*-algebras", *J. Operator Theory* **4**:2 (1980), 289–296.

[Phillips 1987] N. C. Phillips, *Equivariant K-theory and freeness of group actions on C^*-algebras*, Lecture Notes in Mathematics **1274**, Springer, Berlin, 1987.

[Phillips 1989] N. C. Phillips, *Equivariant K-theory for proper actions*, Pitman Research Notes in Mathematics Series **178**, Longman, Harlow, UK, 1989.

[Pimsner 1979] M. Pimsner, "On the Ext-group of an AF-algebra. II", *Rev. Roumaine Math. Pures Appl.* **24**:7 (1979), 1085–1088.

[Pimsner 1985] M. V. Pimsner, "Ranges of traces on K_0 of reduced crossed products by free groups", pp. 374–408 in *Operator algebras and their connections with topology and ergodic theory* (Buşteni, Romenia, 1983), edited by H. Araki et al., Lecture Notes in Math. **1132**, Springer, Berlin, 1985.

[Pimsner 1986] M. V. Pimsner, "KK-groups of crossed products by groups acting on trees", *Invent. Math.* **86**:3 (1986), 603–634.

[Pimsner and Popa 1978] M. Pimsner and S. Popa, "On the Ext-group of an AF-algebra", *Rev. Roumaine Math. Pures Appl.* **23**:2 (1978), 251–267.

[Pimsner and Voiculescu 1980a] M. Pimsner and D. Voiculescu, "Exact sequences for K-groups and Ext-groups of certain cross-product C^*-algebras", *J. Operator Theory* **4**:1 (1980), 93–118.

[Pimsner and Voiculescu 1980b] M. Pimsner and D. Voiculescu, "Imbedding the irrational rotation C^*-algebra into an AF-algebra", *J. Operator Theory* **4**:2 (1980), 201–210.

[Pimsner and Voiculescu 1982] M. Pimsner and D. Voiculescu, "K-groups of reduced crossed products by free groups", *J. Operator Theory* **8**:1 (1982), 131–156.

[Pimsner et al. 1979] M. Pimsner, S. Popa, and D. Voiculescu, "Homogeneous C^*-extensions of $C(X) \otimes K(H)$, I", *J. Operator Theory* **1**:1 (1979), 55–108.

[Pimsner et al. 1980] M. Pimsner, S. Popa, and D. Voiculescu, "Homogeneous C^*-extensions of $C(X) \otimes K(H)$, II", *J. Operator Theory* **4**:2 (1980), 211–249.

[Renault 1980]   J. Renault, *A groupoid approach to C^*-algebras*, Lecture Notes in Mathematics, Springer, Berlin, 1980.

[Rieffel 1974]   M. A. Rieffel, "Induced representations of C^*-algebras", *Advances in Math.* **13** (1974), 176–257.

[Rieffel 1981]   M. A. Rieffel, "C^*-algebras associated with irrational rotations", *Pacific J. Math.* **93**:2 (1981), 415–429.

[Rieffel 1982]   M. A. Rieffel, "Connes' analogue for crossed products of the Thom isomorphism", pp. 143–154 in *Operator algebras and K-theory* (San Francisco, 1981), edited by R. Douglas and C. Schochet, Contemp. Math. **10**, Amer. Math. Soc., Providence, 1982.

[Rieffel 1983a]   M. A. Rieffel, "The cancellation theorem for projective modules over irrational rotation C^*-algebras", *Proc. London Math. Soc.* (3) **47**:2 (1983), 285–302.

[Rieffel 1983b]   M. A. Rieffel, "Dimension and stable rank in the K-theory of C^*-algebras", *Proc. London Math. Soc.* (3) **46**:2 (1983), 301–333.

[Rieffel 1988]   M. A. Rieffel, "Projective modules over higher-dimensional noncommutative tori", *Canad. J. Math.* **40**:2 (1988), 257–338.

[Rørdam 1993]   M. Rørdam, "Classification of certain infinite simple C^*-algebras", Preprint, 1993.

[Rørdam 1998]   M. Rørdam, "On sums of finite projections", Preprint, 1998.

[Rosenberg 1976]   J. Rosenberg, "The C^*-algebras of some real and p-adic solvable groups", *Pacific J. Math.* **65**:1 (1976), 175–192.

[Rosenberg 1982a]   J. Rosenberg, "Homological invariants of extensions of C^*-algebras", pp. 35–75 in *Operator algebras and applications* (Kingston, Ont., 1980), vol. 1, edited by R. V. Kadison, Proc. Sympos. Pure Math. **38**, Amer. Math. Soc., Providence, 1982.

[Rosenberg 1982b]   J. Rosenberg, "The role of K-theory in noncommutative algebraic topology", pp. 155–182 in *Operator algebras and K-theory* (San Francisco, 1981), edited by R. Douglas and C. Schochet, Contemp. Math. **10**, Amer. Math. Soc., Providence, 1982.

[Rosenberg 1983]   J. Rosenberg, "C^*-algebras, positive scalar curvature, and the Novikov conjecture", *Inst. Hautes Études Sci. Publ. Math.* **58** (1983), 197–212.

[Rosenberg 1984]   J. Rosenberg, "Group C^*-algebras and topological invariants", pp. 95–115 in *Operator algebras and group representations, II* (Neptun, Romania, 1980), edited by G. Arsene et al., Monographs Stud. Math. **18**, Pitman, Boston, MA, 1984.

[Rosenberg 1986a]   J. Rosenberg, "C^*-algebras, positive scalar curvature, and the Novikov conjecture, III", *Topology* **25**:3 (1986), 319–336.

[Rosenberg 1986b]   J. Rosenberg, "C^*-algebras, positive scalar curvature and the Novikov conjecture, II", pp. 341–374 in *Geometric methods in operator algebras* (Kyoto, 1983), edited by H. Araki and E. Effros, Pitman Res. Notes Math. Ser. **123**, Longman Sci. Tech., Harlow, 1986.

[Rosenberg 1994]   J. Rosenberg, *Algebraic K-theory and its applications*, Graduate Texts in Mathematics, Springer, New York, 1994.

[Rosenberg 1997]   J. Rosenberg, "The algebraic K-theory of operator algebras", *K-Theory* **12**:1 (1997), 75–99.

[Rosenberg and Schochet 1981] J. Rosenberg and C. Schochet, "Comparing functors classifying extensions of C^*-algebras", *J. Operator Theory* **5**:2 (1981), 267–282.

[Rosenberg and Schochet 1986] J. Rosenberg and C. Schochet, *The Künneth theorem and the universal coefficient theorem for equivariant K-theory and KK-theory*, Mem. Amer. Math. Soc. **348 (62)**, Amer. Math. Soc., Providence, 1986.

[Rosenberg and Schochet 1987] J. Rosenberg and C. Schochet, "The Künneth theorem and the universal coefficient theorem for Kasparov's generalized K-functor", *Duke Math. J.* **55**:2 (1987), 431–474.

[Rosenberg and Stolz 1998] J. Rosenberg and S. Stolz, "Metrics of positive scalar curvature and connections with surgery", Preprint, 1998.

[Sakai 1971] S. Sakai, *C*-algebras and W*-algebras*, Ergebnisse der Mathematik und ihrer Grenzgebiete, Springer, New York, 1971.

[Salinas 1977] N. Salinas, "Homotopy invariance of $Ext(\mathscr{A})$", *Duke Math. J.* **44**:4 (1977), 777–794.

[Salinas 1979] N. Salinas, "Hypoconvexity and essentially n-normal operators", *Trans. Amer. Math. Soc.* **256** (1979), 325–351.

[Salinas 1982a] N. Salinas, "Problems on joint quasitrangularity for N-tuples of essentially commuting, essentially normal operators", pp. 629–631 in *Operator Algebras and Applications, 2*, edited by R. V. Kadison, Proc. Sympos. Pure Math. **38**, Amer. Math. Soc., Providence, 1982.

[Salinas 1982b] N. Salinas, "Some remarks on the classification of essentially n-normal operators", pp. 183–196 in *Operator algebras and K-theory* (San Francisco, 1981), edited by R. Douglas and C. Schochet, Contemp. Math. **10**, Amer. Math. Soc., Providence, 1982.

[Samuel 1997] J. Samuel, "Asymptotic morphisms and E-theory", pp. 291–314 in *Operator algebras and their applications* (Waterloo, ON, 1994/1995), edited by P. A. Fillmore and J. A. Mingo, Fields Inst. Commun. **13**, Amer. Math. Soc., Providence, RI, 1997.

[Schmitt 1991] L. M. Schmitt, "Quotients of local Banach algebras are local Banach algebras", *Publ. Res. Inst. Math. Sci.* **27**:6 (1991), 837–843.

[Schochet 1982] C. Schochet, "Topological methods for C^*-algebras, II: Geometry resolutions and the Künneth formula", *Pacific J. Math.* **98**:2 (1982), 443–458.

[Schochet 1984a] C. Schochet, "Topological methods for C^*-algebras, III: Axiomatic homology", *Pacific J. Math.* **114**:2 (1984), 399–445.

[Schochet 1984b] C. Schochet, "Topological methods for C^*-algebras, IV: Mod p homology", *Pacific J. Math.* **114**:2 (1984), 447–468.

[Schochet 1996] C. Schochet, "The fine structure of the Kasparov groups", Preprint, 1996.

[Schröder 1993] H. Schröder, *K-theory for real C*-algebras and applications*, Pitman Research Notes in Mathematics Series, Longman, Harlow, 1993.

[Segal 1968] G. Segal, "Equivariant K-theory", *Inst. Hautes Études Sci. Publ. Math. No. 34* (1968), 129–151.

[Skandalis 1984] G. Skandalis, "Some remarks on Kasparov theory", *J. Funct. Anal.* **56**:3 (1984), 337–347.

[Skandalis 1988] G. Skandalis, "Une notion de nucléarité en K-théorie (d'après J. Cuntz)", K-*Theory* **1**:6 (1988), 549–573.

[Spanier 1966] E. H. Spanier, *Algebraic topology*, McGraw-Hill, New York, 1966. Corrected reprint, Springer, New York, 1981.

[Takesaki 1979] M. Takesaki, *Theory of operator algebras, I*, Springer, New York, 1979.

[Taylor 1975] J. L. Taylor, "Banach algebras and topology", pp. 118–186 in *Algebras in analysis* (Birmingham, 1973), edited by J. H. Williamson, Academic Press, London, 1975.

[Teleman 1984] N. Teleman, "The index theorem for topological manifolds", *Acta Math.* **153**:1-2 (1984), 117–152.

[Trèves 1980] F. Trèves, *Introduction to pseudodifferential and Fourier integral operators*, Plenum Press, New York, 1980.

[Tu 1998] J. L. Tu, "La conjecture de Baum-Connes pour les feuilletages moyennables", Preprint, 1998.

[Valette 1982] A. Valette, "Extensions of C^*-algebras: a survey of the Brown–Douglas–Fillmore theory", *Nieuw Arch. Wisk.* (3) **30**:1 (1982), 41–69.

[Valette 1983] A. Valette, "A remark on the Kasparov groups $Ext(A, B)$", *Pacific J. Math.* **109**:1 (1983), 247–255.

[Villadsen 1995] J. Villadsen, "Simple C^*-algebras with perforation", Preprint, 1995.

[Villadsen 1997] J. Villadsen, "On the stable rank of simple C^*-algebras", Preprint, 1997.

[Voiculescu 1976] D. Voiculescu, "A non-commutative Weyl–von Neumann theorem", *Rev. Roumaine Math. Pures Appl.* **21**:1 (1976), 97–113.

[Voiculescu 1981] D. Voiculescu, "Remarks on the singular extension in the C^*-algebra of the Heisenberg group", *J. Operator Theory* **5**:2 (1981), 147–170.

[Waldhausen 1978] F. Waldhausen, "Algebraic K-theory of generalized free products, I–IV", *Ann. of Math.* (2) **108**:1–2 (1978), 135–256.

[Wassermann 1983] A. Wassermann, "Equivariant K-theory II: Hodgkin's spectral sequence in Kasparov theory", Preprint, 1983.

[Wassermann 1989] A. Wassermann, "Ergodic actions of compact groups on operator algebras. I. General theory", *Ann. of Math.* (2) **130**:2 (1989), 273–319.

[Wassermann 1994] S. Wassermann, *Exact C^*-algebras and related topics*, Lecture Notes Series **19**, Seoul National University Research Institute of Mathematics Global Analysis Research Center, Seoul, 1994.

[Wegge-Olsen 1993] N. E. Wegge-Olsen, *K-theory and C^*-algebras: A friendly approach*, Oxford Science Publications, Oxford Univ. Press, New York, 1993.

[Whitehead 1941] J. H. C. Whitehead, "On incidence matrices, nuclei and homotopy types", *Ann. of Math.* (2) **42** (1941), 1197–1239.

K-Theory for Operator Algebras
MSRI Publications
Volume **5**, second edition, 1998

INDEX